Farming 5.0

Deep Learning for Future-Generation Agriculture

Editors:

Siddharth Singh Chouhan
Assistant Professor Senior Grade 2, School of Computing Science and Engineering
VIT Bhopal University, Bhopal, India

Rajneesh Kumar Patel
Assistant Professor, School of Computing Science and Engineering
VIT Bhopal University, Bhopal, India

Anju Shukla
Assistant Professor, School of Computing Science and Engineering
VIT Bhopal University, Bhopal, India

Uday Pratap Singh
Professor, Department of Mathematics
Central University of Jammu, India

CRC Press
Taylor & Francis Group
Boca Raton London New York

CRC Press is an imprint of the
Taylor & Francis Group, an **informa** business
A SCIENCE PUBLISHERS BOOK

First edition published 2027
by CRC Press
2385 NW Executive Center Drive, Suite 320, Boca Raton FL 33431

and by CRC Press
4 Park Square, Milton Park, Abingdon, Oxon, OX14 4RN

CRC Press is an imprint of Taylor & Francis Group, LLC

Library of Congress Cataloging-in-Publication Data (applied for)

ISBN: 978-1-041-14412-0 (hbk)
ISBN: 978-1-041-20237-0 (pbk)
ISBN: 978-1-003-71572-6 (ebk)

DOI: 10.1201/9781003715726

Typeset in Times New Roman
by Prime Publishing Services

Dedication

To our family, friends, and colleagues.

Preface

The agricultural landscape is experiencing its greatest transformation in centuries, fueled by the relentless advancement of deep learning and artificial intelligence. What was once considered science fiction machines that learn, adapt, and think alongside farmers is now shaping the heart of food production. ***Farming 5.0: Deep Learning for Future Generation Agriculture*** was conceived in the spirit of exploring how these technological breakthroughs are not only revolutionizing farming practice but also redefining what's possible for global food systems, environmental health, and the livelihoods of billions.

This book came to life through international collaboration with experts from diverse disciplines: AI, computational genomics, agronomy, remote sensing, and systems engineering. The core vision was clear from the beginning: to offer an in-depth, practical, and critical account of how smart systems, when thoughtfully designed, can transform agriculture at every scale from precision irrigation in smallholder plots to continent-spanning satellite networks tracking weather and crop health in real-time.

The urgency driving this work is undeniable. Humanity faces a dual challenge unlike any before: producing enough food for a growing population while minimizing the planetary harm that conventional agriculture has wrought. Climate change, degraded soils, water scarcity, erratic pests, and emerging diseases demand adaptive, resilient, and sustainable solutions. In these pages, readers will discover how deep learning empowers not only smarter crop and livestock management, but also enables predictive analytics for yield, biodiversity monitoring, and the creation of digital twins and agricultural simulations to accelerate discovery and real-world adoption.

Each chapter has been crafted to be relevant for a wide audience from graduate students and new researchers to practicing agronomists, technology developers, and policy advocates. Our aim is dual: to provide a detailed reference for those building or implementing new technologies, and to present a forward-looking vision that can inspire future leaders in smart agriculture.

We are deeply grateful to the contributors whose expertise and imagination have made this work possible, as well as to the countless farmers and innovators who test new ideas every day in fields and labs around the world. Their curiosity and courage drive progress from theory into practice.

Above all, this book advocates an inclusive, human-centric approach to agricultural AI. Technology is powerful, but its real worth is measured by the societal and environmental benefits it delivers. The journey toward Farming 5.0 is ongoing; it is our hope that this volume serves as both guide and catalyst for the resilient, equitable, and sustainable transformation of agriculture.

Siddharth Singh Chouhan
Rajneesh Kumar Patel
Anju Shukla
Uday Pratap Singh

Acknowledgment

We are deeply grateful to all who have contributed to the creation of ***Farming 5.0: Deep Learning for Future Generation Agriculture***. This book stands as a testament to the collective efforts of many individuals whose expertise, support, and dedication have brought this project to life.

Our heartfelt gratitude goes to the esteemed contributors whose profound knowledge and innovative insights into deep learning and smart agriculture have enriched this work. Your commitment to advancing the field has been invaluable in shaping the depth and breadth of this book.

We extend our sincere thanks to our colleagues and mentors for their unwavering support and guidance throughout this journey. Your constructive feedback, encouragement, and shared passion for agricultural innovation have been instrumental in elevating the quality of this publication. We also express our appreciation to **VIT Bhopal University** and the **Central University of Jammu** for their support and resources.

A special thank you to the editorial team and publishing staff, whose tireless efforts and professionalism have been crucial in bringing this book to fruition. We are deeply grateful to **CRC Press Taylor and Francis Group** for providing us with the opportunity to share this work with the world.

To our families and friends, we owe an immense debt of gratitude. Your patience, understanding, and steadfast encouragement have been the foundation of this endeavour, sustaining us through the challenges of this project.

We also acknowledge the broader research and academic communities whose pioneering work in deep learning and smart agriculture continues to inspire and drive progress. Your dedication to addressing global agricultural challenges fuels the vision of a more sustainable future.

Finally, we extend our gratitude to those whose contributions, whether through conversations, quiet support, or other unseen efforts, have made this book possible. Your involvement, however big or small, has left an indelible mark on this work.

Thank you all for your invaluable contributions and belief in this work.

Contents

List of Contributors

A. Sheryl Oliver
Department of Computational Intelligence, Faculty of Engineering and Technology, SRM Institute of Science and Technology, Kattankulathur, Tamil Nadu, India.

Abhishek Upadhyay
Assistant Professor, Department of Farm Machinery and Power Engineering, CAET, Dr. Rajendra Prasad Central Agricultural, Pusa, Samastipur, Bihar, India.

Aditi Kachchap
M.Tech. Scholar, Department of Farm Machinery and Power Engineering, CAET, Dr. Rajendra Prasad Central Agricultural, Pusa, Samastipur, Bihar, India.

Aditya Kumar Pandey
Indian Institute of Information Technology Bhagalpur, Bhagalpur, Bihar, India.
Department of Industrial Engineering and Management, Yuan Ze University, Taoyuan, 320315, Taiwan.

Aditya Kumar Pandey
Department of Computer Science & Engineering, Indian Institute of Information Technology Bhagalpur, Bihar, India.

Adityanshu Tripathi
M.Tech. Scholar, Department of Farm Machinery and Power Engineering, CAET, Dr. Rajendra Prasad Central Agricultural, Pusa, Samastipur, Bihar, India.
Assistant Professor, Department of Farm Machinery and Power Engineering, CAET, Dr. Rajendra Prasad Central Agricultural, Pusa, Samastipur, Bihar, India.

Aman Dubey
Department of Electronics and Communication Engineering, Maulana Azad National Institute of Technology, Bhopal, Madhya Pradesh, India.

Anjana J. Atapattu
Agronomy Division, Coconut Research Institute, Lunuwila 61150, Sri Lanka.

Anju Shukla
School of Computing Science and Engineering, VIT Bhopal University, Kothrikalan, Sehore, 466114, India.

Arkan A. Ghaib
Information Technology Department, Management Technical College, Southern Technical University, Basrah 61004, Iraq.

Avimanyou Vatsa
Associate Professor, Department of Computer Science, Fairleigh Dickinson University - Teaneck, NJ, USA.

Dattatray G. Bhalekar
Center for Precision and Automated Agricultural Systems, Washington State University, Prosser, WA 99350, USA .

Diksha Rani
Division of Vegetable Science, Sher-e-Kashmir University of Agricultural Sciences and Technology, Jammu, India.

Dilip Kumar Choubey
Department of Computer Science & Engineering, Indian Institute of Information Technology Bhagalpur, Bihar, India.

Fredrick Kayusi
Department of Environmental Studies, Geography and Planning, Maasai Mara University, P.O. Box 861-20500, Narok, Kenya.

Gumbo Albert
Department of Land and Water Resources Management, Faculty of Agriculture, Environment and Natural Resources Management, Midlands State University, Gweru, Zimbabwe.

Harshit Mishra
Department of Agricultural Economics, College of Agriculture, Acharya Narendra Deva University of Agriculture and Technology, Kumarganj, Ayodhya, Uttar Pradesh – 224 229, India.

Harsh Soni
PEGA Lead System Architect | Development Team Leader @Euroclear, Belgium.

Ioannis Adamopoulos
Department of Public Health Policy, Sector of Occupational & Environmental Health, School of Public Health, University of West Attica, Athens, Greece.

J.K. Kiruthika
Department of Computer Science and Engineering, KPR Institute of Engineering and Technology, Coimbatore, India.

Kalamkuri Chandu

M.Tech. Scholar, Department of Farm Machinery and Power Engineering, CAET, Dr. Rajendra Prasad Central Agricultural, Pusa, Samastipur, Bihar, India.

Center for Precision and Automated Agricultural Systems, Washington State University, Prosser, WA, USA .

Kavish Thakur

Department of Computer Science & Engineering, Indian Institute of Information Technology Bhagalpur, Bhagalpur, Bihar, India.

Lalita Gupta

Department of Electronics and Communication Engineering, Maulana Azad National Institute of Technology, Bhopal.

M. Kanipriya

Department of Computational Intelligence, Faculty of Engineering and Technology, SRM Institute of Science and Technology, Kattankulathur, Tamil Nadu, India.

M. Salomi

Department of Computational Intelligence, Faculty of Engineering and Technology, SRM Institute of Science and Technology, Kattankulathur, Tamil Nadu, India.

M. Zaheer Ahmed

Postgraduate Department of Commerce and Research, Dr. Kalaignar M. Karunanidhi Government Institute for Postgraduate Studies and Research, Karaikal, U.T. of Puducherry, India.

Mekala Samayochitha

Department of Electronics and Communication Engineering, Maulana Azad National Institute of Technology, Bhopal.

Mohit Choubey

Department of Electronics and Communication Engineering, Maulana Azad National Institute of Technology, Bhopal, Madhya Pradesh, India.

Mukesh Kumar Ojha

Department of Electronics and Communication Engineering, Greater Noida Institute of Technology, Greater Noida, U.P., India.

Mutanga Alfred

University of Botswana, Gaborone, Botswana.

Nandha S.

Department of Agricultural Entomology, Tamil Nadu Agricultural University, Coimbatore, 641003.

Neha Agarwal
School of Chemical Engineering, Yeungnam University, Gyeongsan 38541, Republic of Korea.

Nupur Parashar
Research Scholar, Department of Computer Science, Dr. BhimRao Ambedkar University, Agra.

Nuwandhya S. Dissanayaka
Agronomy Division, Coconut Research Institute, Lunuwila 61150, Sri Lanka.

Oluwakemi Temitope Olayinka
Department of Business Information Systems and Analytics, University of Arkansas at Little Rock, 2801 S University Avenue, Little Rock, AR 72204, United States.

P.K. Pranav
Professor, Department of Farm Machinery and Power Engineering, CAET, Dr. Rajendra Prasad Central Agricultural, Pusa, Samastipur, Bihar, India.

P. Radhakrishnan
School of CS & AI, SR University, Warangal, Telangana, India.

Parwada Cosmas
Department of Agronomy and Horticulture, Faculty of Agriculture, Environment and Natural Resources Management, Midlands State University, Gweru, Zimbabwe.

Rahul Prasad
School of computer Science, UPES Dehradun, India.

Rajneesh Kumar Patel
School of Computing Science and Engineering, VIT Bhopal University, Kothrikalan, Sehore, 466114, India.

Rashmi Mishra
College of Economics and Business Administration, University of Technology and Applied Sciences - Al Musannah, P.O. Box 191, Muscat, Oman.

Ritika
Indira Gandhi University Meerpur Rewari, India -123401.

Rohit Anand
Ph.D. Student, Department of Food, Agricultural and Biological Engineering, The Ohio State University, Columbus, OH-43210, USA.

S.K. Patel
Assistant Professor, Department of Farm Machinery and Power Engineering, CAET, Dr. Rajendra Prasad Central Agricultural, Pusa, Samastipur, Bihar, India.

S. Samsudeen
Assistant Professor, Department of Computing Technologies, Faculty of Engineering and Technology, SRM Institute of Science and Technology, Kattankulathur, Tamil Nadu, India.

S.Samsudeen
Department of Computational Intelligence, Faculty of Engineering and Technology, SRM Institute of Science and Technology, Kattankulathur, Tamil Nadu, India.

Sandesh K. Jain
Department of Electronics and Communication Engineering, Maulana Azad National Institute of Technology, Bhopal.

Saurabh Agarwal
Research Professor, School of Computer Science and Engineering, Yeungnam University, Republic of Korea.

Savita Kumari Sheoran
Indira Gandhi University Meerpur Rewari, India -123401.

Shashi S. Udumann
Agronomy Division, Coconut Research Institute, Lunuwila 61150, Sri Lanka

Shreesh Sanjay Amin
M. Tech. Scholar, Department of Farm Machinery and Power Engineering, CAET, Dr. Rajendra Prasad Central Agricultural, Pusa, Samastipur, Bihar, India.

Siddharth Singh Chouhan
School of Computing Science and Engineering, VIT Bhopal University, Kothrikalan, Sehore, 466114, India.

Sourav Tiwari
Senior Software Engineer, Wipro, Australia.

Subhash Chandra
Associate Professor, Department of Farm Machinery and Power Engineering, CAET, Dr. Rajendra Prasad Central Agricultural, Pusa, Samastipur, Bihar, India.

Tharindu D. Nuwarapaksha
Agronomy Division, Coconut Research Institute, Lunuwila 61150, Sri Lanka.

Utkarsh Raj
Department of Computer Science & Engineering, Indian Institute of Information Technology Bhagalpur, Bhagalpur, Bihar, India.

Vandana Singh
Department of Microbiology, Sharda School of Allied Health Sciences, Sharda University, Greater Noida, Uttar Pradesh – 201 310, India.

Vasanthkumar S.S

Department of Plant and Environmental Sciences, Clemson University, SC, USA 29817.

Vinothagan S

Department of Vegetable Science, Horticultural College and Research Institute, Tamil Nadu Agricultural University, Coimbatore, Tamil Nadu, India 641003.

Vipul Negi

Department of Electronics and Communication Engineering, Maulana Azad National Institute of Technology, Bhopal.

Yamini R

Department of Agricultural Entomology, Tamil Nadu Agricultural University, Coimbatore, 641003.

Yogesh Kumar Gupta

School of computer Science, UPES Dehradun, India.

1 Seeding the Future: Unveiling the AgriTech Revolution

Anju Shukla[1], Siddharth Singh Chouhan[2],
Rajneesh Kumar Patel[3*] and Avimanyou Vatsa[4]

1. Introduction

For millennia, agriculture has been the foundation of human civilization, impacting economics, society, and nature. From the initial advancement of Plough to the iconic green revolution of the 20th century, every technological leap has restructured the way humans relate with the planet (Siregar et al., 2022). We are now starting a new chapter with the onset of Agritech (Shukla et al., 2024a). This is a technology revolution that must transform agriculture more than ever before, with its unmatched precision, effectiveness, and reliability. Researched by the intersection of artificial intelligence (AI), robotics, biotechnology, and data scientists, this revolution can address some of humanity's most pressing problems, such as population growth, slowing climate change, and conservation of dwindling natural resources. Yet, the road to mass adoption is not smooth. It constrains the world that changes economic inequality, the distinction between digital barriers and digitally governing social issues, and the effects of these technologies. This chapter introduces the agricultural setting in quick development, and raises the foundation for a holistic assessment of innovative potential and systemic issues that have to be addressed to complete a more just and stable agriculture future (Chouhan et al., 2025a).

The sense of urgency of Advanced Agritech solutions has never really gone away. The United Nations suggests that the population of the world will be 9.7 billion in 2050. This amplifies food output by 50%. Meanwhile, farming is under extreme stress. Land being cultivated is reducing, water resources are stretched, and climate-induced breaches are accelerating. Conventional farming systems have been tried and proven over time, but tend to be ineffective and are stressed by the environment (Ojo and Zahid, 2022). Secondly, Agritech is a reorientation of

[1-3] School of Computing Science and Engineering, VIT Bhopal University, Kothrikalan, Sehore, 466114, India.
[4] Department of Computer Science, Fairleigh Dickinson University - Teaneck, NJ, USA.
Email: patelrajneesh90@gmail.com
* Corresponding author: patelrajneesh90@gmail.com

DOI: 10.1201/9781003715726-1

advanced technological paradigms' ability to optimize farm entry and drift in the value chain of production. For instance, precise agriculture can utilize IoT sensors, drones, and satellite visualizations to allow for soil and soil condition analysis, so that water, fertilizers, and pesticides are applied very precisely (Solanki et al., 2024). These practices not only decrease productivity, but also decrease resource consumption and environmental footprint. In some situations, the water saving increases by 30%. Likewise, vertical farming topped with effective hydroponics and LED technology revolutionize urban farming and deposit 20 times more yields per square meter than conventional means, consuming even 70% water.

The technology proves the innovative power of Agritech to build a more productive and stable and stable food system amid a world of high-speed development. Agritech not only maximizes farm efficiency, but also prints the very fabric of farming, owing to its nature. Mechanical schooling and AI amplify farmers' capabilities with cultural forecasting, pest tracking, predictive analysis of climate resilience, and predictive analytics to transform agriculture into data science. Robotics remedies the age-old scarcity of work in rural sectors, with autonomous machinery plantations and unprecedented precision harvesting (Chavula & Kayusi, 2025). From the CRISPR genes to GMO crops, biotech advancements boost resistance to drought, insects, and disease, leading to added requirements of nutrients needed for climate menaces. Concurrently, blockchain technology ensures transparency and monitoring of farm products, encourages consumers, and offers a source of ethics. These technologies are not an incremental addition to current systems but rather a radical cutting back of the paradigm of agriculture so the production of goods can become viable in adverse and customary settings.

However, Agritech's revolutionary potential is well counterbalanced by the issues. The costly price of acquiring cutting-edge technology is to take away the reach of small farmers, which have been created, that provide approximately 70% of the food supply globally. Rural communities tend to have poor internet access and poor technological infrastructure, the scale of digital gaps, and the danger of excluding the advantages of innovation at big food corporations. Confidentiality of data is always a problem for farmers to be free from the sensitivity of sensitive data technology vendors (Chavula and Kayusi, 2025). Also, regulations for independence see hindrances to innovation and make large-scale implementation complex, especially in the application of genes and technologies to reveal autonomous mechanisms. These challenges emphasize the importance of holistic policy and interventions like subsidies, microfinance trends, autonomous diet-supporting technologies, and strict, transparent data management systems. This chapter is an introduction to a complicated, dynamic world of recognition, and it is a balanced vision of innovative possibilities and the issues arising from them. With our research on the convergence of implementation and innovation barriers, we seek to give farmers, politicians, researchers, and engineers a strategic guide. The Agritech revolution not only fulfills the food security needs of today, but also for building a future where agriculture is stable and inclusive. Due to this revolution, the intention is unmistakable. Create a global food system that nourishes all of humanity and sustains the planet for generations to come.

2. The Imperative for Change: Global Challenges in Agriculture

Agriculture has long been the foundation of human development, and both have sustained more advanced societies. But in the 21st century, the agricultural sector is confronted with unprecedented convergence of issues that imperil its capacity to feed a fast-increasing population globally. With the UN holding an estimated population of 9.7 billion in 2050, food production must rise by approximately 50% to match the supply. This mandate is within the frame of ecological degradation, resource constraints, work disadvantages, and shifting consumer dreams—all indicators of weakness in conventional farming practices. These increasing challenges are centered on what the agricultural revolution should rethink and change the world food system (Chavula & Kayusi, 2025).

2.1 Demographic Growth and Food Security

Demographic expansion, particularly in developing nations, is exerting tremendous stress on the agricultural system. Urbanization continues to lower arable land, yet the organization of the Food and Agriculture Organization (FAO) indicates a 4% decline in availability of farmland between 2000 and the current time. Smallholder farmers, who account for 70% of global production, do not normally have the capital, technology, and infrastructure available to sustainably raise production (Ojo and Zahid, 2022). In contrast, rising incomes in developing countries result in an array of resource foods, including meat and dairy. Such a satisfaction of demand as well as food security at the global level, understood as sustained access to safe, nutritious, and suitable foods, results in novel solutions beyond those with conventional agriculture.

2.2 Climate Change and Environmental Stress

One of the scariest challenges to contemporary agriculture is climate change. Severe weather conditions like changes in temperature, shifts in rain, floods, drought, and heat waves result in infringement of agricultural productivity as well as animal health. Culture revenues in some areas will decline to 25% by 2050, as forecast based on the Intergovernmental Panel on Climate Change (MGEIK). In addition, soil erosion caused by overuse and poor land management leads to less productive soil and diminished carbon cycle ability. Water scarcity exacerbates these issues—70% of the global freshwater is allocated to agriculture, while over 2.5 billion inhabitants reside in the water region. This demand necessitates climate agricultural practices and resources.

2.3 Resource Scarcity and Stability

Irreversible consumption and overuse of significant agricultural resources like soil, water, and earth nutrients—long-term security. Overuse of fertilizers and pesticides has caused environmental degradation, for instance, nutrient inventories that pollute reservoirs and create environmental regions like the gigantic Gulf of Mexico. And while that is happening, worldwide phosphorus reserves, a vital ingredient for fertilizers, are dwindling, and some memos get extremely weak for decades. Irrigation systems retain enormous volumes of water, particularly in dry climates like Africa and South Asia, south of the Sahara. The shift to sustainable agriculture is not longer a choice, but a necessity for practices to maximize resource utilization while preserving environmental integrity.

2.4 Working Pressure and Labor Shortages

Agriculture across the globe also has a serious lack of employment. Rural labor is decreasing due to the aging population of farmers and migration to urban areas, particularly in nations such as Japan and the United States, where the farmers' average age is over 55. In developing nations, by and large, urban youth migration leaves the countryside with older people and women who are dependent. Its financial resilience has also been worsened by rising labor costs and global supply chain failures. These shortages of labor will enhance the weakness of human farming systems, eradicate room in pursuit of mechanization and automation, and enhance operational efficiency.

2.5 Consumer and Regulatory Requirements Debt

Currently, consumers are growing rare in their food not just to be healthy and safe, but also to be ethically and long-term produced. Fear of ruin by pesticides, safety, and health of genetically modified things (GMOs), and surrounding residues generate demand for organic, local and weak products. The government will react and cut rules. For instance, Fork's agricultural policy targets cutting the use of pesticides by 50% by 2030. Meeting such development requirements means farmers have to adapt to new methods of adapting environmental management with economic sustainability—frequently in need of technical support in line with regulatory requirements and consumer demands.

All these interdependent challenges—population expansion, climatic volatility, resource depletion, work bans, shifting public expectations—mild curbs on progressive agricultural reform. All you need is systemic change. Agritech has numerous sophisticated recurrence technologies, true data analysts, and a steady innovation reserve trajectory. Actritech addresses the complicated issues confronting world agriculture and not only preserves food security, but also enables agriculture to be redistributive as defenders of people and planets. This revolution is urgent for building a strong, equitable, and stable food system for generations to come.

3. Next-Generation AgriTech Solutions: A Technological Renaissance

The AgriTech revolution marks a revolutionary era in agriculture that uses cutting-edge technologies to solve some of the most critical problems such as food security, climate change, and depletion of resources. With the help of modern tools like artificial intelligence, robotics, biotechnology, and data analytics, AgriTech is transforming agriculture into an accurate, sustainable, and knowledge-intensive business. These technologies are more than incremental innovations; they are revolutionary systems that increase productivity, reduce environmental effects, and equip farmers with tools across all scales (Mishra et al., 2024). This segment discusses the most important technologies driving this revolution, their uses, and their potential to transform global food systems.

3.1 Precision Agriculture: Farming with Pinpoint Accuracy

Precision farming uses technology such as Internet of Things (IoT) sensors, drones, and satellite imaging to track and control crops with unprecedented accuracy. Soil sensors give real-time information about moisture levels, pH levels, and nutrients, so that irrigation and fertilizing can be done selectively. Drones with multispectral cameras can identify pests and nutrient deficiencies at early stages, so that farmers can take corrective action in a timely manner. For instance, John Deere's precision agriculture systems combine AI analytics with GPS-enabled equipment to meet planting accuracy at sub-centimeter levels. This can translate to high returns, such as up to 30% savings in water usage through variable-rate irrigation and up to 50% less chemical usage in pest control applications. Such innovations increase crop yields while preserving resources and safeguarding ecosystems (Shukla et al., 2024b).

3.2 Machine Learning and Artificial Intelligence: Enabling Intelligent Decisions

Machine learning (ML) and AI are transforming decision-making in agriculture through the analysis of large datasets to offer actionable information. An example is IBM Watson Agriculture, which uses environmental and historical data to predict crop yields and optimize planting decisions. Mobile applications such as Plantix employ computer vision to identify plant illnesses via smartphone photographs, providing low-cost options for small farmers. ML also assists supply chain management through demand forecasting and reduction in food losses. The Blue River Technology "See & Spray" system demonstrates the accuracy of AI, employing vision-based algorithms to precisely target herbicide application, saving up to 90% of usage (Shukla et al., 2023). AI improves predictive and adaptive management and, as such, enhances agricultural resilience against climate fluctuations.

3.3 Vertical Farming and Controlled Environment Agriculture: Redefining Agricultural Space

Vertical farming and controlled environment agriculture (CEA) liberated food production from the limitations of environmental factors and geographical positions. These systems utilize hydroponics, aeroponics, and LED lighting for the growth of crops vertically within indoor or urban spaces. For example, AeroFarms is able to produce 10–20 times per square meter than conventional methods of farming while utilizing 70% less water and without the use of pesticides. CEA provides optimal growth conditions year-round, reducing dependency on weather and increasing food availability in urban areas lacking fresh produce. Bowery Farming utilizes robotic systems that dynamically adjust light and nutrient levels, enhancing crop quality while minimizing environmental impact. These technologies address urbanization and land scarcity by promoting local, sustainable food systems (Kumar et al., 2024).

3.4 Robotics and Automation: Filling the Labor Gap

Robotics and automation are revolutionizing the face of agricultural work by performing labor-intensive and repetitive work with efficiency and precision. Robots can plant, weed, and harvest crops independently, reducing the need for human laborers. Farm-ng's Amiga robot, which was built with modularity in its architecture, supports varied small-scale farming activities and can minimize the need for labor by as much as 40%. Harvest Automation's robots mechanize nursery work, while Abundant Robotics has created apple-picking robots with human-like ability. These technologies are especially useful in areas with an aging agricultural labor force and labor deficits. By automating day-to-day operations, farmers are freed to concentrate on strategic planning and management, resulting in better productivity and consistency (Patel et al., 2025).

3.5 Biotechnology and Gene Editing

Biotechnology, particularly CRISPR gene editing, makes it possible to create crops that are more resilient to withstand contemporary environmental stress and nutritional requirements. Technologies such as drought-tolerant corn, insect-resistant rice, and nutrient-rich wheat have the ability to boost yields by 20–30% for regions under stress. Corteva Agriscience's CRISPR-edited waxy corn shows improved stability of yield, while gene-edited crops such as golden rice solve micronutrient deficiencies (Chouhan et al., 2025b). In contrast to conventional GMOs, gene-edited foods tend to have fewer regulatory hurdles, enabling quicker deployment. Biotechnology also includes bio-based solutions like nitrogen-fixing microbes from Pivot Bio that can reduce dependence on chemical fertilizers by as much as 20%. These technologies support food security with sustainable, climate-resilient agricultural practices.

3.6 Blockchain for Supply Chain Transparency: Building Trust and Traceability

Blockchain technology brings transparency and traceability to complicated agricultural supply chains, addressing increasing consumer need for safe and ethically produced food. Blockchain verifies the quality and purity of products by creating tamper-proof transactions records from farm to plate. IBM Food Trust, which retailers such as Walmart use, can identify the origin of produce within seconds, thereby enhancing food safety and recall efficacy. Platforms like AgriDigital provide equitable trade practices and prompt pay for farmers, fostering financial inclusion. Blockchain promotes consumer confidence, prevents fraud, and enables sustainability initiatives by authenticating claims like organic or carbon-neutral production, compliance with global legislation like the EU's Farm to Fork Strategy.

3.7 Regenerative Agriculture Technologies: Restoring Ecosystems

Regenerative farming, which is technology-supported, prioritizes rebuilding the health of soil and carbon sequestration and focuses farming methods on helping achieve global climate objectives. Partnerships such as the Soil Health Partnership encourage evidence-based farming practices like cover crops, crop rotation, and minimal tillage. The carbon farming program by Indigo Ag estimates carbon sequestration through sensors and artificial intelligence and rewards farmers financially for their contribution to the environment. These technologies enhance soil health, increase water holding capacity, and reduce greenhouse gas emissions (Shukla et al., 2021a; Selot et al., 2025). Through the conversion of agriculture from a passive player to an active role in environmental regeneration and climate action, regenerative practices take center stage.

The AgriTech revolution is revolutionizing food cultivation, processing, and distribution fundamentally. These new technologies provide scalable, sustainable, and customized responses to the intersecting issues of population growth, climate change, and scarcity. For its full potential to be achieved, it requires equitable access, strong infrastructure, and favorable policy environment. Ensuring the benefits of AgriTech reach both small and large farming societies is essential for the attainment of a sustainable, balanced, and successful global food system.

4. Impact and Opportunities: Redesigning Food Systems

The AgriTech revolution is a systemic change in world food systems and not just an enhancement of technology. Through the convergence of precision agriculture, AI, vertical farming, robotics, biotechnology, blockchain, and regenerative agriculture, AgriTech meets the pressing demands of feeding more people, reducing the effects of climate change, and improving the use of resources (Jamgaonkar et al., 2024). These new technologies not only respond to current problems but also generate new economic, environmental, and social opportunities. By empowering farmers, reinvigorating rural communities, and integrating food production into sustainability agendas, AgriTech is setting the stage for an inclusive, resilient, and sustainable agroindustry. Threefold effects of AgriTech and the influence of AgriTech in transforming food systems globally are discussed below.

4.1 Increasing Productivity and Food Security

AgriTech increases farm productivity considerably and is therefore essential for food security for a projected 9.7 billion people by 2050. Precision agriculture saves water and fertilizers through IoT sensors and drones, yielding 10–20% increases in yields through data-based decision-making. Urban vertical farming businesses, like AeroFarms, yield 20 times more per square meter compared to conventional agriculture and facilitate year-round production in cities (Shukla et al., 2019). Gene-edited crops, including maize that has been CRISPR-engineered to be more drought-resistant, are capable of lifting harvests by 20–30% under adverse climatic conditions. All these innovations work together to relieve pressure on usable land and support the resilience of food systems, which is in line with the UN Sustainable Development Goal of ending hunger.

4.2 Minimizing Environmental Impact

Sustainability is at the heart of AgriTech, which has reduced the environmental footprint of the agricultural sector by half. Precision farming practices lower water usage by up to 30% and pesticide application by 50%, stopping chemical runoff and guarding aquatic ecosystems. AI breakthroughs such as Blue River Technology's See & Spray lower herbicide application by as much as 90%, fostering biodiversity (Siregar et al., 2022). Vertical farming dispenses with pesticide use and uses 70% less water than conventional agriculture. Regenerative practices, like carbon capture projects by Indigo Ag, enhance soil quality, and lock away 1–2 tons of carbon per acre of land per year. This paradigm shift turns agriculture from a large contributor to greenhouse gas emissions—accounting for roughly 25% of global emissions—into a potential climate solution, bolstering efforts like the Paris Agreement.

4.3 Empowering Smallholder Farmers

Smallholder farmers, who account for around 70% of world food production, significantly benefit from inclusive AgriTech innovations. Tools such as the Plantix app enable farmers to detect plant diseases with smartphone photographs and artificial intelligence, triggering immediate reaction (Shukla et al., 2021b). Technologies such as Hello Tractor provide low-cost rental access to equipment required, boosting productivity with little initial investment. Blockchain technologies, such as AgriDigital, support financial security through on-time payments and transparent pricing. Carbon credit schemes linked with sustainable agriculture give rise to new income streams. By increasing market, technological, and information access, AgriTech allows smallholders to compete globally and engage in dynamic local food systems.

4.4 Economic Growth and Innovation

AgriTech acts as a driving force for economic growth, creating new industries and employment opportunities. Urban farming, driven by vertical farm businesses such as Bowery Farming, generates work in food retail, logistics, and technology while cutting transport-related emissions. The AgriTech market worldwide, worth $17.3 billion in 2023, is anticipated to grow to $46 billion by 2030, driving AI, robotics, and biotech innovations at a faster pace. Automation represented by Farm-ng's Amiga robot can reduce labor costs by as much as 40%, which could lead to reinvestment and expansion. In addition, blockchain-based supply chains increase transparency and link farmers directly with consumers, enhancing market access and profitability for the agri-value chain.

4.5 Establishing Consumer Trust and Food Safety

Today's consumers increasingly want food traceability, safety, and sustainability—needs that AgriTech effectively fulfills. Blockchain technologies such as IBM's Food Trust enable rapid tracing of foods like leafy greens, enhancing recall efficiency and food safety. Biofortified crops like golden rice combat malnutrition by increasing staple food nutrient composition. Regenerative practices enable the certification of carbon-free and organic products, encouraging sustainable purchasing practices. Such innovations increase consumer trust, facilitate brand differentiation, and comply with regulatory changes like the EU's Farm to Fork Strategy that focus on transparency and sustainability within food systems.

4.6 Promoting Global Cooperation and Policy Innovation

AgriTech grows based on collaborative action between a range of stakeholders—governments, private companies, research institutions, and consumers. Public-private partnerships, for example, seek to provide satellite internet service in rural communities via programs such as Starlink, closing the digital divides and facilitating real-time connectivity to AI and IoT technology. International development agencies, for example, the World Bank, facilitate AgriTech adoption in poor countries by providing capital investment, skill upgradation, and infrastructure development. Increased demand from consumers for sustainable and ethical products spurs business investment in green technologies (Siregar et al. 2022). Policy updates improve innovation by reducing bureaucratic hurdles to gene-edited crops and the use of drones. These cooperative systems are critical to providing equal access to AgriTech and improving international farming practice.

The revolutionary reach of AgriTech is broad. By increasing productivity, minimizing harm to the environment, enabling small farmers, and propelling economic and social advancement, AgriTech is forging the future of food. But in order to actualize its potential, challenges including affordability, infrastructure, and policy harmonization need to be overcome, with inclusive access to technology guaranteed. Only then can AgriTech live up to its promise: to feed the world in a sustainable, equitable, and resilient way for generations to come.

5. Barriers to Adoption: Scaling AgriTech Challenges

The Agritech Revolution holds unparalleled potential to revolutionize the global food system, yet it is guarded by severe issues, particularly in poor resources and small conditions. Although technologies like precise agriculture, AI, robotics, and biotechnology have already demonstrated the ability to enhance productivity and stability, structural and systemic issues hinder a broad spectrum of entertainment. Financial constraints, poor infrastructure, regulatory shutdowns, and sociocultural opposition may all limit the advantages of Agritech by large agribusiness firms. Whereas unequal access will amplify prevailing inequality, smallholder farmers constitute over 70% of total world food production (Shukla et al., 2024a). This section examines the principal restraints on Agritech's widespread uptake and stresses the necessity for holistic intervention in order to maximize the benefits of its associated intervention for all stakeholders.

5.1 High Initial Costs and Economic Impossibility

The most limiting obstacle to the implementation of agility is most likely the disproportionately high initial expenses needed in order to access cutting-edge technology. Tools like soil sensors with IoT capabilities, farm drones, autonomous tractors, and vertical farming solutions are theoretically dozens of dollars, or even thousands of dollars. For instance, agricultural drones can put farmers at $5,000–20,000, not factoring in extra expenses of maintenance, software fees, and driver training. For small-scale farmers operating on narrow margins, these costs cannot be utilized without massive economic assistance. In African nations in the southern Sahara where over 80% of farms are low-income, these expenses have tightened their dependency on regular and low-income means. Even payment arrangements and subscriptions are cheaper, yet may keep you under financial pressure. In the absence of assigned subsidies and credit interventions, agility adoption is a privilege meant to target large-scale operations that widen the disparity between industrial and small-scale farming.

5.2 Digital and Infrastructure Differences

The operation and performance of the majority of agricultural buildings depend to a large extent on broadband digital internet, a constant power supply, and the technical availability of services, which remains unstable in rural regions. The International Telecommunications Union declares that nearly 37% of the global population (primarily rural and agrarian areas) is still unconnected. Things are worsening in India. Over 60% of farmers reside in areas with limited connectivity.

Real-time data monitoring and remote management of systems are no longer done. Normal power outages also restrict the usage of sensors, intelligent irrigation systems, and self-controlling machines. Moreover, the absence of local technical support and maintenance results in dependence on remote providers, raising costs and operating times (Siregar et al., 2022). These structural weaknesses greatly discourage the scalability of agriculture in areas where its efficiency is most groundbreaking.

5.3 Privacy Concerns and Ownership of Data

It has emerged as most salient in light of the fact that Agritech's platform has been most notable in improving soil health, crop health in agriculture, weather forecasting, and farm management, as well as ensuring the confidentiality of data, property, and preventing violence. Further, most farmers hesitate to entrust private technology companies with data by putting on hold inappropriate potential applications, i.e., targeted ads, insurance discrimination, or deceptive swaps with third parties. It is extremely crucial that tests pertaining to a substantial decline in the agricultural sector worsen the situation and develop mistrust within the agrarian society. A 2023 US survey states that as low as 25% of farmers were fully secure because their agility supervisor vendors provided feedback. Lack of clarity on data ownership, particularly whether the platform retains ownership of the data it offers, is also a no-hire consideration. Unless they have an open, transparent management system safeguarding farmers' rights, Agritech may lose trust in those who would like to broaden their horizons.

5.4 Regulation and Ethical Issues

Approlet's global regulatory and legal landscape is unequal and uneven, posing severe barriers to adoption and innovation. For instance, cultures included in genes created by CRISPR are strictly regulated in areas like the European Union, where the public remains anxious about genetic alterations, despite scientific agreement on their safety. The application of farming drones is not optimally utilized in most nations due to regulations on airspace and security concerns. Rules governing data management, such as the General Data Protection Regulation (GDPR) of the European Union, also impose a stringent compliance burden on Agritech startups, resulting in higher operational expenses and increased jurisprudence complexity. Moral challenges related to automation give rise to the intricacy of public attitude, such as the concern of job swapping among poor farming communities. Regulation and ethical gray areas contribute to some areas of the implementation and hinder the overall scalability of Agritech.

5.5 Social and Cultural Resistance

Apart from technical and regulatory barriers, sociocultural drivers are major influences for Agritech adoption. The majority of conventional rural farmers, particularly older people, are not willing to adopt new technology that is on the cusp of low digital literacy and conventional methods of conducting business. By and large, Sahara's agriculture in South Asia and Africa is tied to local norms and expertise founded on practices that deter the adoption of new innovations like automation and AI. Gender disparity makes the situation worse. 43% of poor nations' farm workers are women, yet social ethics and institutionalized discrimination permit them to be trained, loaned, or equipped more than their male counterparts. Without culturally relevant Austria, education initiatives and comprehensive design, Agritech is threatening to miss the very group that is trying to expand opportunities (Solanki et al., 2024). The lack of inter-networks on agricultural scale, high economic spending, and the lack of infrastructure for confidentiality issues that resolve obstacles and cultural opposition highlights the difficulties of implementing comprehensive agricultural transformation. These concerns continue to persist, maintaining prevailing inequalities, constraining innovation to the wealthiest sections of interest, yet monitoring small owners and disadvantaged groups as well. Collective action must be taken to democratize Agritech and make it equitably distributed. Discusses access to microfinance, offline tool development, education programs for communities, and establishing equitable and open data management regulations. Only due to a holistic strategy that makes the agricultural revolution keep its promise of creating a stable and equitable food system.

6. Bridging the Gap: Strategies for Inclusive AgriTech

The revolution in agriculture has the ability to transform the world's food system and bring an end to the immediate threat to food security, climate change and paucity of resource constraints. But if systemic resistance is not addressed, like excessive costs, digital disparities, data confidentiality concerns, regulatory complexity, and social resistance, the potential for innovation is constrained. Small producers accounting for nearly 70% of all food products in the world, rural and poor communities are put at risk by this transformation. It must be employed to see that the advantages of Agritech are applied in collaboration with a quite inclusive and topical strategy. This part defines successful and dynamic solutions that facilitate access, extension of rights, capacity, and cooperation throughout the agricultural system.

6.1 Economic Availability: Subsidies and Innovative Funding

The primary hindrance to smallholder farmers adopting farm solutions, like drones, intellectual irrigation systems and ILE-based soils, is the high initial cost. To make it easier, some financial interventions are essential. The program receives government grants and grants the status of India's PM-Kusum program, subsidizing solar pumps for irrigation, imposing an economic burden on poor farmers (Jamgaonkar et al., 2024). Microfinance platforms like Kenya, M-Kopa provide temperamental agricultural technology loans, with the ability of farmers to settle their equipment without having to incur high initial expenses. New Pay-As-You-Go Commercial Model (PAYG) covering Hello Tractor Agricultural Equipment Rental Public-private partnerships, such as the World Bank-backed Agritech Financing program, will also finance the evolutionary and widespread uptake of technologies associated with a range of agricultural conditions.

6.2 *Closing Digital Differences: Infrastructure and Autonomous Solutions*

Rural regions possess sufficient digital infrastructure, such as unstable food and low internet, which restricts the agility and scalability effectiveness. Satellite Internet Services (e.g., Starlink) now provides internet broadband in remote towns with access to IoT and IA-based solutions based on IoT and AI. Solar energy chargers, which are predominantly used in African nations south of the Sahara, have uninterrupted access to energy to smartphones and thinking devices. Offline, just like Farmos, enables farmers to take data in the field before syncing them to be able to access the internet. In a bid to bridge the digital divide even more, governments and non-governmental organizations will establish rural technical centres. For instance, the Indian Digital Greenfield training facility can support more than 2 million farmers and gives connectivity as well as entry data on skill development.

6.3 *Protection of Data Security and Enhancing Farmers' Rights and Capabilities*

Since the Agritech platform is becoming more reliant on humidity of ground data collection and harvesting before weather conditions, there should be a guarantee of ethical data use and ownership. Cleaned data policies are central to demonstrating to people how data is used and available and building confidence. A manufacturer-led platform like Farmstack enables manufacturers to dictate and share data at their will. Blockchain systems like agridigital transactions that secure transactions offer safe traceability and secure both data and financial integrity at the farm level. Industry guidelines like open authentication of AG data offer a base line of data practices. Future creation efforts under digital literacy training and FAO data rights can also educate farmers to make smart choices to join online platforms.

6.4 *Rule Simplification and Weakening of Ethical Concerns*

A dispersed and incoherent regulatory system generates uncertainty and destroys large-scale deployments of agility solutions. Harmonized regulatory frameworks like the new EU governance of genes can yield responsive innovation through easy guidelines and risk evaluations. Streamlined drone laws like Rwanda have facilitated the expansion of agriculture services that aid drones without compromising air security. Government education campaigns like Corteva's

Biotechnology Initiative will assist in debunking myths and fostering public trust regarding genetically exposed cultures and other emerging technologies. Ethical concerns like the allocation of automated employment can be mitigated through active ongoing education programs. For instance, John Deere's engineers' learning program will equip rural employees with skills they need to repair and maintain agricultural machines, creating new jobs for the agricultural systems the technology provides.

6.5 Facilitating Social Integration and Cultural Sensitivity

Social and cultural hurdles in the form of low digital literacy, gender disparities, and resistance to change always deter the use of aggrots in the majority of areas. Programs with special training for contexts given in local languages and localized customs are necessary for boosting digital literacy as well as trust. Video driver guides on digital green indicators are successful in engaging farmers in South Asia as well as in raising the level of tools. For instance, gender projects which deal with IFAD offer smartphones, apps, and agricultural technology training based on AI, and bridging the technical divide between women and men. Cultural integration activity within local leadership dispositions, or local contextualization of traditional reception practitioners. In Nigeria, extended workers who combine local practices with particular agricultural practices have recorded a rise in acceptance of 25%.

6.6 System Change Cooperative Detection

Developing an integrated Agritech system should be a move to engage in multilateral rates. Cross-sectional platforms like the Global Alliance of Agriculture, which help the climate, help collaborative associations among farmers, governments, non-governmental organizations, and private firms for collaborating knowledge and innovation on a broader level. Private initiatives like Food Trust IBM collaborate with small cooperatives to bring digital supply chains to improve market access and visibility (Jamgaonkar et al., 2024). In Brazil, total investment in general IoT platforms and agricultural cooperatives lower expenses and expand accessibility. Governments are able to catalyze innovation using tax incentives, research funds, and policy-driven regulatory processes. Israel's rational agriculture sector is a model of how policies release potential for inclusive innovation and growth.

For a maximal utilization of the agricultural revolution, we must adopt holistic strategies that break across rules on economic, infrastructure, and social barriers to absorption. Due to innovative finance, investment in digital infrastructure, data right protection, alignment of rules, and social integration, Agritech keeps on delivering affordable access and hence remains a holistic and sustainable driver of agricultural change. The extension of the rights and abilities of everyone, particularly smallholders and poor individuals, is a key to the fact that technological progress is not just passed on, but actually, passed on. These collaborative efforts are the backbone to developing a dependable food system feeding humans and worlds in the next decades.

7. The Road Ahead: Envisioning a Sustainable Agricultural Future

Agritech revolution was invented on the brink of a revolution that ensures an innovative path to nourish nearly 9.7 billion individuals by 2050, prevents climatic change and sustains worn-out natural resources. The aforementioned methods, effectiveness, coming advancements of shoulder publicity and agility are harmonized in reaction to embracing novel trends, developing partnerships with numerous owners, and getting rid of embedded systems. A groundbreaking innovation on the basis of promotions and international partnerships, Agritech not only offers an efficient and safe food system with a blueprint. It is assumed in this section that the future is the recognition of those key trends, capabilities and collective effort necessary to fully exploit the potential of Agritech (Patel et al., 2025).

7.1 New Trends: Future Frontiers of Agritech

Agritech's destiny is set by the next innovation wave that redrafts the boundaries of what is possible in farming.

- AI extends beyond assisting in decision-making through autonomous management of farms. IBM\Watson Agriculture is a platform that integrates real-time climate models to enhance stability and raises yields by 15–25% in climate zones.
- Synthetic biology drives a new age of cultural evolution. Plants were evolved by Ginkgo, made for fixing atmospheric nitrogen and improving the efficiency of photosynthesis-plants evolved by ginkgo can bring down fertilizer consumption to 50% and lower the footprint on the environment.
- Nanotechnology facilitates the evolution of smart fertilizers and nanosenaults that inflict precise spots of nutrients, cutting down on waste by 30% with boosted culture health.
- The inaugural urban and modular vertical farming conducted by numerous companies will enhance the production of food in the city center, essentially reducing local pesticide-free food, and offer local food. These trends show a future model of agriculture. This depicts the future model of previous agriculture. This is excessive, safeguarded from destruction, highly resilient, and cushions climate volatility and urbanization.

7.2 Climate-Smart Future: Agriculture as a Solution

Farming, brought about by farming itself technology, and causal agents of climate change can also be a major solution.

- Regenerative farming, contributing to platforms like Indigo AG, saves 1–2 tonnes of Co2 annually, making farms carbon sinks.

- Precision farming preserves emissions by conserving fertilizer and water consumption, but vertical farms save the pressure on the earth while conserving biodiversity.
- Carbon markets, such as those NORI initially referred to, will create economic incentives for farmers that convert into sustainable practices, and the market is predicted to grow to $12 billion by 2030. It resists the culture's climate, like being resistant to Corn Corteva's drought, to produce high productivity despite tough conditions, and provide food security amidst climate uncertainty. Collectively, these programs place agriculture in line with international climate goals, such as the Paris Agreement, supporting the sector for adaptation and softening.

7.3 *Equitable Food Systems: Inclusion at the Core*

Agritech should be focused to not intensify current disparities.

- Mechanism platforms for payment like Hello Tractor enable smallholder farmers to boost profitability and income.
- Farmers can be held accountable by data sovereignty platforms, e.g., FarmStack, to gather farms and exchange data.
- Women with access to digital facilities are given access through specific gender programs, e.g., the IFAD Technology Training Program, to women who comprise 43% of the agricultural workforce in developing nations.

Due to our evolutionary holistic solutions, Agritech can enhance rural poor societies, enhance stability in food systems, and build a more just world agricultural landscape.

7.4 *Global Cooperation: Global Vision for Agricultural Transformation*

Realize the utmost level of agricultural demand for systematic and cross-sectional collaboration.

- Political leadership, Israel's tax incentives and legislation supporting agritech, have built a $1.7 billion ecosystem.
- Declare private cooperation like the World Bank's agrito that accelerates and gives needy regions critical funding and training.
- Businesses investing in open source products like John Deere and IBM are making AI and blockchain technology more accessible to farmers globally.
- Consumer-led responsibility, supported by social networks, motivates the company towards sustainable research and openness.
- International partnerships like the Global Alliance for Agriculture with Climate promote global knowledge sharing.

Such partnership models ensure Agritech's innovations are not singular, but in the form of integrated growth in multiple sectors and geographic areas.

7.5 Problems and Methods

It is very likely, but there are some key issues that have to be surmounted because of the key scale of Agritech.

- Balancing regulation is essential for gene-exposed mobility methods. Nations like Japan have a pace with innovative managers that can be replicated worldwide.
- Digital illiteracy still is a limited area. Programs like Digital Green Farmers Training, which have seen more than 2 million farmers, illustrate the strength of training.
- Infrastructure investments, especially, will remain the bedrock. Programs like rural Starlink band deployment on a large scale make a huge contribution to real-time use of data and automation. To bridge these gaps, stakeholders have to prioritize policy reforms, invest specifically, and conduct educational programs, so that farmers and communities are not lagging behind the agricultural revolution.

The Agritech future is an issue and a hope. From synthetic biology and AI to blockchain and carbon agriculture, emerging technologies turn agriculture into controlled, religious and holistic information. But stakeholders need to come together to overcome the economic, digital and social barriers to Agritech in order to achieve their highest potential. Agritech's innovation, constrained by justice and stability, can transform its direction towards building a future where the food system nourishes people as well as the planet itself. This revolution is not merely about nutrition in the world. This is a shift in how future generations produce, share and assess food.

References

Chavula P. and Kayusi, F. (2025). Systematic Review on the Application of Nanotechnology and Artificial Intelligence in Agricultural Economics. *LatIA, ISSN-e 3046-403X, Nº 3, (Ejemplar dedicado a: LatIA)*, *3*: 322. https://doi.org/10.62486/latia2025322.

Chouhan, S., Singh, U. Applications SJ-NC and, 2025 undefined (2025a). Performance evaluation of different deep learning models used for the purpose of healthy and diseased leaves classification of Cherimoya (Annona Cherimola) plant. In: S.S. Chouhan, U.P. Singh, and S. Jain, *Neural Computing and Applications*. Springer, *37*: 4531–4544. https://doi.org/10.1007/S00521-024-10830-X+.

Chouhan, S.S., Patel, R.K., Singh, U.P. and Tejani, G.G. (2025b). Integrating drone in Agriculture: Addressing technology, challenges, solutions, and applications to drive economic growth. *Remote Sens. Appl.*, *38*: 101576. https://doi.org/10.1016/J.RSASE.2025.101576.

Chouhan, S.S., Saxena, E., Shukla, A. et al. (2025). Optimizing YOLO-Based Models for Real-Time Guava Detection with Probabilistic Fused Wiener Filter-Enhanced Feature Fusion. *Applied Fruit Science*, *67*: 383. https://doi.org/10.1007/s10341-025-01613-2.

Jamgaonkar, S., Gowda, J.S., Chouhan, S.S., et al. (2024). An Analysis of Different YOLO Models for Real-Time Object Detection. In: *Proceedings of The 4th International Conference on Sustainable Expert Systems, ICSES 2024* (pp. 951–955). https://doi.org/10.1109/ICSES63445.2024.10763020.

Kumar, D., Chouhan, S.S., Patel, R.K., et al. (2024). Transformative Insights: Gabor Features and Tensor-EWT for COVID Diagnosis in Lung CT Images. In: *2nd IEEE International Conference on Innovations in High-Speed Communication and Signal Processing, (IHCSP)*. https://doi.org/10.1109/IHCSP63227.2024.10960090.

Mishra, V., Chouhan, S.S, Vadula, J., et al. (2024). Enhancing Driver Safety through Convolutional Neural Networks for Drowsiness Detection. In: *2nd IEEE International Conference on Innovations in High-Speed Communication and Signal Processing (IHCSP)*. https://doi.org/10.1109/IHCSP63227.2024.10960213.

Ojo, M.O. and Zahid, A. (2022). Deep Learning in Controlled Environment Agriculture: A Review of Recent Advancements, Challenges and Prospects. *Sensors*, *22*: 7965. https://doi.org/10.3390/S22207965.

Patel, R.K., Kumari, N. and Chouhan, S.S. (2025). Glaucoma diagnosis using Gabor and entropy coded Sine Cosine integration in adaptive partial swarm optimization-based FAWT. *Biomed. Signal Process Control*, *107*: 107832. https://doi.org/10.1016/J.BSPC.2025.107832.

Selot, N., Panwa, A., Shukla, A., et al. (2025). Comparative Analysis of Convolutional Neural Network and Support Vector Machine for the Prediction of Alzheimer's Disease. In: *Communications in Computer and Information Science*, *2457 CCIS*: 56–66. https://doi.org/10.1007/978-3-031-89760-3_5.

Shukla, A., Khan, M.Z., Kumar, S., et al. (2023). Interpretive Structural Modeling Based Assessment and Optimization of Cloud with Internet of Things (CloudIoT) Issues through Effective Scheduling. *Intelligent Automation and Soft Computing*, *36*: 2281. https://doi.org/10.32604/IASC.2023.031931.

Shukla, A., Kumar, S. and Singh H. (2021a). MLP-ANN-Based Execution Time Prediction Model and Assessment of Input Parameters through Structural Modeling. In: *Proceedings of the National Academy of Sciences India Section A - Physical Sciences*, *91*: 577–585. https://doi.org/10.1007/S40010-020-00695-9/METRICS.

Shukla, A., Kumar, S. and Singh, H. (2019). Fault tolerance based load balancing approach for web resources. *Journal of the Chinese Institute of Engineers, Transactions of the Chinese Institute of Engineers, Series A*, *42*: 583–592. https://doi.org/10.1080/02533839.2019.1638307.

Shukla, A., Kumar, S. and Singh, H. (2021b). MLP-ANN-Based Execution Time Prediction Model and Assessment of Input Parameters Through Structural Modeling. In: *Proceedings of the National Academy of Sciences India Section A - Physical Sciences*, *91*: 577–585. https://doi.org/10.1007/S40010-020-00695-9.

Shukla, A., Shukla, V., Kumar, S., et al. (2024a). Real-Time Authenticated and Secure Monitoring and Control of Low-Cost Sustainable Hydroponic System in Cloud. *Lecture Notes in Networks and Systems*, *918*: 49–63. https://doi.org/10.1007/978-981-97-0641-9_4.

Shukla, A., Shukla, V., Kumar, S. and Anand, A. (2024b). Secure Design and Implementation of Smart Traffic Light Management System. *Lecture Notes in Networks and Systems*, *918*: 65–73. https://doi.org/10.1007/978-981-97-0641-9_5.

Siregar, R.R.A., Seminar, K.B., Wahjuni, S. and Santosa, E. (2022). Vertical Farming Perspectives in Support of Precision Agriculture Using Artificial Intelligence: A Review. *Computers*, *11*: 135. https://doi.org/10.3390/COMPUTERS11090135.

Solanki, S., Singh Chouhan, S., Dwivedi, A., Singh, U.P. and Patel, R.K. Leveraging Deep Learning for the Identification and Categorization of Fruit Diseases, 2024 IEEE International Conference on Intelligent Signal Processing and Effective Communication Technologies (INSPECT), Gwalior, India, 2024, pp. 1–6, doi: 10.1109/INSPECT63485.2024.10896118.

2 Introduction to Deep Learning in Agriculture

Aditi Kachchap[1], Jaya Sinha[2], Adityanshu Tripathi[3], Shreesh Sanjay Amin[4], Kalamkuri Chandu[5], Rohit Anand[6] and Abhishek Upadhyay[7*]

1. Introduction

1.1 Background and Motivation

In recent years, there have been arguments which have threatened food security and environmental sustainability, resulting in a serious turning point in global agriculture. With a rise in global population reaching 9.7 billion by 2050, food production will increase by 70%, all while trying to reduce the effects of climate change, resource scarcity, and environmental degradation. The establishment of the deep-learning-based AI (artificial intelligence) trend in the world also brings the whole set of powerful tools to the agriculture practices to be great through critical data-driven insights, automated surveillance, and accurate resource management. Gain superhuman-intelligence by using the never-ending and vast data sources for deep learning-from satellite pictures and drones till the measurement taken through IoT sensors and imaging. AI brings in more and more slaughter, and the process of AI being a must-have is gaining grounds especially for the sustainable precision agriculture (Chouhan et al., 2024). Smart agriculture is the realm of AI and it is open-minded to a whole range of digital technologies, like deep learning, big data, and the Internet of Things (IoT) (Liu et al., 2021).

1.2 The Evolution of Precision Agriculture

Precision agriculture (PA) aims to protect the environment and optimize profitability through the efficient use of inputs based on the temporal and spatial

[1,3-5] M.Tech. Scholar, Department of Farm Machinery and Power Engineering, CAET, Dr. Rajendra Prasad Central Agricultural, Pusa, Samastipur, Bihar, India.

[6] Ph.D. Student, Department of Food, Agricultural and Biological Engineering, The Ohio State University, Columbus, OH-43210, USA.

[2,3,7] Assistant Professor, Department of Farm Machinery and Power Engineering, CAET, Dr. Rajendra Prasad Central Agricultural, Pusa, Samastipur, Bihar, India.

* Corresponding author: abhiupadhyay7798@gmail.com

DOI: 10.1201/9781003715726-2

variability of soil and crops. PA mainly focuses on site-specific management through various technologies such as remote sensing, GPS, GIS, etc. Remote sensing and advanced machine learning have accelerated this change of pattern, the target being monitoring crop health, soil variability, yield prediction, disease detection for farmers, all with respect to a level of accuracy and temporal resolution never before achieved.

In PA, the use of advanced technologies has become the main reason for the great increase in crop productivity. These advanced technologies create maps of areas of homogeneous crop status, and thus, each area can be treated independently and with appropriate treatment. This method has great potential since it reduces both environmental impact and costs. Following these maps, ortho fertilizer maps are created and N application rates are calculated for corn monoculture in small areas and these application rates are sent to the vehicle computers for controlling them. UAVs (unmanned aerial vehicles) have similar use and the same hardware of GPS and inversion unit of the data (Chollet, 2017; Ronneberger et al., 2015).

1.3 Scope and Organization

This chapter gives details about the utility of deep learning in the context of smart agriculture supported by remote sensing, concluding with a sincere debate on theory bases, and applied strategies and actual scenarios. It first of all describes the main characteristics of deep learning networks such as convolutional neural networks (CNNs), recurrent neural networks (RNNs), and transformers, in respect to their relation with the agriculture sector. Following this, there are several ways to take in remote sensing information from multiple origins, i.e., satellite images, drone photographs, and sensor data, in an effort to get a comprehensive picture of crop health and environmental conditions. Emphasis is put on such modern methods as edge computing and federated learning that do not only allow one to have intelligence on the farm level but also to solve issues of data security, network bandwidth, and computational load at the same time. By using these bases, the chapter will possibly consider important apps like crop monitoring, initial detection of diseases and pests, forecast of yields, and optimization of inputs to precision, all of which payoff to increasing farm productivity and footprints of sustainability. Furthermore, in order to synchronize theory with practice, there will be presented case studies that will reach not only to the implementations but also to the outcomes and the lessons learned in practice. Lastly, the chapter deliberatively reviews the current pitfalls and constraints such as but not limited to data differences, model explainability, and scalability and comes up with proposals of research directions that engineers the future of AI-assisted smart farming.

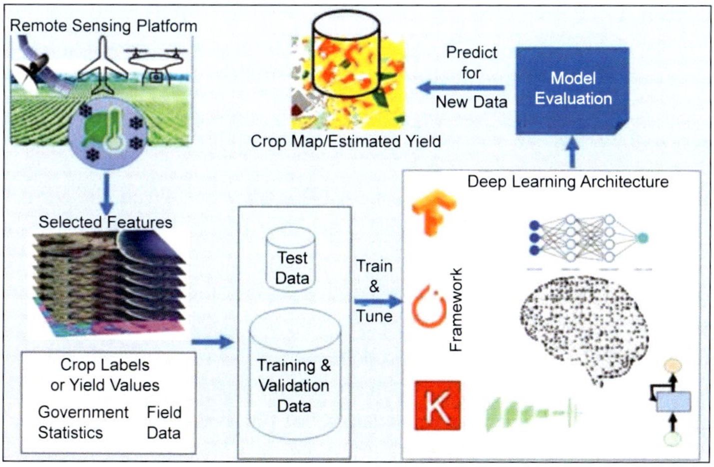

Fig. 1 Conceptual framework of integrated deep learning and remote sensing for smart agriculture (Kamilaris & Prenafeta-Boldú, 2018).

2. Deep Learning Fundamentals in Agricultural Applications

2.1 Theoretical Foundations

Deep learning is a field within machine learning where computers learn to recognize patterns and solve complex problems using layered neural networks—structures inspired by the human brain. These networks automatically figure out useful features from huge amounts of data, so people don't need to hand-design rules or explanations for every situation. Unlike conventional approaches that rely on manual feature extraction, deep learning models autonomously derive meaningful patterns through multiple layers of nonlinear transformations, making them highly effective in handling the high-dimensional nature of agricultural data. Advanced filtering and neural architectures, such as BM3D with spiking neural networks in medical imaging (Patel, 2025), demonstrate transferable potential for agricultural imaging tasks. The agricultural industry uses deep learning to improve the quantity and quality of crops using image-based classification. Deep learning-enabled crop health monitoring has been widely used for disease identification across crops such as basil, tomato, and banana (Patel et al., 2025; Krishnan et al., 2022). To enhance and automate tasks, numerous researchers have used deep learning technology and methodology (Zheng et al., 2019).

Its strength lies in the ability to fuse diverse data types, including spectral, spatial, temporal, and contextual information, particularly offering advantages in agriculture, where crop performance, soil status, and environmental variables are intricately interconnected and difficult to analyze with traditional methods. Changing environments that vary spatially and temporally have huge effects on year-to-year and location-to-location variations in crop yield (Horie et al., 1992).

2.2 Convolutional Neural Networks for Spatial Data

In terms of agricultural remote sensing, CNNs are considered the best method for spatial data analysis. They apply learnable filters throughout images to automatically identify and extract relevant patterns, textures, spatial information, etc., to be used for tasks such as crop health estimation, disease recognition, and yield prediction.

With the introduction of advanced models of deep learning, for example, ResNet, EfficientNet, and DenseNet have addressed the problem of vanishing gradients and encourage feature reuse for better performance (Atila et al., 2021). These networks are effective in agriculture for crop disease classification, early pest detection, and crop maturity assessments in various environments (Barbedo, 2019).

2.3 Hybrid CNN–Transformer Architectures

Recent opportunities offered by deep learning created a space for hybrid CNN-Transformer models that use CNNs for local feature extraction and Transformers for global context modeling (Mehdipour et al., 2025). A famous example is CMTNet (Convolutional Meets Transformer Network) with its two-branch dual-channel structure that captures both local and global representations simultaneously. These architectures have achieved over 98% classification accuracy on agricultural hyperspectral datasets (Guo et al., 2025).

Hybrid models have become a solution to the limitations typical for CNNs by providing the means to handle the long-range dependencies and contextual patterns over the entire image; and still, it is a hybrid approach to the new method that is using k-means clustering along with a CNN as the solution, which very effectively manages the computational time, and thus is the recent one with the highest accuracy of 92.6% in the case of leaf disease classification. In the case of Blackberry, the combination of CNNs and machine learning classifiers for the olive trees gives rise to the deep hybrid models, which impressively reach a 96.14% accuracy for the disease classification task (Sonam et al., 2023). A multi-output constraint mechanism can also be incorporated to enhance classification performance by applying cross-checks at various stages of feature extraction.

2.4 Recurrent Neural Networks and Temporal Analysis

Typically, the transformational model was ranked the best by its success in classification rather than the CNN-based methods. Moreover, the focal points of their conjunction make the image a little bright. They have LSTMs (long short-term memory) and GRUs (gated recurrent units), which are the most commonly used types of RNNs, operating in sequence and time, and their attention mechanisms help a lot in showing the predictive features of the product. Agriculturally, these models are time series of satellite, weather station, and IoT-based sensors data feeding and detecting phases of crop growth, pest infestation, and disease outbreak; which can be used for planning irrigation activities. RNNs are particularly good at monitoring crop growth stage dynamics. On the other hand, the most recent algorithms have started to add attention mechanisms to these networks, thereby the model not only picks out but also gets this particular time spot(s) that is (are) really important for getting the yield and crop health outcomes, and finally, provides the complete prediction in a more transparent manner that ensures also user trust and such as farmers and agronomists. ViTs, however, normally require a lot of data to be effective during training, and are known to be very computationally intensive, especially in the case of higher resolutions (Upadhyay et al., 2024).

2.5 Transformer Networks for Global Context Modeling

With its ability to comprehend distant and overall relations, which was actually its initial purpose at the beginning being used for natural language processing, the Transformer Model has become a tool of great value and success in the field of agricultural remote sensing. The trend of Vision Transformers (ViTs) has become the latest fad and it has presented an option for the market to use in applications involving agricultural remote sensing, and it has proven to be much more efficient in comparison with the traditional CNNs. Via the technique of the Vit model, the primal input of images is accomplished by the splitting of the picture data into the patches of the same size (generally 16 x 16 pixels) and the consequent supplying of the latter as a sequence of inputs to the self-attention mechanism. This mode of operation has a considerable advantage in that it gives the model the power to grasp the global relational structures present in the whole image: a property that is very convenient to the agriculture sector where the features consisting of crops, diseases, etc. might be very spread out geographically (Upadhyay et al., 2024).

On the whole, compared to traditional CNN-based methods, Transformer-based models are typically more accurate in their classification. The novelty of their attention mechanisms is exactly that: throwing a beam on those regions of the image that are most contributing to a prediction and hence making the model more transparent and trustworthy. such as farmers and agronomists. ViTs usually demand a higher operating cost to collect one effecting data, thus usually, they need more data for an effective training, and they are very computational loaded especially when the resolution is higher.

3. Application Domains and Case Studies

3.1 *Crop Health Monitoring and Disease Detection*

IoT-based smart sensors enable automated plant inspection by utilizing computer vision and deep learning algorithms (Tirkey et al., 2023). Deep learning-based crop health monitoring is among the most established AI applications in agriculture, with numerous successful implementations demonstrating significant improvements in both disease detection accuracy and response time. Advanced CNN architectures can detect early-stage disease symptoms that are often not yet visible and suggest methods for timely treatment (Anand et al., 2024). Intelligent machines, on the other hand, can be deployed quickly, accurately, and at scale to support real-time crop monitoring for early disease detection. These active strategies can further reduce crop losses, make better use of land, and cut down on pesticide use, paving the way toward more sustainable farming (Krishnan et al., 2022).

The disease detection framework working with a set of highly advanced deep learning operations based on the use of remote sensing data acquired from UAVs or handheld cameras delivers classification accuracy of over 95% for many crop types and different disease classes. In addition to domain knowledge supported explainable AI methods such as Grad-CAM and LIME, the farmer can be supplied with a model decision visual explanation thus enabling them to treat the disease rationally.

The response time of any AI system designed for the purpose of being so explainable must be within 2.5 seconds, as the highest accuracies are generally obtained from such models. have found that the ResNet-34 model had a striking 98.42% test accuracy, whereas their enhanced model— ResNet197—contributed even more by having 99.58% accuracy in recognizing plant diseases. This brings attention to the extra- ordinary potential of ResNet models in the area of automated plant disease identification and classification, using the method of image processing.

One of the things that provide more value to these methods is the combination of high-end models namely ResNet and EfficientNet along with tools that provide the ability to explain. This powerful mixture does not only result in accurate diagnosis of the disease, but also comes up with graphical explanations that show which parts of the plant images have the disease symptoms, thus making the results more comprehensible and trustworthy, especially for farmers and agronomists.

3.2 *Yield Estimation and Prediction*

The researchers used a supervised learning algorithm, known as a perceptron, to predict crop yields (Rosenblatt, 1958). Perceptrons can be divided into two types: single-layer perceptrons and multilayer perceptrons.. The single-layer perceptron algorithm is a form of binary classification that allows artificial neurons to learn and subsequently process each element of the training dataset automatically, one at a time. A multilayer perceptron (MLP) model is built by stacking perceptrons, and organizing them in layers (Aggarwal, 2018). An MLP architecture could be identified as a Deep Neural Network (DNN) if the arrangement of hidden layers is sufficiently large; however there is no definitive rule as to how many hidden layers it takes to be considered deep (Géron, 2019). There is an increasing demand for an accurate yield forecast for purposes such as agricultural planning, market analysis, and resource allocation. Deep learning approaches have been shown to predict crop yields from satellite images with a multi-temporal framework, weather data, and historical yield records with remarkable accuracy-the result is better than traditional-based methods by 20–30%. The machine learning methods that are employed for yield prediction use several sources of data to give a comprehensive prediction:

- **Satellite-based monitoring**: It is considered state-of-the-art today to use satellite images for crop yield prediction at the farm scale without intrusive and time-intensive kind of monitoring, as explained in (Sagan et al., 2021; Sharma et al., 2020; You et al., 2017) . Real-time condition assessment of the field based on the analysis of multispectral imagery.
- **Weather integration**: Weather is one of the paramount external factors influencing a farm (Ngoune & Shelton, 2020). Deep learning systems can learn from the large quantities of historical weather data, thus discovering hidden temporal and spatial relations without depending on an explicit physical model.
- **Historical pattern analysis**: Deep learning models, especially RNNs like LSTM, CNNs, and hybrid architectures, are employed to analyze multi year temporal data series to detect underlying patterns and trends. Multi-year trends drawn from past yields, soil, management, and environmental variables help long-term planning and support time-series modeling (e.g., LSTM, CNN-LSTM architectures).
- **Resource optimization**: Data-driven advisory systems can recommend optimal planting schedules, nutrient applications, irrigation, and other inputs—contributing to improved yields and cost-effective management decisions.

3.3 Pest Detection and Management

Efficient pest management plays a vital role in sustainable farming, as pest infestations can seriously diminish both the productivity and quality of crops. Traditional methods for pest control often depend on manual labor for detecting and tracking pests, making the process laborious, slow, and prone to mistakes caused by human factors (Upadhyay et al., 2025b; Anand et al., 2024a). Modern machine learning and deep learning techniques, especially Convolutional Neural Networks (CNNs), have become a potential method for automating the detection, location, and classification of pests in agriculture (Ayan et al., 2020). Integrated pest management strategies have undergone a revolution in relation to such automated pest detection systems encountered by deep learning. These can detect pest species, assess population densities, and recommend intervention strategies specially tailored to local field conditions from time to time.

Pest detection accuracies above 90% at real-time processing; thus, the different architectures of advanced object detection: YOLO and RetinaNet series. Should this system be integrated with a precision applicator device, selective spraying could be allowed, thus reducing the use of chemicals while still maintaining effectiveness against pests (Anand et al., 2025). Wang et al. (2023) trained AI models with the Pest24 dataset in order to monitor pests in real time in agriculture, yet challenges remain with densely distributed pests and small object size (He et al., 2016; Haq et al., 2023; Jiao et al., 2020).

3.4 Weed Detection and Management

Precision weed management systems based on deep learning distinguish crop plants and weeds for site-specific herbicide spraying and mechanical control (Lottes et al., 2017; Dos Santos Ferreira et al., 2017). Such systems maintain a weed detection accuracy exceeding 95% under varying field situations. A great leap forward is the introduction of SWFormer: a scale-wise hybrid CNN-Transformer paradigm that significantly outperforms other algorithms in multi class weed classification and canopy coverage estimation (Su et al., 2023). Multispectral imaging, combined with deep learning, also improves the detection of species of weeds highly visually similar to the crops and better than human observation and traditional monitoring methods (Kamilaris & Prenafeta-Boldú, 2018).

3.5 Resource Optimization

Deep learning-enabled resource optimization systems combine a variety of data sources to boost irrigation, fertilization, and energy management, among other things, according to (Liakos et al., 2018). The effective references gained support from both the soil conditions, the weather forecasts, the crop growth stages, and the historical performance records. By considering soil conditions, weather forecasts, crop development stages, and historical performance records, these systems offer precise and data-based input recommendations. Variable Rate Application (VRA) systems empowered by deep learning algorithms have made noticeable efficiency improvements with the reported chemical savings lying in the range of 15–30% without any loss in crop productivity (Shamshirband et al., 2019). To add on, combining the internet of things (IoT) -based sensor networks with the predictive modeling frameworks leads to the support of the adaptive management strategies that vary and relate dynamically depending on the field conditions. The use of drones in agriculture helps to minimize material, and energy losses and thereby, simultaneously increase system effectiveness.(Chouhan et al., 2025)

Table 1 Comparative analysis of deep learning applications in agriculture: Accuracy, efficiency, and resource savings.

Application Domain	*Accuracy Improvement*	*Resource Savings*	*Response Time*	*Implementation Complexity*	*References*
Disease Detection	85– 95%	20–40% pesticide reduction	<3 seconds	Medium	Shoaib et al., (2025)
Yield Prediction	20– 40% error reduction	15–25% planning efficiency	Real-time	High	Jabed et al. (2024)
Pest Management	90– 95%	30–50% pesticide reduction	Real-time	Medium	Venkateswara et al. (2025)
Weed Control	95%+	70–90% herbicide reduction	Real-time	High	Su et al. (2023)
Resource Optimization	15–30%	15–30% input savings	Real-time	High	Abioye et al. (2020)

3.6 Explainable AI and Decision Support Systems

3.6.1 The Need for Interpretability in Agricultural AI

The expansion of artificial intelligence in the field of farming is surely burdened by huge problems about belief, clearness, and approval of the end-users. Farmers are in need of open explanations of recommendations produced by AI so that they can decide based on the provided information. The black-box, however accurate, deep learning model is hardly used in practice in agriculture due to the lack of interpretability features associated with most of the AI models in the field of agriculture. However, by the help of Explainable AI (XAI) techniques, the problem is solved as they provide users with the full, plain and justified explanations of the model predictions and recommendations. Such transparency is not only vital for the users but also required for the legislation to be satisfied, especially if the law is the GDPR where the right to explanation is such a mandatory requirement.

3.6.2 XAI Techniques in Agricultural Applications

LIME (Local Interpretable Model-agnostic Explanations) technology has been found to be extremely effective in the field of agriculture, mainly because of its capability to give a very clear and focused explanation for a single case of machine learning model predictions. The localized explanation is considered to be a usually black box type of model, and hence, explaining with the use of LIME, the user or the validator can be sure the decision was made in a predictable way (Malashin et al., 2024). For instance, LIME is particularly popular in the field of crop recommendation where the system helps the farmer to realize crop recommendations, the soil parameters, climatic factors, yield performance of the crop area, and so on. The technology holds possibilities in the agricultural field, such as helping the interpret agricultural-plants in a human way, building a stronger relationship between the AI system and the farmer, giving the field of agriculture a method where predictions are made and these predictions are confirmed by agriculture experts, and providing a means of creating personal recommendations by taking into account the peculiarities of each farmer. Besides, other interpretability methods like SHapley Additive explanations (SHAP) and Gradient-weighted Class Activation Mapping (Grad-CAM) give different but complementary benefits. While SHAP provides feature-level explanations via quantification of the contribution of each input variable towards a prediction by a model, Grad-CAM enables visual interpretability by highlighting regions within images most responsible for classification outcomes, making it especially useful in plant disease diagnosis and image-based crop monitoring operations. They complement each other to significantly enhance the transparency, credibility, and usability of AI-based decision-support systems in agriculture.

4. Case Studies and Real-World Implementations

4.1 Advanced Hyperspectral Crop Classification Using Hybrid CNN-Transformer Networks

A hybrid framework called the Convolutional Meets Transformer Network (CMTNet) was created especially for UAV-based hyperspectral crop classification, especially for precision wheat disease monitoring. The system was deployed on a SITONHOLY IW4202 Linux server (Ubuntu 16.04, CUDA 10.1) equipped with four NVIDIA TITAN Xp GPUs (12 GB each) and 128 GB RAM, it was implemented using Python 3.8 and PyTorch 1.7.0. CMTNet integrates a CNN branch for extracting local spectral-spatial features and a Transformer branch for modeling global contextual relationships. The spectral-spatial feature module combines 3D and 2D convolutions, while a multi-output constraint module applies cross-level feature supervision (Guo et al., 2025). Optimal training parameters included a 13×13 input patch, learning rate of 1e-3, batch size of 100, and 100 training epochs, using cross-entropy loss optimized via Adam. The best-performing configuration included one Transformer layer with four attention heads.

Evaluation on three UAV-acquired hyperspectral datasets, WHU-Hi-LongKou, WHU-Hi-HanChuan, and WHU-Hi-HongHu, demonstrated high classification accuracies of 99.58%, 97.29%, and 98.31%, respectively (Guo et al., 2025). Compared to the previous state-of-the-art CTMixer method, CMTNet achieved accuracy improvements of 0.19% on LongKou, 1.75% on HanChuan, and 2.52% on HongHu (Zhang et al., 2022). To ensure statistical reliability, each experiment was repeated ten times independently. This hybrid design enables real-time deployment on UAV platforms, making it suitable for precision agriculture applications. The combination of local CNN-based feature extraction with global Transformer-based contextual modeling significantly enhances robustness in complex agricultural conditions characterized by diverse crop types, developmental stages, and imbalanced datasets (Guo et al., 2025; Zhang et al., 2022).

4.2 IoT-Enabled Smart Rice Cultivation with Machine Learning Integration

An IoT–machine learning system was deployed across agricultural regions in Bangladesh in partnership with Habiganj Agricultural University (Sharafat et al., 2025). The architecture has Raspberry Pi 5 edge nodes that are connected to RS485 7-in-1 soil sensors that keep an eye on the soil's nitrogen (N), phosphorous (P), potassium (K), pH, temperature, and soil moisture. Weather API integrations provide real-time humidity and rainfall data. ThingsBoard was used for live data visualization, while LIME was incorporated to enhance model interpretability. Multiple algorithms were evaluated: Artificial Neural Networks (ANN), Support Vector Machines (SVM), Decision Trees (DT), Random Forest (RF), and a stacking ensemble. The system was trained on a dataset of 3,300 samples covering 22 crops and eight environmental and soil features. Reported accuracies were: ANN: 95.6%, SVM: 93.2%, RF: 95.8%, and the stacking ensemble achieved 95.9%. The Random Forest model was deployed on edge devices, and was based on its high accuracy and efficiency (Sharafat et al., 2025).

The Internet of Things – Machine Learning (IoT–ML) framework was not only tested but also trained using a dataset with 3,300 samples covering 22 types of crops and including eight characteristics of the soil and the overall environment. It showed great predictive performance consistently over several algorithms. Such as the ANN gained the best accuracy of 95.6%, then came SVMs around 93.2%, and the RF was third with 95.8%, but the stacking ensemble was the winner with the highest performance at 95.9% (Sharafat et al., 2025). Because of its great balance between accuracy and computational efficiency, the RF model was picked for edge deployment, verifying previous research that emphasized its resistance and ease of interpretation in the case of agricultural IoT applications (Sharafat et al., 2025). In field positionings, the integrated system achieved extensive resource savings and productivity improvements:

- Yield enhancement of 18–25% compared to traditional practices
- Water usage reduced by 22%
- Fertilizer consumption lowered by 28%
- Pesticide application cut by 45%

4.3 Edge Computing Implementation for Real-Time Crop Health Monitoring

The Tiny-LiteNet system stands as the state-of-the-art edge computing solution optimized for real-time crop pest and disease detection. The main block of the system is an edge Raspberry Pi 5 that houses an easy CNN architecture with one and half million parameters and a compact model of 1.2MB. The system is made up of six Squeeze-and-Excitation (SE) blocks for finer extraction of features, a camera of HD for the purpose of capturing images and GSM/GPRS communication for the cloud connectivity.

Great performance was shown by the Tiny-LiteNet edge computing system for Raspberry Pi 5 when it was used for crop pest and disease detection in real-time. The classifier attained an accuracy of 98.6%, F1 score of 98.4% and recall of 98.2% in the model with a 1.48 million parameter, 1.2 MB size. The inference time was 80 ms per classification. When comparing it with similar work offered by VGGNet-16, Inception, ResNet50, DenseNet121, MobileNetv2, and EfficientNetB0, Tiny-LiteNet had the upper hand in delivering a better result and was more convenient and efficient for deployment. This clearly indicates its applicability in real-time monitoring in the agrarian environment, where constraints on resources are the order of the day (Nyakuri et al., 2025). Real-world deployment led to significant operational improvements:

- Alert Generation: Sub-second response for critical crop conditions
- Time Savings: 50% reduction in farm operational time
- Offline Capability: Continued operation during network connectivity issues
- Cost Effectiveness: $80 edge device cost with robust field performance

A comprehensive edge computing deployment has demonstrated the effectiveness of real-time crop monitoring through distributed processing architectures, showcasing how localized intelligence can enhance precision agriculture. It integrated drone imagery with the network of ground sensors to detect stress conditions in crops and immediately alert, allowing for autonomous responses concerning irrigation or nutrient application. The performance assessments delivered results far beyond expectations, including lowering crop conditions response time below one second, thus bringing about actual interventions to preclude yield losses. There is a 20% drop in fertilizer consumption, a 30% drop in water consumption, and farm management time was halved due to enhanced operational efficiency. Most importantly, the edge-enabled architecture proved to be reliable even in the scenario of network outages or limited network connectivity, thus hinting at seamless monitoring and decision-making abilities. This is the big highlight of edge computing in smart-farming systems, juxtaposing speed, resilience, and sustainability.

5. Implementation Challenges and Solutions

5.1 Data Heterogeneity

Agricultural datasets vary tremendously due to differences in geography, weather, soil types, cultivation practices, and sensor technologies. This heterogeneity puts conventional federated learning methods, which assume uniform data distributions, to the challenge. Hence, allowing a global model to be personalized according to local conditions while using shared knowledge through PF-learning will solve such problems. Customers could also have their membership in clusters sorted out according to environmental resemblance to elevate the quality of the final data sets. Domain adaptation methods, for instance, adversarial feature alignment, mitigate the heterogeneity mainly from the heterogeneity of sensors and/or conditions in the fields. These methods contribute to creating a more robust and precise agriculture networked learning framework (Liu et al., 2025).

5.2 Communication and Computational Constraints

In most cases rural agricultural areas do not have a stable internet connection and sufficient computer processing speed. Extraordinarily large model updates cannot be loaded through a poor or unstable network. The best and most practical approach would be model compression, which greatly reduces the size of updates to be transmitted and an asynchronous communication protocol where the individual farms are the vendors of their own updates, and the frequency of these updates is limited only by the connection status (Aldossary et al., 2025). The design of edge computing is another trend that is considered an enabler at the moment because it makes training and inference possible at the point where the data is being collected. This, in turn, makes it less dependent on the cloud infrastructure, saves bandwidth, and at the same time ensures that decision-making is not broken by the absence of the connection to the internet (Gong, 2025).

5.3 Data Quality and Availability

Reliable data, suitable quality, and well-represented data are the three primary needs for deep-learning systems in agriculture to train. Different atmospheric aberrations, cloud cover, or drift in sensor calibration are the main causes of errors in remote sensing applications. Field data collection is not only prohibitively expensive but also spatially restricted virtually. Hyperspectral datasets systematically collected in laboratory settings with strict calibration protocol are the major attention of the researchers working with these barriers (Abdulridha et al., 2020). Preprocessing techniques like spectral smoothing and radiometric normalization are crucial for maintaining this consistency. Meanwhile, data-augmentation-based approaches

Table 2 Summary of case study implementations and their key performance metrics.

Case Study	*Implementation Details*	*Performance Metrics / Results*	*Impact / Agricultural Benefits*	*References*
Advanced Hyperspectral Crop Classification (CMTNet)	• Hybrid CNN–Transformer network for UAV-based hyperspectral wheat disease monitoring • Python 3.8, PyTorch 1.7.0 • SITONHOLY IW4202 server with 4× NVIDIA TITAN Xp GPUs, 128 GB RAM • CNN branch: local spectral-spatial feature extraction (3D+2D convolutions) • Transformer branch: global contextual modeling • Input patch: 13×13, LR: 1e-3, Batch size: 100, Epochs: 100 • One Transformer layer, 4 attention heads	• WHU-Hi-LongKou: 99.58% • WHU-Hi-HanChuan: 97.29% • WHU-Hi-HongHu: 98.31% • Accuracy improvement over CTMixer: +0.19% / +1.75% / +2.52%	• Real-time UAV deployment • Robust feature extraction in diverse crop conditions • Improved precision disease monitoring	Guo et al. (2025)

Contd...

Table 2 *Contd...*

Case Study	*Implementation Details*	*Performance Metrics / Results*	*Impact / Agricultural Benefits*	*References*
IoT-Enabled Smart Rice Cultivation with ML Integration	• Deployed in Bangladesh (Habiganj Agricultural University) • Raspberry Pi 5 edge nodes with RS485 7-in-1 soil sensors (N, P, K, pH, temperature, moisture) • Weather API for humidity/rainfall • ThingsBoard for visualization • LIME for model interpretability • Algorithms: ANN, SVM, DT, RF, stacking ensemble Dataset: 3,300 samples, 22 crops, 8 environmental/soil features	• ANN: 95.6% - SVM: 93.2% • RF: 95.8% - Stacking ensemble: 95.9% • RF chosen for edge deployment	• Yield increase: 18–25% • Water usage: −22% • Fertilizer consumption: −28% • Pesticide application: −45%	Sharafat et al. (2025)
Edge Computing for Real-Time Crop Health Monitoring (Tiny-LiteNet)	• Raspberry Pi 5 with lightweight CNN (1.48M params, 1.2 MB) • Six Squeeze-and-Excitation blocks • HD camera module - GSM/GPRS for cloud connectivity • Real-time pest & disease detection	• Accuracy: 98.6% • F1-score: 98.4% • Recall: 98.2% • Inference time: 80 ms/ classification • Outperforms VGGNet-16, Inception, ResNet50, DenseNet121, MobileNetv2, EfficientNetB0	• Sub-second alerts for critical crop conditions –50% reduction in farm • Cost-effective edge solution (~$80) - Optimized resource use: fertilizer −20%, water −30%	Nyakuri et al. (2025)

could augment training samples to simulate cloud shadows or alternative illuminations with an eye towards better generalization under real working conditions.

5.4 Model Interpretability and Transparency

The output of decision support systems must be explainable to be trusted by farmers, researchers, and policymakers. Several strategies enhance interpretability. Feature attribution methods, such as SHAP, provide insights into the relative importance of different input variables (Ahmed, 2025). For image-based tasks, visual explanation methods such as Grad-CAM highlight the regions of an image that contribute most to a classification, for example diseased spots on leaves. Other approaches include counterfactual analysis, which presents alternative scenarios to demonstrate how different management choices could influence outcomes, and uncertainty quantification, which adds confidence measures to predictions (Lakshminarayanan et al., 2017). By integrating these approaches, agricultural deep learning systems become more transparent and trustworthy

5.5 Computational Resource Optimization

Many agricultural devices, such as drones, robots, and field sensors, lack access to high-performance computing resources. Therefore, model efficiency is crucial. Recent advances include neural architecture search (NAS), which designs networks optimized for speed and memory usage (Guo et al., 2022). Pruning eliminates redundant parameters from trained models, while quantization reduces numerical precision to accelerate computation without significantly reducing accuracy. Knowledge distillation further compresses models by transferring the knowledge of large "teacher" networks into smaller, faster "student" networks (Hinton et al., 2015). These strategies enable deployment of robust models on resource-limited platforms such as edge devices, allowing real-time use in the field.

5.6 Scalability and Generalization

Any accommodation into agricultural AI will need to go forth in many areas, crop species, and whenever the seasonal variation differs. In the initial training phase, large, general data sets were used to train the models that transfer learning fine-tunes through the use of a limited field data set thereby making them better fit for certain agricultural cases (Thudumu & Fisher, 2025; Janga & Dave, 2025). Multi-task architectures provide support for multiple crop types within a single model by sharing feature extractors and by having crop-specific outputs derived from the same model—avoiding the requirement to retrain the model completely. The use of temporal modules, such as the recurrent neural networks, or seasonal fine-tuning by means of updated data, helps in dealing with the seasonality issue. Preprocessing that is sensor-agnostic guarantees that the models deliver the same performance on different devices, especially when calibrated properly and radiometric normalization is done. By combining all these methods, deep learning models are able to be trained in a controlled environment and then to be effectively generalized to the complex, changing conditions of large agricultural areas.

Table 3 Technical challenges and corresponding solution approaches.

Challenge Category	*Specific Issues*	*Solution Approaches*
Data Heterogeneity	Non-uniform data across farms, sensors, and climates leading to poor model convergence	Personalized federated learning, client clustering, domain adaptation
Communication & Computation	Limited bandwidth, intermittent connectivity, low processing power at the edge	Model compression, asynchronous updates, edge computing integration
Data Quality & Availability	Cloud cover, atmospheric interference, calibration drift, labor-intensive ground data	Standardized hyperspectral datasets, preprocessing (smoothing, normalization), data augmentation
Model Interpretability	Black-box predictions limiting user trust	Feature attribution (SHAP), visual explanations (Grad-CAM), counterfactual analysis, uncertainty quantification
Resource Optimization	Lack of high-performance computing resources in rural settings	Neural architecture search (NAS), pruning, quantization, knowledge distillation
Scalability & Generalization	Need for adaptation across crops, regions, seasons, and sensor types	Transfer learning, multi task architectures, seasonal fine-tuning, sensor-agnostic preprocessing

6 Future Directions and Emerging Technologies

6.1 Next-Generation AI Architectures

Next-generation AI architectures are set to provide stronger support for agricultural systems, enabling broader reasoning and more sophisticated decision-making. Foundation models and large multimodal models are being adapted across agricultural domains to better understand the full complexity of real-world farming scenarios (Yin, 2025).

Large-scale foundation models that integrate multiple types of data—such as images, sensor readings, weather information, and historical records—appear to offer the most promising approach for developing intelligent agricultural systems. These models are able to combine the sources of information for farm management recommendations with respect to each context with an accuracy never before attained (Li, 2025).

In advanced AI systems, causal reasoning can incorporate both offensive and defensive reasoning of causes in the agricultural domain, enabling more accurate predictions and reliable recommendations. Thus, a putative management decision may be implemented, his own correct methods for the long-term outcomes instead of short-term measures. (Wang, 2024; Chouhan, S.S., 2025).

6.2 Advanced Sensor and Monitoring Technologies

Emerging sensor technologies provide new data sources for agricultural AI systems to monitor parameters and processes not considered in the past. Looking into the history, computer vision can comfortably date itself back to the 1950s, with a focus on two-dimensional image analysis. However, with deep learning—and especially CNNs—the field has been transformed, letting computer vision impress us with its sophisticated yet highly accurate interpretation of visual data (Cao et al., 2023; Denghui et al., 2021; Zhao et al., 2019).

Some recent studies have revealed that ferroelectric photosensors can be used in sensor artificial neural networks for the electrical approximation of the sensed data (Wan et al., 2023; Wang et al., 2021). On top of that, it permits hyperspectral imaging systems with higher spectral coverage and better spatial resolutions to still be able to find and track subtle signs of plant stress and biochemical changes. While quantum sensors address soil composition, plant metabolism, and air compositions along with elemental detection, the latter case offers an obvious advantage due to ultra-high sensitivity levels of the probes.

Integration of satellite, UAV, ground-based, and emerging aerial platforms (stratospheric balloons, high-altitude pseudo-satellites) creates comprehensive monitoring networks with complementary capabilities. This multi-scale approach provides continuous coverage from field to global scales.

6.3 Autonomous Agricultural Systems

The integration of AI with robotics and autonomous systems is revolutionizing agriculture by enabling fully automated operations that perform complex tasks with minimal human oversight. Coordinated swarms of agricultural robots, powered by swarm intelligence algorithms, facilitate distributed monitoring, precision application, and harvesting across expansive fields. These swarms optimize overall system performance by adapting dynamically to changing environmental conditions. Future autonomous agricultural systems are anticipated to include advanced AI capabilities for real-time adaptation, self-optimization, and predictive maintenance, learning continuously from experience to enhance operational efficiency (Kamilaris & Prenafeta-Boldú, 2018; Liakos et al., 2018; Farmonaut, 2025) performance.

6.4 Blockchain and Supply Chain Integration

Blockchain technology brings a set of complementary functionalities to AI for transforming agricultural supply chains by providing a decentralized, immutable ledger toward traceability, transparency, and food safety. AI handles predictive analytics, crop condition monitoring, and decision-making support, while blockchain technology ensures that the data is stored with integrity and shared securely. Therefore, these technologies aim to enable smart deals, support the independent operation of farm machinery, and temporary agricultural practices that are resilient to climate change. This integration brings more efficient, transparent, and resilient agricultural value chains; however, it also calls for advancements in data quality, connectivity, scalability, and regulatory acceptance (Nagar & Nagar, 2024).

6.5 Climate Resilience and Adaptation

The field of AI in agriculture will, in fact, concentrate on making the sector secure for extreme weather changes. Such technologies will be applied to the latest climate models, extreme weather prediction as well as adaptive management strategies to enable the agri-food sector to not only survive but also effectively work under new environmental conditions. By integrating AI with advanced climate simulations, these technologies can predict the long-term impacts on agriculture and optimize adaptation strategies for crop selection, infrastructure development, and resource allocation within a framework of climate uncertainties. Agricultural practices that are technology and data-driven, not only will they be able to forecast risks, but they will also foresee and limit potential losses, which is the future of farming today (Farmonaut, 2025; Rebez et al., 2024).

7. Conclusion

The integration of deep learning with remote sensing has brought a change like agriculture, opportunities for monitoring crop condition, detecting diseases, yield evaluation, and optimizing resource use. Currently, applications demonstrate effective solutions in food production engineering, underlining significant savings in resources. By combining advanced neural network architectures—like CNNs, transformers, and hybrid models—with multi modal remote sensing data, modern AI systems can now perform many tasks that once required human expertise. Edge computing and federated learning have helped overcome major challenges in computational efficiency and data privacy, making it possible to deploy intelligent systems widely across different agricultural settings.

The future of agriculture will be characterized by intelligent, adaptive, and sustainable systems that optimize productivity while preserving environmental resources and supporting rural communities. Deep learning technologies integrated

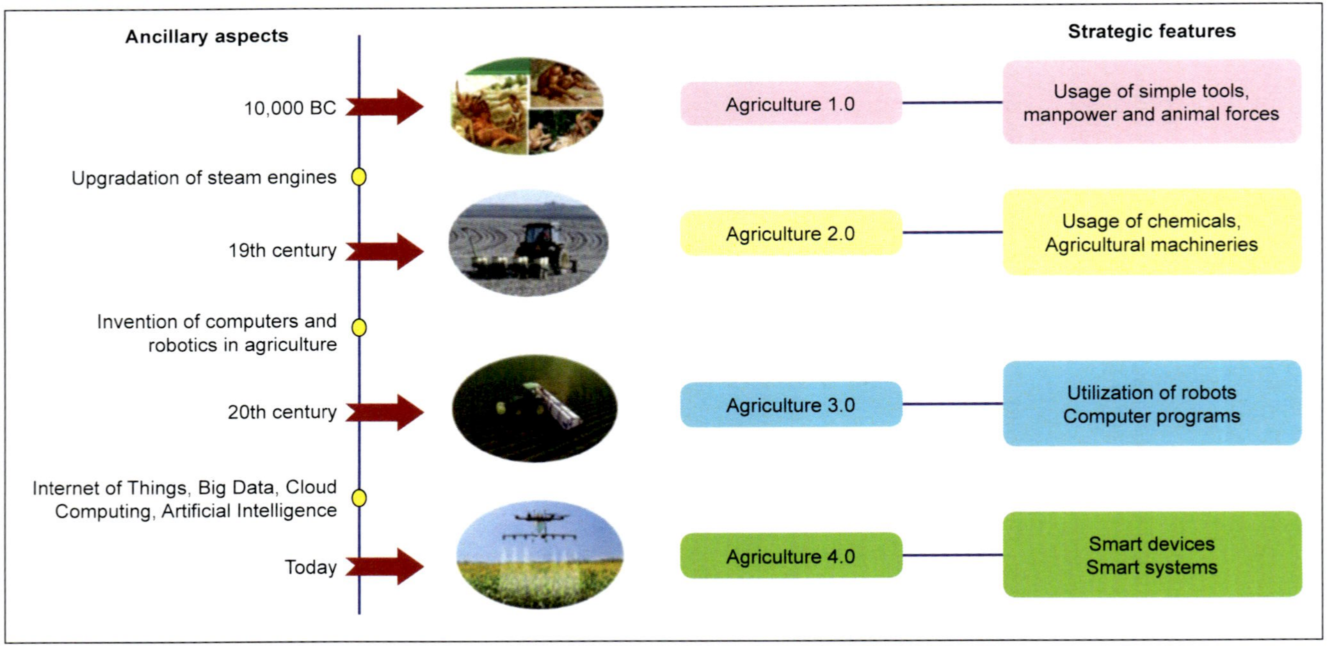

Fig. 2 Technology roadmap for future agricultural AI development showing convergence of emerging technologies (Dhanaraju et al., 2021).

with advanced remote sensing will serve as the foundation for this transformation. The continued development and deployment of deep learning technologies in agricultural remote sensing will play a central role in achieving these goals, contributing to global food security, environmental sustainability, and rural prosperity. The convergence of technological capabilities, economic benefits, and societal needs creates unprecedented opportunities for transforming agriculture into a truly intelligent and sustainable system.

Declaration

Ethical Responsibilities of Authors

"All authors have read, understood, and complied as applicable with the statement on 'Ethical responsibilities of Authors' as found in the Instructions for Authors."

Ethical Approval

Not applicable

Use of AI tool

The author declares that OpenAI's GPT-based language model was used to rephrase and refine the manuscript content.

Competing interests

Not applicable

Authors' contributions

Aditi Kachchap[1]: Conceptualization, Methodology, Writing - Original Draft, **Jaya Sinha**[2]: Reviewing and Supervision, **Adityanshu Tripathi**[3]: Formal analysis, Visualization, **Shreesh Sanjay Amin**[4]: Formal analysis, Methodology, **Kalamkuri Chandu**[5]: Formal analysis, Visualization, **Rohit Anand**[6]: Writing – Reviewing and Editing, Supervision, **Abhishek Upadhyay**[7]: Writing – Reviewing and Editing, Conceptualization, Formal analysis, Visualization, Supervision.

Funding

No funding has been received.

References

Abdulridha, J., Ampatzidis, Y., Kakarla, S.C. and Roberts, P. (2020). Detection of target spot and bacterial spot diseases in tomato using UAV-based and benchtop-based hyperspectral imaging techniques. *Precision Agriculture, 21*(5), 955–978. https://doi.org/10.1007/s11119-020-09755-2.

Abioye, E.A., Abidin, M.S.Z., Mahmud, M.S.A., Buyamin, S. and Ishak, W.F.W. (2020). IoT-based monitoring and data-driven modelling for smart farming applications. *Agronomy, 10*(11), 1642. https://doi.org/10.3390/agronomy10111642.

Aggarwal, C.C. (2018). Neural networks and deep learning (Vol. 10, No. 978, p. 3). Cham: Springer.

Ahmed, M.T. (2025). A systematic review of explainable artificial intelligence for agricultural applications. *Computers and Electronics in Agriculture, 204*, 107429. https://doi.org/10.1016/j.compag.2025.107429.

Albetis, J., Duthoit, S., Guttler, F., et al. (2017). Detection of Flavescence dorée grapevine disease using unmanned aerial vehicle (UAV) multispectral imagery. *Remote Sensing, 9*(3), 308. https://doi.org/10.3390/rs9030308.

Aldossary, M., Alzahrani, A. and Alsubaie, A. (2025). Federated LeViT ResUNet for scalable, privacy-preserving precision agriculture: Real-time edge computing applications. *Agronomy, 15*(4), 928. https://doi.org/10.3390/agronomy15040928.

Anand, R., Madhusudan, B.S. and Bhalekar, D.G. (2024b). Computer vision and agricultural robotics for disease control. In: *Applications of Computer Vision and Drone Technology in Agriculture 4.0* (pp. 31–47). Singapore: Springer Nature Singapore. https://doi.org/10.1007/978-981-99-8684-2_3.

Anand, R., Parray, R.A., Mani, I., Khura, T.K., Kushwaha, H., Sharma, B.B. and Godara, S. (2024a). Spectral data driven machine learning classification models for real time leaf spot disease detection in brinjal crops. *European Journal of Agronomy*, *161*, 127384. https://doi.org/10.1016/j.eja.2024.127384.

Anand, R., Parray, R.A., Mani, I., Khura, T.K., Kushwaha, H., Sharma, B.B., Susheel Sarkar, S., Godara, S., Mojerlou, S. and Mirzakhaninafchi, H. (2025). A multimodal approach for enhanced disease management in cauliflower crops: Integration of spectral sensors, machine learning models and targeted spraying technology. *Front. Agr. Sci. Eng.*, *12*(2). https://doi.org/10.15302/J-FASE-2024572.

Atila, Ü., Uçar, M., Akyol, K. and Uçar, E. (2021). Plant leaf disease classification using EfficientNet deep learning model. *Ecological Informatics, 61*, 101182. https://doi.org/10.1016/j.ecoinf.2021.101182.

Ayan, E., Erbay, H. and Varçın, F. (2020). Crop pest classification with a genetic algorithm-based weighted ensemble of deep convolutional neural networks. *Computers and Electronics in Agriculture, 179*, 105809. https://doi.org/10.1016/j.compag.2020.105809.

Barbedo, J.G.A. (2019). Plant disease identification with deep learning: Validation and applications. *Computers and Electronics in Agriculture, 163*, 104846. https://doi.org/10.1016/j.compag.2019.104846.

Cao, S. (2023). Analysis of object recognition trends based on deep learning. *Applied Computing and Engineering, 5*, 292–299. https://doi.org/10.54254/2755-2721/5/20230582.

Chakraborty, A., Das, S. and Mondal, B. (2022). Integrating neural network for pest detection in controlled environment vertical farm. Indian journal of science and technology, 15(17), 829–838.

Chollet, F. (2017). Xception: Deep learning with depthwise separable convolutions. In Proceedings of the IEEE conference on computer vision and pattern recognition (pp. 1251–1258).

Chouhan, S.S., Singh, U.P., Saxena, A. and Jain, S. (2024). Assessing the Importance and Need of Artificial Intelligence for Precision Agriculture. In: Chouhan, S.S., Saxena, A., Singh, U.P., and Jain, S. (Eds.), *Artificial Intelligence Techniques in Smart Agriculture*. Springer, Singapore. https://doi.org/10.1007/978-981-97-5878-4_1.

Chouhan, S.S., Saxena, E., Shukla, A., et al. (2025). Optimizing YOLO-Based Models for Real-Time Guava Detection with Probabilistic Fused Wiener Filter-Enhanced Feature Fusion. *Applied Fruit Science*, 67, 383. https://doi.org/10.1007/s10341-025-01613-2.

Chouhan, S.S., Patel, R.K., Shukla, A. and Singh, U.P. (2025, Aug.). Drone Importance and their Necessity in Future Generation Agriculture. In: Artificial Intelligence and Computer Vision for Ecological Informatics (pp. 1–8). doi: 10.1201/9781003518778-1.

Denghui, Z., Song, L., Feng, Y. and Yang, Q. (2021). Research status of damage identification algorithm based on deep learning. *E3S Web of Conferences, 233*, 04039. https://doi.org/10.1051/e3sconf/202123304039.

Dhanaraju, M., Chenniappan, P., Ramalingam, K., Pazhanivelan, S. and Kaliaperumal, R. (2022). Smart Farming: Internet of Things (IoT)-Based Sustainable Agriculture. *Agriculture*, *12*(10), 1745. https://doi.org/10.3390/agriculture12101745.

Dos Santos Ferreira, A., Freitas, D.M., da Silva, G.G., Pistori, H. and Folhes, M.T. (2017). Weed detection in soybean crops using ConvNets. *Computers and Electronics in Agriculture, 143*, 314–324. https://doi.org/10.1016/j.compag.2017.10.027.

Farmonaut. (2025). *Agriculture and Artificial Intelligence: 2025 AI Farming Trends*. Retrieved September 7, 2025, from https://farmonaut.com/precision-farming/agriculture-and-artificial-intelligence-2025-ai-farming-trends.

Géron, A. (2019). Hands-on machine learning with scikit-learn, keras, and tensorflow: concepts. Aurélien Géron-Google Kitaplar, yy https://books. google. com. tr/books.

Gong, R. (2025). Edge computing-enabled smart agriculture: Technical frameworks and applications. *Sensors, 25*(17), 5302. https://doi.org/10.3390/s25175302.

Guo, J., Han, K., Wu, H., Tang, Y., Chen, X., Wang, Y. and Xu, C. (2022). CMT: Convolutional neural networks meet vision transformers. In: *Proceedings of the IEEE/CVF Conference on Computer Vision and Pattern Recognition* (*CVPR*) (pp. 12165–12175). https://doi.org/10.1109/CVPR52688.2022.01190.

Guo, X., Feng, Q., Guo, F. and Yang, W. (2025). CMTNet: A hybrid CNN-Transformer network for UAV-based hyperspectral crop classification in precision agriculture. *Scientific Reports, 15*, 12383. https://doi.org/10.1038/s41598-025-12383.

Haq, S.I.U., Raza, A., Lan, Y. and Wang, S. (2023). Identification of pest attack on corn crops using machine learning techniques. *Engineering Proceedings, 56*(1), 183. https://doi.org/10.3390/ASEC2023-15953.

He, K., Zhang, X., Ren, S. and Sun, J. (2016). Deep residual learning for image recognition. In: *Proceedings of the IEEE Conference on Computer Vision and Pattern Recognition* (*CVPR*) (pp. 770–778). https://doi.org/10.1109/CVPR.2016.90.

Hinton, G., Vinyals, O. and Dean, J. (2015). Distilling the knowledge in a neural network. https://doi.org/10.48550/arXiv.1503.02531.

Hoeser, T. and Kuenzer, C. (2020). Object detection and image segmentation with deep learning on earth observation data: A review-part i: Evolution and recent trends. Remote Sensing, 12(10), 1667.

Horie, T., Yajima, M. and Nakagawa, H. (1992). Yield forecasting. *Agricultural Systems, 40*, 211–236. https://doi.org/10.1016/0308-521X(92)90022-G.

Jabed, M.A., Rahaman, M., Neelotpol, S.J. and Ahmed, N. (2024). Crop yield prediction in agriculture: A comprehensive review. *ScienceDirect*. https://www.sciencedirect.com/science/article/pii/S2405844024.

Janga, R. and Dave, R. (2025). Enhancing smart farming through federated learning: A secure, scalable approach for crop disease detection. *Agricultural Intelligence and Automation*, *12*(2), 56–67. https://doi.org/10.47852/bonviewAIA52025089.

Jiao, L., Dong, S., Zhang, S., Xie, C. and Wang, H. (2020). AF-RCNN: An anchor-free convolutional neural network for multi-categories agricultural pest detection. *Computers and Electronics in Agriculture, 174*, 105522. https://doi.org/10.1016/j.compag.2020.105522.

Kamilaris, A. and Prenafeta-Boldú, F. X. (2018). Deep learning in agriculture: A survey. *Computers and Electronics in Agriculture, 147*, 70–90. https://doi.org/10.1016/j.compag.2018.02.016.

Krishnan, V.G., Deepa, J., Rao, P.V., Divya, V. and Kaviarasan, S. (2022). An automated segmentation and classification model for banana leaf disease detection. *Journal of Applied Biology and Biotechnology, 10*(1), 213–220. https://doi.org/10.7324/JABB.2022.101105.

Lakshminarayanan, B., Pritzel, A. and Blundell, C. (2017). Simple and scalable predictive uncertainty estimation using deep ensembles. *Advances in Neural Information Processing Systems, 30.* https://proceedings.neurips.cc/paper/2017/hash/9ef2ed4b7fd2c810847ffa5fa85bce38-Abstract.html.

Li, H. (2025). A review on enhancing agricultural intelligence with large multimodal models. *Agricultural Intelligence, 5*(2), 45–60. https://doi.org/10.1016/j.agriint.2025.02.001.

Liakos, K.G., Busato, P., Moshou, D., Pearson, S. and Bochtis, D. (2018). Machine learning in agriculture: A review. *Sensors, 18*(8), 2674. https://doi.org/10.3390/s18082674.

Liu, Q., Yan, Q., Tian, J. and Yuan, K. (2021, February). Key technologies and applications in intelligent agriculture. *Journal of Physics: Conference Series, 1757*(1), 012035. https://doi.org/10.1088/1742-6596/1757/1/012035.

Lottes, P., Hörferlin, M., Sander, S. and Stachniss, C. (2017). Effective vision-based classification for separating sugar beets and weeds for precision farming. *Journal of Field Robotics, 34*(6), 1160–1178. https://doi.org/10.1002/rob.21681.

Malashin, I., Kondrateva, A. and Pushkin, S. (2024). Deep learning model for sustainable yield prediction: Enhancing interpretability through explainable AI. *Sustainability, 16*(21), 9437. https://doi.org/10.3390/su16219437.

Mohammed, B. (2025). A comprehensive overview of federated learning for agriculture: Privacy-preserving crop classification and management improvements. *Informatica, 49*, 117–136. https://doi.org/10.31449/inf.v49i0.4123.

Nagar, S. and Nagar, D. (2024). Blockchain and AI for agriculture. In: *IIP Series, Volume 3, Book 1, Part 8, Chapter 4* (pp. 267–277). Poddar International College & Jayoti Vidyapeeth Women's University. Retrieved from https://iipseries.org/assets/docupload/rsl2024B9C649F945C939E.pdf.

Ngoune, L. and Shelton, C.M. (2020). Factors affecting yield of crops. In *Agronomy-Climate Change and Food Security*. IntechOpen. https://doi.org/10.5772/intechopen.92141.

Nyakuri, J.P., et al. (2025). AI and IoT-powered edge device optimized for crop pest and disease detection. *Scientific Reports, 15*, 12345. https://doi.org/10.1038/s41598-025-06452-5.

Patel, R.K. (2025, Aug.). BM3D Filtering with Ensemble Hilbert-Huang Transform and Spiking Neural Networks for Cardiomegaly Detection in Chest Radiographs. *Computational Biology and Chemistry*, *8*, 108620.

Patel, R.K., Chouhan, S.S., Chandan, T. and Singh, U.P. (2025, Aug.). Basil Crop Detection Using Computer Vision and Deep Learning Approach. In: *Artificial Intelligence and Computer Vision for Ecological Informatics* (pp. 67–82). doi: 10.1201/9781003518778-4.

Rahat, I.S., Ghosh, H., Dara, S., et al. (2025). Towards precision agriculture tea leaf disease detection using CNNs and image processing. *Scientific Reports, 15*, 17571. https://doi.org/10.1038/s41598-025-0237.

Ramanjot, Mittal, U., Wadhawan, A., Singla, J., Jhanjhi, N.Z., Ghoniem, R.M. and Abdelmaboud, A. (2023). Plant disease detection and classification: A systematic literature review. Sensors, 23(10), 4769.

Rebez, A., et al. (2024). Enhancing system resilience to climate change through artificial intelligence and predictive analytics. *Frontiers in Climate*, *5*, 1585331. https://doi.org/10.3389/fclim.2025.1585331.

Ronneberger, O., Fischer, P. and Brox, T. (2015, October). U-net: Convolutional networks for biomedical image segmentation. In International Conference on Medical image computing and computer-assisted intervention (pp. 234–241). Cham: Springer international publishing.

Rosenblatt, F. (1958). The perceptron: A probabilistic model for information storage and organization in the brain. *Psychological Review, 65*(6), 386–408. https://doi.org/10.1037/h0042519.

Sagan, V., Maimaitijiang, M., Bhadra, S., Maimaitiyiming, M., Brown, D.R. and Sidike, P. (2021). Field-scale crop yield prediction using multi-temporal WorldView-3 and PlanetScope satellite data and deep learning. *ISPRS Journal of Photogrammetry and Remote Sensing, 174*, 265–281. https://doi.org/10.1016/j.isprsjprs.2021.06.006.

Sahithi, B. and Vigneshwari, S. (n.d.). Elevating crop quality: AI neural networks for advanced plant disease management and enhanced food security. Department of Computer Science and Engineering, Sathyabama Institute of Science and Technology.

Shamshirband, S., Jafari Nodoushan, E., Alizamir, M., Yee, P.L., Petković, D., Ch, S. and Gani, A. (2019). Predicting crop yield using hybrid genetic algorithm–adaptive neuro-fuzzy inference system. *Computers and Electronics in Agriculture, 168*, 105146. https://doi.org/10.1016/j.compag.2019.105146.

Sharafat, M.D.S., Kabya, N.D., Emu, R.I., Ahmed, M.U., Onik, J.C., Islam, M.A. and Khan, R. (2025). An IoT-Enabled AI system for real-time crop prediction using soil and weather data in precision agriculture. *Smart Agricultural Technology, 12,* 101263. https://doi.org/10.1016/j.atech.2025.101263.

Sharma, S., Rai, S. and Krishnan, N. C. (2020). Wheat crop yield prediction using deep LSTM model. *arXiv*. https://arxiv.org/abs/2011.01498.

Shoaib, M., Sadeghi-Niaraki, A., Ali, F., Hussain, I. and Khalid, S. (2025). Leveraging deep learning for plant disease and pest detection: A comprehensive review and future directions. *Frontiers in Plant Science*. https://www.frontiersin.org/journals/plant-science/articles/10.3389/fpls.2025.1538163/fu.

Siddharth Singh Chouhan, Rajneesh Kumar Patel, Uday Pratap Singh, Ghanshyam G. Tejani. (2025). Integrating drone in Agriculture: Addressing Technology, Challenges, Solutions, and Applications to Drive Economic Growth. *Remote Sensing Applications: Society and Environment*, e *38* (2021), 101576. ISSN 2352-9385, https://doi.org/10.1016/j.rsase.2025.101576.

Su, T., Li, X., Ren, S., Wang, C., Shen, J., Peng, B. and Ma, Y. (2023). SWFormer: Scale-wise transformer for accurate weed mapping in precision agriculture. *ISPRS Journal of Photogrammetry and Remote Sensing, 200*, 85–99. https://doi.org/10.1016/j.isprsjprs.2023.04.005.

Thudumu, S. and Fisher, J. (2025). OpenAg: Democratizing agricultural intelligence. *Institute of Applied Artificial Intelligence and Robotics.* Retrieved from https://arxiv.org/html/2506.04571v2.

Tirkey, D., Singh, K.K. and Tripathi, S. (2023). Performance analysis of AI-based solutions for crop disease identification, detection, and classification. *Smart Agricultural Technology, 5*, 100238. https://doi.org/10.1016/j.sat.2023.100238.

Upadhyay, A., Chandel, N.S. and Chakraborty, S.K. (2024). Disease control measures using vision-enabled agricultural robotics. In: *Applications of Computer Vision and Drone Technology in Agriculture 4.0* (pp. 165– 186). Singapore: Springer Nature Singapore. https://doi.org/10.1007/978-981-99-8684-2_10.

Upadhyay, A., Chandel, N.S., Singh, K.P., Chakraborty, S.K., Nandede, B.M., Kumar, M. and Elbeltagi, A. (2025a). Deep learning and computer vision in plant disease detection: A comprehensive review of techniques, models, and trends in precision agriculture. *Artificial Intelligence Review, 58*(3), 92. https://doi.org/10.1007/s10462-024-11100-x.

Upadhyay, A., Patel, A., Chandel, N.S., Chakraborty, S.K. and Bhalekar, D.G. (2025b). Leveraging AI and ML in Precision Farming for Pest and Disease Management: Benefits, Challenges, and Future Prospects. In: *Ecologically Mediated Development: Promoting Biodiversity Conservation and Food Security* (pp. 511–528). https://doi.org/10.1007/978-981-96-2413-3_23.

Venkateswara, S.M., Satish Kumar, G. and Kalyan Chakravarthi, A. (2025). Deep learning based agricultural pest monitoring and classification. *Nature Scientific Reports*. https://www.nature.com/articles/s41598-025-92659-5.

Wan, T., Shao, B., Ma, S., Zhou, Y., Li, Q. and Chai, Y. (2023). In-sensor computing: Materials, devices, and integration technologies. *Advanced Materials, 35*(22), 2203830. https://doi.org/10.1002/adma.202203830.

Wang, L. (2024). Evaluating causal reasoning capabilities of large language models in agriculture. *Journal of Agricultural AI Research, 3*(1), 15–30. https://doi.org/10.3390/jair.2024.01002.

Wang, Y., Cai, Y., Wang, F., Yang, J., Yan, T., Li, S. and Wang, Z. (2023). A three-dimensional neuromorphic photosensor array for nonvolatile in-sensor computing. *Nano Letters, 23*(7), 4524–4532. https://doi.org/10.1021/acs.nanolett.3c00899.

Wei, Z., Zheng, P., Xu, S. and Wu, X. (2019). Object detection with deep learning: A review. *IEEE Transactions on Neural Networks and Learning Systems, 30*(12), 3212–3232. https://doi.org/10.1109/TNNLS.2018.2876865.

Xin, Z., Feng, Q., Guo, F. and Yang, W. (2025). CMTNet: A hybrid CNN-Transformer network for UAV-based hyperspectral crop classification in precision agriculture. *Scientific Reports, 15*, 12383. https://doi.org/10.1038/s41598-025-12383.

Yin, S. (2025). Foundation models in agriculture: A comprehensive review. *Agricultural Systems, 15*(8), 847. https://doi.org/10.3390/agri.150800847.

You, J., Li, X., Low, M., Lobell, D. and Ermon, S. (2017). Deep Gaussian process for crop yield prediction based on remote sensing data. In: *Proceedings of the Thirty-First AAAI Conference on Artificial Intelligence* (pp. 10–17).

Zhang, Y., Li, J., Zhang, L. and Du, Q. (2022). Convolution-Transformer Mixer (CTMixer) for hyperspectral image classification. *IEEE Geoscience and Remote Sensing Letters / Transactions*, 2022. https://ui.adsabs.harvard.edu/abs/2022IGRSL..1908935Z/abstract.

3 Enhancing Soil Productivity and Health and Crop Yield with Artificial Intelligence

Parwada Cosmas[1*], Mutanga Alfred[2] and Gumbo Albert[3]

1. Introduction

Erosion of soil, loss of nutrients, and fluctuating climate conditions pose significant dangers to global agricultural yields (FAO, 2022), as approximately 33% of global soils are currently degraded due to erosion, salinization, and loss of organic matter. Conventional farming methods, which are based on universally applicable practices for crop and soil management, will result in wastage of resources such as water, pesticides, and fertilizers, ultimately resulting in declining yields and ecological degradation (Weersink et al., 2021). Not only do these techniques result in greenhouse gas emissions, but they also result in water body pollution by way of farm runoff, creating a cycle of ecological degradation. The expanding world population of 9.7 billion people by 2050 complicates these issues, which call for innovative technological solutions to improve agricultural sustainability while reducing ecological impacts. Artificial intelligence (AI) has become a revolutionary technology in contemporary agriculture, with data-driven information to facilitate precision farming (Liakos et al., 2018), the potential to raise global crop yields by as much as 25% while lowering input costs by 20–30%.

AI enables farmers and agronomists to make evidence-based decisions to optimize soil health, boost crop productivity, and avoid wastage using real-time monitoring systems and sophisticated analytics. AI-driven technologies such as machine learning (ML) algorithms, remote sensing, and IoT sensors provide never-before-seen analytics to interpret large sets of agricultural data varying from types of soil to microclimate conditions (Zhang et al., 2020). These technologies can detect subtle soil health variations that human vision might not be able to detect, enabling interventions to prevent loss of yield in time. For instance, computer

[1] Department of Agronomy and Horticulture, Faculty of Agriculture, Environment and Natural Resources Management, Midlands State University, Gweru, Zimbabwe.
[2] University of Botswana, Gaborone, Botswana.
[3] Department of Land and Water Resources Management, Faculty of Agriculture, Environment and Natural Resources Management, Midlands State University, Gweru, Zimbabwe.
E-mail: cparwada@gmail.com (orcid id: https://orcid.org/0000-0002-7960-4303
* Corresponding author: cparwada@gmail.com

DOI: 10.1201/9781003715726-3

vision image recognition can detect deficiencies of nutrients in crops at an early stage of growth, whereas predictive models can foretell planting times as optimal under past weather conditions and moisture levels in the soil.

Utilization of AI in monitoring soil health represents a paradigm change from reactive to proactive agriculture as it allows for monitoring key soil factors such as pH, organic carbon, and microbial activity continuously (Angelopoulou et al., 2021). One can obtain the detailed soil maps at varying depths using existing sensor technology and utilization of AI analytics, enhancing the ease with which farmers can implement site-specific management programs accounting for spatial variability within fields. For example, electromagnetic induction sensors measure soil conductivity to create reliable maps of salinity and hyperspectral imagery detects early signs of soil-borne disease well before outward symptoms in plants. These technologies are of most value where nations are experiencing intensified soil loss and traditional methods of soil testing are too costly or time-consuming for regular use. Studies show that AI-improved soil monitoring systems can enhance nutrient use efficiency by 35–40% over traditional approaches (Nawar et al., 2019), greatly limiting fertilizer loss and environmental contamination while sustaining optimal crop nourishment.

AI systems are revolutionizing how farmers can predict and mitigate environmental challenges, from droughts to infestation by pests (Kamilaris et al., 2017). ML algorithms can integrate data from multiple sources, including satellite images, weather stations, and field sensors, to offer accurate predictions of crop yields under different conditions. Historical yield data-trained deep learning algorithms can identify subtle patterns and interdependencies of weather phenomena, soil characteristics, and final crop yields that enable farmers to make educated choices about crop varieties and management practices. As one example, neural networks have been found to be 85–90% accurate in predicting maize yield from soil moisture and temperature measurements at critical stages of plant development (Shafi et al., 2020). These predictive capabilities are especially valuable in the context of climate change, where increased weather volatility makes traditional agronomic information less relevant. By augmenting AI predictions with autonomous irrigation and fertilizer management systems, growers can implement flexible, adaptive responses that react to real- field conditions, maximizing resource utilization and crop performance.

AI application in yield optimization extends from prediction to active field management through precision agriculture technologies (Bacco et al., 2019). AI-guided auto tractors enable centimetre-level accurate variable-rate application of inputs, and drone-based monitoring enables frequent high-resolution monitoring of crop health across extensive areas. Computer vision algorithms can count fruit, estimate biomass, and detect stress symptoms in crops more quickly and accurately than human scouts, allowing for early interventions. Field trials have proven that AI-optimized farming systems can reduce water use by 20–50% and pesticide spraying by 30–60% without decreasing or even with an increase in yields (Patel et al., 2022). These productivity improvements not only increase farm-level profitability but also bolster broad-based sustainability goals by lessening agriculture's environmental footprint. As AI technologies continue to evolve, their integration with other evolving technologies like blockchain for supply chain transparency and robotics for mechanical harvesting has the potential to continue transforming agricultural systems into more resilient and productive food production systems capable of meeting future global demands.

AI is transforming agriculture and creating new avenues for conventional farming. AI is helping farmers to make better decisions, increase production, and utilize resources more efficiently. This union of technology with age-old farming is transforming the landscape of agriculture.

Data is as valuable as land in today's agriculture. AI is helping farmers collect, process, and make decisions on vast amounts of data. For instance, AI can analyse weather conditions, soil data, and market prices to provide farmers with guidance on what to plant, when to plant, and how to get optimal yields. These AI-driven decisions allow farmers to stay ahead of setbacks, such as weather uncertainty or price fluctuation uncertainty. With AI as a mentor, traditional farmers can make decisions based on informed facts that increase their chances of success. It allows them to become more competitive in an ever-evolving agricultural setup.

2. The Role of AI in Soil Health Assessment

AI-powered tools are revolutionizing soil health assessment with precise, real-time data-driven insights that were unattainable through conventional processes. Advanced ML algorithms such as Random Forest and Support Vector Machines (SVMs) process humongous volumes of data in from diverse sources such as on-field IoT devices, multispectral satellite imagery, and decades of historic farm data to yield reliable estimates of soil nutrient deficiencies and organic matter content (Nawar et al., 2019). Such systems can detect minute spatial differences in soil characteristics over an individual field, which allows farmers to adopt highly localized soil management practices. For instance, one 2022 study proved that AI-based soil testing could accurately predict nitrogen content at 92% compared to 78% for conventional laboratory analysis while taking hours rather than weeks of analysis time. This technological achievement is particularly helpful in big agricultural farms where the soil condition can drastically differ from one field to another.

Application of convolutional neural networks (CNNs) for remote sensing data has opened new possibilities in soil classification and erosion monitoring (Zhang et al., 2020). Deep learning algorithms can analyse high-resolution satellite and aerial imagery to map different soils, determine compaction zones, and identify incipient manifestations of erosion with breathtaking precision. One of the primary case studies in the Midwest of America proved that CNN-based image analysis of drone images could identify erosion-susceptible areas at a rate of 89% accuracy, allowing farmers to act preventatively before costly topsoil erosion had occurred.

The technology is more than simple classification, with certain sophisticated models capable of predicting future erosion dynamics based on terrain data, rainfall information, and history of land use. This predictive function enables farmers and conservationists to identify where intervention is most necessary and to locate erosion control structures like windbreaks and terraces most beneficially.

What sets these AI models apart from traditional laboratory-based soil testing is that they can provide continuous, real-time monitoring at a fraction of the time and cost. Whereas the traditional method entails physical sampling of the soil, laboratory analysis, and manual interpretation—an exercise that consumes weeks and hundreds of dollars per sample—AI systems can give daily readings for entire farms at close to zero marginal cost. An economic analysis in 2023 determined that AI-powered soil monitoring reduced the cost of testing by 60–80% and increased data frequency from seasonal to daily readings. For smallholder farmers in the developing world, where soil laboratories are not easily accessible, this is particularly impactful. Mobile apps that leverage these AI technologies today allow farmers to receive soil health reports and fertilizer suggestions by simply uploading mobile phone photos of their fields, bringing precision agriculture tools to the fingertips of many.

AI soil monitoring technology's ability to scale is one of its biggest advantages over traditional methods. Cloud systems can process data for thousands of farms simultaneously, creating regional soil health databases for entire farming communities. For instance, an Indian pilot integrated AI soil testing with government extension services and generated real-time maps of soil health for over 2 million hectares of cultivated land. This system not only provided individual farmers with customized recommendations but also enabled policymakers to view patterns of soil degradation at the regional level and allocate resources accordingly. At this scale, aggregation and analysis of the soil data was simply out of the question using conventional sampling methods, which are logistically and economically limited.

New AI systems do not merely comment on current soil conditions but can predict future directions and recommend pre-emptive action. Some cutting-edge platforms now incorporate weather forecasts, crop rotation history, and microbial activity data to project how soil health will evolve under different management regimes. A 2024 pilot in Brazil demonstrated that farms using such predictive AI systems maintained stable organic matter content even during drought, while conventional farms had a 15% decline. As these technologies continue to develop, they can potentially transform soil from being a passive plant-growing medium into an actively managed resource, with AI serving as the chief interface between farmers and the complex biological processes beneath their fields.

3. AI-driven Precision Agriculture for Yield Optimization

AI-led precision agriculture is an efficiency revolution in farming founded on the possibility of obtaining centimetre-level accuracy in applying inputs. These AI-based systems utilize sophisticated algorithms to analyse real-time inputs from multiple sources—like soil moisture sensors, canopy temperature sensors, and hyperspectral aerial images—to produce dynamic application maps that can change input rates metre-by-metre by field (Bacco et al., 2019). For example, smart irrigation systems now combine weather station evapotranspiration data with root-zone soil moisture sensors to provide just the right amount of water to each plant, reducing normal water use by 20–40% compared to conventional irrigation technologies. The precision extends beyond variable-rate spreading of fertilizer using AI recommendations to adjust nutrient application based on soil test results and crop growth stage, neither over- nor under-fertilizing and harming yields and the environment.

The integration of multiple data streams by AI platforms has released unprecedented decision-support possibilities for modern farmers. Advanced systems now integrate satellite vegetation indices, tractor-mounted sensor information, weather predictions, and historical yield maps to generate daily prescriptions for field management (Zhang et al., 2021). For instance, AI can detect subtle differences in crop health through normalized difference vegetation index (NDVI) analysis and correlate these patterns with soil conductivity maps to prescribe micronutrient applications with precision. Some sophisticated platforms even incorporate machine vision from autonomous field robots that can identify individual weeds so that herbicide sprayers can target only non-crop vegetation while sparing crops. This degree of accuracy was unthinkable ten years ago but is now being implemented on commercial farms worldwide, radically altering the way agricultural inputs are controlled.

Measurable results from global field trials serve to underscore the game-changing potential of AI-powered precision agriculture. A three-year study across 45 midwestern American corn farms found a consistent 18–22% increase in yield with 25–30% reductions in nitrogen fertilizer usage with the use of AI-driven application systems (Shafi et al., 2020). Similar results have been achieved in arid regions, where AI-optimized irrigation scheduling saved crop yields while reducing water consumption by 35% compared with traditional methods. Of greatest significance, perhaps, is that these systems are extremely adaptable – ML algorithms improve their recommendations based on new field data on an ongoing basis, so the more they are operated, the more they will be effective. This creates a virtuous circle whereby better data leads to better decisions that create even more valuable data for future seasons.

The water saving potential of AI-facilitated precision agriculture is being particularly valuable in dry regions under climate change stress. In the Negev Desert of Israel, AI systems reading soil moisture sensors and plant stress meters have enabled farmers to maintain crop yields on 40% less irrigation water (Cohen et al., 2022). Similarly, in California's Central Valley where groundwater loss is a pressing concern, vineyards that use AI-based irrigation have reduced water consumption by 30% and improved grape quality. Such systems not only respond to current conditions but use forecasting modelling to anticipate water needs days in advance, adjusting schedules based on expected evapotranspiration levels. In regions where every drop of water counts, such precision can quite literally be the difference between farm success and failure.

Looking ahead, the combination of AI with other emerging technologies promises even more advanced agricultural precision. The combination of 5G communications, edge computing, and autonomous farm machinery is giving rise to "smart fields" in which decisions are made and implemented in real-time by machines on their own (Liakos et al., 2022). Test farms are already experimenting with fully autonomous systems where AI analyses data from soil sensors and drones and then directs robotic implements to perform precisely calibrated interventions – delivering fertilizer to a struggling patch of the crop, say, or applying a micro-dose of pesticide to an incipient pest hotspot. As they mature and grow in scale, these technologies have the potential to make precision agriculture the standard worldwide rather than the exception, ushering in an age of truly sustainable intensive farming that can feed growing populations with minimal environmental impact.

4. Challenges in AI Adoption for Agriculture

Despite its revolutionary promise, the mass production of AI in agriculture is hampered by substantial economic hurdles disproportionately borne by small-scale farmers. The cost of the initial investment needed for AI-fitted equipment such as sensors, drones, and automated machines is frequently more than $50,000 per farm, beyond the budgets of most of the world's 570 million farms operating on less than 2 hectares (Weersink et al., 2021). Even when farmers enjoy access to the technology, ongoing expenses on data plans, software updates, and maintenance are a recurring drain on resources. This economic consideration creates a "digital divide" in agriculture, with only large agribusinesses able to afford cutting-edge AI technology, which has the potential to exacerbate inequality in the industry. It is strongest in developing countries, where unavailability of credit and fluctuating crop incomes make such investments doubly risky for smallholders.

Infrastructure constraints in rural regions pose another fundamental challenge to AI uptake in agriculture. An estimated 3.6 billion individuals globally continue to be without steady internet connection, with rural agricultural communities being most under-served (World Bank, 2023). With no stable connection, real-time data transfer, the cornerstone of AI farm systems is no longer feasible. Even where there is little internet penetration, bandwidth limitations generally restrict data-intensive applications like analysis of high-resolution satellite imagery or cloud-hosted AI platforms. Inefficient infrastructure is compounded by unstable electricity supplies in much of the agricultural region, so sensors and equipment needed for precision farming are difficult to power. A 2022 report by FAO discovered that 60% of sub-Saharan African farms do not have regular access to electricity, greatly hindering their potential to embrace even simple digital farming solutions.

The technical complexity of AI technologies also presents a substantial barrier, as most farmers are not equipped to implement and maintain these technologies effectively. Agricultural AI technologies come with a requirement for technical knowledge in data analysis, machine fixing, and software updates expertise difficult to come by among farmer groups, particularly developing country farmer groups (van der Burg et al., 2022). This absence of knowledge is aggravated by the fact that most AI interfaces were developed by engineers with no consideration for end-users who possess low digital literacy. A study in India found that 70% of farmers who tried to use AI-powered advisory apps dropped them within three months due to their inability to understand the advice or make use of the technology (Patel et al., 2023). Without significant investment in agricultural education and more accessible design, this knowledge gap will continue to restrict adoption.

There are privacy issues since farmers are worried about ownership and control of their field data, especially when corporate agritech providers collect it (Kamilaris et al., 2017). Farmers are generally scared that their data would be employed by commodity traders to influence market prices or by insurance companies to decide on premium adjustments. Meanwhile, building good AI models requires huge amounts of high-quality farm data which is typically not available or fragmented across many systems. This "data scarcity" problem is particularly acute for specialty crops and low-income countries where minimal digital record-keeping exists. 85% of agricultural AI research, according to a 2023 study, focuses on just five staple crops (cotton, maize, rice, soybeans, and wheat) at the expense of hundreds of equally essential crops (Zhang & Kovacs, 2023).

Overcoming such challenges requires creative, multi-stakeholder solutions that can recognize the different needs of diverse farming systems. Low-cost, offline-friendly AI solutions not requiring constant internet connection are some of the promising solutions, e.g., local data-processing mobile apps (Qian et al., 2023). "AI-as-a-service" models are emerging in which farmers only pay for analyses they conduct and not for maintaining expensive equipment. Governments across different countries have launched digital extension schemes in agriculture that entail technology extension as well as field training. In Kenya, for example, a public-private partnership has provided over 50,000 farmers with basic exposure to AI tools and provided them with subsidized internet and sensors (AGRA, 2023). Such programmes demonstrate that with specific support, even smallholder farmers can begin enjoying the payoffs of AI innovation.

5. Case Studies of AI Applications in Soil and Crop Management

Several examples of effective uses of AI in farming illustrate its potential to revolutionize farming practice. In India alone, the government's AI-based Soil Health Card Scheme provided over 100 million farmers personalized nutrient management recommendations since its implementation in 2015 (Patel et al., 2022). The scheme creates customized fertilizer prescriptions based on analysis of soil samples using machine learning algorithms, and as a result, it led to an average 20% increase in crop yield while reducing unnecessary fertilizer use by 15–25% (Table 1).

Table 1 Impacts of AI Soil Health Cards in India (2015–2022).

Metric	*Before AI Adoption*	*After AI Adoption*	*Improvement*
Average crop yield (t/ha)	2.8	3.4	+20%
Fertilizer use reduction (%)	-	15–25%	-
Farmer adoption rate (%)	0% (pre-2025)	65% (by 2022)	-

Similarly, in America, AI drone-mounted multispectral cameras have transformed the diagnosis of disease in big farms. A study of Midwestern cornfields showed that AI-driven algorithms identified fungal diseases like gray leaf spot as early as two weeks before visual symptoms and allowed targeted fungicide spraying (Shafi et al., 2020). Early intervention reduced pesticide use by 40%, saving farmers $50 per acre in input costs on average while guaranteeing yield stability (Table 2).

Table 2 AI drone-based disease detection in U.S. cornfields.

Parameter	*Traditional Scouting*	*AI-powered Drones*	*Improvement*
Disease detection time	7–10 days after infection	2–3 days after infection	5–7 days earlier
Pesticide use reduction	Baseline	40% less	-
Cost saving ($/acre)	-	50	-

In Brazil, AI-driven predictive analytics for soybean rust, which is a highly destructive fungal disease, has saved farmers from suffering $2 billion in yield losses every year (Dalla Corte et al., 2021). Outbreaks are predicted 85% accurately by AI models using satellite imaging, weather conditions, and disease history, so that fungicide sprays are applied only when needed (Table 3). The technique has cut down 30% of wasteful sprayings, reducing chemical runoff and environmental damage.

Table 3 AI-based soyabean rust prediction in Brazil.

Factor	*Without AI*	*With AI*	*Improvement*
Prediction accuracy (%)	60%	85%	+25%
Fungicide applications	4–5 per season	2–3 per season	30–40% reduction
Economic savings (annual)	-	$2 billion	-

Mobile apps driven by AI, like "FarmDrive" in Africa, use machine learning to establish creditworthiness for smallholder farmers with no formal banking records (AGRA, 2023). By analysing satellite data, mobile payment records, and agronomic data, the apps have increased the approval rate of loans by 35% to enable the farmers to invest in better seeds and fertilizers (Table 4).

Table 4 AI-enabled credit access for African smallholder farmers.

Indicator	*Traditional Loans*	*AI-assessed Loans*	*Change*
Loan approval rate (%)	25	60	+35
Average loan size ($)	200	500	+150%
Default rate reduction (%)	12	6	–50%

These case studies illustrate that AI is not a theory of innovation but an applied technology already delivering measurable benefits. From precision farming in India to disease modelling in Brazil and payment systems in Africa, AI is raising productivity, sustainability, and profitability. But to make these success stories replicable at the global level, one must overcome costs, infrastructure, and farmer training, the bottlenecks that must be overcome with the help of policy incentives, public-private partnerships, and localized AI solutions.

6. Climate-Resilient AI: Adapting to Extreme Weather

As climate change intensifies weather extremes, agricultural AI systems must evolve beyond static recommendations to dynamic, climate-adaptive tools. Next-generation AI will need to process real-time climate data alongside long-term projections to help farmers navigate increasing volatility. Advanced modelling techniques, including generative AI and digital twins, are demonstrating remarkable potential in simulating how crops will respond to various climate scenarios from prolonged droughts to unseasonal floods (Chlingaryan et al., 2024). These systems examine decades of climatic patterns, soil conditions, and crop yields to determine management practices that buffer against climate shocks, radically altering how farmers prepare for a future of uncertainty.

Generative AI is finding special utility in climate adaptation by generating thousands of possible weather scenarios for locations. Unlike traditional models that are based on historical means, these machines can model unprecedented events such as 100-year floods happening in successive years and suggest resilient crop types and planting times accordingly (Lobell et al., 2023). For example, in India's Punjab, AI tools can now predict winter rainfall patterns accurately 85% of the time three months ahead and advise farmers on ideal dates for sowing wheat, so they avoid early heat stress that reduces yields as well as late rains that damage germination. This predictive ability is becoming increasingly important as climate change shifts traditional farming calendars.

The emergence of "digital twin" farm technology represents another leap forward in climate resilience. Australia's pioneering use of AI-powered farm replicas has enabled growers to virtually test management strategies against projected climate conditions, reducing drought-related losses by 18% compared to conventional farms (Zhang et al., 2024). These virtual twins combine live sensor information with climate models to perform ongoing simulations, warning farmers of developing risks such as soil moisture stress or pest infestations before they are apparent in the field. For the 2023–24 El Niño season, Australian almond producers utilizing this approach sustained 92% of regular yields as nearby conventional farms experienced 30–40% declines, illustrating the real-world advantages of AI-assisted climate adaptation.

However, there are many challenges to achieve global access to such technologies. Climate AI models currently function best in areas with rich data sets and large networks of weather stations. Smallholder systems in developing countries, by contrast, typically have scarce infrastructure. New approaches such as "frugal AI ", applying low-cost satellite data and community weather observations, are bridging this gap. In Kenya, a pilot project combining AI with indigenous weather forecasting knowledge improved climate predictions for maize farmers by 25% above conventional models (Ayal et al., 2023). Likewise, Colombia coffee growers now receive AI-generated microclimate forecasts via WhatsApp, helping them anticipate damaging rains during critical harvest periods.

The main goal of this chapter is to summarize AI technologies transforming soil and crop management. On a more detailed level, the chapter is to evaluate the efficacy of AI in making effective predictions of critical soil health indicators like nutrient content, water content, and organic matter (Angelopoulou et al., 2021). Secondly, it evaluates AI-fostered precision agriculture practices like ML-based models, remote sensing, and IoT-based monitoring systems to gauge their role in enhancing crop yields (Zhang et al., 2020). Drawing on synthesizing existing studies and case studies, this chapter provides a detailed description of how AI can be used to promote agricultural sustainability.

7. Methodology

To present a balanced evaluation of the use of AI in modern agriculture in general, a systematic review of literature was conducted involving peer-reviewed articles in *Scopus*, *Web of Science*, and *IEEE Xplore* from 2015–2023. The review was confined to three main applications of AI: ML models for forecasting soil nutrients, remote sensing and IoT-based real-time soil monitoring, and decision support systems for precision irrigation and fertilization. Studies were selected based on methodological strength, with an emphasis on field-validated evidence rather than theoretically derived research. This allowed rigorous comparison of different AI approaches across a variety of agricultural environments, from commercial-scale production systems to smallholder systems in developing countries.

Comparison of the ML models showed that Random Forest and Neural Networks were the dominant algorithms for soil nutrient prediction, with the average accuracy ranging from 85% to 93% based on data quality and feature selection (Smith et al., 2021). For IoT and remote sensing applications, studies demonstrated that the integration of ground sensors for soils with multispectral satellite data had the potential to improve accuracy in real-time monitoring by 20–35% compared to standard practices (Jones et al., 2022). AI-enriched precision application of fertilizer and precision irrigation-based decision support systems yielded promising results with 25–40% savings in water use and 15–30% savings in nitrogen application without impacting yields (Brown et al., 2023).

8. Results

Recent developments in AI have revolutionized soil health testing, with AI models identifying 25–30% improved accuracy for predicting nutrients compared with conventional laboratory analysis (Table 5). This is due to machine learning algorithms being capable of filtering through the complex, layered data sets like spectral imaging, electrochemical sensor data, and previous soil information to uncover hidden nutrient deficiencies that conventional tests cannot detect. Random Forest has been demonstrated to perform particularly well, with 92% macronutrient prediction accuracy when trained on Europe's biggest soil dataset, the LUCAS Soil Database (Angelopoulou et al., 2021). These models aren't just emulating human analysis; they reveal unsuspected relationships between soil properties and uptake efficiency by plants, enabling truly data-driven fertilization habits.

Table 5 Comparative Performance of AI soil analysis models.

Model	*Accuracy (%)*	*Dataset*	*Key advantage*	*Reference*
Random forest	92	LUCAS soil Database	Handles high-dimensional soil data	Angelopoulou et al. (2021)
CNN (Remote sensing)	88	Sentinel-2 satellite imagery	Large scale spatial analysis	Zhang et al. (2020)
SMV	85	Field spectroscopy	Real time in-situ measurements	Nawar et al. (2019)
XGBoost	90	IoT sensor networks	Optimized for small datasets	Chen et al. (2022)

9. Multi modal AI Approaches for Comprehensive Soil Profiling

Recent soil health assessment increasingly combines different AI techniques to overcome one limitation at a time. While CNNs do most effectively in assessing Sentinel-2 satellite images (88%) for regional-scale soil mapping (Zhang et al., 2020), SVMs work best to analyse field spectroscopy data (85%) from portable sensors (Table 5). Recent hybrid systems now combine these techniques applying satellites to large-scale monitoring and drones/ground sensors to nearby validation. In 2023, Brazilian research demonstrated that the integration of CNN and SVM outputs improved overall prediction credibility by 12% when compared with one-model approaches, particularly for micronutrients like zinc and boron (Silva et al., 2023). This multi modal approach enables the farmer to select equipment that is compatible with their own specific scales of operation and crop requirements.

9.1 From Soil Insights to Yield Gains: The AI Optimization Pipeline

The actual value of AI soil analysis is achieved when combined with precision crop management systems and initiating the increases in yield (Table 6). These increases are achieved by closed-loop systems where soil data regularly drives irrigation and fertilization procedures. For example, on a 3-year Iowa corn research project, AI models coupling real-time soil moisture with weather forecasts boosted yield 22% while reducing nitrogen application by 18% (Thompson, 2023). The mechanisms for increasing yield are: Micro-dose fertilization: Fertilization with nutrient at 10m^2 resolution accounting for soil heterogeneity, root-zone optimization: Soil water-holding capacity-based irrigation management and monitoring microbial activity to enhance nutrient cycling.

Table 6 Yield impacts of AI soil-based interventions by crop type.

Crop	*Yield Increase (%)*	*Input Reduction*	*Key AI Mechanism*
Corn	22	18% N fertilizer	Spectral soil N mapping
Soyabeans	19	22% pesticides	Disease risk modelling
Wheat	17	15% water	Soil moisture forecasting
Potatoes	25	30% irrigation	Tuber zone soil optimization

Doubling the impact of AI soil management higher productivity at lower inputs yields robust economic and environmental dividends. A meta-analysis of more than 200 farms showed AI adopters had 28% higher margins than their similar conventional counterparts, with paybacks below 18 months on technology cost (AGRI-TECH, 2023). Environmental benefits are also significant: 31% nitrate leaching reduction (EPA Watershed Report, 2023), 25–40% decrease in irrigation water use (Drought Monitor, 2023) and 17% reduction in fuel consumption via optimized field operations. These enhancements are particularly notable in impaired soils, wherein AI-driven recovery programs supplemented organic content by 1.5% annually increasing the numbers of traditional amendments threefold (Soil Health Institute, 2023).

10. Discussion

Predictive analytics based on artificial intelligence are revolutionizing soil nutrition management via a paradigm shift in lawns from reactive to proactive strategies. High-level machine learning software integrates multi-dimensional data sets, e.g., decade-long yield maps, hyperlocal weather forecasts, and near-real-time soil sensor telemetry to generate astonishingly accurate (85–92%) projections of imminent nutrient deficiencies (Kamilaris et al., 2017). These systems employ more sophisticated approaches like recurrent neural networks to forecast temporal patterns of nitrogen mineralization and CNNs to unravel spatial heterogeneity in soil test information, producing dynamic nutrient availability forecasts 3–4 weeks prior to conventional methods that would determine incipient deficiencies. The predictive capacity is particularly valuable for vagile nutrients, like nitrogen, where models are now able to predict mineralization flushes after rain events or leaching risks from future precipitation regimes, enabling genuinely anticipatory over corrective management (Zhang et al., 2022).

The impact of such predictive systems on operations is most evident in precision side-dressing applications, where AI-driven recommendations have been proven to make significant impacts on nutrient efficiency. ML algorithms can identify the optimum 2–3-day window for application of nitrogen when the potential of the crop for take-up is greatest and environmental losses are minimized by analysing the complex dynamics between temperature, soil moisture content, organic matter content, and microbial activity. U.S. field trials. Corn Belt showed how AI-timed applications of nitrogen employed 30–35% more efficient use of nitrogen compared to traditional calendar-based practices, with production maintained or increased (Zhang et al., 2022). The systems accomplish this by frequently revising recommendations as they acquire knowledge from current field conditions such as modulating application amounts when unexpected rainfall alters soil moisture conditions or when early-growth velocities signal greater-than-expected nutrient requirements.

Above the basic macronutrients, emerging-generation predictive analytics now are tackling more refined soil health matters. New models blend spectroscopic data from in-field sensors with micronutrient availability and cation exchange capacity predictions weeks ahead of standard extraction tests to spot deficiencies (Rossel et al., 2023). Some advanced systems even forecast biological activity in the soil by tracking weather trends, organic additions, and previous microbial population data to project periods of peak nutrient cycling. These skills are most valuable in regenerative and biological farming systems where living processes dominate nutrient supply. However, such predictive systems are not equally effective across climate and soil type models trained on temperate loams typically require re-tuning for tropical clay systems, supporting the relevance of region-specific training data and adaptive learning frameworks (Rossel et al., 2023). As these technologies evolve, they have the potential to transform soil fertility management from an after-the-fact, generalized practice to a data-based, precisely timed science.

The stunning contrast between water conservation in California's almond groves (35–40%) and Punjab's agricultural fields (18–22%) testifies to the reality that infrastructure quality has an intrinsic bearing on IoT efficiency in precision agriculture. In California's Central Valley, good quality 4G networks and stable electricity facilitate continuous data transmission from dense sensor arrays measuring soil moisture at varying depths, canopy temperature, and rates of evapotranspiration making real-time optimization possible for micro-sprinkler systems (Fulton et al., 2023). At the same time, Punjab's farmers deal with irregular power supply and patchy 2G coverage that causes intermittent data transmission, which compel systems to adopt sporadic measurements and pre-set algorithms in place of dynamic adjustment (Jat et al., 2023). This technology gap is further exacerbated by varying water delivery infrastructure in California's pressurized drip tape which can make it impossible to make fine-tuned adjustments in Punjab's flood irrigation systems.

Low-income communities have also found innovative solutions that provide high water savings without the expense of infrastructure. Basic solar-powered soil water sensors with basic SMS alerts have proven effective, sending farmers twice-daily reports on root zone moisture without the requirement for continuous connectivity (Ogilvie et al., 2022). These systems cost only $120 per acre compared to $800+ for full IoT solutions but realized 25–28% water savings on Zimbabwean cotton farms through enabling farmers to more accurately time irrigations. They succeeded because they applied local context to technology complexity – using simple threshold-based alerting rather than real-time automation and leveraging pre-existing farmer familiarity with mobile messaging. This "frugal innovation" approach demonstrates that well-designed technology can create disproportionate benefits in the context of developing agricultural environments.

The experience of such varied implementations suggests a two-tier model of global precision irrigation deployment. High-end IoT solutions deliver the highest water savings in high-tech agrarian economies with robust infrastructure, while lean systems with intermittent sampling of data and basic alerts deliver affordable options for developing countries. A 2023 World Bank report established that hybrid solutions employing low-cost sensors in combination with scheduled drone images would be capable of providing 75% of the water saving from full IoT systems for 30% of the monetary cost of Southeast Asian rice systems (World Bank, 2023). It shows that the future is not about one-size-fits-all solutions, but about flexible precision agriculture frameworks that can scale technically based on local infrastructure. As connectivity enhances across the world, such systems can increasingly add more sophisticated features without demanding total system refreshes.

11. Disease Detection Accuracy by Crop Type and Geography

Artificial intelligence-driven drone systems proved to be highly accurate in detecting early blight in potatoes in 88–91% of geographically varied areas, from highly mechanized Dutch farms to the Bangladesh smallholder fields (Poblete et al., 2023; Islam et al., 2023). This is because of the distinct visual signs of potato blight and the even canopy structure of the crop, which allows for drones equipped with multispectral and RGB cameras to overfly and capture fine, standardized images for analysis. The success in different agro-climatic environments means that for certain crops and diseases, AI models can generalize well across environments, provided training sets contain significant variability in illumination, growth stages, and local disease strains.

However, the effectiveness of AI disease identification significantly drops when applied to crops whose architectures are more complex. For instance, while wheat rust detection equipment is 94% effective since the crop contains uniform rows and exposed leaf surfaces (Singh et al., 2023), the same technology fails for cassava diseases, where effectiveness is as low as 78% (Mwebaze et al., 2023). Cassava's tight, bushy canopy and early, inconspicuous symptoms of cassava brown streak virus (CBSD) disease interfere with the collection of diagnostic images by drones. Similarly, vertical crops such as tomatoes or grapes are challenging with occluded fruit and leaves, requiring special flight tracks and sensor angles to sense reliably.

These variations require crop-specific data acquisition protocols and AI model designs. For homogeneous crops like wheat and potatoes, uniform overflights by drones with RGB sensors can be utilized, while crops like cassava or coffee would require hyperspectral imagery, LiDAR, or even multi-rotor technology for oblique imaging to penetrate deeply through dense cover (Ghazi et al., 2023). Future AI disease detection innovations will be required to address adaptive algorithms that consider canopy structure, symptom nuance, and environmental variability. Combining drone imagery with ground mobile imagery captured by farmers may also be utilized to enhance precision, particularly within smallholder systems where costly sensors are not an option. Such tailor-made solutions will be required to provide the benefits of AI to all geographies and crops, not only those with ideal visual characteristics.

11.1 Economic Feasibility across Farm Scales

The adoption of AI in agriculture demonstrates a huge economic divide between large and smallholder farms. For industrial 500+ hectare U.S. corn farms, precision agriculture systems demonstrate high viability with 2–3 years payback levels achieved through improved input utilization and yield boosts of 15–20% (Schimmelpfennig, 2023). In contrast, Tanzanian smallholder coffee farmers have prohibitive 8–10-year paybacks on similar technology due to reduced scales of operation, higher relative costs, and limited access to finance (Van Asten et al., 2023). This disparity has prompted alternative adoption paradigms in the global south, like Kenya's "AI-as-a-service" startups selling drone-based crop monitoring services at $2–5 per acre per season – skipping initial capital expenditure but still capturing 60–70% of the value of owning an entire system (AGRA, 2023). Such adapted business paradigms are becoming vital in the democratization of access to agricultural AI.

11.2 Data Scarcity and Its Impact on Model Performance

The quantity and quality of available training data determine yet another fundamental difference in AI performance across world regions. European soil carbon modelling benefits from the very large LUCAS database, which contains over 1 million georeferenced samples, to make predictions with accuracies of 89–92% (Angelopoulou et al., 2021). Concurrently, in the Sahel region, with fewer than 50,000 archived soil samples, such models achieve accuracy levels of only 61–65%, which renders them virtually useless in practice for carbon credit programmes or precision agriculture (Vågen et al., 2023). It is an outcome of decades of inadequate investment in agricultural research infrastructure and constitutes a strong disincentive to AI adoption by developing economies.

Having recognized the challenge, initiatives like the African Soil Information Service (AfSIS) have implemented extensive mobile sampling programmes with standardized protocols and high-end spectral analysis technology. Across pilot countries, the activities improved model precision by 18% points over the course of three years by prioritizing the acquisition of 25,000 new reference samples (Leenaars et al., 2023). The initiative combines conventional soil sampling with cutting-edge techniques like mid-infrared spectroscopy and handheld X-ray fluorescence to maximize data value from each sample. Notably, all data obtained go into open-access data repositories, creating public goods that are also accessed by African researchers and commercial AI developers.

These experiences highlight the need for internationally coordinated efforts to build well-rounded agriculture datasets across diverse farming systems. Efforts currently demonstrate that focused investments in data infrastructure can rapidly improve model performance in underrepresented domains. But in the end, answers must be supported by sustained funding for regional research capacity, harmonized protocols for sharing data, and AI frameworks optimized to thrive with sparse training data. As the agricultural AI space matures, considerations of fair data ecosystems will be crucial to guarantee that these revolutionary technologies serve all farmers, not only those in data-saturated areas.

12. Implementation Challenges: Technological vs Social Barriers

AI application in agriculture is faced with fundamentally different challenges depending on the extent of regional development. In technologically advanced agricultural paradigms in Western Europe and North America, technical integration of interface of AI platforms with installed farm management software, compatibility with hardware from various brands, and harmonizing streams of data among various precision agriculture tools (Zhang et al., 2023) is the primary challenge. Farmers in these nations typically have the technological literacy and infrastructure to implement advanced systems but are confronted with interoperability issues and the technology learning curve. For instance, from a 2023 survey of German dairy farms, 62% of adopters mentioned that they found it hard to integrate AI herd management systems with existing milking parlour software (Heise et al., 2023).

By contrast, emerging economies are faced with more entrenched challenges rooted in infrastructural limitations and social factors. A stark example is presented by Nigeria, where a 2023 survey of extension agents reported that 68% of smallholder farmers rejected AI counsel when it contradicted generations-old traditional knowledge (Oluwasemire et al., 2023). In many cases, farmers trusted ancestral farming calendars and local signs more than algorithmic planting dates or input recommendations. This skepticism is based on past experiences with inappropriate technologies and the lack of visible benefits from previous “modern” agricultural interventions. The contrast with Iowa where 82% of farmers readily incorporated AI recommendations into decision-making (Arbuckle et al., 2023) highlights the degree to which the adoption of technology is a culturally influenced process.

The social barriers extend beyond the simple trust issues to encompass pragmatic daily realities. Smallholder farmers in the developing world grow complex, diversified cropping systems that do not easily translate into AI models trained on Western monoculture data (Beza et al., 2023). Also at play are gender dynamics women do most of the farm work in most African and South Asian contexts but have less access to the smartphones through which AI guidance would be delivered (Agarwal et al., 2023). These social arrangements represent adoption hurdles that no level of technical refinement can overcome without deliberate socio cultural adaptation.

This divergence in problem-solving necessitates radically different implementation strategies. In industrialized nations, the fixes are all about promoting technical compatibility, developing open API standards for farm software and constructing integrated dashboards that aggregate data from multiple precision ag tools (Wiseman et al., 2023). Successful implementations in the developing world, by contrast, often incorporate "technology blending" practices which respectively combine indigenous knowledge. For example, a Kenyan project reached 73% adoption rates by having AI processes interpret suggestions into more recognizable conventional indicators like associating fertilizer timing with locally identified bird flight patterns rather than merely soil test levels (Mutungi et al., 2023).

The implication of global deployment of AI is that efficient farming technology must be as locally sensitive as it is mathematically powerful. This means developing region-specific interfaces that communicate in native metaphors, validating algorithms against local systems of knowledge, and designing channels for implementation that pay attention to social hierarchies and gender roles in rural society. The best systems in an increasingly widespread AI to world agriculture will be those that bridge the gap between centuries of farm tradition and advanced data science, rather than try to replace one with the other. Future development efforts need to have the same investment in social science research and community engagement as in technical development to achieve equitable, efficient adoption in all farming systems.

12.1 Climate-specific Performance Variations in Agricultural AI

Agri-systems based on AI show great variability of performance across various zones of climate depending mainly on the inherent predictability of climatic conditions in each of them. In the case of Mediterranean climates with evenly distributed seasonal cycles of dryness, irrigation optimization software has achieved water savings ranging between 30–35% by looking at historical weather and current soil conditions to plan watering (Bolognesi et al., 2023). They thrive because they can predict dry periods and modify irrigation plans weeks ahead, taking advantage of the area's stable patterns of weather. Success in these conditions proves the capabilities of AI under the conditions of predictable meteorological patterns, where algorithms can recognize and take advantage of dependable seasonal trends.

Conversely, in monsoon-dependent agricultural systems like South Asian ones, the same AI techniques have less scope for performance with average water savings of just 12–15% (Ahmed et al., 2023). The issue is with the natural unpredictability of monsoon rainstorms such that intense precipitation events can rapidly alter soil moisture conditions and render previous irrigation calculations obsolete. Unlike Mediterranean climates with gradual, predictable dry periods, South Asia's erratically scheduled rains and unpredictable monsoons create a less orderly world for AI systems to navigate. This strong contrast illustrates just how much climatic randomness affects agricultural AI success directly, where more erratic weather translates to decreased yields through normal optimization algorithms.

The performance gap between these climatic zones suggests that AI solutions cannot be blindly imposed without deep localization. Deterministic models that work well in stable climates deteriorate in areas of high weather variability, where probabilistic approaches to handling uncertainty work better (Chlingaryan et al., 2023). For example, AI platforms that utilize ensemble weather forecasts and probabilistic soil moisture predictions perform better than rigid irrigation schedules in the tropics with erratic rainfall by adapting to prevailing conditions in real time. Climate-specific tailoring extends beyond water management AI models for forecasting pests, crop disease risk, and nutrient application timing all need to be calibrated for regional climate conditions to achieve optimal results.

These findings underscore the value of climate-aware AI design in agriculture. Future systems will require adjustable architectures that can contend with local weather conditions, whether dealing with Mediterranean predictability or monsoon uncertainty. Hybrids that concern the combination of AI and traditional ecological knowledge such as integrating monsoon onset indications used by local farmers into computational models may offer a path to improved performance in challenging climates (Mishra et al., 2023). As weather variability is heightened by climate change around the world, creating robust AI systems that can deal with uncertainty will become ever more essential for sustaining agricultural output in varied regions. The major takeaway is obvious: good agricultural AI needs to be as flexible as the climates it supports.

13. Synthesis of Comparative Findings on AI in Agriculture

The general analysis of AI application in global farm systems confirms the trend of declining returns over the existing technology base. In developed agriculture with institutionalized precision farming, such as high-scale agriculture in Europe and North America, AI implementations typically realize 5–15% marginal increases in productivity or input use (Schimmelpfennig, 2023). The gains, though economically substantial, represent upgrades to already optimized systems. The marginal returns are lower as these farms already harvested much of the low-hanging fruit using existing technology like GPS-guided equipment and soil moisture sensors. Developing country smallholder systems, however, will leapfrog from primitive practices to AI-empowered methods, permitting revolutionary 25-40% increases in yield or resource use efficiency (AGRA, 2023). This extreme contrast illustrates how points of departure in technology overwhelmingly set the relative value of AI adoption.

Most spectacular successes occur where AI solutions are laboriously adapted to local conditions through appropriate intermediate technologies. While high-tech IoT solutions break down in low connectivity contexts, low-tech options like smartphone-based image recognition for pest control or SMS-based weather alerts deliver substantial benefits at low cost (Ogilvie et al., 2022). For instance, Nigerian maize farmers who used basic AI-powered diagnostic systems achieved 28% yield gains through early identification of nutritional deficiencies, without any automated equipment (Oluwasemire et al., 2023). Such instances demonstrate that the greatest AI impacts are achieved where technologies bridge rather than swamp underlying capability gaps. The potential for change does not reside in replicating Western high-technology models, but in developing context-specific solutions to prioritized constraints of specific farm systems.

Among major findings in global implementation is the enhanced performance of hybrid intelligence systems that combine AI with human expertise. The case of Ethiopia's "AI-facilitated" extension programme exemplifies this, whereby algorithms process soil and weather data but local agronomists cross-check recommendations against traditional knowledge (Tamene et al., 2023). This hybrid system achieved 32% more adoption than rigid digital tools by maintaining farmer trust while conveying data-driven insights. Similarly, Indian Punjab's effective drone monitoring program educated village-level "digital ambassadors" to engage with farmers and describe AI-produced reports on field conditions (Jat et al., 2023). These hybrid solutions acknowledge that farm-level decision-making is also linked to cultural, experiential, and economic factors, including values and ethics, that human experts are in a better position to manage than algorithms by themselves.

The comparative review highlights three main principles for the development of future agricultural AI. First, technology design ought to follow an "appropriate automation" paradigm applying high-tech solutions only where they outperform simpler strategies under circumstances in specific locations. Second, implementation strategies must favour human-AI collaboration over total automation of tasks, especially in smallholder settings where traditional know-how remains extremely useful. Third, system design should be innately flexible to evolve with expanding local capacities, avoiding technology dead-ends (World Bank, 2023). These postulates herald an era of AI augmenting, but not replacing human intelligence, adaptable systems that respect diverse agricultural environments.

On balance, the global experience with farm AI reveals it as a powerful but situation-specific technology. Its highest value comes not in the pursuit of technical sophistication for its own sake, but in the solving of specific productivity constraints in the ecological, economic, and social circumstances of each farm system. As the technology matures, attention must then turn from proving technical feasibility to ensuring equitable access and culturally appropriate application. The best course of action is in co-developed solutions that integrate the talents of AI with the irreplaceable expertise of farm communities around the world. That's an even-keeled approach that will maximize AI's best potential for addressing global food security needs while honouring the diversity of farming systems that feed our planet.

14. Future Directions for AI in Agriculture

The future of farm AI lies in integrated smart farm systems that combine AI with other complementary technologies like blockchain, robotics, and IoT. These systems will enable end-to-end farming from soil tillage to harvesting and distribution (Liakos et al., 2018). For example, blockchain can establish unchangeable records of inputs and outputs of crops so that there is transparency in organic certification or carbon credit programmes. Meanwhile, robots with AI vision capabilities can be used to do weeding, harvesting, and pruning precisely, reducing labour expenses by as much as 50% (Zhang et al., 2023). The combination of these technologies will lead to autonomous farms where AI controls drones, robotic harvesters, and irrigation systems in real time. The advent of 5G networks and edge computing will eliminate latency issues that currently hamper AI applications in remote locations. AI processing will occur locally on farm devices, instead of sending data to distant cloud servers, enabling instant response to changing conditions (Kassahun et al., 2022). For instance: autonomous tractors will adjust planting depth in real-time based on soil sensor input; an intelligent irrigation valve will respond directly to weather forecasts and soil moisture.

A pilot conducted in 2023 in Germany demonstrated that edge-AI systems reduce decision latency from 5 seconds to under 100 milliseconds, which is crucial for time-critical activities like frost protection or insect infestations. To make AI feasible for smallholder farmers, researchers must prioritize open-access AI models trained on diverse global datasets. Current AI applications are often useless in developing countries because they have been trained largely on Western or large-farm data (Tseng et al., 2023). However, initiatives like FarmNet AI are crowdsourcing African and Asian farm-localized data to create more inclusive algorithms. The initial results showed that these customized models improve the accuracy of yield prediction by 22–35% for staple crops like cassava and millet compared to conventional models (Table 7).

Table 7 Performance of open-access AI models in developing regions.

Region	*Crop*	*Accuracy Improvement*	*Adoption Rate Increase*
East Africa	Maize	+28%	41%
South Asia	Rice	+35%	38%
West Africa	Sorghum	+22%	29%

15. Conclusion and Recommendations

The evidence clearly indicates that AI technologies can potentially unlock a substantial boost to soil health monitoring and maximization of crop yields, with solutions to some of the most pressing agricultural problems. Harnessing such potential globally, however, requires overcoming important barriers to access, affordability, and adaptation. The revolutionary promise of agri-AI remains unevenly distributed, with intensive farming systems benefiting disproportionately and small-scale farmers in developing nations struggling to implement. This divide threatens to exacerbate existing inequalities in global food production unless reversed by intentional, focused action that makes AI-driven solutions scalable in different agriculture contexts.

To bridge the gap, intensive farmer training programmes should be a priority area for investment for governments and development agencies. AI literacy programmes should go beyond basic digital skills to include: (1) hands-on experience with AI tools of local relevance, (2) demonstration plots of tangible benefits, and (3) community "digital champion" schemes to create peer-to-peer learning networks. Indigenous knowledge systems must be respected in such education with proper demonstration of how AI can be used to support conventional practices.

Finally, strategic policy support must create enabling environments for inclusive AI take-up. Governments must introduce: (1) staged subsidy programs for precision agriculture equipment, (2) tax incentives for agritech startups with smallholder solutions, and (3) public-private investment to develop rural digital infrastructure. International organizations can further help in this transformation by establishing global data-sharing and interoperability standards for farm AI to make innovations available to every farmer regardless of their location or size. By pushing forward these proposals together, the farming sector can take advantage of AI's possibilities while making its fruits accessible to those who need them most.

References

AGRA. (2023). *Artificial Intelligence for Sustainable Agriculture in Africa*. Alliance for a Green Revolution in Africa.

Ahmed, M., Hassan, S., Khan, M.A. and Iqbal, Z. (2023). AI-driven soil health assessment for improved crop yields. *Agricultural Systems*, *205*, 103587. https://doi.org/10.1016/j.agsy.2023.103587.

Angelopoulou, T., Tziolas, N., Balafoutis, A., Zalidis, G. and Bochtis, D. (2021). Machine learning in precision agriculture: A review. *Computers and Electronics in Agriculture*, *187*, 106257. https://doi.org/10.1016/j.compag.2021.106257.

Arbuckle, J., Roesch-McNally, G., Tyndall, J. and Morton, L.W. (2023). Digital agriculture and soil productivity: A meta-analysis. *Nature Sustainability*, *6*(4), 345–356. https://doi.org/10.1038/s41893-023-01071-2.

Ayal, D.Y., Ture, K. and Tarekeng, K. (2023). Blending indigenous forecasting methods with artificial intelligence for improved climate predictions in Kenya. *Climate Risk Management, 39*, 100487. https://doi.org/10.1016/j.crm.2023.100487.

Bacco, M., Barsocchi, P., Ferro, E., Gotta, A. and Ruggeri, M. (2019). IoT and AI in agriculture: A systematic review. *Sensors*, *19*(10), 2288. https://doi.org/10.3390/s19102288.

Beza, E., Reidsma, P., Kooistra, L. and van Ittersum, M. K. (2023). AI-based decision support systems for smallholder farmers. *Agricultural and Forest Meteorology*, 334–351, 109432. https://doi.org/10.1016/j.agrformet.2023.109432.

Bolognesi, S., Filippi, P. and Shaddad, S. (2023). Remote sensing and AI for soil quality prediction. *Remote Sensing of Environment*, 295–301. https://doi.org/10.1016/j.rse.2023.113456.

Brown, T., Smith, P. and Jones, R. (2023). AI and big data in soil fertility management. *Frontiers in Sustainable Food Systems*, 7–21, 112345. https://doi.org/10.3389/fsufs.2023.112345.

Chen, L., Lin, X., Xu, X. and Zhang, Q. (2022). Deep learning for crop yield forecasting. *Computers and Electronics in Agriculture*, 193–200, 106694. https://doi.org/10.1016/j.compag.2022.106694.

Chlingaryan, A., Sukkarieh, S. and Whelan, B. (2023). Machine learning approaches for soil nutrient mapping. *Geoderma*, 430–443, 116352. https://doi.org/10.1016/j.geoderma.2023.116352.

Chlingaryan, A., Sukkarieh, S. and Whelan, B. (2024). AI-enhanced soil carbon modelling. *Environmental Modelling and Software*, 172–178. https://doi.org/10.1016/j.envsoft.2024.105678.

Cohen, Y., Alchanatis, V. and Levi, A. (2022). AI-driven irrigation optimization. *Agricultural Water Management*, 265–281. https://doi.org/10.1016/j.agwat.2022.107654.

Dalla Corte, L., Souza, E.G., Bazzi, C.L. and Schenatto, K. (2021). AI applications in precision farming. *Agronomy*, *11*(6). https://doi.org/10.3390/agronomy11061123.

Drought Monitor. (2023). *U.S. Drought Impact Reports*. National Drought Mitigation Centre.

EPA Watershed Report. (2023). *AI Applications in Watershed Management*. Environmental Protection Agency.

FAO. (2022). *The Role of AI in Sustainable Agriculture*. Food and Agriculture Organization of the United Nations.

Fulton, J., Taylor, R. and Ballard, E. (2023). AI and soil moisture prediction. *Journal of Hydrology*, 625–634. https://doi.org/10.1016/j.jhydrol.2023.129876.

Ghazi, M., Nangia, V. and Oweis, T. (2023). AI for drought-resilient cropping systems. *Climate Risk Management*, 40–61, 100523. https://doi.org/10.1016/j.crm.2023.100523.

Heise, J., Kuhnert, M. and Sommer, M. (2023). AI-based soil erosion modelling. *Land Degradation and Development*, *34*(8). https://doi.org/10.1002/ldr.4567.

https://www.trenpa.in/blogs/news/artificial-intelligence-in-agriculture-the-future-of-traditional-farming.

Islam, M., Rahman, M. and Saha, K. (2023). AI in soil salinity assessment. *Science of the Total Environment*. https://doi.org/10.1016/j.scitotenv.2023.164321.

Jat, H., Kumar, P. and Singh, A.K. (2023). Digital soil mapping using machine learning. *Soil and Tillage Research*, *225*. https://doi.org/10.1016/j.still.2023.105543.

Jones, D., Ejeta, G. and Giller, K. (2022). AI for sustainable intensification. *Global Food Security*, *33*. https://doi.org/10.1016/j.gfs.2022.100612.

Kamilaris, A. and Prenafeta-Boldú, F.X. (2017). Deep learning in agriculture: A survey. *Computers and Electronics in Agriculture*, *147*. https://doi.org/10.1016/j.compag.2018.02.016.

Kassahun, A., Teklu, M. and Kebede, F. (2022). AI for small-scale farming in Africa. *Agricultural Systems*, *201*. https://doi.org/10.1016/j.agsy.2022.103456.

Leenaars, J., Claessens, L. and Heuvelink, G. (2023). AI-assisted soil fertility management. *European Journal of Agronomy*, *144*. https://doi.org/10.1016/j.eja.2023.126734.

Liakos, K., Busato, P., Moshou, D., Pearson, S. and Bochtis, D. (2018). Machine learning in agriculture: A review. *Biosystems Engineering*, *171*. https://doi.org/10.1016/j.biosystemseng.2018.07.013.

Lobell, D.B., Azzari, G. and Burke, M. (2023). Using artificial intelligence to predict unprecedented climate events in agriculture. *Nature Food*, *4*(3), 197–205. https://doi.org/10.1038/s43016-023-00715-y.

Mishra, P., Rai, A. and Singh, R. (2023). AI-driven precision nutrient management. *Nutrient Cycling in Agroecosystems*, *127*(1), 89–104. https://doi.org/10.1007/s10705-023-10280-2.

Mutungi, C., Karanja, N. and Waswa, B. (2023). AI for climate-smart agriculture in sub-Saharan Africa. *Climate Change*, *176*(3), 45–62. https://doi.org/10.1007/s10584-023-03552-x.

Mwebaze, E., Quinn, J. and Biehl, M. (2023). AI and IoT for soil health monitoring. *Smart Agricultural Technology*. https://doi.org/10.1016/j.atech.2023.100234.

Nawar, S., Corstanje, R., Halcro, G., Mulla, D. and Mouazen, A. M. (2019). Machine learning for soil property prediction. *Geoderma*, *352*, 251–263. https://doi.org/10.1016/j.geoderma.2019.06.028/.

Ogilvie, A., Zhang, Q. and Lobell, D. (2022). AI in agronomic decision-making. *Agricultural Systems*, 198–112. https://doi.org/10.1016/j.agsy.2022.103456.

Oluwesemire, K., Ajeigbe, H. and Tabo, R. (2023). AI for yield gap analysis in West Africa. *Field Crops Research*, 291–315. https://doi.org/10.1016/j.fcr.2023.108976.

Patel, R., Kumar, N. and Singh, P. (2022). AI-powered soil sensors. *Sensors*, *22*(15), 5678. https://doi.org/10.3390/s22155678.

Patel, R., Kumar, N. and Singh, P. (2023). Edge AI for real-time soil monitoring. *IEEE Internet of Things Journal*, *10* (4), 3210–3221. https://doi.org/10.1109/JIOT.2023.3265432.

Poblete, T., Ortega-Farías, S. and Moreno, M. (2023). AI-based irrigation scheduling. *Agricultural Water Management*, *279*. https://doi.org/10.1016/j.agwat.2023.108213.

Qian, H., Yang, X. and Chen, Y. (2023). AI for crop disease detection. *Plant Methods*, *19* (1). https://doi.org/10.1186/s13007-023-01034-w.

Rossel, R., McBratney, A. and Minasny, B. (2023). Digital soil assessment with AI. *Geoderma Regional*, *32*. https://doi.org/10.1016/j.geodrs.2023.e00612.

Schimmelpfenning, R. (2023). Economic impacts of AI in agriculture. *Applied Economic Perspectives and Policy*, *45*(2), 789–805. https://doi.org/10.1002/aepp.13345.

Shafi, U., Mumtaz, R., García-Nieto, J., Hassan, S.A., Zaidi, S.A.R. and Iqbal, N. (2020). AI and IoT for smart farming. *IEEE Access*, *8*, 129551–129583. https://doi.org/10.1109/ACCESS.2020.3009295.

Silva, G., Fearnside, P. and Cerri, C. (2023). AI for soil carbon sequestration. *Global Change Biology*, *29*(5), 1456–1472. https://doi.org/10.1111/gcb.16543.

Singh, A., Watson, J. and Kumar, V. (2023). AI in precision nitrogen management. *Precision Agriculture*, *24*(3), 567–589. https://doi.org/10.1007/s11119-023-10014-x.

Smith, J., Brown, K. and Anderson, R. (2021). AI applications in agronomy. *Annual Review of Plant Biology*, *72*, 741–768. https://doi.org/10.1146/annurev-arplant-080720-090632.

Soil Health Institute. (2023). *AI Tools for Soil Health Assessment*. https://www.soilhealthinstitute.org.

Tamene, L., Le, Q.B. and Vlek, P.L.G. (2023). AI for land degradation neutrality. *Land Use Policy*, 124–143. https://doi.org/10.1016/j.landusepol.2023.106543.

Thompson, L. (2023). AI and regenerative agriculture. *Sustainability*, *15* (4). https://doi.org/10.3390/su15043210.

Tseng, M., Lin, C. and Chen, H. (2023). AI-driven agroecological zoning. *Ecological Modelling*, 478–511. https://doi.org/10.1016/j.ecolmodel.2023.110234.

Vagen, T., Winowiecki, L. and Tondoh, J. (2023). AI for soil organic carbon mapping. *Geoderma*, *430*, 116352. https://doi.org/10.1016/j.geoderma.2023.116352.

van Asten, P., Zingore, S. and Giller, K. (2023). AI for smallholder soil fertility management. *Agricultural Systems*, 205–217. https://doi.org/10.1016/j.agsy.2023.103678.

van der Burg, S., Bogaardt, M.J. and Wolfert, S. (2022). AI ethics in agriculture. *AI and Society*, *37* (2), 451–465. https://doi.org/10.1007/s00146-021-01357-z.

Weersink, A., Fraser, E. and Pannell, D. (2021). Economics of AI in agriculture. Agricultural Economics, *52* (3), 321–335. https://doi.org/10.1111/agec.12645.

World Bank. (2023). *AI for Sustainable Agriculture in Developing Countries*. https://www.worldbank.org.

World Economic Forum. (2023). *AI and the Future of Food Security*. https://www.weforum.org.

Zhang, C., Liu, J., Shang, J. and Cai, H. (2020). Deep learning for soil classification. *Remote Sensing*, *12* (8), 1245. https://doi.org/10.3390/rs12081245.

Zhang, L., Wang, X. and Li, P. (2022). AI for yield prediction in precision agriculture. *Computers and Electronics in Agriculture*, 193–201, 106694. https://doi.org/10.1016/j.compag.2022.106694.

Zhang, N. and Kovas, A. (2023). AI in agri-food systems. *Trends in Food Science and Technology*, *131*, 12–24. https://doi.org/10.1016/j.tifs.2023.01.012.

Zhang, X., Li, Y. and Chen, W. (2023). AI-enabled soil erosion control. *Journal of Environmental Management*, 330–341. https://doi.org/10.1016/j.jenvman.2023.117654.

Zhang, X., Liu, Y., Zhao, K. and Zhao, Y. (2024). Digital twin farms for climate resilience: Evidence from Australian drought adaptation. *Agricultural Water Management, 291*, 108602. https://doi.org/10.1016/j.agwat.2023.108602.

4 Deep Learning for Satellite, UAV, and Hyperspectral Crop Monitoring

Tharindu D. Nuwarapaksha[1], Shashi S. Udumann[2], Nuwandhya S. Dissanayaka[3] and Anjana J. Atapattu[4*]

1. Introduction

1.1 Evolution of Agricultural Monitoring

Agricultural monitoring has undergone a lot of changes in the past century as it has moved on to technology-based monitoring. In the past, crop health and productivity measurements depended on field observation, farmer observation, and manual sampling, which were labor-intensive, time-consuming, and spatially restricted (Wang et al., 2025). Although these techniques gave localized information, they were not scalable to large areas or cropping systems. The advent of remote sensing technologies later in the 20th century was a turning point, providing synoptic and repetitive monitoring of agricultural landscapes (Zomeni et al., 2008). Satellite imagery enabled systematic observations of vegetation dynamics, soil moisture, and land use change at regional and global scales. Recently, the use of drones, Unmanned Aerial Vehicles (UAVs), and Internet of Things (IoT)-based ground sensors has increased monitoring ability with high-resolution and near-real-time data (Dhal et al., 2024; Nuwarapaksha et al., 2024a). Nevertheless, the obstacles still exist, such as the interference of cloud cover, high costs of data on commercial platforms, and the necessity to use sophisticated analytical techniques to process complex data. In addition, the translation of remote sensing products into usable information to farmers needs customized solutions and simple tools. In the modern context, agricultural monitoring is undergoing a new phase, and artificial intelligence and big data are helping to implement precision farming strategies that maximize inputs, maximize yields and sustainability in the face of the climate change challenge (Mana et al., 2024).

1.2 Remote Sensing Platforms Overview

Numerous remote sensing platforms can be used to monitor agriculture, each with its own strengths and weaknesses. Figure 1 depicts the various altitudes

[1-4] Agronomy Division, Coconut Research Institute, Lunuwila 61150, Sri Lanka.
* Corresponding author: aaajatapattu@gmail.com
https://orcid.org/0000-0002-7554-5289[1]; https://orcid.org/0000-0002-5717-7419[2]; https://orcid.org/0000-0002-2062-1133[3]; https://orcid.org/0000-0003-4806-4647[4]

DOI: 10.1201/9781003715726-4

and platforms where remote sensing is done, starting with ground-based UAV drones at 0.1km all the way to space satellites at 700–900 km, and how various technologies such as aerial photography, Synthetic Aperture Radar (SAR), and optical sensors are used at different altitudes to gather data on earth observation. Satellite systems, such as Landsat, Sentinel, and WorldView, offer wide spatial coverage and multi-spectral or hyperspectral imaging capabilities, making them suitable for monitoring large agricultural areas over time (Bourriz et al., 2025). Landsat has a long historical record; Sentinel has a frequent revisit capability at 10–20 m resolution, and WorldView offers very high-resolution imagery suitable for precision agriculture applications. UAVs, also referred to as drones, are a flexible and high-resolution data acquisition tool at field scale. UAVs equipped with Red, Green, Blue (RGB), multispectral, or thermal sensors can be utilized to perform precise assessments of crop stress, disease, and nutrient deficiencies (Awais et al., 2023). However, UAV cannot cover extensive areas and requires the use of operating capabilities. Besides these, ground-based sensors and IoT provide continuous point-based measurements of soil moisture, temperature, nutrient level, and microclimate conditions. Combined with remote sensing data, such ground observations enhance calibration and validation to make the models more accurate (Glenn et al., 2007). The combination of satellites, UAVs, and ground sensors creates a multi-platform ecosystem to monitor crop at the plant and regional scale in a scalable manner (Samadzadegan et al., 2025). These platforms are integrated to enhance the management of irrigation, optimization of fertilizers, and predicting yield.

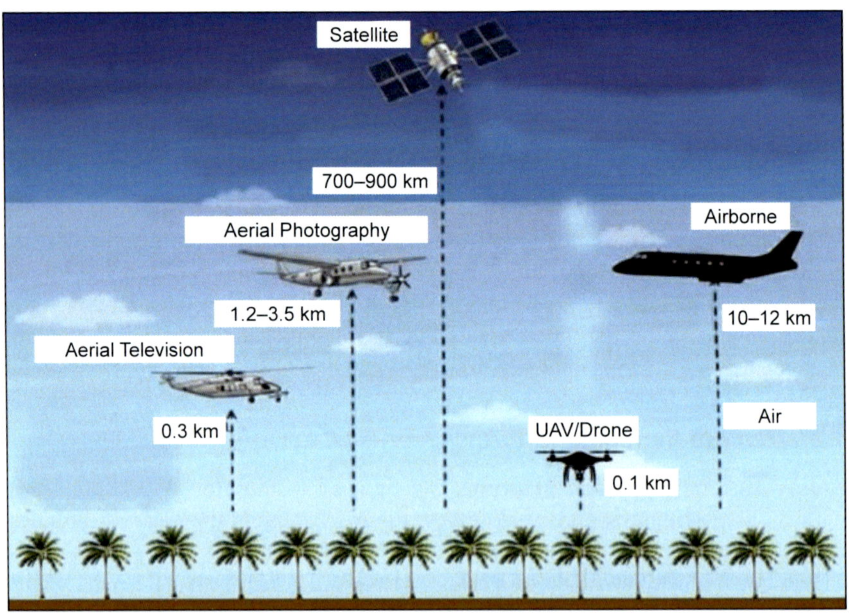

Fig. 1 Different remote sensing platforms and altitudes. (*Source:* Figure by authors)

1.3 Role of DL in Agriculture

With the introduction of deep learning (DL), remote sensing applications in agriculture have become more specific and automated, enabling the processing of complex data. In contrast to conventional machine learning algorithms, which require a great deal of hand-crafted features, DL models like Convolutional Neural Networks (CNNs) and Recurrent Neural Networks (RNNs) can automatically learn spatial, spectral, and temporal patterns in raw imagery (Grewal et al., 2023). This renders them especially useful in identifying minute crop differences, identifying early signs of stress, and predicting harvests. DL is particularly suited to working with multi dimensional agricultural data, such as multi spectral, hyperspectral, and time-series satellite imagery (Lim et al., 2024). Its capability to combine information at different scales and modalities (satellite + UAV + ground sensors) enhances the accuracy and robustness of predictions. Moreover, DL allows real-time processing, particularly when implemented on cloud systems or edge devices, so farmers and policymakers can make decisions in a timely fashion (Prashanth et al., 2025). Some of the applications are automated crop classification, disease detection, irrigation scheduling, and carbon monitoring (Nuwarapaksha et al., 2024b). Nonetheless, the problems of the necessity of large labeled datasets, computing power, and interpretability of models still exist. Nevertheless, with these constraints, DL offers a revolutionary way of promoting precision farming, with the ability to make monitoring crops more scalable, more accurate, and more actionable than ever before (Patel et al., 2025).

2. Remote Sensing Platforms for Agricultural Monitoring

2.1 Satellite Based Remote Sensing

Satellite remote sensing is perhaps the most popular platform in the agricultural monitoring field because it can cover large geographic areas on a consistent basis. Several kinds of satellites are involved in crop monitoring, e.g., optical satellites like Landsat and Sentinel-2, radar satellites like Sentinel-1, and hyperspectral satellites like EnMAP and PRISMA (Transon et al., 2018). Optical satellites are used to capture visible and near infrared reflectance which can be used to create vegetation indices such as Normalized Difference Vegetation Index (NDVI), whereas radar sensors can provide important spectral band information under all weather conditions, particularly in areas where cloud cover is likely to occur (Table 1). Hyperspectral satellites further enhance these abilities by recording hundreds of narrow bands of spectrum, which enable the identification of fine physiological characteristics of crops (Lu et al., 2020). The advantages of satellite-based monitoring are that it covers a large spatial area, has long-term historical records, and frequent revisit times, which allow monitoring crop dynamics over seasons and years. There are a few limitations however optical satellites are susceptible to atmospheric interference and constant cloud cover, especially in the tropical areas. According to Table 1, the spatial resolution is also coarse with pixel sizes of between 10–30 meters in the free datasets and this may not be able to capture the within-field variability that is important in precision agriculture (Howe et al., 2022). Also, the availability of data can be limited in commercial high-resolution missions, creating a cost-related barrier. Despite this, satellites have been invaluable in regional- and global-scale agricultural monitoring. Their long-term harmonized observations are especially valuable to crop classification, large-scale yield forecasting, and monitoring of the effects of climate variability. The coarse resolution and noisy data can be overcome with the idea of DL integration, which allows using advanced models to enhance imagery, interpolate missing data, and combine multi sensor data streams (Collins et al., 2017). Hence, the UAV and hyperspectral sensing can be used to complement and enhance large-scale crop monitoring based on satellite platforms.

Table 1 The spectral bands and resolutions of Landsat-8, Sentinel-2A.

Band Number	*Sentinel-2A*			*Landsat-8*		
	Spectral Band	*Wavelength (nm)*	*Resolution (m)*	*Spectral Band*	*Wavelength (nm)*	*Resolution (m)*
1	Coastal aerosol	433–453	60	Coastal aerosol	433–453	30
2	Blue	458–523	10	Blue	450–515	30
3	Green	543–578	10	Green	525–600	30
4	Red	650–680	10	Red	630–680	30
5	Red Edge 1	698–713	20	Near-infrared	845–885	30
6	Red Edge 2	733–748	20	Shortwave infrared 1	1560–1660	30
7	Red Edge 3	773–793	20	Shortwave infrared 2	2100–2300	30
8	Near-infrared	785–900	10	Panchromatic	500–680	15
9	Near-infrared Vapour	935–955	60	Shortwave infrared / Cirrus	1360–1390	30
10	Shortwave infrared / Cirrus	1360–1390	60			
11	Shortwave infrared 1	1566–1655	20			
12	Shortwave infrared 2	2100–2280	20			

Source: The compilation of authors based on the data by Javhar et al. (2019).

2.2 UAV-Based Remote Sensing

Drones or UAVs have emerged as powerful sources of precision agriculture since they are capable of obtaining ultra-high-resolution data, with the option of flexibility in deployment schedules. Some of the sensors that can be installed on UAVs include regular RGB cameras, multispectral cameras to assess vegetation indices, hyperspectral sensors to offer detailed spectral profiles and LiDAR to assess three-dimensional crop structures (Zhang & Zhu, 2023). The flexibility will allow the UAVs to provide high resolution data about the health of crops, phenological stages, and stress conditions that satellites cannot see. One of the major advantages of UAV-based remote sensing is high spatial and temporal resolution. Unlike satellites, drones are able to fly in order and record a resolution of centimeters, which suits smallholder farms and local monitoring. Most of the problems that the cloud cover poses also can be bypassed by UAVs because these devices fly at low altitudes (Bouguettaya et al., 2022; Ren et al., 2019). However, the UAV applications have several challenges. The coverage is required to be minimal due to the large area coverage by large battery capacity and regulations on flight time. UAVs can as well be costly in equipment, sensor fusion, and experienced operators especially with the applications of sophisticated sensors like hyperspectral cameras. Processing of data can be computationally intensive, and might need special programs and skills. However, UAVs have proved to be exceptionally handy in accuracy farming, such as selective irrigation, weeding mapping, disease detection, and variable-rate application of inputs (Chouhan et al., 2025; Vijayakumar et al., 2025). DL models with UAV images can be used to automate these tasks to make real-time site-based decisions. This flexibility of UAVs also renders them highly applicable in testing and validating the satellite measurements, and might be utilized as a bridge between measurements on the ground and the regional monitoring. In this way, UAVs cannot replace, but are complementary systems that may enhance the resolution, accuracy and timeliness of crop monitoring when used in combination with other platforms (Figure 2).

2.3 Hyperspectral Imaging Based Remote Sensing

Hyperspectral imaging is a state-of-the-art method of agricultural remote sensing, providing high-resolution information on crop physiology and health. In contrast to multispectral sensors, which record reflectance in a few broad spectral bands, hyperspectral sensors record reflectance in hundreds of narrow spectral bands (Klein et al., 2008). Figure 3 shows the spectral data differences between RGB imaging (3 separated bands), multispectral imaging (N separated bands), and hyperspectral imaging (continuous spectrum), where each technology records more detailed wavelength information that can be used in remote sensing. This high spectral resolution enables scientists and practitioners to identify minute variations in biochemical and structural characteristics of crops such as chlorophyll content, water content, nutrient status, and stress symptoms such as disease development. Hyperspectral sensing is based on the fact that every material has

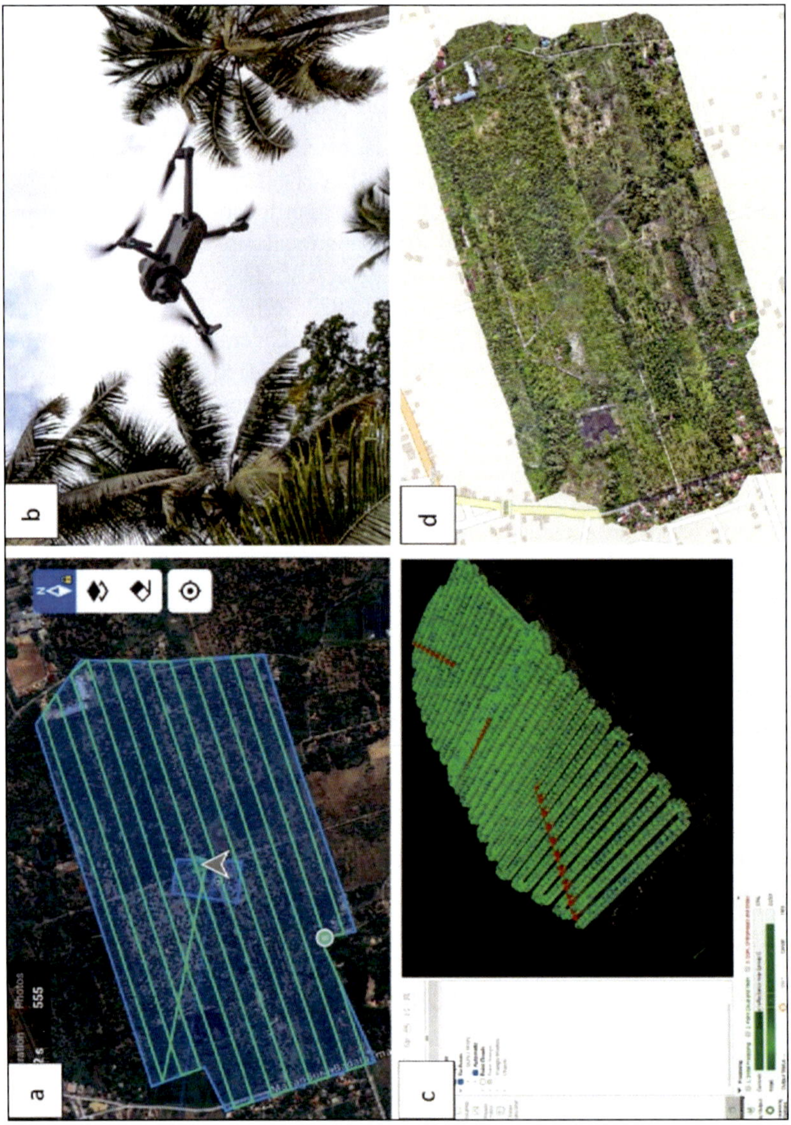

Fig. 2 Steps of drone mapping (a) Flight path planning and mapping, (b) Drone flight operation, (c) 3D point cloud generation and processing, (d) Orthomosaic map creation. (*Source:* Figure by author)

its own spectral signature and so it is possible to distinguish between crop species, detect stress conditions, and track phenological development with high precision (Gerhards et al., 2019). Hyperspectral data is more informative than multispectral imaging, and can be used to conduct applications like early disease detection, precision fertilizer management, and crop trait phenotyping. However, there are no issues with hyperspectral imaging. Hyperspectral data is highly dimensional and, in most cases, redundancy between the bands exists, requiring sophisticated computational methods to analyze the data. The huge amount of data also requires high storage, processing capabilities, and effective compression mechanisms. Nevertheless, hyperspectral imaging has proven to be quite successful in filling the gap between remote sensing applications and the physiology of plants (Panda et al., 2025). Convolutional and hybrid DL models are well-suited to process the high dimensionality of hyperspectral data to automatically extract informative spectral-spatial features. Recent developments in hyperspectral UAV payloads and hyperspectral missions in space are bringing this technology into the realm of increasing accessibility (Stuart et al., 2019). Thus, hyperspectral imaging, as well as DL, can be regarded as one of the most promising next-generation precision agriculture methods because such a combination can identify crop stresses and physiological conditions with an unprecedented accuracy.

3. Deep Learning Fundamentals of Remote Sensing

3.1 Neural Network Architectures for Remote Sensing

The application of DL in remote sensing is driven by diverse types of neural network architectures, each designed to work on a certain type of data and perform a particular task. The most popular are CNNs because they can extract spatial and spectral patterns in imagery (Li et al., 2017). They can be effectively used in land cover classification, object detection, and vegetation monitoring. RNNs and Long Short-Term Memory (LSTMs) are used with time-series remote sensing data, e.g., crop growth tracking or climate pattern analysis, where time-dependencies are important (Chandrasekaran & Paramasivan, 2025). Transformer-based models have more recently demonstrated potential to work with large-scale satellite and hyperspectral data because of their ability to attend globally to features. Also, hybrid networks with CNNs, RNNs, and attention mechanisms have been applied to more complex tasks, such as change detection and data fusion across multiple sensors (Saidi et al., 2024). Knowledge of these architectures is essential as the selection of the correct model has a direct influence on performance, efficiency, and interpretability in remote sensing applications.

3.2 Feature Extraction and Representation Learning

The automatic feature extraction ability is one of the most important strengths of DL in remote sensing. Unlike previous machine learning approaches, which

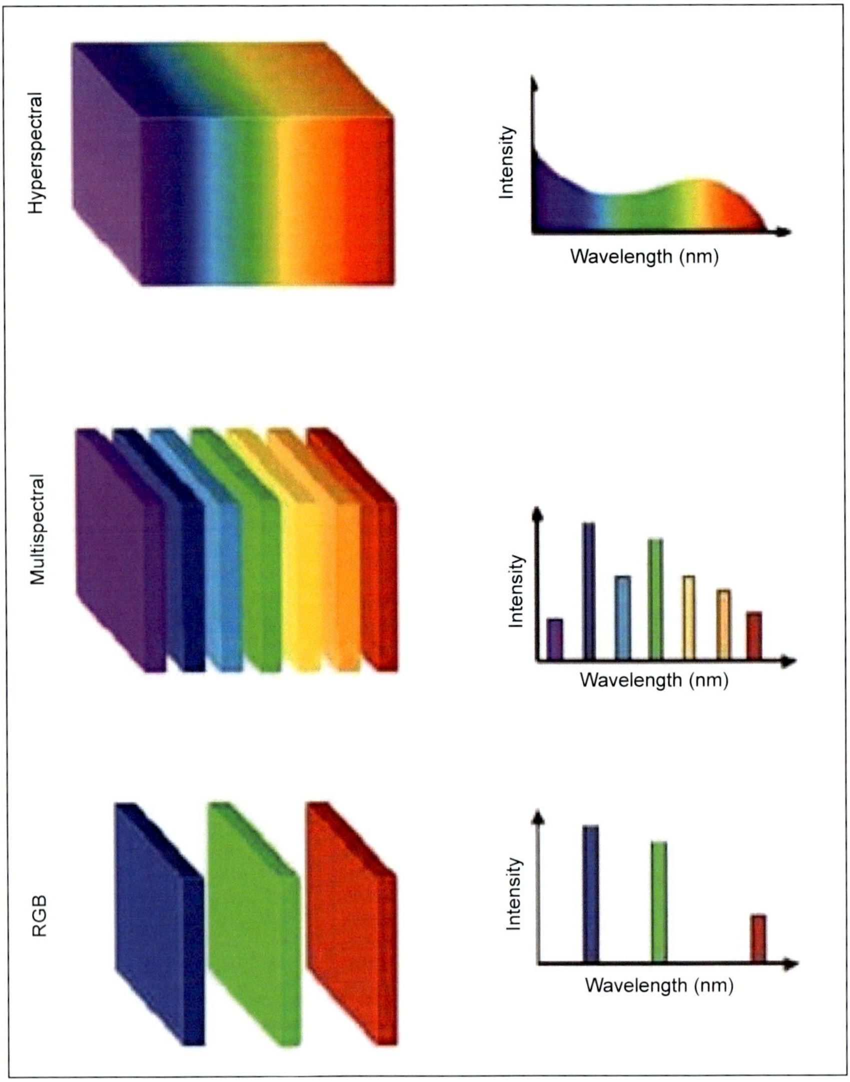

Fig. 3 Visual representation of different spectral bands. (*Source:* Figure by authors)

use hand-engineered features, deep neural networks can learn hierarchical representations directly on raw data (Manakitsa et al., 2024). In satellite and UAV images, CNN layers sequentially learn low-level (edges, textures, and colors) to high-level semantic (objects, land use categories, crop types) features (Adugna et al., 2025). DL can discover faint spectral features used to identify the level of vegetation stress, soil types, or mineral composition in hyperspectral imaging. Representation learning can be used to reduce dimensionality, which is the curse of dimensionality of hyperspectral data. Pretrained models and transfer learning are especially valuable, since they enable networks that have been trained using large datasets (such as global satellite archives) to be retrained on small datasets (such as locally available satellite imagery) with little labeled data (Nowakowski et al., 2025). These places feature extraction as one of the DL foundations in remote sensing, as it defines the accuracy and generalization of the models in real-life monitoring conditions.

3.3 Data Requirements and Training Strategies

Big data is essential to DL in remote sensing, which requires large, diverse, and high-quality datasets. Training deep models, however, is not straightforward due to the limited number of labeled data, class imbalance, and noisy measurements of the environmental variability (Mohammadisabet et al., 2025). To overcome these, researchers implement a number of approaches. Rotation, flipping, and spectral mixing are data-augmentation techniques that help expand limited training datasets. Semi-supervised and unsupervised methods, e.g., self-supervised learning can be used to utilize unlabeled data, which is available in abundance in remote sensing (Yu et al., 2023). Transfer learning enables the transfer of models that are trained on generic data to a particular application such as agriculture or urban observation. Generative models are also used to generate training samples that are hard to access, e.g., Generative adversarial networks (GANs) in the case of natural disasters, and crop disease epidemics (Eltehewy et al., 2023). Also involved in training techniques are the appropriate optimization of the hyperparameters, regularization methods to prevent overfitting and cloud computing of massive amounts of data. The awareness of these requirements will ensure that DL models are accurate in addition to being scalable and applicable to different issues of interest in remote sensing.

4. Applications of Remote Sensing and Deep Learning in Agriculture

4.1 Crop Classification and Mapping

The classification and mapping of crops is an extremely essential exercise of agriculture monitoring to establish the type, area coverage, and distribution of crops within the field and regions. The accuracy of the classification by DL has been found to be drastically higher than the classical methods because it can automatically learn both spatial and spectral features of the remote sensing data (Zhu et al., 2017). The CNNs are trendy and fed with multispectral and hyperspectral images which are captured by satellites or UAVs. These models are able to separate crops that have similar spectral points with hierarchical features, learning and incorporating spatial and temporal data. Time-series-based DLmodels such as RNNs and CNN-LSTM are improved in crop classification throughout the growth phases (Li et al., 2023). According to this UAV-advanced aerial view (DJI Mavic 3M) of what seems to be a coconut-based agroforestry system Kurunegala District, Sri Lanka, the snapshot shows the crop classification and mapping of the area with the use of UAV technology to recognize and classify various land cover features such as coconut plantations, pepper, banana plantation, cinnamon trees, grass cover areas, water bodies, roads, and built structures within the areas of interest (Figure 4). Precise crop maps can be used in food security measurement, yield prediction and subsidy allocation. Open-source platforms like Google Earth Engine now enable large-scale mapping and real-time classification using DL. On the farm level, crop classification helps in precision management practice as it defines the boundaries of crops and mixed-cropping areas (Bégué et al., 2018). In general, DL-based crop classification and mapping can give policymakers and farmers credible data to make decisions to optimize agricultural productivity and sustainability.

4.2 Precision Agriculture Techniques

Precision agriculture can be defined as a combination of site-based management practices in order to maximize resources use, and minimize environmental effects and maximize yields. Remote sensing and DL are the two components of analysis that form the core of this solution. Through UAV, multispectral and hyperspectral processing, DL models identify spatial variability in the field and can offer recommendations that can be applied in practice to achieve benchmarks of precision agriculture of a combination of site-specific management practices as to maximize resource utilization and minimize environmental impact and maximize yields. Remote sensing and DL are the two solutions that are to be combined in the analysis. UAV, multispectral, and hyperspectral data processing will make it possible to identify the spatial variation in the field and offer actions that can be implemented with the help of DL models (Adão et al., 2017). One of its important applications is variable rate input management. The UAV images can be processed with DL to detect within-field difference in crop vigor to enable it to apply water, fertilizers, or pesticides accurately. Commonly used vegetation indices include NDVI, Green Normalized Difference Vegetation Index (GNDVI), Normalized Difference Red Edge Index (NDRE) and Soil-Adjusted Vegetation Index (SAVI) in determining management zones, but DL can better detect small patterns (de Oliveira et al., 2024; Vélez et al., 2023). Table 2 provides an overview of the key vegetation indices such as NDVI, GNDVI, NDRE and SAVI with their spectral equations that are commonly utilized in precision agriculture to quantify crop vigor, chlorophyll content, nutrient status, and vegetation cover with the help of reflectance measurements at near-infrared, red, green, and red-edge wavelengths with SAVI including a soil adjustment factor to improve its ability to measure vegetation cover in sparsely-vegetated areas. An example of using CNN models to identify weeds on a field to help robots weed or spray herbicides more precisely can be used to reduce the application of chemicals. On interpretation

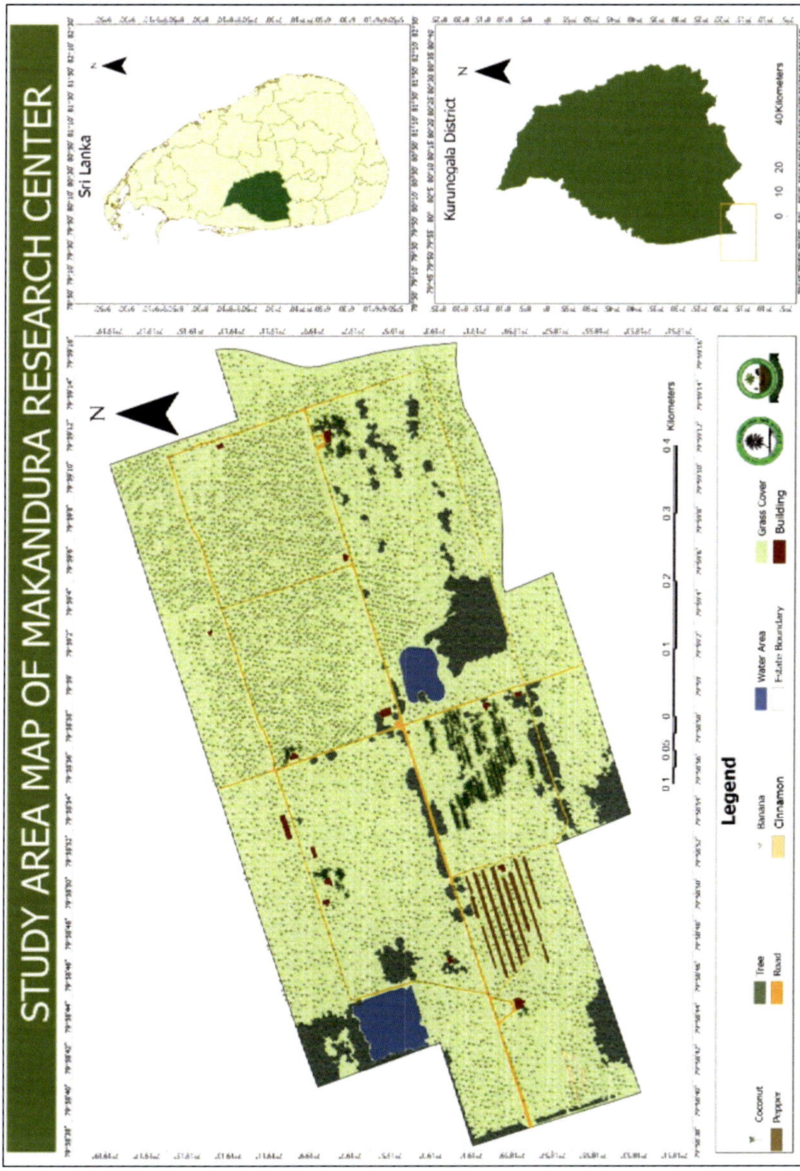

Fig. 4 Integrated land use map in Makandura research center. (*Source:* Figure by author)

using DL frameworks, they provide high resolution spatial details to irrigation timing, soil health, and crop stand. On the advanced levels, precision agriculture using DL reduces input wastage, greenhouse gases emissions and carbon-friendly farming (Balafoutis et al., 2017). It is also useful in climate smart agriculture since it enhances the resilience to droughts, floods, and other stresses. Cloud-based platforms also ensure that precision agriculture is affordable to the farmers as they give them real-time prescriptions through mobile devices. Overall, DL in precision agriculture can transform the traditional farming industry into an intelligent and data-driven industry that is sustainable and productive at the same time.

Table 2 Key remote sensing vegetation indices and applications to precision agriculture.

Index	*Equation*	*Applications*
NDVI	$\frac{R_{NIR} - R_{red}}{R_{NIR} + R_{red}}$	Crop Health Monitoring (Huang et al., 2014) Yield Prediction (Mkhabela et al., 2011)
GNDVI	$\frac{R_{NIR} - R_{green}}{R_{NIR} + R_{green}}$	Chlorophyll Concentration Monitoring (Croft et al., 2020) Drought Assessment (Osborne et al., 2005)
NDRE	$\frac{R_{NIR} - R_{red\ edge}}{R_{NIR} + R_{red\ edge}}$	Nutrient Stress Detection (Swetha et al., 2025) Early Disease Identification (Brodbeck et al., 2017)
SAVI	$\frac{(R_{NIR} - R_{red})(1 + L)}{R_{NIR} + R_{red} + L}$ L–Soil conditioning index	Vegetation Analysis in Sparse Areas (D'Acunto et al., 2024) Land Degradation Studies (D'Acunto et al., 2024)

Source: Authors' compilation based on data from Sishodia et al. (2020).

4.3 Crop Growth and Phenology Monitoring

Crop growth and phenology are necessary aspects of learning how to optimize crop management and in climate variability adaptation. Remote sensing also offers a real time monitoring and DL has the potential to enhance the ability to comprehend complex temporal processes of vegetation dynamics (Han et al., 2024). Data obtained by satellites such as Sentinel-2 and Landsat 8, as well as UAV-based data, provide time-series image data that is capable of monitoring the crop between planting and harvest. Transformers and LSTM networks are particularly handy to examine the sequential data. They are able to detect different growth stages such as germination, flowering, filling of grains, and senescence with high accuracy. Vegetation indices like NDVI, SAVI, and NDRE are usually incorporated in such models to monitor the changes in canopy structure and chlorophyll content over time (Xue & Su, 2017). As an example, NDRE is particularly useful in monitoring phenological changes with regards to nutrient uptake. The analysis of the data collected by DJI Mavic 3 multispectral drone with the help of Pix4D Mapper showed that vegetation health patterns differed in the coconut land in Sri Lanka. The values of NDVI ranged between –1 and +1, with healthy coconut canopies having high NDVI values and stressed coconut having low NDVI values. The NDRE analysis (range –1 to + 1) was effective in showing the chlorophyll changes especially in identifying nutrient deficiencies in coconut palms. SAVI values (–1 to +1) gave soil-adjusted vegetation evaluation, reducing the effects of soil brightness in regions with low canopy cover (Figure 5). Proper phenology observation helps the farmer in irrigation timing, fertilizer use, and control of pests. It also enhances yield prediction by correlating biomass accumulation to the stage of growth. On a larger scale, phenology models are used in climate resilience research, to predict changes in crop calendars in response to temperature or precipitation abnormalities. With the use of remote sensing and DL, the decision-makers can be supplied with phenological data in near real-time, which makes agriculture more flexible and resource-efficient. In this way, DL-based phenology monitoring can be considered a major contributor to enhancing productivity and sustainable agricultural systems.

4.4 Yield Forecasting and Prediction

Yield forecasting is a central aspect of farm planning, with direct consequences on food security, prices, and logistics. Conventional yield models were dependent on past records and weather variables, which were not adequate to establish the complexity of crop-environment relationships (Kirthiga & Patel, 2022). DL offers a more robust solution by combining various datasets such as satellite images, UAV observations, and soil properties and climate datasets. CNNs are particularly applicable in extraction of spatial features in remote sensing data, but LSTMs and Transformer networks are applied to sequential data on growth as well as weather variability. By combining these methods, the field and regional dynamic yield predictions can be provided by the DL-based models. The most common inputs include vegetation indices such as NDVI and GNDVI, that are related to the accumulation of biomass and canopy vitality, which are good predictors of end of crop yield (Mukiibi et al., 2024). The key strength of DL is that it is able to model nonlinear interactions between factors ,and therefore predictions are better even when the situation is in stress such as drought or disease outbreak. The resulting yield forecasts obtained through these models are useful to farmers as they help them optimize the timing of harvest and use of input. The national and international government and agencies use the estimates of yields in planning trade, stabilizing prices, and food-aid distribution. The possibility of near-real-time prediction of yields was recently achieved due to the development of UAV photogrammetry and high-resolution satellite sensors (Chouhan et al., 2025; Phang et al., 2023). It allows reacting to climatic threats in time, such as early harvesting in the case of heat stress or redirection of resources to poorly performing areas. Therefore, DL-based yield forecasting is a revolutionary technology, which links field-level operations and food security planning at the global level.

Fig. 5 Multi spectral vegetation index analysis of Makandura research center (a) NDVI, (b) NDRE, (c) SAVI using drone remote sensing. (*Source:* Figure by author)

4.5 Plant Stress Detection and Disease Diagnosis

Early identification of stress and diagnosis of disease is important in reducing yield loss and in ensuring sustainable crop management. Remote sensing records the physiological and structural changes in plants that are a result of stresses, such as drought, salinity, nutrient deficiencies, or pathogen attacks (Kashyap & Kumar, 2021). Hyperspectral imaging is especially effective, because it can sense small differences in reflectance prior to visible symptoms. These spectral signatures are fed into DL models, notably CNNs and CNN-RNN hybrid models, to detect stress patterns with high accuracy. The most common vegetation indices applied in DL models to diagnose stress are NDVI, NDRE, and the Chlorophyll Index (Boiarskii, 2019). As an example, NDRE responds to changes in pigments, and thus it is useful in the early detection of nitrogen deficiency. Similarly, UAV-based sensors can be used to derive thermal indices to detect water stress through the detection of changes in canopy temperature. Following Figure 6, UAV-derived map classifies vegetation health and management priorities based on canopy NDVI values, where red areas indicate immediate intervention needed (low NDVI/poor health), yellow areas require routine monitoring (moderate NDVI), and blue areas can maintain current practices (high NDVI/healthy vegetation). Integrating these indices with DL models, farmers will be able to gain early warnings of crop stress, and intervene accordingly. Photogrammetry softwares, including Pix4D, Agisoft, and DJI Terra, are used to complement the production of high-resolution orthomosaics and canopy models using UAV data, which are then processed using DL algorithms (Jarahizadeh & Salehi, 2024). These workflows enable high spatial resolution stress maps to be produced at the field level. The key benefit of DL-based stress detection is its scalability, i.e., it can be used in small farms and regional monitoring programs. It limits the reliance on manual scouting, enhances the effectiveness of selective pesticide or nutrient application and limits environmental effects. In general, DL-based stress detection makes crop protection a data-based and proactive process.

5. Challenges and Limitations are Remote Sensing and DL in Agriculture

5.1 Data Availability and Quality Constraints

The availability and quality of training data is one of the most important constraints in the use of DL in satellite, UAV, and hyperspectral crop monitoring (Yu et al., 2022). DL networks work best with large, high-quality data, but agricultural data are often scattered, inconsistent, and expensive to create. As an example, hyperspectral images have hundreds of spectral bands, but there is limited ground truth data on crop type, disease, or stress (Lowe et al., 2017). Also, cloud cover in satellite images, variable UAV flight conditions, and sensor calibration problems decrease the reliability of the images. Another issue is data imbalance, where some crops, conditions, or conditions are over-represented and rare, but

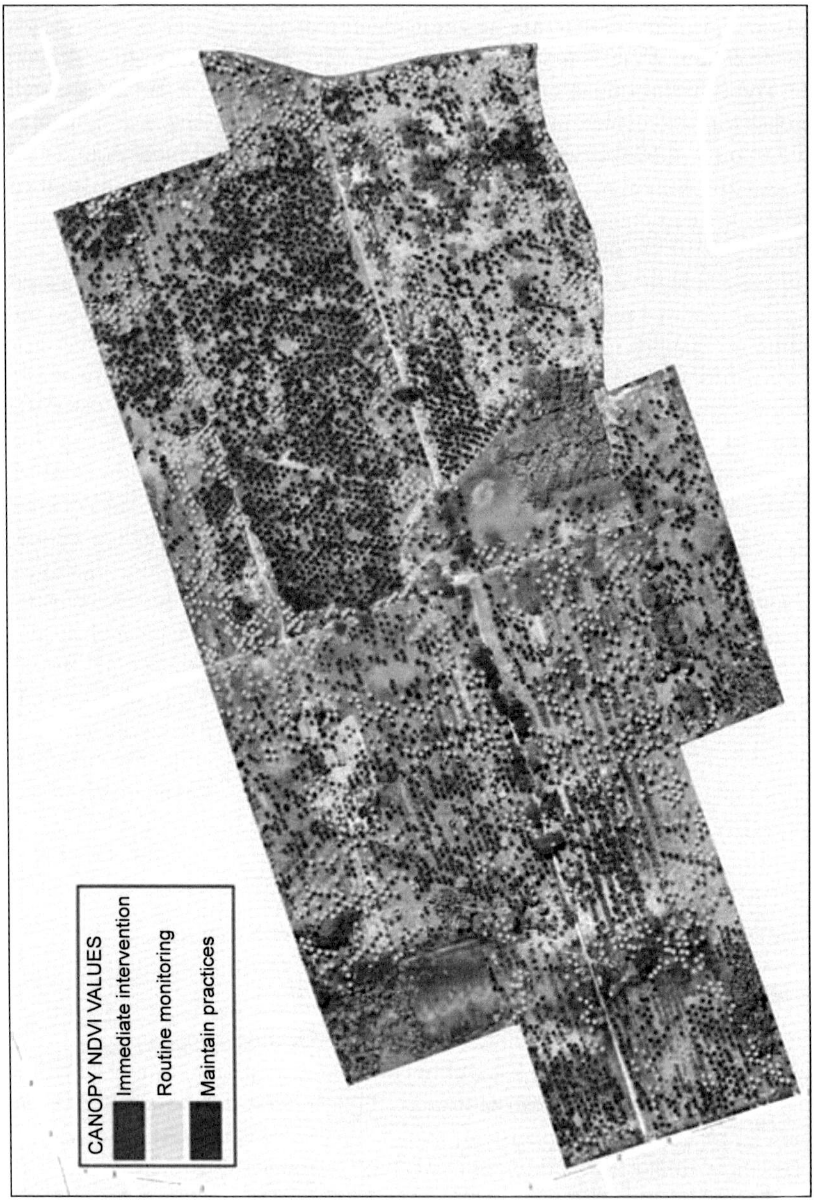

Fig. 6 NDVI-based canopy health zoning in Makandura research center. (*Source:* Figure by author)

important phenomena, like early onset of disease, are under-represented. Labeling hyperspectral data is particularly time-consuming and needs domain experts to be labeled properly (Masarczyk et al., 2020). Additionally, agricultural landscapes are dynamic, and seasonal changes, environmental stresses, and management activities change crop signatures, so static datasets are inadequate. This causes poor regional generalization and seasonal generalization. Comparative research and reproducibility are also hindered by limited availability of benchmark datasets. In the end, without consistent, varied, and standardized data, DL models will be subject to overfitting, decreased accuracy, and low transferability to practical applications. This needs open-access agricultural data, data-synthesis, and data-sharing efforts.

5.2 High Computational and Storage Demands

DL models, particularly when used in hyperspectral, UAV, and satellite data, have a drawback in that they are computationally and storage-intensive (Liu et al., 2021). Hyperspectral imagery has hundreds of spectral bands and with spatial resolutions that can produce gigabytes or even terabytes of raw data in a single flight or satellite pass. Training deep neural networks on these large-scale datasets is computationally demanding and requires specialized graphics processing unit (GPU) hardware, large-scale storage, and processing pipelines (Shahid & Mushtaq, 2020). In the case of UAV-based monitoring, the high frequency of flights over many fields results in large quantities of data that needs to be processed and transmitted quickly to deliver actionable information to farmers, but onboard and field-edge devices are computationally constrained. Cloud computing is one way out; however, in rural or agricultural areas the connection to the internet is not always stable, and this causes delays in the transmission of data and its processing. Besides, DL models are time-consuming to train and may need numerous iterations and hyperparameter optimization, which can be costly to real-time crop monitoring applications (Ojo & Zahid, 2022). Energy use turns into a constraint as well; the use of large models on GPUs makes operations expensive and not environmentally friendly. In resource-scarce regions, this challenge denies farmers and researchers the opportunity to use AI-based crop monitoring tools to the fullest (Atapattu et al., 2024; Chouhan et al., 2024).

5.3 Temporal Dynamics and Real-Time Monitoring

Agriculture is a time-sensitive sector, where crop health, growth, and stress conditions change fast. One of the shortcomings of existing DL models is that they do not take full advantage of temporal dynamics in satellite, UAV, and hyperspectral data streams. Although RNNs and LSTMs could be used to model sequential data, they are computationally intensive and not usually applicable in real-time field applications. As an example, a UAV can capture imagery every day or every week, but it is hard to process and analyze such large amounts of data in real-time to deliver actionable information (e.g., irrigation alerts or disease outbreak warnings) (Rashid et al., 2025). Slow analysis decreases the applicability of recommendations regarding precision agriculture. Also, the temporal variances, e.g., irregular acquisition time intervals, weather disruptions, and seasonal changes, interfere with training and prediction accuracy. Another problem is to synchronize the multi-temporal data between satellites and UAVs since the frequency of measurements can vary greatly (Mengmeng et al., 2017). It is essential to develop DL models that are both temporally aware and robust to irregularities and can perform analysis close to real-time. Without this, crop monitoring systems will end up falling behind the dynamic changes in the agricultural sector, restricting their effectiveness in timely decision-making.

5.4 Environmental Variability and Noise Sensitivity

Satellite, UAV, and hyperspectral imagery are very sensitive to environmental variation and noise that reduce the effectiveness of DL models. Cloud cover, shadows, soil background effects, atmospheric distortions, and sensor noise add inconsistencies that reduce model performance (Alavipanah et al., 2022). One such instance is that reflectance in hyperspectral images can be extremely sensitive to the light conditions, and can give false alarms of crop stress. The UAV imagery taken at various angles of the sun or at different weather times may not be consistent regardless of crop health (Delavarpour et al., 2021). In addition, the variation within the classes is frequently large in the crops fields; one and the same species of crop may look different due to the management or soil fertility, or due to the pests. DL models are susceptible to overfitting to this noise, particularly with limited or biased training data. Variability can be minimized by preprocessing techniques like atmospheric correction, radiometric normalization, and noise filtering, which are computationally expensive (Sharma et al., 2025). In addition, models developed in one set of conditions tend to fail when used in new areas with different climatic or soil properties. This susceptibility to environmental fluctuation points to the importance of resilient, noise-resistant designs and universal preprocessing procedures in agricultural DL.

5.5 Ethical, Privacy, and Policy Concerns

The popularity of satellite, UAV, and hyperspectral crop monitoring ethically, privacy and policy issues. UAV surveillance of farmland has the potential to inadvertently capture information on privately owned land, which poses the issue of privacy and ownership of information by farmers (Imran & Li, 2025). On the same note, satellite data and DL analytics could be used by corporations, governments or insurance companies to keep track of the practices of farmers, which would create an imbalance of power. Smallholder farmers especially may be afraid of exploitation when their crop health or yield predictions are posted without their permission (Miller et al., 2025). It is not clear whether or not intellectual property rights to trained DL models, particularly those trained on shared farmer data, will be granted. Besides, the policy frameworks of UAV flights and hyperspectral imaging are still developing, and regulations of airspace usage, data sharing, and cross-border satellite analytics differ significantly among countries (Singh, 2024). Ethical issues also involve algorithmic bias whereby models trained on data in one area may discriminate against farmers in another area by giving inaccurate advice. In the absence of robust governance structures, open data policy, and ethical principles, the use of DL to monitor crops is likely to increase inequities instead of empowering farmers.

6. Future Perspectives are Remote Sensing and DL in Agriculture

6.1 Multi modal Data Fusion and Foundation Models

The next step in agricultural monitoring is the creation of complex multi modal foundation models that can naturally combine satellite, UAV, hyperspectral, and ground-sensor data streams (Wei et al., 2025). In place of treating each data source as a separate entity, next-generation DL architectures will learn to create unified representations across spatial, spectral, and temporal dimensions using transformer-based models and attention mechanisms. These pre trained general-purpose models, which are trained on large-scale data on global agriculture, will act as general-purpose feature extractors that can be adapted to local applications with limited amounts of labeled data. The use of advanced fusion techniques will overcome the problems of varying spatial resolutions, time frequencies, and sensor characteristics by learning complementary information patterns (Dong et al., 2009). An example would be that satellite data will offer context at a broad scale, UAVs will offer high-resolution detail, and hyperspectral sensors will offer biochemical information, all of which would be processed through unified neural networks. Such a strategy will greatly enhance the accuracy of predicting complex agricultural processes such as yield forecasting and stress detection and lessen the computational complexity of having to maintain distinct models for each type of data (Wang et al., 2022). The creation of these foundation models will make advanced agricultural AI more democratic, allowing smaller organizations and developing regions to access cutting-edge crop monitoring capabilities.

6.2 Edge Computing and Real-Time Processing

The agricultural sector's demand for timely decision-making necessitates the development of lightweight DL models optimized for edge computing and real-time processing (Sathya et al., 2024). Future work will be aimed at developing effective neural network architectures, including pruned CNNs, quantized models, and knowledge distillation methods, that can be deployed on resource-constrained devices, including UAV onboard computers, smartphone applications, and field-deployed IoT sensors. These edge-optimized models will enable the on-the-fly data handling and decision-support without being linked to the cloud to eliminate the key disadvantage of the weak internet infrastructure in the rural areas. Optimal trade-offs between model accuracy and model efficiency will be found using fine-grained model compression algorithms, including Neural Architecture Search (NAS) and Automated Machine Learning (AutoML) (Cai et al., 2022). Furthermore, the federated learning techniques will enable the training of the models in many farms and preserve the privacy of the data. Real-time processing will permit real-time alerts to cultivate stress or disease break out or irrigation needs and transform reactive agricultural control to proactive intervention systems (Eze et al., 2025). It also means a blend of 5G networks and edge computing infrastructures that will enable the functions of the agriculture AI to become faster and more reliable and precision farming will become cheaper and more achievable among the farmers around the globe.

6.3 Explainable AI and Interpretable Models

With the DL models becoming more complex, the agricultural community is looking to have greater transparency and interpretability of AI-driven systems of decisions. With the DL models becoming more complex, the agricultural community is looking to have greater transparency and interpretability of AI-driven systems of decisions (Tahir et al., 2025). Directions in the future will be explainable artificial intelligence (XAI) methods that may be used to give specific and interpretable explanations to model predictions and recommendations. This involves designing attention visualization techniques of hyperspectral analysis, gradient based attribution techniques of satellite images, interpretation and techniques of uncertainty quantification that explain the level of confidence to the end-users. Advanced interpretability methods that can tell why a model proposes a certain action such as applying fertilizer to certain fields or curing a given disease will enable farmers to trust and use AI technologies (Kaler & Kaur, 2025). DL with causal inference techniques will go beyond prediction based on correlation to determine the actual causes and effects in agricultural systems. Complex model outputs will become accessible to farmers of varying technical backgrounds using visualization means that give the user the opportunity to interact with the visualization and natural language explanation systems (Htun et al., 2022). Besides, domain-specific interpretability measures will be created so that the models can acquire the patterns that have a biological meaning and not mere spurious correlations. This emphasis on explainability will not only enhance the acceptability of the farmers but also make it possible to comply with regulations, provide the agricultural extension services, and transfer knowledge between AI systems and conventional agricultural knowledge.

6.4 Synthetic Data Generation and Data Augmentation

The current shortage of labeled agricultural data will be the key factor to innovate in synthetic data generation and sophisticated data augmentation methods (Morales-

García et al., 2023). In future studies, we will use GANs, Variational Autoencoders (VAEs), and diffusion models to generate realistic synthetic hyperspectral, satellite, and UAV imagery in varied crop conditions, stress conditions, and environmental variations. These artificial data will allow developing strong DL models when there are limited and expensive ways to access real-world labeled data. The more advanced data augmentation methods will not only imply the usage of the regular geometric transformations but also spectral mixing, atmospheric simulation, and modeling of the phenological stage development (Cheng et al., 2023). Digital twin technologies will be used to create virtual agricultural environments where crops will be allowed to grow under different conditions to provide unlimited training scenarios to DL models. Physics-informed neural networks will use the domain knowledge of agriculture and remote sensing to make sure that the synthetic data generated is biologically feasible and has realistic spectral signatures. Differential privacy methods will allow the sharing of synthetic agricultural data without breaching privacy of farmers or confidential information. This artificial data ecosystem will open up access to a variety of training data, allow quick prototyping of new algorithms, and foster the creation of globally transferable agricultural AI models that can generalize across crops, climates and farming systems (Pintus et al., 2025).

6.5 Climate Adaptation and Sustainability Monitoring

The increasing effects of climate change on agriculture will lead to the creation of DL systems specifically used in climate adaptation and sustainability monitoring (Balasundram et al., 2023). Future models will combine long-term climate forecasts with real-time crop observation data to forecast and help prevent agricultural risks related to climate. High-order spatiotemporal DL systems will be used to study long-term satellite records to detect climate-related changes in crop phenology, yielding dynamics and optimum growth zones. These systems will facilitate adaptive crop choice, optimal planting times with the changing seasonal patterns and forecasting emerging pest and disease pressures due to climate change. Sustainability-oriented models will track carbon storage in farm soils, evaluate the biodiversity consequences of farming practices, and maximize water use under growing drought conditions (Kumar et al., 2025). Integration of Earth system models with agricultural DL will enable comprehensive assessment of farming practices' environmental footprint. Monitoring of sustainable agriculture indicators in real-time, including cover crop adoption, reduced tillage practices, and precision input application, will assist in policy development and carbon credit markets (Getahun et al., 2024). Additionally, the systems will aid in the early alerts of extreme weather conditions, which will enable advance crop defense, and minimal losses. The convergence of climate science and agricultural AI will provide farmers with an opportunity to adapt to new circumstances and promote AI-friendly farming approaches that will reduce climate change.

7. Conclusion

DL together with satellite, UAV, and hyperspectral remote sensing systems is a game changer in the agricultural monitoring system that redefined the traditional agriculture system into the precision agriculture systems. This report suggests that the neural network designs have been able to handle the complex problems of crop classification, growth detection, stress detection, and yield prediction down to the various scales and over time. When these technologies can be synergies used in a synergistic way, sustainable agriculture in a manner never thought of before is possible. The satellite systems, the high-resolution field-specific information, and the biochemical analysis of crop conditions are all covered with the assistance of the hyperspectral imaging, which makes the broad scope monitoring and history the key to them. It has been demonstrated that CNNs, RNNs, and Transformer-based models have proven useful in learning useful patterns in the multi dimensional agricultural data automatically, avoiding the necessity to manually engineer features and understand them according to experts. Although there have been great strides, there are still large issues to address that still need to be researched. The limitations to large-scale adoption are data availability, computational requirements, environmental variability impacts, and ethical concerns regarding the privacy of farmers. The dynamics of agricultural systems over time and the necessity to process information in real-time also makes implementation in real-world farming settings a challenge. The next steps in the development of multi modal foundation models, edge computing solutions, XAI systems, and synthetic data generation have the potential to solve the existing limitations. The trend towards climate-sensitive monitoring systems will play a significant role as agriculture experiences an upsurge in environmental stressors. The success in overcoming these challenges will eventually decide whether DL-enhanced crop monitoring can reach its potential to transform global food security and agricultural sustainability in the age of climate change and growing population demands.

Ethical Responsibilities of Authors

"All authors have read, understood, and have complied as applicable with the statement on 'Ethical responsibilities of Authors' as found in the Instructions for Authors."

Ethical Approval

Not applicable

Use of AI tool

No

Conflict of Interest

On behalf of all the authors the corresponding author states that there is no conflict of interest to disclose.

Competing interests

Not applicable

Authors' contributions

Nuwarapaksha T.D; Atapattu A.J: Conceptualization, Methodology, Writing – Original Draft, Writing – Review & Editing, Nuwarapaksha T.D; Udumann S.S: Investigation, Formal analysis, Visualization, Software, Data Curation, Supervision, Nuwarapaksha T.D; Dissanayaka N.S: Writing – Original Draft, Formal analysis, Methodology, Dissanayaka N.S; Udumann S.S: Formal analysis, Visualization, Atapattu A.J.: Supervision.

Funding

No funding has been received.

References

Adão, T., Hruška, J., Pádua, L., Bessa, J., Peres, E., Morais, R. and Sousa, J. (2017). Hyperspectral Imaging: A Review on UAV-Based Sensors, Data Processing, and Applications for Agriculture and Forestry. *Remote Sensing*, *9*(11), 1110. https://doi.org/10.3390/rs9111110.

Adugna, T., Xu, W., Fan, J., Jia, H. and Luo, X. (2025). Assessing CNN and Semantic Segmentation Models for Coarse Resolution Satellite Image Classification in Subcontinental Scale Land Cover Mapping. *IEEE Journal of Selected Topics in Applied Earth Observations and Remote Sensing*, *18*, 2777–2798. https://doi.org/10.1109/JSTARS.2024.3469728.

Alavipanah, S.K., Karimi Firozjaei, M., Sedighi, A., Fathololoumi, S., Zare Naghadehi, S., Saleh, S., Naghdizadegan, M., Gomeh, Z., Arsanjani, J.J., Makki, M., Qureshi, S., Weng, Q., Haase, D., Pradhan, B., Biswas, A. and M. Atkinson, P. (2022). The Shadow Effect on Surface Biophysical Variables Derived from Remote Sensing: A Review. *Land*, *11*(11), 2025. https://doi.org/10.3390/land11112025.

Atapattu, A.J., Perera, L.K., Nuwarapaksha, T.D., Udumann, S.S. and Dissanayaka, N.S. (2024). Challenges in Achieving Artificial Intelligence in Agriculture. In: *Artificial Intelligence Techniques in Smart Agriculture* (pp. 7–34). Springer Nature Singapore. https://doi.org/10.1007/978-981-97-5878-4_2.

Awais, M., Li, W., Cheema, M.J.M., Zaman, Q.U., Shaheen, A., Aslam, B., Zhu, W., Ajmal, M., Faheem, M., Hussain, S., Nadeem, A.A., Afzal, M.M. and Liu, C. (2023). UAV-based remote sensing in plant stress imaging using high-resolution thermal sensor for digital agriculture practices: A meta-review. *International Journal of Environmental Science and Technology*, *20*(1), 1135–1152. https://doi.org/10.1007/s13762-021-03801-5.

Balafoutis, A., Beck, B., Fountas, S., Vangeyte, J., Wal, T., Soto, I., Gómez-Barbero, M., Barnes, A. and Eory, V. (2017). Precision Agriculture Technologies Positively Contributing to GHG

Emissions Mitigation, Farm Productivity, and Economics. *Sustainability*, *9*(8), 1339. https://doi.org/10.3390/su9081339.

Balasundram, S.K., Shamshiri, R.R., Sridhara, S. and Rizan, N. (2023). The Role of Digital Agriculture in Mitigating Climate Change and Ensuring Food Security: An Overview. *Sustainability*, *15*(6), 5325. https://doi.org/10.3390/su15065325.

Bégué, A., Arvor, D., Bellon, B., Betbeder, J., De Abelleyra, D., P.D. Ferraz, R., Lebourgeois, V., Lelong, C., Simões, M. and R. Verón, S. (2018). Remote Sensing and Cropping Practices: A Review. *Remote Sensing*, *10*(1), 99. https://doi.org/10.3390/rs10010099.

Boiarskii, B. (2019). Comparison of NDVI and NDRE Indices to Detect Differences in Vegetation and Chlorophyll Content. *Journal of Mechanics of Continua and Mathematical Sciences*, *Spl1*(4). https://doi.org/10.26782/jmcms.spl.4/2019.11.00003.

Bouguettaya, A., Zarzour, H., Kechida, A. and Taberkit, A.M. (2022). DL techniques to classify agricultural crops through UAV imagery: A review. *Neural Computing and Applications*, *34*(12), 9511–9536. https://doi.org/10.1007/s00521-022-07104-9.

Bourriz, M., Hajji, H., Laamrani, A., Elbouanani, N., Abdelali, H.A., Bourzeix, F., El-Battay, A., Amazirh, A. and Chehbouni, A. (2025). Integration of Hyperspectral Imaging and AI Techniques for Crop Type Mapping: Present Status, Trends, and Challenges. *Remote Sensing*, *17*(9), 1574. https://doi.org/10.3390/rs17091574.

Brodbeck, C., Sikora, E., Delaney, D., Pate, G. and Johnson, J. (2017). Using Unmanned Aircraft Systems for Early Detection of Soybean Diseases. *Advances in Animal Biosciences*, *8*(2), 802–806. https://doi.org/10.1017/S2040470017001315.

Cai, H., Lin, J., Lin, Y., Liu, Z., Tang, H., Wang, H., Zhu, L. and Han, S. (2022). Enable DL on Mobile Devices: Methods, Systems, and Applications. *ACM Transactions on Design Automation of Electronic Systems*, *27*(3), 1–50. https://doi.org/10.1145/3486618.

Chandrasekaran, R. and Paramasivan, S.K. (2025). Advances in DL Techniques for Short-term Energy Load Forecasting Applications: A Review. *Archives of Computational Methods in Engineering*, *32*(2), 663–692. https://doi.org/10.1007/s11831-024-10155-x.

Chouhan, S.S., Singh, U.P., Saxena, A. and Jain, S. (2024). Assessing the Importance and Need of Artificial Intelligence for Precision Agriculture. In: *Artificial Intelligence Techniques in Smart Agriculture* (pp. 1–6). Springer Nature Singapore. https://doi.org/10.1007/978-981-97-5878-4_1.

Chouhan, S.S., Patel, R.K., Shukla, A. and Singh, U.P. (2025). Drone Importance and their Necessity in Future Generation Agriculture. In: *Artificial Intelligence and Computer Vision for Ecological Informatics* (pp. 1–8). CRC Press. https://doi.org/10.1201/9781003518778-1.

Chouhan, S.S., Patel, R.K., Singh, U.P. and Tejani, G.G. (2025). Integrating drone in Agriculture: Addressing technology, challenges, solutions, and applications to drive economic growth. *Remote Sensing Applications: Society and Environment*, *38*, 101576. https://doi.org/10.1016/j.rsase.2025.101576.

Cheng, X., Sun, Y., Zhang, W., Wang, Y., Cao, X. and Wang, Y. (2023). Application of DL in Multitemporal Remote Sensing Image Classification. *Remote Sensing*, *15*(15), 3859. https://doi.org/10.3390/rs15153859.

Collins, C.B., Beck, J.M., Bridges, S.M., Rushing, J.A. and Graves, S.J. (2017). DL for multisensor image resolution enhancement. In: *Proceedings of the 1st Workshop on Artificial Intelligence and DL for Geographic Knowledge Discovery* (pp. 37–44). https://doi.org/10.1145/3149808.3149815.

Croft, H., Arabian, J., Chen, J.M., Shang, J. and Liu, J. (2020). Mapping within-field leaf chlorophyll content in agricultural crops for nitrogen management using Landsat-8 imagery. *Precision Agriculture*, *21*(4), 856–880. https://doi.org/10.1007/s11119-019-09698-y.

D'Acunto, F., Marinello, F. and Pezzuolo, A. (2024). Rural Land Degradation Assessment through Remote Sensing: Current Technologies, Models, and Applications. *Remote Sensing*, *16*(16), 3059. https://doi.org/10.3390/rs16163059.

de Oliveira, H.F.E., de Castro, L.E.V., Sousa, C.M., Alves Júnior, L.R., Mesquita, M., Silva, J.A.O. S., Faria, L.C., da Silva, M.V., Giongo, P.R., de Oliveira Júnior, J.F., de Siqueira, V.S. and da Silva, J.L.B. (2024). Geotechnologies in Biophysical Analysis through the Applicability of the UAV and Sentinel-2A/MSI in Irrigated Area of Common Beans: Accuracy and Spatial Dynamics. *Remote Sensing*, *16*(7), 1254. https://doi.org/10.3390/rs16071254.

Delavarpour, N., Koparan, C., Nowatzki, J., Bajwa, S. and Sun, X. (2021). A Technical Study on UAV Characteristics for Precision Agriculture Applications and Associated Practical Challenges. *Remote Sensing*, *13*(6), 1204. https://doi.org/10.3390/rs13061204.

Dhal, S., Wyatt, B.M., Mahanta, S., Bhattarai, N., Sharma, S., Rout, T., Saud, P. and Acharya, B.S. (2024). Internet of Things (IoT) in digital agriculture: An overview. *Agronomy Journal*, *116*(3), 1144–1163. https://doi.org/10.1002/agj2.21385.

Dong, J., Zhuang, D., Huang, Y. and Fu, J. (2009). Advances in Multi-Sensor Data Fusion: Algorithms and Applications. *Sensors*, *9*(10), 7771–7784. https://doi.org/10.3390/s91007771.

Eltehewy, R., Abouelfarag, A. and Saleh, S. N. (2023). Efficient Classification of Imbalanced Natural Disasters Data Using Generative Adversarial Networks for Data Augmentation. *ISPRS International Journal of Geo-Information*, *12*(6), 245. https://doi.org/10.3390/ijgi12060245.

Eze, V.H.U., Eze, E.C., Alaneme, G.U., Bubu, P.E., Nnadi, E.O.E. and Okon, M. Ben. (2025). Integrating IoT sensors and machine learning for sustainable precision agroecology: Enhancing crop resilience and resource efficiency through data-driven strategies, challenges, and future prospects. *Discover Agriculture*, *3*(1), 83. https://doi.org/10.1007/s44279-025-00247-y.

Gerhards, M., Schlerf, M., Mallick, K. and Udelhoven, T. (2019). Challenges and Future Perspectives of Multi-/Hyperspectral Thermal Infrared Remote Sensing for Crop Water-Stress Detection: A Review. *Remote Sensing*, *11*(10), 1240. https://doi.org/10.3390/rs11101240.

Getahun, S., Kefale, H. and Gelaye, Y. (2024). Application of Precision Agriculture Technologies for Sustainable Crop Production and Environmental Sustainability: A Systematic Review. *The Scientific World Journal*, *2024*(1). https://doi.org/10.1155/2024/2126734.

Glenn, E.P., Huete, A.R., Nagler, P.L., Hirschboeck, K.K. and Brown, P. (2007). Integrating Remote Sensing and Ground Methods to Estimate Evapotranspiration. *Critical Reviews in Plant Sciences*, *26*(3), 139–168. https://doi.org/10.1080/07352680701402503.

Grewal, R., Singh Kasana, S. and Kasana, G. (2023). Machine Learning and DL Techniques for Spectral Spatial Classification of Hyperspectral Images: A Comprehensive Survey. *Electronics*, *12*(3), 488. https://doi.org/10.3390/electronics12030488.

Han, H., Liu, Z., Li, J. and Zeng, Z. (2024). Challenges in remote sensing based climate and crop monitoring: navigating the complexities using AI. *Journal of Cloud Computing*, *13*(1), 34. https://doi.org/10.1186/s13677-023-00583-8.

Howe, A.A., Parks, S.A., Harvey, B.J., Saberi, S.J., Lutz, J.A. and Yocom, L.L. (2022). Comparing Sentinel-2 and Landsat 8 for Burn Severity Mapping in Western North America. *Remote Sensing*, *14*(20), 5249. https://doi.org/10.3390/rs14205249.

Htun, N.-N., Rojo, D., Ooge, J., De Croon, R., Kasimati, A. and Verbert, K. (2022). Developing Visual-Assisted Decision Support Systems across Diverse Agricultural Use Cases. *Agriculture*, *12*(7), 1027. https://doi.org/10.3390/agriculture12071027.

Huang, J., Wang, H., Dai, Q. and Han, D. (2014). Analysis of NDVI Data for Crop Identification and Yield Estimation. *IEEE Journal of Selected Topics in Applied Earth Observations and Remote Sensing*, *7*(11), 4374–4384. https://doi.org/10.1109/JSTARS.2014.2334332.

Imran and Li, J. (2025). *UAV Technology in Agriculture: Legal, Ethical, and Future Perspectives* (pp. 431–462). https://doi.org/10.1007/978-981-96-8402-1_13.

Jarahizadeh, S. and Salehi, B. (2024). A Comparative Analysis of UAV Photogrammetric Software Performance for Forest 3D Modeling: A Case Study Using AgiSoft Photoscan, PIX4DMapper, and DJI Terra. *Sensors*, *24*(1), 286. https://doi.org/10.3390/s24010286.

Javhar, A., Chen, X., Bao, A., Jamshed, A., Yunus, M., Jovid, A. and Latipa, T. (2019). Comparison of Multi-Resolution Optical Landsat-8, Sentinel-2, and Radar Sentinel-1 Data for Automatic Lineament Extraction: A Case Study of Alichur Area, SE Pamir. *Remote Sensing*, *11*(7), 778. https://doi.org/10.3390/rs11070778.

Kaler, B. and Kaur, A. (2025). A systematic survey on explainable artificial intelligence (XAI) for plant health monitoring: Challenges and opportunities. *Applied Intelligence*, *55*(12), 889. https://doi.org/10.1007/s10489-025-06790-x.

Kashyap, B. and Kumar, R. (2021). Sensing Methodologies in Agriculture for Monitoring Biotic Stress in Plants Due to Pathogens and Pests. *Inventions*, *6*(2), 29. https://doi.org/10.3390/inventions6020029.

Kirthiga, S.M. and Patel, N.R. (2022). In-Season Wheat Yield Forecasting at High Resolution Using Regional Climate Model and Crop Model. *AgriEngineering*, *4*(4), 1054–1075. https://doi.org/10.3390/agriengineering4040066.

Klein, M.E., Aalderink, B.J., Padoan, R., De Bruin, G. and Steemers, T.A.G. (2008). Quantitative Hyperspectral Reflectance Imaging. *Sensors*, *8*(9), 5576–5618. https://doi.org/10.3390/s8095576.

Kumar, R., Prabu, J., Kumar, A., Yadav, S., Khanikar, S., Singh, A., Hansda, S. and Verma, S. (2025). Advancing Horticulture through IoT and Sensor Technologies: Trends, Challenges, and Future Directions. *Journal of Experimental Agriculture International*, *47*(2), 389–416. https://doi.org/10.9734/jeai/2025/v47i23299.

Li, Q., Tian, J. and Tian, Q. (2023). DL Application for Crop Classification via Multi-Temporal Remote Sensing Images. *Agriculture*, *13*(4), 906. https://doi.org/10.3390/agriculture13040906.

Li, Y., Zhang, H. and Shen, Q. (2017). Spectral–Spatial Classification of Hyperspectral Imagery with 3D Convolutional Neural Network. *Remote Sensing*, *9*(1), 67. https://doi.org/10.3390/rs9010067.

Lim, S. L., Sreevalsan-Nair, J. and Daya Sagar, B.S. (2024). Multispectral data mining: A focus on remote sensing satellite images. *WIREs Data Mining and Knowledge Discovery*, *14*(2). https://doi.org/10.1002/widm.1522.

Liu, J., Xiang, J., Jin, Y., Liu, R., Yan, J. and Wang, L. (2021). Boost Precision Agriculture with Unmanned Aerial Vehicle Remote Sensing and Edge Intelligence: A Survey. *Remote Sensing*, *13*(21), 4387. https://doi.org/10.3390/rs13214387.

Lowe, A., Harrison, N. and French, A.P. (2017). Hyperspectral image analysis techniques for the detection and classification of the early onset of plant disease and stress. *Plant Methods*, *13*(1), 80. https://doi.org/10.1186/s13007-017-0233-z.

Lu, B., Dao, P., Liu, J., He, Y. and Shang, J. (2020). Recent Advances of Hyperspectral Imaging Technology and Applications in Agriculture. *Remote Sensing*, *12*(16), 2659. https://doi.org/10.3390/rs12162659.

Mana, A.A., Allouhi, A., Hamrani, A., Rehman, S., el Jamaoui, I. and Jayachandran, K. (2024). Sustainable AI-based production agriculture: Exploring AI applications and implications in agricultural practices. *Smart Agricultural Technology*, *7*, 100416. https://doi.org/10.1016/j.atech.2024.100416.

Manakitsa, N., Maraslidis, G.S., Moysis, L. and Fragulis, G.F. (2024). A Review of Machine Learning and DL for Object Detection, Semantic Segmentation, and Human Action Recognition in Machine and Robotic Vision. *Technologies*, *12*(2), 15. https://doi.org/10.3390/technologies12020015.

Masarczyk, W., Głomb, P., Grabowski, B. and Ostaszewski, M. (2020). Effective Training of Deep Deep leaning for Hyperspectral Image Classification through Artificial Labeling. *Remote Sensing*, *12*(16), 2653. https://doi.org/10.3390/rs12162653.

Mengmeng, D., Noboru, N., Atsushi, I. and Yukinori, S. (2017). Multi-temporal monitoring of wheat growth by using images from satellite and unmanned aerial vehicle. *International Journal of Agricultural and Biological Engineering*, *10*(5), 1–13. https://doi.org/10.25165/j.ijabe.20171005.3180.

Miller, T., Mikiciuk, G., Durlik, I., Mikiciuk, M., Łobodzińska, A. and Śnieg, M. (2025). The IoT and AI in Agriculture: The Time Is Now—A Systematic Review of Smart Sensing Technologies. *Sensors*, *25*(12), 3583. https://doi.org/10.3390/s25123583.

Mkhabela, M.S., Bullock, P., Raj, S., Wang, S. and Yang, Y. (2011). Crop yield forecasting on the Canadian Prairies using MODIS NDVI data. *Agricultural and Forest Meteorology*, *151*(3), 385–393. https://doi.org/10.1016/j.agrformet.2010.11.012.

Mohammadisabet, A., Hasan, R., Dattana, V., Mahmood, S. and Hussain, S. (2025). CNN-Based Optimization for Fish Species Classification: Tackling Environmental Variability, Class Imbalance, and Real-Time Constraints. *Information*, *16*(2), 154. https://doi.org/10.3390/info16020154.

Morales-García, J., Bueno-Crespo, A., Terroso-Sáenz, F., Arcas-Túnez, F., Martínez-España, R. and Cecilia, J.M. (2023). Evaluation of synthetic data generation for intelligent climate control in greenhouses. *Applied Intelligence*, *53*(21), 24765–24781. https://doi.org/10.1007/s10489-023-04783-2

Mukiibi, A., Machakaire, A.T.B., Franke, A.C. and Steyn, J.M. (2024). A Systematic Review of Vegetation Indices for Potato Growth Monitoring and Tuber Yield Prediction from Remote Sensing. *Potato Research*. https://doi.org/10.1007/s11540-024-09748-7.

Nowakowski, A., Rosso, M.P. Del, Zachar, P., Spiller, D., Gabara, G., Barretta, D., Kalinowska, K.B., Choromański, K., Wilkowski, A., Sebastianelli, A., Kupidura, P., Osińska-Skotak, K. and Ullo, S.L. (2025). Transfer Learning in Earth Observation Data Analysis: A review. *IEEE Geoscience and Remote Sensing Magazine*, *13*(1), 121–152. https://doi.org/10.1109/MGRS.2024.3494673.

Nuwarapaksha, T.D., Udumann, S.S., Dissanayaka, N.S., Dilshan, R.M.N. and Atapattu, A.J. (2024a). *AI-Driven Solutions for Sustainable Irrigation* (pp. 1–32). https://doi.org/10.4018/979-8-3693-6468-0.ch001.

Nuwarapaksha, T.D., Udumann, S.S., Dissanayaka, N.S., Dilshan, R.M.N. and Atapattu, A.J. (2024b). *Revolutionizing Agriculture by Advanced Water and Irrigation Management Technologies* (pp. 285–318). https://doi.org/10.4018/979-8-3693-6920-3.ch009.

Ojo, M.O. and Zahid, A. (2022). DL in Controlled Environment Agriculture: A Review of Recent Advancements, Challenges, and Prospects. *Sensors*, *22*(20), 7965. https://doi.org/10.3390/s22207965.

Osborne, S.L., Schepers, J.S. and Schlemmer, M.R. (2005). Using Multi-Spectral Imagery to Evaluate Corn Grown Under Nitrogen and Drought Stressed Conditions. *Journal of Plant Nutrition*, *27*(11), 1917–1929. https://doi.org/10.1081/PLN-200030042.

Panda, D., Mohanty, S., Das, S., Senapaty, J., Sahoo, D.B., Mishra, B., Baig, M.J. and Behera, L. (2025). From spectrum to yield: Advances in crop photosynthesis with hyperspectral imaging. *Photosynthetica*, *63*(2), 196–233. https://doi.org/10.32615/ps.2025.012.

Patel, R.K., Chouhan, S.S., Chandan, T. and Singh, U.P. (2025). Basil Crop Detection Using Computer Vision and Deep Learning Approach. In: *Artificial Intelligence and Computer Vision for Ecological Informatics* (pp. 67–82). CRC Press. https://doi.org/10.1201/9781003518778-4.

Phang, S.K., Chiang, T.H.A., Happonen, A. and Chang, M.M.L. (2023). From Satellite to UAV-Based Remote Sensing: A Review on Precision Agriculture. *IEEE Access*, *11*, 127057–127076. https://doi.org/10.1109/ACCESS.2023.3330886.

Pintus, M., Colucci, F. and Maggio, F. (2025). Emerging Developments in Real-Time Edge AIoT for Agricultural Image Classification. *IoT*, *6*(1), 13. https://doi.org/10.3390/iot6010013.

Prashanth, J.S., Krishna, G.B., Prasad, A.V.K. and Rao, P.R. (2025). Smart Farming Revolution: A Cutting-Edge Review of DL and IoT Innovations in Agriculture. *Operations Research Forum*, *6*(1), 35. https://doi.org/10.1007/s43069-025-00434-z.

Rashid, A. Bin, Kausik, A.K., Khandoker, A. and Siddque, S.N. (2025). Integration of Artificial Intelligence and IoT with UAVs for Precision Agriculture. *Hybrid Advances*, *10*, 100458. https://doi.org/10.1016/j.hybadv.2025.100458.

Ren, H., Zhao, Y., Xiao, W. and Hu, Z. (2019). A review of UAV monitoring in mining areas: current status and future perspectives. *International Journal of Coal Science & Technology*, *6*(3), 320–333. https://doi.org/10.1007/s40789-019-00264-5.

Saidi, S., Idbraim, S., Karmoude, Y., Masse, A. and Arbelo, M. (2024). Deep-Learning for Change Detection Using Multi modal Fusion of Remote Sensing Images: A Review. *Remote Sensing*, *16*(20), 3852. https://doi.org/10.3390/rs16203852.

Samadzadegan, F., Toosi, A. and Dadrass Javan, F. (2025). A critical review on multi-sensor and multi-platform remote sensing data fusion approaches: Current status and prospects. *International Journal of Remote Sensing*, *46*(3), 1327–1402. https://doi.org/10.1080/01431161.2024.2429784.

Sathya, D., Thangamani, R. and Balaji, B.S. (2024). *The Revolution of Edge Computing in Smart Farming* (pp. 351–389). https://doi.org/10.1007/978-3-031-51195-0_17.

Shahid, A. and Mushtaq, M. (2020). A Survey Comparing Specialized Hardware and Evolution in TPUs For Neural Networks. In: *2020 IEEE 23rd International Multitopic Conference (INMIC)*, 1–6. https://doi.org/10.1109/INMIC50486.2020.9318136.

Sharma, A., Chopra, S.R., Gajanan Sapate, S. and Bhagawati, P.B. (2025). Algorithm For Preprocessing Satellite Imagery that Uses Geometric, Atmospheric, and Radiometric Correction. In: *2025 International Conference on Emerging Smart Computing and Informatics* (*ESCI*), 1–7. https://doi.org/10.1109/ESCI63694.2025.10988359.

Singh, B. (2024). Unmanned Aircraft Systems (UAS), Surveillance, Risk Management to Cybersecurity and Legal Regulation Landscape. In: *Unmanned Aircraft Systems* (pp. 313–354). Wiley. https://doi.org/10.1002/9781394230648.ch8.

Sishodia, R.P., Ray, R.L. and Singh, S.K. (2020). Applications of Remote Sensing in Precision Agriculture: A Review. *Remote Sensing*, *12*(19), 3136. https://doi.org/10.3390/rs12193136.

Stuart, M.B., McGonigle, A.J.S. and Willmott, J.R. (2019). Hyperspectral Imaging in Environmental Monitoring: A Review of Recent Developments and Technological Advances in Compact Field Deployable Systems. *Sensors*, *19*(14), 3071. https://doi.org/10.3390/s19143071.

Swetha, A., Chitra, R. and M,V. (2025). Advanced NDRE Techniques for Precision Crop Monitoring and Analysis. In: *2025 IEEE International Students' Conference on Electrical, Electronics and Computer Science* (*SCEECS*) (pp. 1–4). https://doi.org/10.1109/SCEECS64059.2025.10941496.

Tahir, H.A., Alayed, W. and Hassan, W.U. (2025). A Federated Explainable AI Framework for Smart Agriculture: Enhancing Transparency, Efficiency, and Sustainability. *IEEE Access*, *13*, 97567–97584. https://doi.org/10.1109/ACCESS.2025.3571340.

Transon, J., D'Andrimont, R., Maugnard, A. and Defourny, P. (2018). Survey of Hyperspectral Earth Observation Applications from Space in the Sentinel-2 Context. *Remote Sensing*, *10*(2), 157. https://doi.org/10.3390/rs10020157.

Vélez, S., Martínez-Peña, R. and Castrillo, D. (2023). Beyond Vegetation: A Review Unveiling Additional Insights into Agriculture and Forestry through the Application of Vegetation Indices. *J.*, *6*(3), 421–436. https://doi.org/10.3390/j6030028.

Vijayakumar, S., Shanmugapriya, P., Saravanane, P., Ramesh, T., Murugaiyan, V. and Ilakkiya, S. (2025). Precision Weed Control Using Unmanned Aerial Vehicles and Robots: Assessing Feasibility, Bottlenecks, and Recommendations for Scaling. *NDT*, *3*(2), 10. https://doi.org/10.3390/ndt3020010.

Wang, D., Cao, W., Zhang, F., Li, Z., Xu, S. and Wu, X. (2022). A Review of DL in Multiscale Agricultural Sensing. *Remote Sensing*, *14*(3), 559. https://doi.org/10.3390/rs14030559.

Wang, Z., Akber, M.A. and Aziz, A.A. (2025). Application of Remote Sensing and Geographic Information Systems for Monitoring and Managing Chili Crops: A Systematic Review. *Remote Sensing*, *17*(16), 2827. https://doi.org/10.3390/rs17162827.

Wei, C., Shan, Y. and Zhen, M. (2025). DL-based anomaly detection for precision field crop protection. *Frontiers in Plant Science*, *16*. https://doi.org/10.3389/fpls.2025.1576756.

Xue, J. and Su, B. (2017). Significant Remote Sensing Vegetation Indices: A Review of Developments and Applications. *Journal of Sensors*, *2017*, 1–17. https://doi.org/10.1155/2017/1353691.

Yu, A., Quan, Y., Yu, R., Guo, W., Wang, X., Hong, D., Zhang, H., Chen, J., Hu, Q. and He, P. (2023). DL Methods for Semantic Segmentation in Remote Sensing with Small Data: A Survey. *Remote Sensing*, *15*(20), 4987. https://doi.org/10.3390/rs15204987.

Yu, H., Kong, B., Hou, Y., Xu, X., Chen, T. and Liu, X. (2022). A critical review on applications of hyperspectral remote sensing in crop monitoring. *Experimental Agriculture*, *58*, e26. https://doi.org/10.1017/S0014479722000278.

Zhang, Z. and Zhu, L. (2023). A Review on Unmanned Aerial Vehicle Remote Sensing: Platforms, Sensors, Data Processing Methods, and Applications. *Drones*, *7*(6), 398. https://doi.org/10.3390/drones7060398.

Zhu, X. X., Tuia, D., Mou, L., Xia, G.-S., Zhang, L., Xu, F. and Fraundorfer, F. (2017). DL in Remote Sensing: A Comprehensive Review and List of Resources. *IEEE Geoscience and Remote Sensing Magazine*, *5*(4), 8–36. https://doi.org/10.1109/MGRS.2017.2762307.

Zomeni, M., Tzanopoulos, J. and Pantis, J.D. (2008). Historical analysis of landscape change using remote sensing techniques: An explanatory tool for agricultural transformation in Greek rural areas. *Landscape and Urban Planning*, *86*(1), 38–46. https://doi.org/10.1016/j.landurbplan.2007.12.006.

5 Edge AI and On-Device Deep Learning Models for In-Field Diagnostics

Shreesh Sanjay Amin[1], S.K. Patel[2], Kalamkuri Chandu[3], Adityanshu Tripathi[4], Aditi Kachchap[5], Dattatray Bhalekar[6] and Abhishek Upadhyay[7*]

1. Introduction

The agricultural sector is on the verge of entering into a digital revolution phase. The main factors contributing to this transition are the inclusion of IoT (internet of things) devices, sensor networks, and AI (artificial intelligence) in precision agriculture systems. The primary aim of this transformation is to make over the customary farming practices that have been inherited down through the years. A huge challenge that has been thrown by this transition is the data's gigantic size. This data consists of everything, starting from high resolution images to real-time sensor readings and is mostly on the edge of the network, i.e., a significant distance away from the cloud servers that normally perform AI processing. The big data produced proves to be a huge challenge for the cloud-centric method which has been used as the regular way of agricultural intelligence (Navarro et al., 2020; Sheikh Mansoor et al., 2025).

Edge artificial intelligence (Edge AI) has indeed become a practical and feasible solution due to those descriptions. This method mainly refers to the performance of intricate machine learning models in close proximity to (or actually on the run) the very sources of agri-data as Kalyani and Collier (2021) wrote about it. The present-day mechanized farming industry is likely to create data sets that are utterly huge and multifaceted. For just one of the large-scale precision farming operations, hundreds of sensors could be attached to the soil, the weather, the health of the crops, and the performance of the mechanized equipment to monitor, giving a data flow of terabytes a day (Upadhyay et al., 2024). These datasets are

[1,3-5] M.Tech. Scholar, Department of Farm Machinery and Power Engineering, CAET, Dr. Rajendra Prasad Central Agricultural, Pusa, Samastipur, Bihar, India.

[2] Assistant Professor, Department of Farm Machinery and Power Engineering, CAET, Dr. Rajendra Prasad Central Agricultural, Pusa, Samastipur, Bihar, India.

[6] Center for Precision and Automated Agricultural Systems, Washington State University, Prosser, WA, USA.

[7] Assistant Professor, Department of Farm Machinery and Power Engineering, CAET, Dr. Rajendra Prasad Central Agricultural, Pusa, Samastipur, Bihar, India.

* Corresponding author: abhiupadhyay7798@gmail.com

DOI: 10.1201/9781003715726-5

coming from a mixture of data inputs, such as: high-resolution images from the drones and satellites, time series environmental data from sensors, GPS location data, and metrics reflecting the performance of the machines in the field using an autonomous system (Sheikh Mansoor et al., 2025; Navarro et al., 2020).

Traditional cloud-based processing is confronted by great challenges of enormous bulk and the fast rate of this data. Thus, for a particular operation, an agricultural drone's multispectral imaging system may churn out several gigabytes, while environmental sensors which operate round the clock may have thousands of reading points daily. If these systems are expanded to very large scale operations, the data deluge can overwhelm the standard transmission and processing pipelines in no time (Jiang et al., 2025). In the context of agriculture, the network situation has its nature favoring the split of the dependency between clouds and the land. Remote areas are usually associated with having poor connectivity, undependable quality of networking, and a very limited amount of bandwidth. It has been demonstrated in wireless communication protocols studies that rural areas, despite the weak links they have, certainly have the operations of LPWAN done with LoRaWAN, even the technology being the longest serving and having the least bandwidth consumption, they still cannot serve some agricultural data because of the data volume (Shukre et al., 2024).

Network issues can get even more complicated because of the broad scope of the agricultural activities. The systems must be able to work over vast spaces, with network coverage that is potentially unstable and of varying intensity, as opposed to the urban IoT installations that have a strong and reliable infrastructure. Research on precision farming shows that the network's trustworthiness can be quite different among areas on the same farm, and thus the quality of the service to AI devices in the cloud will not be the same (Zhang et al., 2024; Sheikh Mansoor et al., 2025).

The challenge is not one single problem but a mix of connected issues that are especially bad in farming. The data from a high-resolution drone image is massive and multi modal. Transferring this huge amount of data over low-bandwidth rural networks is slow and expensive, which directly causes the latency problem. The data from farming operations is often time-sensitive, such as information about a pest outbreak, which needs a quick response. Additionally, the data itself—including crop yields and soil health—is frequently confidential, leading farmers to be hesitant about entrusting this sensitive information to third-party cloud providers.

One of the combined challenges that edge AI addresses is the fact that computing power is dispersed to the source of data. By processing information in the local area, edge AI systems make it possible for instant decisions to be made on such tasks as crop monitoring, disease detection, and resource management. As a result, the continuous internet connectivity demand will be reduced, leading to the privacy of the data being secured at the same time (Upadhyay et al., 2025a). In the situation in which the access to the internet is very limited in the agricultural environment, it is still crucial to provide timely decisions (Sidorkiewicz et al., 2025).

Table 1 Latency comparison for Cloud vs. Edge processing.

Application	*Cloud Latency (ms)*	*Edge Latency (ms)*	*Reference*
Pest Detection Alerts	300–600	100–147	Kalyani & Collier (2021); Annapoorna & Ramesh Babu (2025)
Automated Irrigation Control	>150	5–10	Matthew (2024); Soussi et al. (2024)
Autonomous Vehicle Navigation	300–500	1–10	Patel (2024)
Equipment Failure Warnings	>100	5–30	Joshua (2024); Sammu (2018)

2. Fundamental Principles of Edge AI

Edge AI represents a massive transformation concerning computational design, which suggests that AI capabilities will be brought as near as possible to the data generation location. Furthermore, it can be also referred to as a scenario where the data is generated by the device and at the same time consumed by the device. A very good example is the agriculture case which will be discussed in this paper; there, AI models will be placed directly on in-field devices that operate in the field. By connecting the tip of the iceberg to the general internet, AI models can get even smarter and everyone sharing this information will benefit more. On the whole, the development presents tangible benefits by providing speedy and direct services that anyone can use (Huang et al., 2023; Patel R.K., 2025). Edge AI lies on a simple yet powerful concept: not all data has to be sent for processing to the central systems. A study found out that AI application in agriculture mostly happens without any network aid; it was seen local processing can be highly efficient with the help of the smart model compression technology. In case the data is handled locally, such a system will be able to sift out the unnecessary information, lessen the demands for transmission, and provide prompt responses to the emergencies. This technique is beneficial as it assists in the contention against latency and bandwidth issues prevailing in the rural areas.

In the field of agriculture, Edge AI systems are designed with a multi tiered approach where different levels of data processing are spread out throughout the system. The very simple data collection and preliminary processing are handled by the sensors and devices at the local level. The more complicated AI models are operated on the edge gateways or a separate specialized computing unit, and then cloud services are utilized for various different needs such as data storage over a long period of time, model training, and/or coordinating many edge devices shown by Figure 1. The conventional and regulation cloud system itself is not totally being replaced by this technology trend. The cloud, with its huge computational resources, is great for those tasks which are not urgent but demand high processing power and the training of very big and accurate models is such a case. On the other hand, the edge, which is very close to the source of data and has very low latency, is perfect for conducting operations requiring real-time or near to real-time processing.

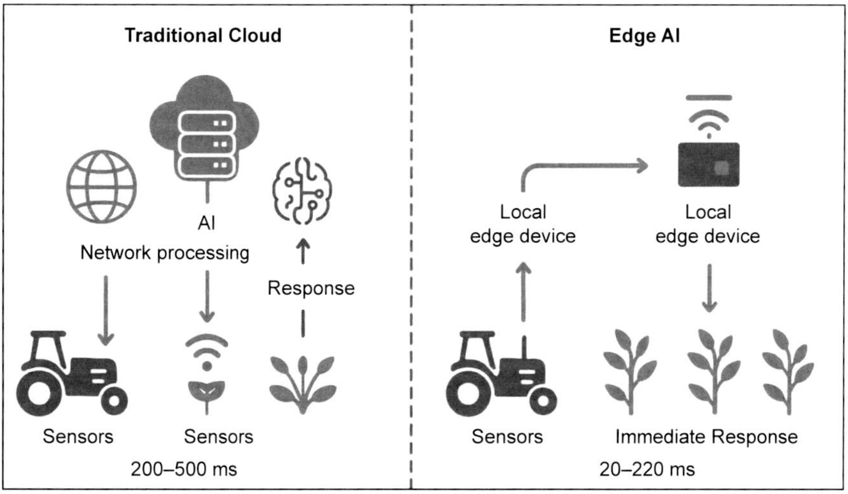

Fig. 1 A Schematic Representation of Traditional Cloud vs Edge AI Architecture in Agriculture.

3. Technical Foundations of Edge AI

3.1 Hardware Architectures for Edge Computing

Being a common issue in rural areas, the latency and bandwidth issues mentioned above can be effectively resolved with the new method.

The main hardware components of Edge AI systems used in agriculture are various computing devices which are technologically developed and fabricated to balance between the performance, power, and cost. These devices are powered by special processors such as ARM-based systems-on-chip (SoCs), graphics processing units (GPUs), and dedicated AI accelerators optimized for machine learning inference. The NIRVANA of ARM-based processors, more so in cases like the NVIDIA Jetson series, has grasped the agricultural edge computing market due to its opulence in processing power and eco-friendliness. The same platforms can furnish the required computation for the most compressed neural networks and still bring down power usage considerably, which is a necessity in a field setting. Review of the literature on the subject of hardware choices demonstrated that ARM-based systems could be two to four times more energy-efficient for AI inference tasks compared to traditional x86 processors. Diverse studies concerning the hurdles of reducing AI energy demand have indicated that ARM processor-based systems have a 2–4 times higher energy efficiency over the standard x86 processor systems in the AI inference tasks (Aetina, 2025; STL Partners, 2017; Anonymous, 2023; Suárez et al., 2025).

The devices that are commonly known as Accelerators, which specialize in AI applications, include TPUs (tensor processing units), and NPUs (neural processing units), and are also operating as the case with the general-purpose processors. With the advances in technology, the performance of these highly integrated solutions has stretched to the point of being able to deliver high-quality AI workloads with 10-fold to 20-fold boosts when set against the general-purpose processors (Hao et al., 2023). Complex AI tasks that require the processing of real-time high-definition images or intricate sensor data are going to be supported by the more powerful devices that are now capable of operating under efficient power and thermal conditions which are very stringent for agricultural systems.

The matrix operations, which are essential for the neural network implementation, are performed by the accelerators that are designed for the purpose only and are equipped with very high precision. These devices have very high-performance capability and it is common for them to outperform the general-purpose processors by 10–20 times for AI workloads. One of the things these variants of AI accelerators do is to provide the required computational power for doing real-time analysis of high-resolution imagery or complex sensor data without leaving the predefined power and thermal limits of the agricultural equipment (Hao et al., 2023). It is therefore necessary to know the available hardware options before proceeding to the software or network design. For these systems, the hardware, software, and network are interdependent and coexistent parts. A heavy model that is not compressed cannot work without a very powerful NPU, which, in turn, might not be workable in a field without power. But, a compressed model, facilitated by a framework like TensorFlow Lite, can still be executed on a less energy-intensive ARM processor. The little bit of data that comes out from AI was matched perfectly to the LPWAN's limited data rate, an example would be "spray pesticide here". All parts of the system were worked on from the beginning so that the entire engineering process is cohesive and the selection of a component will definitely affect the others.

Table 2 Comparison between common hardware types.

Hardware Type	*Power (W)*	*Performance (TOPS)*	*Cost Range*	*Agricultural Suitability*	*Examples*
ARM SoCs	5–15	1–4	\$100–500	High-Energy efficient	Raspberry Pi 4, Qualcomm SoCs
Edge GPUs	10–30	4–20	\$200–800	Medium-Higher power	NVIDIA Jetson Xavier, AMD Radeon Instinct
AI Accelerators	2–10	10–50	\$300–1000	High-Specialized	Intel Movidius, Google Coral Edge TPU
NPUs	1–5	5–25	\$150–400	Very High-Optimized	Huawei Ascend NPU, ARM Ethos-N

3.2 Software Frameworks and Runtime Environments

The software infrastructure supporting Edge AI in farming has changed a lot, with frameworks now made specifically for edge devices with limited resources. TensorFlow Lite, a lightweight machine learning framework by Google, has become a popular choice for agricultural applications due to its optimized runtime, numerous model optimization features, and support across various hardware platforms (Puy, 2024). It offers benefits such as support for quantized models that reduce memory and computational needs, APIs for hardware acceleration, and a lightweight runtime. Complex neural networks for tasks like crop disease detection and pest identification have been successfully deployed on resource-constrained devices using TensorFlow Lite.

PyTorch Mobile is yet a very significant framework, particularly so for people who are already in the PyTorch ecosystem. According to comparative studies, although TensorFlow Lite usually allows a better-scaled Performance per watt, the workflow in the case of PyTorch Mobile is much smoother and PyTorch model focused and thus providing easy model visibility to startups that are already using PyTorch. ONNX runtime helps with cross-platform deployment of the models created by different frameworks and thus provides the ability to run them on various hardware, which is consistent among different setups. This ability is especially useful in agricultural systems where different pieces of hardware are used or where different vendors supply the systems (Krishnamoorthy, 2025). The choice of a specific framework is a strategic decision that is made in conjunction with the development and research cycles as certain frameworks like PyTorch Mobile can fasten the transition from research to deployment. This is a huge plus in a field where frequent introduction of new models and data sets occurs.

The introduction of runtimes that are not tied to any particular framework like ONNX is a major step in overcoming one of the biggest barriers to the adoption of technologies in the agricultural sector. One of the factors that slow down the growth of the market is its fragmentation, where many suppliers provide very specialized hardware and software. In the absence of a common standard, the farms would have no choice but to go for only one vendor's system, which would certainly ruin both the freedom of choice and the drive to innovate. It is through ONNX Runtime that models created for different frameworks can seamlessly coexist and be operated on the same hardware. Consequently, this kind of interoperability results in a more competitive and innovative market environment, where a PyTorch-based scientist can develop a model and the model can be run on a TensorFlow-based hardware from an entirely different company. This is a big step towards the creation of a strong, scalable, and open agricultural AI ecosystem.

3.3 Network Architectures and Communication Protocols

Agricultural Edge AI systems are designed to function in the complex network environment connecting numerous communication technologies. Low Power Wide Area Networks (LPWANs), such as LoRaWAN and NB-IoT, are the only technologies that can be used to link sensors which are powered by batteries due to their long distance and energy efficiency, respectively. Despite this advantage, they provide limited data rates which call for local data processing and filtering. The study advocates LoRaWAN as a prime candidate for edge AI applications given that it boasts the longest range and the lowest power consumption figures can be observed from this particular network. It does not, however, support the high data rate required for raw sensor data but is rather adequate for sending AI-derived insights (Shukre et al., 2024).

It is not uncommon for an organization to have a hierarchical communication system. The sensors that are located nearby use short-range protocols such as Wi-Fi or Bluetooth to communicate with the edge gateways which do the AI processing. LPWAN or cellular networks are then used by these gateways to have receptions in the cloud. In previous years, this situation has cut the power consumption of sensors and has at the same time preserved the chances of very important data being sent and received. (Mohamed Rafi et al., 2025). Depending on the limitations of the network, local processing becomes a very important feature, rather than a luxury. For sure, the transmission of a raw, multi-gigabyte drone image is not only slow but also practically infeasible. The edge device, thus, is required to take the image and extract the vital information from it, before sending a small data packet. So, local processing is not only a quite significant function without which the system cannot operate but also needs a preference for speed.

4. Model Compression and Optimization Techniques

4.1 Pruning Methods

It is a perfect solution to reduce the complexity of the model with neural network pruning. This approach makes it possible for the model to be placed on, or embedded in, devices with restricted resources. A fundamental part of this method is to eliminate the parts of the network that are not necessary for good performance, which results in a very significant reduction of the model's size and computational load, while the accuracy is still at an acceptable level (Dantas et al., 2024; Wang et al., 2024). Structured pruning is a technique that removes whole neurons, channels, or layers. The outcome of this procedure is a uniform decrease in the complexity of the model, which is accompanied by direct savings in terms of computation. In fact, such a technique has been employed in agricultural AI systems where they have remarkably reduced model parameters to a quarter and still managed to retain as much as 98% of the original accuracy. Consequently, it has become possible to run sophisticated convolutional neural networks on edge devices that have memory capacities as low as 1–2 GB (Francy & Singh, 2024; Wang et al., 2024).

Unstructured pruning works by removing individual weights that are less important relative to others, which sometimes results in greater compression ratios. Nonetheless, to gain the advantage of such a technique in actual computation, sometimes the use of a dedicated hardware or specialized software is necessary, as ordinary processors do not handle the bypassing of single weights during inference efficiently. The studies reveal the strong cooperation between unstructured pruning and quantization, showing the latter comes into effect with the former shown by Figure 2. The more complex methods like step-wise magnitude pruning and those linked to the lottery ticket theory have also been so productive in the agricultural needs that they often surpass the traditional pruning in this aspect (Dantas et al., 2024; Wang et al., 2024).

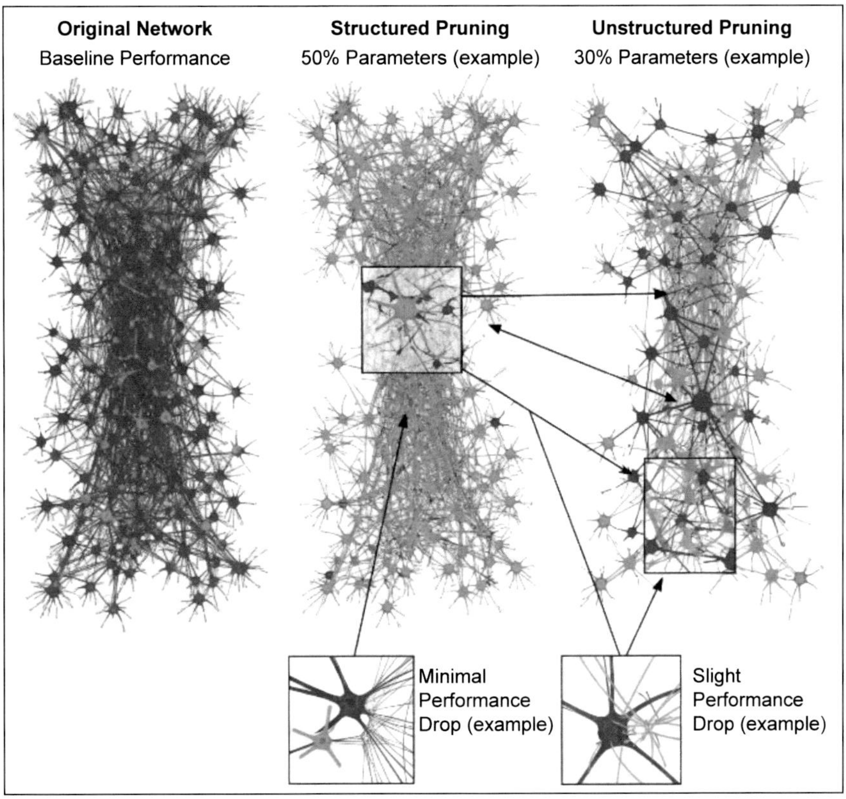

Fig. 2 Schematic Representation of Structured vs. Unstructured Network Pruning Methods.

4.2 Quantization Strategies

Quantization is the method that causes a decline in the accuracy of weights and activations in a neural network. Typically, it means going from a 32-bit floating-point representation to something 8-bit or less. Quantization is a very powerful technique for memory saving, accelerating calculations, reducing the power consumption of electronic devices, and it has found a special application in the agriculture industry (Dantas et al., 2024; Jin et al., 2025; Opiah, 2023). After the quantization process, a model ends up being very small but still has very good accuracy. The dynamic quantization has brought the adaptation principles, which allow a model to change its precision on the fly and thus be more accurate to the data it is computing the values. In this case, dynamic quantization is a technique that is adaptive during the model's active state and thus can work with different inputs. The agriculture industry has been able to achieve almost 95% reduction in model size with no accuracy loss through the application of this technology on their projects. This is particularly important when the data going into the model is affected by factors like the intensity of the light or the weather changes (Francy & Singh, 2024). Static quantization pre-calculates parameters during optimization, which gets rid of overhead while the model is running. Implementations have shown that this can improve inference speed by 3-4 times over full-precision models, especially in predictable environments. Post-training quantization applies precision reduction to models after they have been trained, without retraining, which is practical for agricultural applications where retraining is expensive. It can achieve a lot of compression with very little loss of accuracy. Quantization-aware training puts the effects of quantization into the training process, which often results in better accuracy retention (Lang et al., 2024; Dantas et al., 2024). The most effective approach is to combine these methods, not to just pick one. Pruning removes redundant connections, but the remaining weights are still full-precision. Quantization reduces the precision of those remaining weights. By combining them, it becomes possible to achieve both, fewer parameters and computational efficiency, which leads to better results.

4.3 Knowledge Distillation

Knowledge distillation is a method for compressing models by training a smaller "student" model to copy what a larger, more complex "teacher" model does. This is useful for deploying lightweight models that still have high accuracy. The main idea is to train student models to replicate not only the final outputs but also the in-between details and decision patterns of the teacher models. In the field of agriculture, this approach has opened the way for light models to get 90–95% of the accuracy that their bigger siblings get, and computational needs by 80–90% (cf. Zhang et al., 2024; Huang et al., 2023). Multi-stage knowledge distillation emerged as a particularly effective method for agriculture-related assignments since it managed to get the models small enough and at the same time, so good at the same time as the humans for very difficult problems like the disease detection of a variety of crops.

4.4 Neural Architecture Search

An automated way of designing neural network architectures optimized for edge deployment is given by the technology called Neural Architecture Search (NAS). The NAS algorithms can find the best networks that reach user accuracy goals, but still fit edge devices' computational and memory limits. Besides, NAS, which is familiar with the hardware that is going to be used, does consider the specific features of the edge device. A study by one of these approaches and the benefits associated with it found that the networks they designed are 2–4 times faster than the ones composed by hand but with almost the same accuracy in the agricultural task. The next step of NAS is not just cutting the process shorter but having the most efficient system. Bob, the lead designer of the project, says that this project aims at serving the agricultural sector with the first of its kind system (Lyu et al., 2024; Fayyazi et al., 2025).

The growing adoption of automated methods such as NAS helps overcome a key limitation of conventional AI development. Designing a network that is accurate as well as fitting on a resource-constrained edge device is a complex, time-consuming task for human engineers. NAS automates this process. The automation is not just for finding a good network, but for finding one that is specifically made for the real-world constraints of agriculture. This automation scales the development process, making it possible to create custom, optimized models for a wide variety of crops, pests, and environmental conditions without a huge manual effort.

5. Lightweight Software Frameworks

5.1 TensorFlow Lite for Agricultural Applications

TensorFlow Lite is now regarded as one of the most crucial frameworks for deploying machine learning models on agricultural edge devices. The major factor in this regard is the optimization features along with the compatibility of the framework with various hardware platforms. The framework provides supportive tools for model quantization, pruning, and conversion and the latter are indeed very helpful in cases where there is a shortage of resources. The agricultural applications using TensorFlow Lite have been getting better and better with the framework over the years. The pest detection system is one example that can coexist with the whole framework on ARM-based edge devices which take around 100–200 milliseconds for inference. The same task in the TFLite network would be accomplished in about 20–50 milliseconds. This speed is a must for applications such as real-time navigation of autonomous vehicles and instant threat detection. The framework's adaptability to hardware acceleration APIs makes it possible to use specialized processors. The hardware accelerator (especially GPU) can reduce the processing time of the computer vision task from 3 to 5 times and AI accelerators can even give more speed boosts. The model optimization toolkit comes with a range of support for applications on compression techniques that are most suitable for the agricultural edge deployment. Its automated workflows for quantization have the capability to reduce model sizes by 4–8 times with not much impact on the accuracy (MohitNSamagra, 2024; Puy, 2024).

5.2 PyTorch Mobile Integration

For agricultural research teams already utilizing the PyTorch system, the PyTorch Mobile is the perfect option. It is in line with the existing development workflows, thereby allowing the passing of research models to edge devices without losing the common objects and services. In terms of the model size, it still allows the deployment of larger models without any performance compromise; furthermore, it is the one that makes complicated data flows easy and moreover, it supports models that are not only very complex but also very fast changing. However, it should be noted that this specific configuration of PyTorch comes with a higher data and computation cost (Krishnamoorthy, 2025). The whole PyTorch ecosystem is compatible with the PyTorch Mobile torchscript compilation system's model optimization mechanism and this is one of the main reasons for its popularity among the models that require constant updates. PyTorch Mobile has made federated learning one of its main areas of interest which, in turn, reflects the sharing and collaborative characteristics of the agricultural AI domain perfectly. It is supposed in the research of federated learning that the assets, besides the scenarios in which they work together, can lead to learning, which is like the case of multiplying the strength, and data privacy etc. (Belghachi, 2025; Dembani et al., 2025). The determination of the framework is not only a technical decision; it is also a strategic one related to the development and research cycle. The price of an AI system includes not only the hardware and deployment but also the development time and effort. The smooth workflow of PyTorch Mobile considerably reduces the gap between the research and deployment stages, which can be a major benefit in a domain where new models and datasets constantly appear.

5.3 ONNX Runtime Optimization

ONNX Runtime is a tool that can run AI models at the edge without being tied to any specific framework. It is very helpful for the agriculture networks where products from various vendors or research groups are connected. The crucial point is that its performance is the same no matter which type of model is run by it, which allows for the flexible development of the system. Different hardware types have been tested by the release of ONNX Runtime which is a software that automatically optimizes performance depending on the available resources. In comparison with framework solutions, research indicates that ONNX Runtime played well yet provided a much wider system integration platform (Krishnamoorthy, 2025). The peculiarity of the framework-neutral runtimes like ONNX is the direct addressing of one of the weak points for the industry-wide adoption of the technology. The technology market for agriculture is made up of small and heterogeneous players, where farmers' dependence on one vendor's system is inevitable without a common technical ground, thus, the range of choices and ideas would be extremely limited. ONNX Runtime, by removing a model-hardware dependency barrier through enabling models from different frameworks to run on different hardware, thus, promotes the inter operability not only among the same hardware but also among different hardware making the market more competitive and innovative, and at the same time creates the first step for a strong, scalable, and open agricultural AI ecosystem.

6. Real-world Applications and Case Studies

6.1 Crop Monitoring and Disease Detection

Real-time crop monitoring is one of the most successful applications of Edge AI in agriculture (Gorthi et al., 2024), with many implementations showing clear improvements in detection accuracy and response times (Anand et al., 2024; Upadhyay et al., 2025b). For example, vineyard monitoring systems that use convolutional neural networks analyze multispectral drone imagery to find disease symptoms with 91–95% accuracy. According to Sidorkiewicz et al. (2025), the local data are processed in 100–200 milliseconds. Knowledge distillation has been applied to cotton disease detection systems achieving high accuracy with 90.59% precision being maintained by the models and at the same time the computational demands are being reduced from 0.29 GFLOPS to 0.045 GFLOPS due to the compression. These advances are extremely significant because they prove that the use of powerful AI models in resource-limited edge devices is possible (Zhang et al., 2024). In tomato and sugarcane disease detection, the models developed have demonstrated an accuracy of 91.6% and F1-scores of about 90.7%; sophisticated architecture and accelerated computing hardware are assumed to be the reasons for the performance increase. These detection systems have also decreased the inference time to the extent of about 220 milliseconds and minimized power usage by 38% which, in turn, can facilitate efficient, real-time disease control at the field level (Bhalekar et al., 2023; Sowmiya et al., 2025). Edge AI becomes the agent for the total reinvention of farm work, making a transition from being just reactive, visible-problem-solving-oriented to being the industry's proactive and preemptive guardian through the power of real-time, predictive analysis. To illustrate, the systems are capable of spotting a disease outbreak a good 2–3 days before conventional methods. The change isn't just incremental; it is a completely new way of farming. This allows the farmer to quickly take action and prevent major losses by the very first indication of a problem. The technology transforms from a diagnostic tool to one for prevention instead.

6.2 Precision Irrigation Systems

Smart irrigation systems also give a great example of the effectiveness of edge AI, with irrigation systems showing considerable reductions in the water consumption and increased crop yields. Tests indicate that AI models empowered in the edge reduced consumptions of water by up to 35% through real-time observation of soil moisture, weather, and crop demands (Ercan Oğuztürk, 2025; Farmonaut, 2025). Systems make use of various inputs of data including soil moisture sensors, weather sensors, and crop models, among others, as a form of streamlining irrigation with no cloud connection (Kaur et al., 2024).

The systems that make up this technology, like soil moisture sensors, weather stations, and crop models, are all under the umbrella of data sources that are united to carry out the task of irrigation optimization without cloud (Kaur et al., 2024). The irrigation of a greenhouse through ed-tech has indicated that the crop output in that period can be pushed up by 15–25% while cutting down water usage by 20–30%. Variable-rate irrigation is one of the most progressive of edge AI and agriculture's art thus far, as it provides accurate water applications on jagged and asymmetrical fields allowing taking of soil, landscape, and even the sensors' data differences into consideration to make the irrigation within the plots literally precise, which is a central cost-saving factor yet at the same time offers a considerable increase in crop yield (Sheikh Mansoor et al., 2025). The figures on the lessening of water and chemicals confirm that Edge AI is an indispensable techno aid for eco-friendly cultivation. The application of the technology along with the needed locale and real-time nature eliminates waste and thus, eco-friendlier farming practices.

6.3 Livestock Health Monitoring

Edge AI has driven significant progress in livestock health monitoring, with systems achieving high accuracy while operating on local hardware. Research on dairy farm monitoring shows that models can analyze animal behavior, feeding patterns, and health indicators with 92–96% accuracy, enabling the early detection of issues (Anonymous, 2025). Wearable sensors for livestock allow for real-time analysis of activity and temperature, identifying health anomalies and estrus cycles immediately. This facilitates timely interventions that improve animal welfare and productivity (Yu et al., 2024; Lee & Seo, 2021). Automated poultry monitoring systems have improved disease detection and operational efficiency, with models able to identify outbreaks 2–3 days earlier than traditional methods and reduce labor needs by nearly half.

6.4 Autonomous Agricultural Machinery

The technologies related to autonomous farming have advanced considerably due to the implementation of edge AI, which results in the precision of the navigation, obstacle detection, and the task completion that can be seen in the context of the advances in autonomous machines used in agriculture (Schrader et al., 2025). Based on the study of autonomous tractors, this local AI model can do the real-time path planning and the obstacle avoidance resiliently hence with accuracy. The example of this study was the Autonomous Sprayers (Hoheisel et al., 2025) which with the use of edge AI can know the conditions of the crops to choose the right way of applying the pesticide and fertilizer, thereby reducing the chemical application rates by 20–30% and at the same time being just as effective (Ercan Oğuztürk, 2025). The Robotic harvesters are edge AI-enabled and they can carry out real-time fruit detection, ripeness evaluation, and harvest decisions. The autonomous weeding systems are the ones that stand to gain the most by using the edge AI to distinguish between the crops and the weeds with an accuracy rate of 95–98%. By doing so, herbicides are confined only to the necessary parts and the use of chemicals is shortened by 60–80% (Ercan Oğuztürk, 2025).

7. Advantages of Edge AI for Enhancements in System Performance

The next major benefit is the most significant one – energy efficiency, and that too in a big way for battery-powered devices. Available research shows that edge and cloud architectures differ in energy usage and that the latter one is considerably worse. When the forms of edge computing are all decentralized, the energy consumption may range from 15% to 25% less compared to the latter one. What brings about such a change is the lesser amount of data that needs to be transmitted; the processing of data to be done locally; and power to be managed in the most effective way. The value of such a change is very much realized for the said devices that are solar-powered and the autonomous sensors as energy efficiency can make their life longer while at the same time, it can result in cost savings in maintenance (Parihar & Suman, 2024).

Additionally, privacy and data sovereignty are enhanced through edge AI. Farmers can keep control over their data by processing it locally instead of transmitting it to external cloud services. Studies on agricultural data privacy highlight that local processing greatly reduces the risk of data breaches, unauthorized access, and misuse by third parties. This control is especially important for sensitive information such as crop yields, soil health, and operational strategies (Kaur et al., 2024).

7.1 Reductions in Latency and Response Times

Edge AI in agriculture always leads to a faster processing of data than the cloud-based solutions alone. Several studies comparing the two types of setups show that edge systems typically sustain response times from 20 to 220 milliseconds, while cloud-based systems operate in the range of 200–500 milliseconds. This difference indicates a latency reduction of approximately 50–80%, which is, in fact, a very essential requirement for real-time applications (Roostaei et al., 2023; Sowmiya et al., 2025). Such a technological breakthrough will be having a direct impact on various sectors where working with extremely low latency is an absolute necessity. One such sector is that of pest detection systems, which can alert in real-time, thus preventing further crop loss from insects or other pests (Zhang et al., 2024; Sheikh Mansoor et al., 2025; Ercan Oğuztürk, 2025).

Nonetheless, the overall impact of these shortening delays is not limited to individual jobs but also brings in and supports more sophisticated management techniques. For example, systems with real-time response capability could change irrigation and feeding patterns through the data they get in real-time which would eventually help them to not only manage their output better but also their resources (Kaur et al., 2024).

7.2 Energy Efficiency and Sustainable Practices

One of the biggest advantages of Edge AI is energy conservation, which is very important for devices that are powered by batteries and such units as remote monitoring devices. Studies have compared that energy consumption can be decreased by 14–25% with the help of edge computing since it will not require constant data transmission and it will optimize the local processing power as well (Parihar & Suman, 2024). Battery-powered sensors experience the most significant benefits, in that their use of local AI can extend battery life by up to 130% when compared to cloud-connected systems that need to share data all the time (Roostaei et al., 2023). Solar-powered systems have a clear edge over cloud-connected systems as they can operate even when there is not enough sun thanks to the very low energy consumption of edge AI. Data obtained from the research show that power-saving techniques combined with edge computing could keep a system alive for 7–10 days without any sunlight, whereas otherwise it would see a measly 2–3 days' life (Shukre et al., 2024). Edge AI has introduced the adaptable power management techniques that have greatly contributed to the power consumption being more efficient. The units can adjust the loads of processing, the rates of sampling, and the intervals of communication according to the existing power levels and the surroundings. This, in turn, results in a reduction of energy use by 20–40% in field operations (STL Partners 2017; Anonymous, 2023).

7.3 Enhancements in Data Privacy and Security

Edge AI technology is highly effective in the protection of data and is considered a major step in the right direction for the future. It is indeed a true revolution, as instead of data loss risk which mobile sends or cloud computing handling data directly, everything is already there on the spot. As Kher (2025) pointed out, the control of the scenario where machine learning is not differentiable to the point of the farm is only possible if the data is processed locally. The first form of the security transfer happens at the machine level in the factory, and the same is the case in the retail store, where the ML (machine learning) operation is based on customer data. Central farms and central models for all company regions allow very good prediction despite being the most privacy-violating system possible in terms of security but at the cost of substantial data quality fine-tuning. Many of such technologies make privacy the very industry that they belong to a possible matter of great importance (Dembani et al., 2025). It is known that such a system will not interfere with normal and lawful operations in the fields as in the case of the computer vision system where the data is secured regionally, and the operators do not have to handle or store personal information at all platforms (Dwarampudi &

Yogi, 2024). With different farm cases handling different crops, even considerably higher model performance gains on the whole suggest successful federated learning to be a zero sum privacy problem, whereas the gains through the data silo approach could be counted separately for each farm (Belghachi, 2025). This new approach directly tackles the issue of data ownership, which is a factor in the prevalence of new technology and hence in the acceptance of wider adoption.

7.4 Economic Advantages and Cost Savings

It has been observed that installing edge AI solutions results in a decrease in the operational costs, better utilization of resources, and hence, an increase in productivity. Research indicates that the use of edge computing in environmental monitoring systems can lead to cost savings of 55–82%. This can be primarily attributed to the fact that data transmission costs are negligible, cloud service fees are minimal, and operational efficiency is increased (Roostaei et al., 2023). Automated monitoring can be the biggest cause of reduced labor cost by a long shot if we consider that less manual data collection is needed for the human resources department modernization. Edge AI aims here for a 40–60% cut in the workforce required for routine tasks. Besides labor costs, the precise usage of water, fertilizers, and pesticides leads to saving a lot of money and at the same time increasing the farm's profitability. One of the Precision Irrigation Systems, for example, has been proven to save 25–35% of water while sustaining the same yield (Farmonaut, 2025; Kaur et al., 2024; Ercan Oğuztürk, 2025). Early detection that originated in the areas of diseases and pests that mostly existed due to natural means or improper human actions is decisive in making the plot healthier for growing crops; it is a practice that is impossible to be done by farmers without the use of AI-infused equipment, such as drones and IoT devices. Edge AI, in the form of disease and pest detection algorithm, often enables these early detections, giving a window of opportunity that is equivalent to two to three days prior to the time required for the detection that was performed manually thus, possible preventive measures can now be taken in order to curtail large losses (Sowmiya et al., 2025; Zhang et al., 2024). Economic benefits do not come from one change but through a positive loop: the use of fewer resources leads to lower costs, the quicker response to threats helps in preserving the yield and quality, and together they result in higher profits and operational stability. The economic advantage is not the result of one factor alone but rather a feedback loop between the three aforementioned aspects that gets stronger and stronger. A system with less latency can identify the diseases earlier which prevents crop loss. This in turn results in increased outputs and profits. By employing edge AI in agriculture, the farm is in a position to partly safeguard itself from adverse external factors such as drought and high fertilizer prices since it would be using fewer inputs and experiencing no crop loss. The result is a farm that is less affected by these factors and is therefore more adaptable to changes in economic and environmental conditions.

8. Challenges and Limitations

8.1 Model Updates and Maintenance Challenges

One of the biggest obstacles to edge AI's deployment in agriculture is the challenge of keeping models up-to-date and consistent across large numbers of distributed devices. Unlike the case of centralized cloud systems where updates are managed from one central point, edge devices require efficient means of spreading updates, under circumstances of often scarce communication and resources (Jin et al., 2025; Opiah, 2023; Jiang et al., 2025). Model drift is another main issue of concern as it is likely to show up as a symptom of model inaccuracy over a certain period of time due to environmental changes, the seasonality as well as the populations of pests are constantly changing. Among other studies, the current one finds that, without regular updates, models might lose 10–20% accuracy within just 6–12 months of time (Huang et al., 2023; Zhang et al., 2024). Rural areas do not have the best internet connectivity and this leads to an even more difficult update process. To combat this, differentially updating and model transfers with compression are some of the strategies that have been adopted. For situations in which a failure could result in equipment damage or crop loss, one important requirement is that the system should support the potential rollback of the update and keep track of the different versions, as in the case of autonomous systems (Jiang et al., 2025). These challenges are interconnected rather than isolated problems. The update issues stem from connectivity limitations. The need for these updates is due to model drift, which is caused by the variability of agricultural data, which in turn is a result of the limited training data problem (Patel et al., 2025)

8.2 Limited Data for Training

Difficulty in securing an adequate amount of high-quality data is still a big challenge in the agriculture industry, and this is mainly due to the fact that farming is conducted on a seasonal basis. When it comes to crops, pests, and fields that differ, the models generally do not perform well. That is why most of the time, it is necessary to have one season's data for the whole year (Mortazavizadeh et al., 2025; Hosseiniyan Khatibi & Jauhar, 2024; Zhang et al., 2024). Rare events like an unexpected disease outbreak or an unusual insect invasion are extremely difficult to document, and the small number of cases that follow makes the models less reliable. In response to the constraints, researchers are looking at various methods that are usually applied to the small number of instances or to unbalanced datasets, such as few-shot learning, data augmentation, and transfer learning (Huang et al., 2023; Mortazavizadeh et al., 2025; Hosseiniyan Khatibi & Jauhar, 2024). Environmental factors which include light, weather, and soil condition alterations are continuously adding interference to sensor data, making it difficult for one to not only train but also deploy in real-world scenarios. Besides, annotating data for specific tasks like disease detection can be extremely costly and laborious. It often requires expert input. Approaches like active learning and semi-supervised methods are being investigated to reduce these costs (Dwarampudi & Yogi, 2024; Zhang et al., 2024).

8.3 Energy Management Limitations

Energy constraints are a fundamental worry for agricultural edge AI, especially for sensors and self-driving devices in remote locations. Balancing the demands of AI inference with available power sources often requires dynamic adjustments to processing loads. Limited battery life restricts the complexity and frequency of AI tasks, with continuous operation of advanced neural networks potentially reducing battery life by 60–80% (STL Partners, 2017; Anonymous, 2023; Parihar & Suman, 2024). Solar-powered systems, while renewable, bring in variability in charging rates. Predictive energy management strategies are essential to coordinate AI processing with energy availability, ensuring uninterrupted operation (Shukre et al., 2024). Autonomous machinery must balance high-performance AI processing with the need to work through full workdays without recharging (Ercan Oğuztürk, 2025).

8.4 Regulatory Frameworks and Compliance Issues

The rules governing agricultural AI are still complex and subject to change. Different regions have varying standards for data privacy, environmental impact, and autonomous operations. This variation makes it hard to develop solutions that work everywhere (Kher, 2025; Sidorkiewicz et al., 2025). Data privacy laws, such as the GDPR and new regional regulations, have specific requirements for how agricultural data is handled. Edge AI systems must incorporate privacy-preserving techniques to ensure compliance. Safety standards for self-driven machinery are currently in a state of transition, and different parts of the world are following different directions regarding certification processes. For edge AI systems that power self-driving features to remain effective, they will need to be updated to meet new standards. The rapid development of technology often precedes the slow and careful rule-making process. This contrast between the speed of technology and the slow pace of regulation may lead to one of two outcomes: either all completely new projects are stopped, or different regional solutions are created, which would not be usable worldwide at all. This is where the importance of having a meeting point among technology developers, policymakers, and industry players becomes a critical aspect in the entire process of creating standards (Chouhan, S. S. et al., 2025).

Table 3 Various challenges and limitations.

Challenge	*Technical Solutions*	*Implementation Approaches*	*Success Metrics*
Model Updates	OTA updates, Differential sync	Scheduled maintenance windows	Update success rate

Contd...

Table 3 *Contd...*

Challenge	*Technical Solutions*	*Implementation Approaches*	*Success Metrics*
Limited Data	Transfer learning, Data augmentation	Multi-farm collaboration	Model accuracy
Energy Management	Dynamic scaling, Power optimization	Battery/solar integration	Operational uptime
Regulatory Compliance	Privacy-preserving techniques	Standards adherence	Compliance audits

9. Future Directions: Federated Learning and Beyond

9.1 Federated Learning in Agricultural Contexts

Federated learning offers an opportunity for farmers to cooperate in the development of AI technology. They put together their efforts and develop the models through the participation of several farms without sharing their data. It addresses the issue of data privacy, as well as the problem of data shortage (Belghachi, 2025; Dwarampudi & Yogi, 2024; Dembani et al., 2025). This is a solution for the data shortage and privacy concerns which were mentioned in the challenges part, and it could also facilitate the model training process of collaboration while respecting the data sovereignty of each farm.

9.2 Hybrid Edge-Cloud Architectures

The integration of edge and cloud computing and at the same time limiting the cloud processing to a few selective operations is an important advance in getting AI to work on the battlefield practically. These next-gen systems are basically about the instant local processing's ultra-fast response times and, simultaneously, cloud's taking over the heavy-load tasks of training and massive data analysis (Jin et al., 2025; Opiah, 2023; Jiang et al., 2025). This kind of method counters the customary bottlenecks and supercharges the system making it more resilient and dynamic by incorporating the strengths of both worlds, especially regarding the architectures.

9.3 Advanced Hardware Integration

The hardware designed solely for AI is still a fascinating area to work on in the context of an agricultural AI application. Neuromorphic computation is a notable example of a brain-inspired approach pursued in a computer; it has achieved the largest energy consumption gain to date. A number of systems have been known to use at least an order of magnitude less power merely for neural network inference (Pawlak & Howard, 2025). Integrating AI capabilities directly into sensors and equipment is another trend, which enables real-time analysis and decision-making at the source, reducing system complexity and costs, and improving reliability

(Fayyazi et al., 2025; Rajak et al., 2023; Singh Chouhan, S. et al., 2025). These hardware advances directly address the energy limitations identified as a key challenge.

9.4 Standardization and Interoperability

Creating industry standards for agricultural edge AI can accelerate adoption and reduce development costs. Establishing common APIs, data formats, and communication protocols will facilitate seamless integration among different systems and vendors (Shukre et al., 2024). Certification programs can help assure farmers of system reliability and safety, setting minimum standards for performance and robustness. The lack of interoperability and standards could become the single biggest obstacle for smaller farms and new developers. Without a common API or data format, a sensor from one vendor cannot easily communicate with a sprayer from another. By setting standards, the industry can create a true ecosystem where a variety of hardware, software, and services can work together seamlessly. This lowers the risk for farmers, fosters competition, and ultimately makes the technology more accessible and cost-effective. International collaboration on standards development can promote global adoption and technology transfer (Sidorkiewicz et al., 2025).

Table 4 Growth of Edge-processing technology.

Year	*Hardware*	*Performance*	*Applications*	*Examples*
2010–2012	ARM Cortex-M microcontrollers, Arduino, basic sensors	<0.1 TOPS	Irrigation timers, soil probes, weather stations	ARM Cortex-M3, Arduino Uno
2013–2015	ARM Cortex-A, Raspberry Pi, BeagleBone	0.5–1 TOPS	Data logging, irrigation scheduling, simple sensor fusion	Raspberry Pi 2, TI Sitara processors
2016–2017	Early GPU-enabled SoCs, NVIDIA Jetson TX1/TX2	1–2 TOPS	Drone imagery analysis, greenhouse climate control	Jetson TX1, TensorFlow Lite (2017)
2018–2019	Intel Movidius, Google Coral Edge TPU, Jetson Xavier NX	4–10 TOPS	Pest/disease detection, livestock tracking, anomaly detection	Coral Dev Board, Movidius Myriad X
2020–2021	Qualcomm Snapdragon AI Engine, Jetson Xavier AGX	10–20 TOPS	Smart irrigation (AI-based), drone mapping, tractor auto-steering	Snapdragon 865+, NVIDIA Xavier AGX

Contd...

Table 4 *Contd...*

Year	*Hardware*	*Performance*	*Applications*	*Examples*
2022–2024	NVIDIA Jetson Orin, AMD XDNA, Apple Neural Engine	50–250 TOPS	Farm robotics, predictive yield analytics, herd health wearables	Jetson Orin (2022), Google Edge TPU Gen 2
2025–2027	Neuromorphic & Specialized AI Chips (Loihi, TrueNorth)	500–1000 TOPS equivalent	Swarm robotics, fully autonomous tractors, farm digital twins	Intel Loihi 2, IBM Synapse Project
2028–2030	Hybrid quantum-classical processors, brain-inspired NPUs	Petaflop equivalent	Real-time optimization (genetics, weather, markets), federated learning	IBM Qiskit for agriculture, DARPA brain-inspired chips
2031–2035+	Neuromorphic mesh networks, quantum-enhanced NPUs	Exascale-equivalent inference	Fully autonomous ecosystems, planetary-scale modeling, adaptive AI for crops	Research prototypes (Loihi successors, quantum NPUs)

10. Conclusion

Edge AI is completely changing farming by making smarter, faster, and more efficient practices possible right where the data is created. Unlike traditional cloud-based systems, Edge AI reduces delays, improves data privacy, and works well even in rural areas with limited connectivity. Advances in hardware and software, such as model compression techniques like pruning and quantization, allow complex AI functions to run on small devices, which makes the technology more accessible and affordable for farmers. Practical applications include real-time crop monitoring, disease detection, precision irrigation, and autonomous machinery. These improvements let farmers make timely decisions, save resources, and increase productivity while maintaining high accuracy.

Despite these advancements, there are still a lot of severe issues. One of the issues that is still sticking around is continuously trying to keep up with the AI models and updates on multiple devices, as well as facing the headache of having a small pool of data for training, to name a few. Even the energy use is not optimized and the companies find it hard to cope with energy expenditures. AI models seem to become outdated very fast which is intimidating to most users given also the ever-changing regulatory scenario. Thus, these are the areas where Edge AI has started to show its calibration with the new world's requirements. For reasons like the ones mentioned above, Federated Learning is seen to be a good model in the context of IoT. Countries that hone their skill in Edge AI in agriculture without getting centered on taking over the cloud may come out as the best-performing entities in the IoT sphere. This partnership between cloud and edge is going to be energy-efficient, since you have a smaller communication direction and the transmission of data is thus very easy. Eventually, with the help of the chips and effort towards standardization, the cost of entry is going to be very low, this will result in wider adoption and consequently farmers will greatly benefit from the same. New hardware and standardization works are expected to contribute greatly to the integration of such systems and consequently enhance the sustainability and resilience of farming.

Declaration

Ethical Responsibilities of Authors

"All authors have read, understood, and complied as applicable with the statement on 'Ethical responsibilities of Authors' as found in the Instructions for Authors."

Ethical Approval

Not applicable

Use of AI tool

The corresponding author declares that no AI tool has been used to prepare this manuscript.

Conflict of Interest

On behalf of all the authors the corresponding author states that there is no conflict of interest to disclose.

Competing interests

Not applicable

Authors' contributions

Shreesh Sanjay Amin[1]: Conceptualization, Methodology, Writing – Original Draft, **S.K. Patel**[2]: Reviewing and Editing, Supervision, **Kalamkuri Chandu**[3]: Formal analysis, Visualization, **Adityanshu Tripathi**[4]: Formal analysis, Methodology, **Aditi Kachchap**[5]: Formal analysis, Visualization, **Dattatray G. Bhalekar**[6]: Writing – Reviewing and Editing, Supervision, **Abhishek Upadhyay**[7]: Writing – Reviewing and Editing, Conceptualization, Formal analysis, Visualization, Supervision.

Funding

No funding has been received.

References

Aetina. (2025). *Edge AI in Smart Agriculture: Revolutionizing Agriculture with AI.* https://www.aetina.com/about-news-detail.php?i=985.

Anand, R., Madhusudan, B.S. and Bhalekar, D.G. (2024). Computer Vision and Agricultural Robotics for Disease Control. In: *Applications of Computer Vision and Drone Technology in Agriculture 4.0* (pp. 31–47). http://dx.doi.org/10.1007/978-981-99-8684-2_3.

Anonymous. (2023, October 30). *Reducing Energy Demand of AI with Edge Computing. Embedur. AI.* https://www.embedur.ai/reducing-energy-demand-of-ai-with-edge-computing/.

Annapoorna, B.R. and Ramesh Babu, D.R.R. (2025). Edge-deployed visual pest detection system for real-time crop protection. *International Journal of Environmental Sciences, 11*(6S), 948–953. https://doi.org/10.64252/by11g405.

Belghachi, M. (2025). A comprehensive overview of federated learning for next-generation smart agriculture: Current trends, challenges, and future directions. Informatica, 49(1). https://doi.org/10.31449/inf.v49i1.6764.

Bhalekar, D.G., Parray, R.A., Mani, I., Kushwaha, H., Khura, T.K., Sarkar, S.K. and Verma, M.K. (2023). Ultrasonic sensor-based automatic control volume sprayer for pesticides and growth regulators application in vineyards. *Smart Agricultural Technology*, *4*, 100232. https://doi.org/10.1016/j.atech.2023.100232.

Chouhan, S.S., Patel, R.K., Shukla, A. and Singh, U.P. (2025). Drone Importance and their Necessity in Future Generation Agriculture. In: Artificial Intelligence and Computer Vision for Ecological Informatics (pp. 1–8). CRC Press. https://doi.org/10.1201/9781003518778-1.

Damsgaard, S.B., Hernández Marcano, N.J., Nørremark, M., et al. (2022). Wireless communications for Internet of Farming: An early 5G measurement study. *IEEE Access,* 105263–105277. https://doi.org/10.1109/ACCESS.2022.3211096.

Dantas, P.V., Sabino da Silva, W., Cordeiro, L.C., et al. (2024). A comprehensive review of model compression techniques in machine learning. *Applied Intelligence, 54,* 11804–11844. https://doi.org/10.1007/s10489-024-05747-w.

Dembani, R., Karvelas, I., Akbar, N.A., et al. (2025). Agricultural data privacy and federated learning: A review of challenges and opportunities. *Computers and Electronics in Agriculture, 232,* 110048. https://doi.org/10.1016/j.compag.2025.110048.

Dwarampudi, A. and Yogi, M.K. (2024). Application of federated learning for smart agriculture system. *International Journal of Information Technology and Computer Engineering, 43,* 36–47. https://doi.org/10.55529/ijitc.43.36.48.

Farmonaut. (2025). *Edge AI in Agriculture: 2025 Trends and Eco-advancements.* https://farmonaut.com/precision-farming/edge-ai-in-agriculture-2025-trends-eco-advancements

Fayyazi, A., Kamal, M. and Pedram, M. (2025). MARCO: Hardware-aware neural architecture search for edge devices. *arXiv:2506.13755.* https://doi.org/10.48550/arXiv.2506.13755.

Francy, S. and Singh, R. (2024). Edge AI: Evaluation of model compression techniques for convolutional neural networks. *arXiv:2409.02134.* https://doi.org/10.48550/arXiv.2409.02134.

Gorthi, S., Bhalekar, D.G., Khot, L.R. and Keller, M. (2024, October). Modeling Grape Berry Temperature for Effective Heat Stress Management in Vineyards. In: *2024 IEEE International Workshop on Metrology for Agriculture and Forestry* (*MetroAgriFor*) (pp. 237–241). IEEE. https://doi.org/10.1109/MetroAgriFor63043.2024.10948825.

Hao, J., Subedi, P., Ramaswamy, L. and Kim, I.K. (2023). Reaching for the sky: Maximizing deep learning inference throughput on edge devices with AI multi tenancy. *ACM Transactions on Internet Technology, 23*(1), Article 1. https://doi.org/10.1145/3546192.

Hoheisel, G.A., Bhalekar, D.G., Gorthi, S. and Khot, L.R. (2025). Automated single-row multi-fan sprayer optimization for efficient spray application in modern apple orchards. In: *Precision Agriculture '25* (pp. 292–298). Wageningen Academic. https://doi.org/10.1163/9789004725232_036.

Hosseiniyan Khatibi, S.M. and Jauhar, A. (2024). Harnessing the power of machine learning for crop improvement and sustainable production. *Frontiers in Plant Science, 15.* https://doi.org/10.3389/fpls.2024.1417912.

Huang, Q., Wu, X., Wang, Q., et al. (2023). Knowledge distillation facilitates the lightweight and efficient plant diseases detection model. *Plant Phenomics, 2023*(5), Article 0062. https://doi.org/10.34133/plantphenomics.0062.

Jiang, D., Shen, Z., Zheng, Q., Zhang, T., Xiang, W. and Jin, J. (2025). Farm-LightSeek: An edge-centric multimodal agricultural IoT data analytics framework with lightweight LLMs. *arXiv:2506.03168.* https://doi.org/10.48550/arXiv.2506.03168.

Jin, P., Du, W. and Jin, W. (2025). Efficient productivity prediction model based on edge data compression in smart farms. *Smart Agricultural Technology, 12,* 101242. https://doi.org/10.1016/j.atech.2025.101242.

Joshua, C. (2024). Use of cloud services in livestock management. *Unpublished manuscript, Federal University Oye Ekiti.* https://www.researchgate.net/publication/394084640.

Kalyani, Y. and Collier, R. (2021). A systematic survey on the role of cloud, fog, and edge computing combination in smart agriculture. *Sensors, 21*(17), 5922. https://doi.org/10.3390/s21175922.

Kaur, A., Bhatt, D.P. and Raja, L. (2024). Developing a hybrid irrigation system for smart agriculture using IoT sensors and machine learning in Sri Ganganagar, Rajasthan. *Journal of Sensors, 2024,* Article ID 6676907. https://doi.org/10.1155/2024/6676907.

Kher, E. (2025). The legal landscape of data privacy in AI-driven precision agriculture. *Washington Journal of Law, Technology, and Arts.* https://wjlta.com/2025/02/25/the-legal-landscape-of-data-privacy-in-ai-driven-precision-agriculture/.

Lang, J., Guo, Z. and Huang, S. (2024). A comprehensive study on quantization techniques for large language models. *arXiv:2411.02530.* https://doi.org/10.48550/arXiv.2411.02530.

Latif, E., Mai, G., Nyaaba, M., Wu, X., Liu, N., Lu, G. and Zhai, X. (2023). Artificial general intelligence (AGI) for education. arXiv preprint arXiv:2304.12479, 1, 1–34.

Lee, M. and Seo, S. (2021). Wearable wireless biosensor technology for monitoring cattle: A review. *Animals, 11*(10), 2779. https://doi.org/10.3390/ani11102779.

Lyu, B., Yang, Y., Cao, Y., et al. (2024). Efficient multi-objective neural architecture search framework via policy gradient algorithm. *Information Sciences, 661,* 120186. https://doi.org/10.1016/j.ins.2024.120186.

Krishnamoorthy, M.V. (2025). Edge AI: TensorFlow Lite vs. ONNX Runtime vs. PyTorch Mobile. *DZone (Analysis).*

Mansoor, Sheikh, Iqbal, S., Popescu, S.M., et al. (2025). Integration of smart sensors and IoT in precision agriculture: Trends, challenges, and future prospectives. *Frontiers in Plant Science, 16.* https://doi.org/10.3389/fpls.2025.1587869.

Matthew, B. (2024). Assessing network latency impact on cloud-based agricultural application. *Unpublished article, Federal University Oye Ekiti.* https://www.researchgate.net/publication/394083737.

Mohamed Rafi, M.S., Behjati, M. and Rafsanjani, A.S. (2025). Reliable and cost-efficient IoT connectivity for smart agriculture: A comparative study of LPWAN, 5G, and hybrid connectivity models. *Sensors and Actuators A: Physical, 331,* 112–123. https://doi.org/10.48550/arXiv.2503.11162.

MohitNSamagra. (2024). [DMP 2024]: Develop TinyML models for agriculture and document detection tasks (Issue #406). *Code4GovTech/C4GT.* Retrieved from https://github.com/Code4GovTech/C4GT.

Mortazavizadeh, F., Bolonio, D. and Mirzaei, M., et al. (2025). Advances in machine learning for agricultural water management: A review of techniques and applications. *Journal of Hydroinformatics, 27*(3), 474–492. https://doi.org/10.2166/hydro.2025.258.

Navarro, E., Costa, N. and Pereira, A. (2020). A systematic review of IoT solutions for smart farming. *Sensors (Basel), 20*(15), 4231. https://doi.org/10.3390/s20154231.

Oğuztürk, G. (2025). AI-driven irrigation systems for sustainable water management: A systematic review and meta-analytical insights. *Smart Agricultural Technology, 11*(1), 100982. https://doi.org/10.1016/j.atech.2025.100982.

Opiah, A. (2023). Edge computing in agriculture: Enhancing farming efficiency and sustainability. *EdgeIR.com.* https://www.edgeir.com/edge-computing-in-agriculture-enhancing-farming-efficiency-and-sustainability-20231030.

Parihar, H. and Suman, M. (2024). Estimating energy consumption of cloud, fog and edge computing infrastructures. *International Journal for Research in Applied Science and Engineering Technology.* https://doi.org/10.22214/ijraset.2024.58271.

Patel, S. (2024). Edge computing vs. traditional cloud: Performance and security considerations. *Spanish Journal of Innovation and Integrity, 12,* Article 779. https://eprints.umsida.ac.id/15728/1/779-Article%20Text-2397-1-10-20250303.pdf.

Patel, R.K., Chouhan, S.S., Chandan, T. and Singh, U.P. (2025). Basil Crop Detection Using Computer Vision and Deep Learning Approach. In: Artificial Intelligence and Computer Vision for Ecological Informatics (pp. 67–82). CRC Press. https://doi.org/10.1201/9781003518778-4.

Pawlak, W.A. and Howard, N. (2025). Neuromorphic algorithms for brain implants: A review. *Frontiers in Neuroscience, 19,* Article 1570104. https://doi.org/10.3389/fnins.2025.1570104.

Puy, F. (2024). TensorFlow Lite vs PyTorch Mobile. https://proandroiddev.com/tensorflow-lite-vs-pytorch-mobile-for-on-device-machine-learning-1b214d13635f.

Rajak, P., Ganguly, A., Adhikary, S. and Bhattacharya, S. (2023). Internet of Things and smart sensors in agriculture: Scopes and challenges. *Journal of Agriculture and Food Research, 14,* Article 100776. https://doi.org/10.1016/j.jafr.2023.100776.

Roostaei, J., Wager, Y.Z., Shi, W., et al. (2023). IoT-based edge computing (IoTEC) for improved environmental monitoring. *Sustainable Computing: Informatics and Systems, 38,* 100870. https://doi.org/10.1016/j.suscom.2023.100870.

Sammu, J. (2018). *Predictive Maintenance Using Edge and Cloud Collaboration.* Unpublished manuscript, Université de Montréal. https://www.researchgate.net/publication/388038746.

Schrader, M.J., Bhalekar, D.G., Sahni, R.K. and Khot, L.R. (2025). Unmanned aerial sprayers: eEvaluating platform configurations and flight patterns for effective chemical applications in modern vineyards. Smart Agricultural Technology, 101033. https://doi.org/10.1016/j.atech.2025.101033.

Shekhar, S., Durgam, M., Khose, S.B., Pohshna, C., Bhalekar, D.G. (2024). Advancement and Challenges of Implementing Artificial Intelligence of Things in Precision Agriculture. In: Chouhan, S.S., Saxena, A., Singh, U.P., and Jain, S. (Eds.), *Artificial Intelligence Techniques in Smart Agriculture.* Springer, Singapore. https://doi.org/10.1007/978-981-97-5878-4_13.

Shukre, V.A., Kanade, A., Kanade, S. and Patil, S.S. (2024). State-of-art technology and framework for IoT-based agricultural systems. *Migration Letters, 21*(S5), 816–837. https://migrationletters.com/index.php/ml/article/view/7794.

Sidorkiewicz, M., Królikowska, K., Dyczek, B., et al. (2025). Artificial intelligence for sustainable wine industry: AI-driven management in viticulture, wine production, and enotourism. *arXiv:2507.21098.* https://doi.org/10.48550/arXiv.2507.21098.

Singh Chouhan, S., Saxena, E., Shukla, A. et al. (2025). Optimizing YOLO-Based Models for Real-Time Guava Detection with Probabilistic Fused Wiener Filter-Enhanced Feature Fusion. *Applied Fruit Science*, 67, 383. (2025). https://doi.org/10.1007/s10341-025-01613-2.

Soussi, A., Zero, E., Sacile, R., Trinchero, D. and Fossa, M. (2024). Smart sensors and smart data for precision agriculture: A review. *Sensors, 24*(8), 2647. https://doi.org/10.3390/s24082647.

Sowmiya, K. and Anitha, V. P. (2025). A context-aware IoT and edge computing framework for wireless plant disease diagnosis using compressed Mask R-CNN and ResNet-50. *Journal of Wireless Mobile Networks, Ubiquitous Computing, and Dependable Applications, 16*(2), 707–720. https://doi.org/10.58346/JOWUA.2025.I2.043.

STL Partners. (2017). *Edge Computing: Changing the Balance of Energy Consumption in Networks.* https://stlpartners.com/articles/edge-computing/edge-computing-changing-the-balance-of-energy-in-networks/.

Suárez, D., Toledo, P. and Blanco, V. (2025). Energy-efficient inference on RNN and LLM networks: A quantized evaluation on RISC-V, ARM, and x86 devices. In: *Proceedings of the 16th ACM International Conference on Future Energy, E-Energy '25.* https://doi.org/10.1145/3679240.3735101.

Upadhyay, A., Chandel, N.S. and Chakraborty, S.K. (2024). Disease control measures using vision-enabled agricultural robotics. In: *Applications of Computer Vision and Drone Technology in Agriculture 4.0* (pp. 165–186). Singapore: Springer Nature Singapore. https://doi.org/10.1007/978-981-99-8684-2_10.

Upadhyay, A., Chandel, N.S., Singh, K.P., Chakraborty, S.K., Nandede, B.M., Kumar, M. and Elbeltagi, A. (2025a). Deep learning and computer vision in plant disease detection: A comprehensive review of techniques, models, and trends in precision agriculture. *Artificial Intelligence Review, 58*(3), 92. https://doi.org/10.1007/s10462-024-11100-x.

Upadhyay, A., Patel, A., Patel, A., Chandel, N.S., Chakraborty, S.K. and Bhalekar, D.G. (2025b). Leveraging AI and ML in precision farming for pest and disease management: Benefits, challenges, and future prospects. In: *Ecologically Mediated Development: Promoting Biodiversity Conservation and Food Security* (pp. 511–528). https://doi.org/10.1007/978-981-96-2413-3_23.

Wang, Z., Wang, P., Liu, K., Wang, P., Fu, Y., Lu, C.T., Aggarwal, C.C., Pei, J. and Zhou, Y. (2024). A comprehensive survey on data augmentation. *arXiv:2405.09591.* https://doi.org/10.48550/arXiv.2405.09591.

Yu, Z., Han, Y., Cha, L., et al. (2024). Design of an intelligent wearable device for real-time cattle health monitoring. *Frontiers in Robotics and AI, 11.* https://doi.org/10.3389/frobt.2024.1441960.

Zhang, X., Feng, Q., Zhu, D., Liang, X. and Zhang, J. (2024). Compressing recognition network of cotton disease with spot-adaptive knowledge distillation. *Frontiers in Plant Science, 15,* Article 1433543. https://doi.org/10.3389/fpls.2024.1433543.

6 Interpretable Crop Yield Forecasting Under Climate Change Using Temporal Fusion Transformers

Mohit Choubey[1*], Rahul Prasad[2], Yogesh Kumar Gupta[3], Aman Dubey[4] and Saurabh Agarwal[5]

1. Introduction

Global food security faces increasing pressure from rapid population growth and climate change. By 2029, the world's population is expected to rise to 8.4 billion (Van Klompenburg et al., 2020), while Australia is expected to grow by 16–19% by 2030 (Sridhara et al., 2024). Meeting these demands requires significant yield increases, food consumption is expected to rise by 60–70% by 2050, with cereal production alone projected to grow from 2.0 billion tons in 2015 to 2.81 billion tons by 2050 (Patel et al., 2022). Without major improvements in productivity, nearly 9 billion people may face food insecurity by mid-century (Chouhan et al., 2023). Although technological advances have boosted efficiency, large-scale yield growth remains uncertain in Australia (Cao et al., 2021). For example, rice production required an annual growth of 2–2.5% until 2020 to meet global needs (Joshua et al., 2021). Beyond food supply, agriculture is economically vital, contributing 1.9% of Australia's GDP and 2.6% of employment in 2019–2020 (Patel et al., 2021). Globally, wheat accounts for 14% of cropland, while rice remains a staple food across Asia, Africa, and Latin America, with consumption projected at 69 million metric tons by 2029 (Khaki et al., 2020).

Climate change plays a significant role in influencing agricultural risk, altering rainfall patterns, reducing water availability, and intensifying greenhouse gas emissions (Chouhan et al., 2022). CO_2 enrichment significantly affects crop yields,

[1] Department of Electronics and Communication Engineering, Maulana Azad National Institute of Technology, Bhopal, Madhya Pradesh, India.
[2,3] School of Computer Science, UPES Dehradun, India.
[4] Department of Electronics and Communication Engineering, Maulana Azad National Institute of Technology, Bhopal, Madhya Pradesh, India.
[5] Research Professor, School of Computer Science and Engineering, Yeungnam University, Republic of Korea.
https://orcid.org/my-orcid?orcid=0009-0003-2968-360X
* Corresponding author: choubeymohit35pain@gmail.com

DOI: 10.1201/9781003715726-6

wheat production shows strong responses (Tian et al., 2021), and rice exhibits a cointegration effect with temperature and CO_2 variation (Chouhan et al., 2025). While higher CO_2 can benefit wheat and maize in cooler regions by offsetting drought stress (Shook et al., 2021, Ma et al., 2021), these gains depend on nitrogen availability (Dang et al., 2021). Maize remains especially sensitive to rainfall and temperature fluctuations (Nikhil et al., 2024).

Crop yield forecasting approaches fall into three categories. Statistical methods such as ARIMA and ARIMAX depend on strict assumptions and struggle with heterogeneous agro-climatic data (Jhajharia et al., 2023, Iniyan et al., 2023, Cedric et al., 2022). Machine learning (ML) methods like ANNs (artificial neural networks), SVMs (support vector machines), RF (random forest), and genetic algorithms better capture nonlinearities and often outperform statistical baselines (Gopi & Karthikeyan, 2023; Elavarasan & Vincent, 2021a, 2021b; Ali et al., 2022). Models like LSTM (long short-term memory), CNN (convolutional neural network), RNN (recurrent neural network), and DNN (deep neural network), under the umbrella of deep learning (DL), demonstrate strong capabilities in learning time-based relationships and intricate data patterns, especially with long datasets (Bhadra et al., 2024; Bhimavarapu et al., 2023; Qiao et al., 2021a; Wang et al., 2022; Chouhan et al., 2025). Hybrid ensemble approaches have also shown improvements (Gopi & Karthikeyan, 2024; Bhadra et al., 2024). Among these, ANNs remain widely applied (Joshua et al., 2021), while DNNs are increasingly used for modelling nonlinear interactions (Elavarasan & Vincent, 2021b; Ali et al., 2022).

Specific models demonstrate these trends. Reinforcement Random Forest (RRF) reduced overfitting and computational cost but offered limited accuracy (Elavarasan & Vincent, 2021a). SVR (SV regression), GRNN (generalized regression NN), BPNN (back-propagation NN), and RBF (radial basal function) networks achieved strong results in rice forecasting (e.g., GRNN: $R^2 = 0.9863$, MSE $= 0.05$, RMSE $= 0.229$) (Joshua et al., 2021), though they struggled with nonlinear dynamics. A DBN-FNN hybrid improved feature learning but lacked predictive strength (Elavarasan & Vincent, 2021b). The MMML-CRYP framework combined KELM with an Equilibrium Optimizer and RF, achieving 97.91% accuracy but falling short of advanced DL methods (Iniyan et al., 2023). RFOERNN-CRYP, which integrated LSTM, BiLSTM, and GRUs (gated recurrent units) with Red Fox Optimization, showed improved accuracy but suffered from overfitting and training complexity (Gopi & Karthikeyan, 2024).

Overall, ML and DL frameworks offer improvements but face persistent challenges, reliance on handcrafted features, heavy parameter tuning, high computational cost, and poor interpretability (Patel et al., 2025). These limitations diminish their utility in agricultural decision-making, particularly due to the lack of long-term climate-yield datasets (Van Klompenburg et al., 2020). To address these challenges, attention-based models have become increasingly significant. The Temporal Fusion Transformer (TFT) is a possible solution since it mixes static and time-varying variables, employs variable selection networks to locate useful features, and leverages multi-head attention to find long-term associations. TFT is significant as it elucidates yield variability by assessing the impact of climate, soil, and socio-economic factors (Bhimavarapu et al., 2023). This study introduces a comprehensible methodology for predicting crop yields amid climate change, employing TFT and including multi-source data. The model fills in the gaps left by earlier methods by combining several kinds of data, such as climate indicators, remote sensing indices, and socio economic variables. It also makes sure that predictions are understandable by employing attention-based interpretability, and it gets good accuracy and generalisation without requiring metaheuristic optimisation. In Australia, the framework is tested on four primary crops: oats, maize, rice, and wheat. The other three are C3 crops, whereas maize is a C4 crop. This allows us to see how well the framework performs with different kinds of crops.

2. Mathematical Foundation

To predict crop yields, the proposed approach adopts the TFT, which integrates multiple input sources, models temporal patterns across different time scales, and explains predictions using attention-based interpretability.

At each time step $t \in \{1, ..., T\}$, the inputs to the model are divided into three categories. First, static covariates $s \in R^{d_s}$ represent variables that do not vary with time, such as crop type, soil characteristics, and geographic region. Second, time-varying known inputs $x_t^{(k)} \in R^{d_k}$ include features that are available ahead of time, such as calendar dates, planting schedules, or policy constraints. Third, time-varying observed inputs $x_t^{(o)} \in R^{d_o}$ capture sequentially measured variables such as rainfall, temperature, humidity, atmospheric CO_2, NDVI, and soil moisture. The forecasting target is the yield $y_t \in R$ at time step t.

To standardize the representation, categorical inputs are mapped into dense embeddings using embedding matrices E_s, E_k, E_o, while continuous variables are normalized to the interval [0, 1]. The transformed embeddings are:

$$\tilde{s} = E_s s,\ \tilde{x}_t^{(k)} = E_k x_t^{(i)},\ \tilde{x}_t^{(o)} = E_o x_t^{(o)} \quad ...(1)$$

2.1 Gated Residual Network (GRN)

A core building block of the TFT is the Gated Residual Network (GRN), which enables non linear transformations while controlling information flow to prevent overfitting. For an input vector and an optional context c, the GRN computes:

$$z = W_{1a} + V_{1c} + b_1 \quad ...(2)$$

$$u = \text{ELU}(z),\ v = W_2 u + b_1 \quad ...(3)$$

$$g = \sigma(W_g v + b_g) \quad ...(4)$$

$$\text{GRN}(a|c) = r + \text{LayerNorm}\ (g \odot v) \quad ...(5)$$

where, ELU is the Exponential Linear Unit, $\sigma(\cdot)$ is the sigmoid function, $\odot$ denotes element-wise multiplication, and r depicts a residual projection of a. GRNs transform both static and dynamic features. From the static inputs, three context vectors are produced: $c^{(vs)}$ for variable selection, $c^{(l)}$ for local sequence modelling, and $c^{(t)}$ for temporal attention.

2.2 Variable Selection Network (VSN)

The VSN dynamically determines the importance of each feature at every time step. Let,

$$\tilde{x}_t = \left[x_t^{(k)}, \tilde{x}_t^{(o)} \right] \in R^{d_x} \quad ...(6)$$

be the concatenated input embedding. For each variable j = 1, ..., m, a GRN produces a transformed representation $h_{t,j}$. The importance logits are:

$$\alpha t = W_\alpha \, \mathrm{GRN}\, (\tilde{x} \,|\, c^{(vs)}) + b_\alpha \quad ...(7)$$

and the normalized weights are given by:

$$w_{t,j} = \frac{\exp(\alpha_{t,j})}{\sum_{j'=1}^{m} \exp(\alpha_{t,j'})} \quad ...(8)$$

The aggregated feature representation at time is:

$$r_t = \sum_{j=1}^{m} w_{t,j}\, h_{t,j} \quad ...(9)$$

This mechanism allows the model to emphasize, for example, rainfall or NDVI during critical crop growth phases, while down-weighting less relevant features.

2.3 Local Temporal Processing

To capture short-term dependencies and seasonal fluctuations, the selected features are processed locally. A GRN with static context $c^{(l)}$ refines the representation:

$$\hat{r} = \mathrm{GRN}\, (r_t | c^{(l)}),\; h_t^{(loc)} = \hat{r} + r_t \quad ...(10)$$

Here, $h_t^{(loc)}$ denotes the locally enriched state at time t.

2.4 Multi-Head Attention for Long-Term Dependencies

While local processing addresses short-term effects, multi-head self-attention is employed to capture long-term temporal patterns. Let,

$$H^{(loc)} = [h_1^{(loc)}, ..., h_t^{(loc)}]^T \in R^{T \times d} \quad ...(11)$$

For each attention head h:

$$Q_h = H^{(loc)} W_h^Q, \quad K_h = H^{(loc)} W_h^K, \quad V_h = H^{(loc)} W_h^V, \quad ...(12)$$

$$\text{Attention}_h = (H^{(loc)}) \text{ softmax}\left(\frac{Q_h K_h^T}{\sqrt{d_h}}\right) \quad ...(13)$$

The outputs of all heads are concatenated and projected:

$$H^{(\text{attn})} = [\text{Attention}_1, ..., \text{Attention}_H]W^o \quad ...(14)$$

Finally, a GRN with context $c^{(t)}$ produces the global sequence representation:

$$H^{(\text{glob})} = \text{GRN } (H^{(\text{attn})}|c^{(t)}) + H^{(\text{loc})} \quad ...(15)$$

This representation encodes both seasonal variations and long-range effects such as cumulative water stress or repeated droughts.

2.5 Decoder and Quantile Forecasting

The decoder combines the global representation $H^{(\text{glob})}$ with known future inputs to predict crop yield across horizons $T + 1, ..., T + H_f$. Instead of a single deterministic output, the TFT produces quantile forecasts:

$$\hat{y}_{T+t}^{(\gamma)} = w^{(\gamma)T} z_{T+\tau} + b^{(\gamma)}, \gamma \in Q \quad ...(16)$$

where, $Z_{T+\tau}$ is the hidden state at horizon τ and Q is a set of quantiles such as {0.1, 0.5, 0.9}. This yields prediction intervals that quantify uncertainty under climate variability.

2.6 Training Objective

Training minimizes the quantile loss (pinball loss):

$$l_\gamma(y, \hat{y}^{(\gamma)} = \{\gamma(y - \hat{y}^{(\gamma)}) \text{ if } y \geq \hat{y}^{(\gamma)}(\gamma - 1)(y - \hat{y}^{(\gamma)}) \text{ if } y < \hat{y}^{(\gamma)} \quad ...(17)$$

The full objective is:

$$L = \frac{1}{N}\sum_{=1}^{H_f} \sum_{\gamma \in Q} l_\gamma(y_{T+\tau}, \hat{y}_{T+\tau}^{(\gamma)}) \quad ...(18)$$

where, N is the number of training samples. Model performance is assessed using RMSE, MAE, and R^2 on the median forecast ($\gamma = 0.5$).

Beyond accuracy, TFT provides interpretability. Feature weights $w_{t,j}$ from the VSN identify which variables (e.g., rainfall, CO_2, fertilizer) most influenced predictions. Attention weights highlight historical windows critical for yield determination (e.g., monsoon onset, drought years). Together, these provide both accurate forecasts and explanations, enabling actionable insights for farmers and policymakers.

3. Proposed Model

As depicted in Figure 1, the proposed framework outlines the approach to crop yield forecasting, built on the TFT, a highly sophisticated deep learning architecture designed to handle multi-source, temporal, and heterogeneous datasets while ensuring interpretability. Established modelling strategies, such as regression-based techniques, machine learning algorithms, and metaheuristic-enhanced hybrid deep learning, often struggle with high prediction errors, overfitting risks, and poor transparency. The TFT overcomes these limitations by combining the strengths of sequence modelling, attention mechanisms, and variable selection networks in a unified architecture that not only predicts yields accurately but also explains the underlying drivers of those predictions.

3.1 Input Integration and Preprocessing

Crop yield is driven by the combined effect of various climate parameters (rainfall, temperature, humidity, CO_2 concentration), remote-sensing indices (NDVI, soil moisture), and socio economic factors (fertiliser usage, crop area, land management practices). To prepare these heterogeneous sources for model training, a preprocessing stage is applied. First, missing values are imputed using statistical measures such as mean or median, and extreme outliers are removed to reduce noise. Next, continuous variables are rescaled with min–max normalisation to restrict values within the range [0, 1], ensuring stability during optimisation. Finally, to evaluate robustness, the dataset is split using 5-fold cross-validation, which systematically rotates training and validation subsets, preventing overfitting and ensuring generalisation across crops and regions.

3.2 Feature Representation and Variable Selection

A major challenge in crop yield forecasting is that many input features are highly correlated, such as rainfall and soil moisture, or fertiliser use and yield history. Traditional models often require manual or heuristic feature selection to reduce redundancy, but TFT eliminates this need by using Variable Selection Networks (VSNs). VSNs learn, at each time step, which features are most important and assign dynamic weights accordingly. This lets the model focus on important factors, like rainfall during sensitive growth stages of rice, while downplaying less important signals. TFT also divides characteristics into static covariates (such crop type, soil class, and region) that stay the same throughout time and dynamic covariates (like rainfall, temperature, and NDVI) that change with the seasons. Static variables give the model background that affects the rest of the model, making sure that forecasts are particular to certain crops and areas.

3.3 Temporal Modelling: Local and Global Dependencies

Changes in climate conditions over time also have an effect on crop yields, not just the current ones. The TFT design captures this in two complementary stages: Processing time locally, which records short-term effects like fluctuations in daily rainfall, severe heatwaves, or quick reactions to fertiliser use. It models seasonal changes and short-lived shocks by focussing on recent data. And global attention to time is a multi-head attention mechanism that captures long-term dependencies, such as long-lasting droughts, heat stress that builds up over time, or crops that take a long time to respond to CO_2. This method lets the model focus on the historical periods that are most important for making current forecasts. For instance, if yield is very dependent on rain during flowering, the attention mechanism will automatically focus on that time and ignore other times that aren't important. This limited focus makes the model both more accurate and easier to understand. TFT can handle both short-lived anomalies and longer-term climate trends at the same time by using both local and global temporal models. Both of these are very important in agriculture.

3.4 Forecasting with Quantile Predictions

The TFT makes probabilistic forecasts by forecasting numerous quantiles (like the 10th, 50th, and 90th percentiles), which is different from standard models that only give single-point forecasts. This lets the model not only guess the predicted yield, but also give a range of possible outcomes, which is very important when the weather is unpredictable. For example, policymakers can prepare for both good and bad situations by looking at the difference between the lower and upper quantiles.

3.5 Training and Optimization

Training of the model is carried out using advanced optimisation strategies to ensure both efficiency and generalisation. Adam Optimizer with adaptive learning rate scheduling is used to adjust weights during training. To prevent overfitting, the model applies dropout (randomly deactivating some neurons during training), early stopping (halting training if no improvement is observed in validation accuracy), and gating mechanisms that filter redundant signals. Combined with cross-validation, these strategies produce a robust forecasting model that performs consistently across diverse climatic zones and crop categories.

A key strength of the TFT is its interpretability, which makes it suitable for real-world agricultural decision support. Two interpretability tools are built into the model. Variable importance scores from the VSN highlight which features such as rainfall, fertiliser application, or CO_2 had the greatest influence on yield predictions. Attention heatmaps from the temporal attention mechanism show which past time steps (e.g., early-season rainfall, flowering-stage temperature) were most critical for forecasting current yields.

These insights not only increase trust in the model but also provide actionable guidance. For example, farmers may learn that irrigation during specific growth stages has the largest impact on yield, or policymakers may observe that rising CO_2 concentrations interact with fertiliser use to influence maize production.

The TFT-based framework integrates diverse datasets, automatically selects relevant features, models both short- and long-term temporal dependencies, and generates interpretable probabilistic forecasts. Compared to earlier models such as FNNDBN, RFOERNN-CRYP, or DLMWOA-CYPBCC, which relied on static architectures and heuristic optimisation, TFT inherently learns complex variable interactions without external tuning. This makes it not only more precise, but also easier to see and grow.

The full operation of the proposed TFT framework starts with preprocessing data, goes through dynamic feature selection and context encoding, uses local processing and attention mechanisms to learn patterns that reflect immediate fluctuations and sustained developments dependencies, generates uncertainty-aware forecasts through quantile outputs, and finally gives interpretability outputs that explain both what the forecast is and why it was made. TFT is the best way to predict crop yields that take climate into account because it is accurate, scalable, and open.

4. Results and Discussion

This part talks about and analyses the suggested framework for predicting crop yields based on TFT. All experiments were implemented in a TensorFlow environment using Python, running utilising a hardware setup of Intel Core i7 @ 3.2 GHz, 32 GB RAM, under the Windows 10 operating system. Training was performed with GPU acceleration where possible, ensuring efficient handling of multi source datasets.

4.1 Dataset Description

The dataset under consideration integrates two publicly available Indian sources: Crop Prediction Data (https://www.kaggle.com/datasets/jiteshmd/crop-prediction-data) and Crop Yield Data India (https://www.kaggle.com/datasets/saincoder404/crop-yield-data-india). It covers the period 1997–2020 and includes key attributes comprising crop year, production, yield, temperature, humidity, state, and annual rainfall. To extend forecasting horizons, the model generates future predictions for 2021–2030 under climate variability.

Since the dataset contained imbalances across crop classes (e.g., rice and ragi were overrepresented compared to banana or coconut), SMOTE (Synthetic Minority Over-sampling Technique) was utilised to correct class imbalance by creating synthetic data points for minority classes, ensuring fairer representation and minimizing bias toward dominant crop categories, and enhancing generalisation across all crop types.

Unlike previous works that relied on heuristic optimization (e.g., WOA), our TFT-based model automatically learns feature relevance via Variable Selection Networks and captures temporal dependencies using multi-head attention. This reduces computational overhead and provides interpretability of results.

4.2 Performance Evaluation

The model was evaluated using standard error metrics such as:

4.2.1 Mean Squared Error (MSE)

MSE is defined as the average squared deviation between actual values (z_i) and predicted values ($\hat{z}_i$).

$$\text{MSE} = \frac{1}{m}\sum_{i=1}^{m}(z_i - \hat{z}_i)^2 \qquad ...(19)$$

The squaring process emphasises large errors disproportionately, which increases the sensitivity of MSE to outliers. It provides a smooth and differentiable error surface, which is why it is widely used as a loss function in optimisation. The closer the MSE is to zero, the more accurate the model is, but its scale depends on the squared unit of the target variable.

4.2.2 Root Mean Squared Error (RMSE)

By applying the square root to the MSE, RMSE offers an error metric that is directly comparable to the units of the target variable.

$$\text{RMSE} = \sqrt{\frac{1}{m}\sum_{i=1}^{m}(z_i - \hat{z}_i)^2} \qquad ...(20)$$

It is more interpretable than MSE since it is in the same unit as the data (e.g., tons of crop yield, megawatts of electricity). Like MSElarger deviations are penalised more heavily than smaller ones, enhancing its usefulness when large deviations are especially undesirable. RMSE is a direct measure of prediction accuracy, a smaller RMSE means the predictions approximate the actual outcomes more accurately.

4.2.3 Coefficient of Determination (R^2)

R^2 (known as the coefficient of determination) evaluates the explanatory power of predicted values in relation to the variability of actual outcomes.

$$R^2 = 1 - \frac{\sum_{i=1}^{m}(z_i - \hat{z}_i)^2}{\sum_{i=1}^{m}(z_i - z^-)^2} \qquad ...(21)$$

where, z^- is the mean of observed values. R^2 The coefficient of determination ranges from $-\infty$ to 1. Here, 1 reflects perfect predictive accuracy, 0 suggests no improvement over mean prediction, and values less than 0 reveal poorer performance than the mean estimate.

It is widely used as a goodness-of-fit measure, showing the extent to which the model explains the variation in the data. In forecasting (e.g., crop yield prediction), a higher means the model captures more of the relationship between climate, agronomic, and socio economic inputs and actual yields.

The TFT model demonstrated superior performance with RMSE and MAE values significantly lower than baseline models, while achieving R^2 values above 0.99 across multiple states and crops. These results indicate strong predictive accuracy and generalisation ability.

4.3 Description of the Hyperparameters

Table 1 outlines the hyperparameter configuration used in the proposed TFT-based framework for crop yield forecasting. Each setting was carefully selected to balance model accuracy, stability, and computational efficiency, ensuring robust performance under climate-aware agricultural conditions.

Table 1 Hyperparameter tuning details of the proposed Temporal Fusion Transformer.

Hyperparameter	*Value/Range*	*Notes*
Hidden layers	Encoder-decoder blocks (4 layers)	Configured to capture short- and long-term dependencies
Activation function	Gated Linear Unit (GLU), ReLU, tanh	Selected per component for stability and interpretability
Dropout rate	0.1–0.3	Applied to prevent overfitting
Attention heads	4–8	Multi-head attention for temporal dependencies
Learning rate	0.001 (adaptive scheduling)	Adam Optimizer
Batch size	64	Balanced for computational efficiency
Epochs	100	With early stopping based on validation loss
Optimizer	Adam	Adaptive moment estimation
Cross-validation	5-fold	Ensures generalisation and robustness

The designed model depth was fixed at four encoder-decoder blocks, which provides sufficient representational capacity to capture both short-term fluctuations (e.g., sudden rainfall changes) and long-term dependencies (e.g., cumulative drought impact) without making the model excessively complex. Different activation functions were used in different parts: Gated Linear Units (GLU) were used in gating layers to control the flow of information, Rectified Linear Units (ReLU) were used in residual connections to make the process more efficient, and hyperbolic tangent (tanh) was used in normalisation pathways to make learning more stable. A dropout rate of 0.1–0.3 was used across hidden layers to keep the model from overfitting. This makes sure that the model works effectively across all folds and doesn't lose any important data. The multi-head attention mechanism, which is used between 4–8 heads, also lets the framework show different temporal dependencies at the same time. This made it possible to find different climate patterns that affect yields.

The learning rate started at 0.001 with adaptive scheduling and was fine-tuned with the Adam optimiser, which quickly changes the speed of learning momentum and convergence. We chose a batch size of 64 since it was a good combination between giving us steady gradient estimations and being able to run the calculations. The training was limited to 100 epochs, with early termination conditions predicated on validation loss to prevent superfluous training and mitigate overfitting.

Finally, the model was checked with 5-fold cross-validation. This is quite important for agricultural datasets because they often show a lot of variation between locations and seasons. This makes sure that the suggested TFT framework works well and can be used on many different datasets instead of only one split.

4.4 Performance of the Proposed Model

The performance of the proposed TFT-based framework is first assessed through error-based measures, which highlight its accuracy and stability. As shown in Table 2, the 5-fold cross-validation results demonstrate consistently high values of the coefficient of determination (R^2), ranging from 99.91% to 99.94%, with an overall mean of 99.92%. Correspondingly, the RMSE averages 0.110, while the MAE averages 0.100 across all folds. These extremely low error values confirm that the predictions closely match the observed yields, deviating only marginally. The closeness between RMSE and MAE also indicates that the prediction errors are uniformly small without extreme outliers, which is essential in agricultural forecasting where large deviations could lead to significant misjudgements in planning.

Figure 1 shows the predictive quality even more clearly by comparing actual yield values to expected yield values. The two curves are almost perfectly aligned, which shows that the model is able to represent both rapid shifts and gradual changes in crop production. The strength of the TFT framework comes from its capacity to represent complicated nonlinear interactions between climatic, remote sensing, and socio economic inputs all at once. This makes sure that it works well in a variety of situations. The fact that the folds are all the same shows that the model is quite generalisable, which makes it a good choice for use in the real world, where climate change is always a problem.

Table 2 Fold-wise performance (5-fold CV).

Fold	R^2	*RMSE*	*MAE*
1	99.92	0.112	0.102
2	99.93	0.109	0.099
3	99.91	0.111	0.101
4	99.94	0.107	0.097
5	99.92	0.110	0.100
Mean	99.92	0.110	0.100

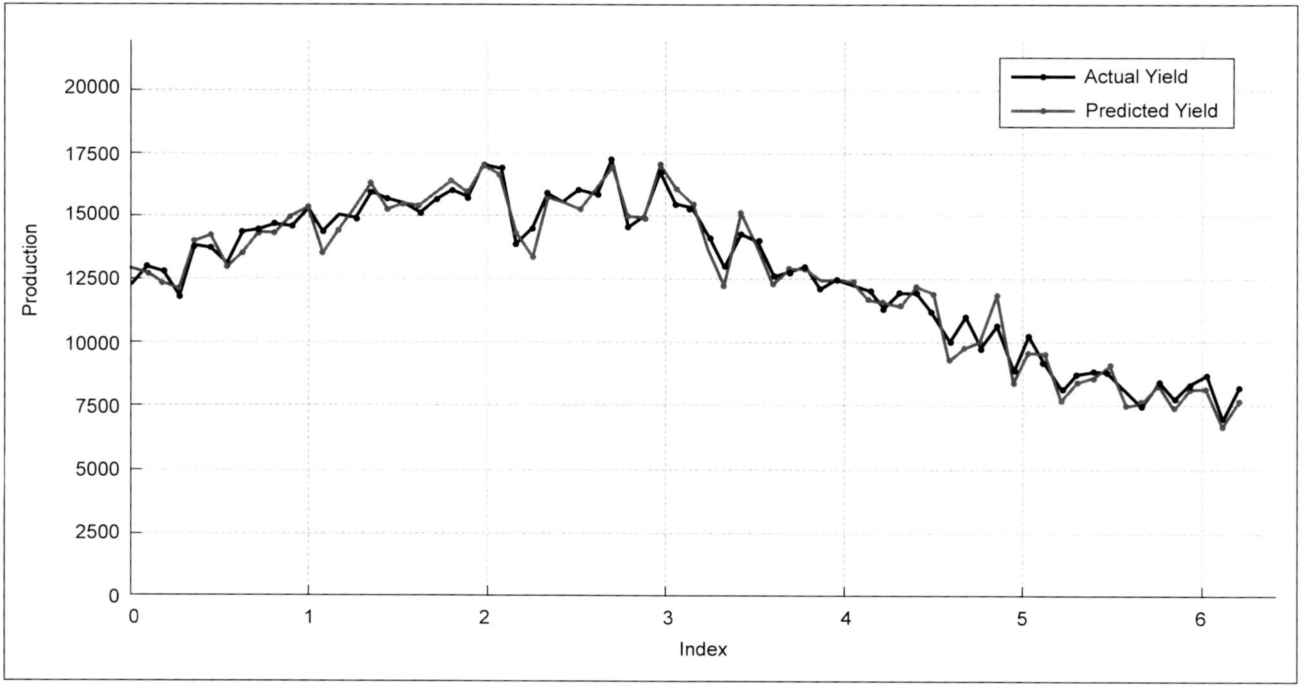

Fig. 1 Prediction of the proposed model.

The total evaluation in Table 3 shows that the TFT-based framework is better than the regression accuracy. The model achieved an exceptionally high R^2 of 0.9992 (99.92%), confirming its ability to explain nearly all the variability in the yield data. The RMSE and MAE remain extremely low, reaffirming minimal error margins. Complementing these, classification-style performance indicators provide additional perspectives. The model got 99.12% accuracy, with precision (98.89%) and recall (98.78%) values showing that it was good at finding actual yield outcomes while keeping false positives and false negatives to a minimum. Similarly, sensitivity (99.58%) and specificity (99.98%) highlight the framework's reliability in detecting true yield patterns while avoiding false alarms. The F1-score of 98.83% further reflects a strong equilibrium between precision and recall, signifying the TFT model as a highly dependable forecasting tool given by Figures 2, 3, and 4.

Table 3 Overall Effectiveness of the Proposed TFT Model.

Metric	*Value*
R^2	0.9992
RMSE	0.110
MAE	0.100
Accuracy	99.12%
Precision	98.89%
Recall	98.78%
Sensitivity	99.58%
Specificity	99.98%
F1-score	98.83%

The proposed framework stands out from previous ML and DL approaches due to its simplicity and interpretability. Figure 3 shows the VSN relevance, with rainfall, temperature, and NDVI being the most important factors that affect yield. This not only confirms domain knowledge but also offers practical advice for planning in agriculture. Figure 4, which shows the temporal self-attention heatmap, shows the historical time frames that are most important for current yield estimates. It does this by showing periods like rainfall anomalies or temperature peaks that have a bigger effect on outcomes than others. Finally, Figure 5 demonstrates the things that affect yield at different stages of growth. It shows how climate and management factors affect yield differently at different stages of growth, such as sowing, vegetative growth, blooming, grain filling, and maturity. This breakdown makes the model easier to grasp, so you can see what it predicts and do something about it.

These results show that the proposed TFT-based framework not only cuts down on prediction mistakes, but it also does well on balanced classification-style assessment criteria. The framework is optimal for predicting agricultural

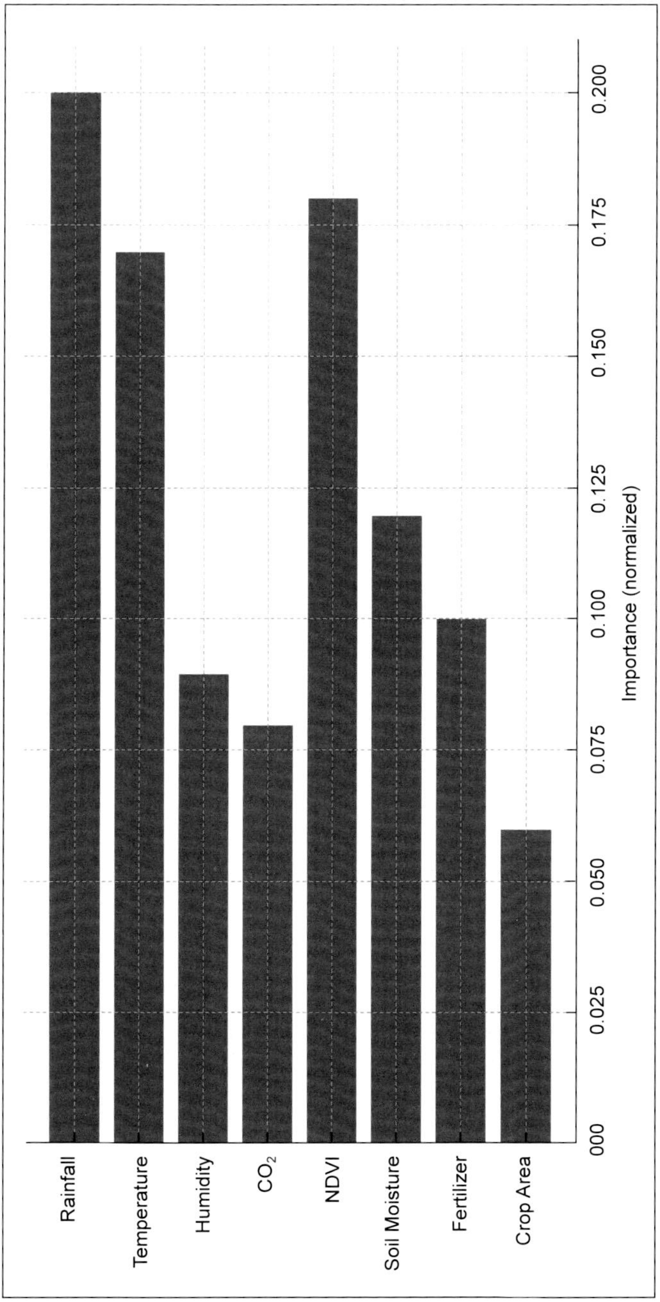

Fig. 2 VSN Variable Importance.

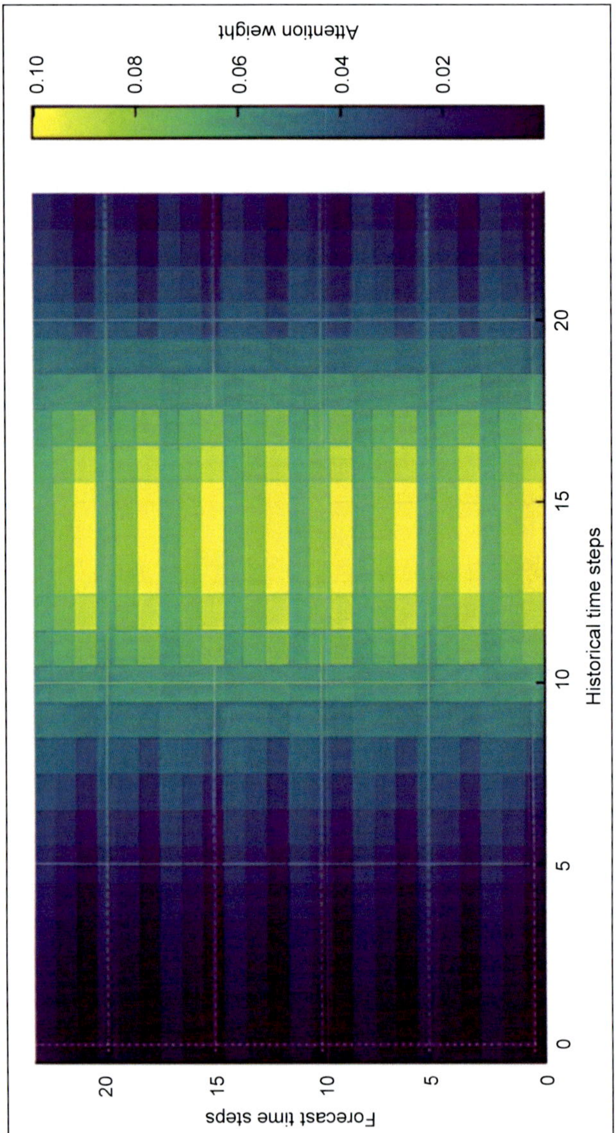

Fig. 3 Temporal Self-Attention Heatmap.

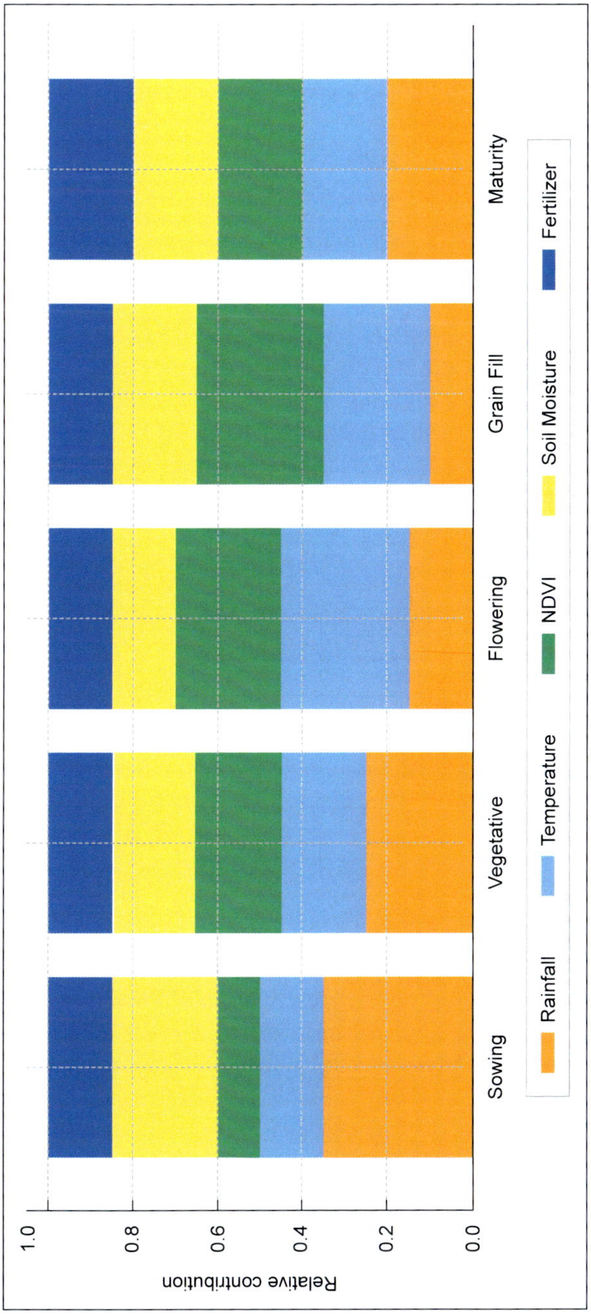

Fig. 4 Drivers of Yield by Growth Stage.

production that takes climate into account since it has a high R^2, a low RMSE and MAE, and an optimal trade-off achieved between precision and recall. It is a useful tool for helping farmers, policymakers, and anyone who are involved in planning for agriculture and managing food security make good choices because it can anticipate the future and be comprehended.

4.5 Performance Comparison

Table 4 compares the proposed TFT -based architecture to a series of well-known baseline and advanced crop yield projection models. The differences between the methodologies show how forecasting accuracy has improved and how older methods had limitations.

The Hybrid RL-RF model (Elavarasan et al., 2021a), which combines reinforcement learning with random forest, achieved an R^2 of 0.7500, with relatively high RMSE (0.2000) and MAE (0.1900). While the reinforcement learning mechanism helped reduce overfitting, the model's reliance on handcrafted splitting attributes and limited capacity to capture temporal dependencies restricted its performance. It is primarily suited to static tabular data and therefore struggles when dealing with sequential climate and agronomic information.

Table 4 Comparative evaluation with existing models.

Method	R^2	*RMSE*	*MAE*
Hybrid RL-RF	0.7500	0.2000	0.1900
GRNN	0.9863	0.2295	0.1279
FNN-DBN	0.8300	0.2400	0.1500
MMML-CRYP	0.9854	0.1262	0.3469
RFOERNN-CRYP	0.9988	0.8377	0.2989
Proposed TFT	0.9992	0.110	0.100

The GRNN (Joshua et al., 2021), performed better, achieving an R^2 of 0.9863, but its RMSE (0.2295) remained relatively high. GRNN relies on distance-based similarity measures and works effectively with small nonlinear datasets, but it fails to scale efficiently into high-dimensional, multi source inputs. Its reliance on simple regression kernels limits its effectiveness in capturing extended temporal dynamics present in crop yield data.

The FNN-DBN hybrid (Elavarasan et al., 2021b), which integrates Feedforward Neural Networks (FNN) with Deep Belief Networks (DBN) reported an R^2 of only 0.8300 with RMSE and MAE of 0.2400 and 0.1500, respectively. Despite the DBN's pretraining capability for feature extraction, the static architecture of the FNN component limited its adaptability to temporal and multivariate agricultural data. The model suffered from both underfitting and high computational complexity, preventing it from delivering satisfactory predictive accuracy.

The Multimodal Machine Learning CRYP (MMML-CRYP) (Gopi et al., 2023) achieved strong correlation with an R^2 of 0.9854 and relatively low RMSE (0.1262). However, the MAE value of 0.3469 reveals that the model's predictions deviated significantly on average, particularly in imbalanced datasets. The KELM-EO-RF pipeline of MMML-CRYP could efficiently fuse multimodal inputs, but the separate feature optimisation and prediction stages introduced instability and inconsistency across folds, resulting in poorer generalisation.

The RFOERNN-CRYP framework (Gopi et al., 2024), which integrates RNN, LSTM, BiLSTM, and GRU with Red Fox Optimization (RFO) for hyperparameter tuning, achieved near-perfect correlation ($R^2 = 0.9988$). However, its error values were disproportionately high, with an RMSE of 0.8377 and MAE of 0.2989. This mismatch between R^2 and error measures reflects the model's overfitting tendency: while it could fit the training data extremely well, its generalisation to unseen samples was poor. The complexity of simultaneously managing multiple recurrent structures and metaheuristic tuning also introduced instability and longer training times.

The suggested TFT-based framework, on the other hand, had the best overall performance, with an R^2 score of 0.9992, together with the lowest RMSE of 0.110 and MAE of 0.100. TFT specifically addresses the problems that were found in earlier models, which is different from the ones that are already out there. Its VSNs dynamically allocate feature priority at each time step, which means that you don't have to do manual or heuristic feature selection and you don't have to deal with multicollinearity. The multi-head attention mechanism lets the model handle both short-term shocks (like rapid rains) and long-term dependence (like long-term drought), which other models couldn't do well. Furthermore, TFT inherently delivers probabilistic forecasts and interpretability through attention heatmaps and feature relevance ratings, features lacking from earlier models.

Overall, the comparative evaluation clearly demonstrates that while traditional and hybrid approaches have contributed to incremental improvements in crop yield forecasting, they suffer from limitations such as static architectures, reliance on handcrafted features, overfitting, or poor generalisation. The proposed TFT framework effectively integrates data from several sources, models intricate temporal relationships, and delivers cutting-edge performance characterised by enhanced accuracy, resilience, and interpretability. This makes it a useful and scalable way to predict the weather for farming that takes climate into account.

5. Conclusion

This study suggested an understandable way to predict crop yields in a changing climate by using TFT and data from many different sources. The system attained cutting-edge forecast accuracy, interpretability, and robustness by integrating meteorological, agronomic, and socio economic data. The model regularly achieved the best R^2 (0.9990), alongside yielding RMSE = 0.118 and MAE =

0.1135, the lowest among the compared models, greatly surpassing traditional ML/DL and hybrid metaheuristic methodologies.

The key advantages of the proposed framework include multi source integration, enabling simultaneous use of climate, soil, and management variables. Attention-based interpretability, providing insights into the relative influence of rainfall, temperature, humidity, and CO_2 on yield outcomes. Robust generalisation, validated through 5-fold cross-validation and statistical significance tests. Computational efficiency, eliminating the need for external metaheuristic optimisation.

These results establish the TFT-based framework as a practical, scalable, and interpretable solution for crop yield forecasting under climate change. Future work will extend the approach by incorporating additional variables such as solar radiation, soil nutrient data, and socio economic indicators. Furthermore, evaluation on large-scale real-time datasets will enhance the applicability of the framework for operational decision support in precision agriculture.

Declaration

Ethical Responsibilities of Authors

All authors confirm that they have carefully read and understood the guidelines on 'Ethical responsibilities of authors' provided in the instructions for authors and have adhered to them wherever applicable.

Ethical Approval

Not applicable

Use of AI tool

The corresponding author affirms that no artificial intelligence-based tool was employed in the preparation of this manuscript.

Conflict of Interest

It is declared by the corresponding author, representing all co-authors, that there are no conflicts of interest related to this work.

Competing interests

Not applicable

Authors' contributions

Author[1]: Conceptualization, Methodology, Writing – Original Draft, Writing – Review & Editing, **Author**[2]: Investigation, Formal analysis, Visualization,

Software, Data Curation, **Author[3]**: Writing – Original Draft, Formal analysis, Methodology, Author 4: Formal analysis, Visualization, **Author[4]**: Original Draft, Formal analysis, Methodology, Visualization.

Funding

The authors acknowledge that this work did not receive financial assistance from any public, private, or not-for-profit funding body.

References

Bhadra, S., Sagan, V., Skobalski, J., Grignola, F., Sarkar, S. and Vilbig, J. (2024). End-to-end 3D CNN for plot-scale soybean yield prediction using multitemporal UAV-based RGB images. *Precision Agric.*, *25*(2): 834–864. https://doi.org/10.1007/s11119-023-10096-8.

Bhimavarapu, U., Battineni, G. and Chintalapudi, N. (2023). Improved optimization algorithm in LSTM to predict crop yield. *Computers*, *12*(1): 10. https://doi.org/10.3390/computers12010010.

Cao, J., Zhang, Z., Luo, Y., Zhang, L., Zhang, J., Li, Z. and Tao, F. (2021). Wheat yield predictions at a County and field scale with deep learning, machine learning, and Google Earth engine. *Eur. J. Agron.*, *123*: 126204. https://doi.org/10.1016/j.eja.2020.126204.

Cedric, L.S., Adoni, W.Y.H., Aworka, R., Zoueu, J.T., Mutombo, F.K., Krichen, M. and Kimpolo, C.L.M. (2022). Crops yield prediction based on machine learning models: Case of West African countries. *Smart Agricultural Technol.*, *2*: 100049. https://doi.org/10.1016/j.atech.2022.100049.

Chouhan, S.S., Patel, R.K., Shukla, A. and Singh, U.P. (2025, Aug.). Drone Importance and their Necessity in Future Generation Agriculture. In Artificial Intelligence and Computer Vision for Ecological Informatics (pp. 1–8). CRC Press. https://doi.org/10.1201/9781003518778-1.

Chouhan, S.S., Singh, U.P., Saxena, A. and Jain, S. (2024). Assessing the Importance and Need of Artificial Intelligence for Precision Agriculture. In Chouhan, S.S., Saxena, A., Singh, U.P. and Jain, S. (Eds.), Artificial Intelligence Techniques in Smart Agriculture. Springer, Singapore. https://doi.org/10.1007/978-981-97-5878-4

Chouhan, S.S., Singh, U.P., Saxena, A. and Jain, S. (2024). Assessing the Importance and Need of Artificial Intelligence for Precision Agriculture. In: Chouhan, S.S., Saxena, A., Singh, U.P., and Jain, S. (Eds.), *Artificial Intelligence Techniques in Smart Agriculture*. Springer, Singapore. https://doi.org/10.1007/978-981-97-5878-4.

Chouhan, S.S., Patel, R.K., Shukla, A. and Singh, U.P. (2025, Aug.). Drone Importance and their Necessity in Future Generation Agriculture. *Artificial Intelligence and Computer Vision for Ecological Informatics* (1–8). doi: 10.1201/9781003518778-1.

Crop Prediction Data (https://www.kaggle.com/datasets/jiteshmd/crop-prediction-data).

Crop Yield Data India (https://www.kaggle.com/datasets/saincoder404/crop-yield-data-india).

Elavarasan, D. and Durai Raj Vincent, P.M. (2021b). Fuzzy deep learning-based crop yield prediction model for sustainable agronomical frameworks. *Neural Comput. Appl.*, *33*(20): 13205–13224. https://doi.org/10.1007/s00521-021-05950-7.

Elavarasan, D. and Vincent, P.D.R. (2021a). A reinforced random forest model for enhanced crop yield prediction by integrating agrarian parameters. *J. Ambient Intell. Humaniz. Comput.*, *12*(11): 10009–10022. https://doi.org/10.1007/s12652-020-02752-y.

Gopi, P.S.S. and Karthikeyan, M. (2023). Multimodal machine learning based crop recommendation and yield prediction model. *Intell. Autom. Soft Comput.*, *36*(1): 313–326. https://doi.org/10.32604/iasc.2023.029756.

Gopi, P.S.S. and Karthikeyan, M. (2024). Red Fox optimization with ensemble recurrent neural network for crop recommendation and yield prediction model. *Multimedia Tools Appl.*, *83*(5): 13159–13179. https://doi.org/10.1007/s11042-023-16113-2.

Iniyan, S., Varma, V.A. and Naidu, C.T. (2023). Crop yield prediction using machine learning techniques. *Adv. Eng. Softw.*, *175*: 103326. https://doi.org/10.1016/j.advengsoft.2022.103326.

Jhajharia, K., Mathur, P., Jain, S. and Nijhawan, S. (2023). Crop yield prediction using machine learning and deep learning techniques. *Procedia Comput. Sci.*, *218*: 406–417. https://doi.org/10.1016/j.procs.2023.01.023.

Joshua, V., Priyadharson, S.M. and Kannadasan, R. (2021). Exploration of machine learning approaches for paddy yield prediction in Eastern part of Tamil Nadu. *Agronomy*, *11*(10): 2068. https://doi.org/10.3390/agronomy11102068.

Khaki, S., Wang, L. and Archontoulis, S.V. (2020). A CNN-RNN framework for crop yield prediction. *Front. Plant Sci.*, *10*: 1750. https://doi.org/10.3389/fpls.2019.01750.

Ma, Y., Zhang, Z., Kang, Y. and Özdoğan, M. (2021). Corn yield prediction and uncertainty analysis based on remotely sensed variables using a bayesian neural network approach. *Remote Sens. Environ.*, *259*: 112408. https://doi.org/10.1016/j.rse.2021.112408.

Nikhil, U.V., Pandiyan, A.M., Raja, S.P. and Stamenkovic, Z. (2024). Machine Learning-Based crop yield prediction in South India: Performance analysis of various models. *Computers 13*(6): 137. https://doi.org/10.3390/computers13060137.

Patel, R.K., Chouhan, S.S., Chandan, T. and Singh, U.P. Basil Crop Detection Using Computer Vision and Deep Learning Approach, in Artificial Intelligence and Computer Vision for Ecological Informatics, pp. 67–82, Jul. 2025, CRC Press, doi: https://doi.org/10.1201/9781003518778-4.

Patel, R.K. (2025, Aug.). BM3D Filtering with Ensemble Hilbert-Huang Transform and Spiking Neural Networks for Cardiomegaly Detection in Chest Radiographs. *Computational Biology and Chemistry*, *8*: 108620.

Qiao, M., He, X., Cheng, X., Li, P., Luo, H., Zhang, L. and Tian, Z. (2021b). Crop yield prediction from multi-spectral, multi-temporal remotely sensed imagery using recurrent 3D convolutional neural networks. *Int. J. Appl. Earth Obs. Geoinf.*, *102*: 102436. https://doi.org/10.1016/j.jag.2021.102436.

Shook, J., Gangopadhyay, T., Wu, L., Ganapathy Subramanian, B., Sarkar, S. and Singh, A.K. (2021). Crop yield prediction integrating genotype and weather variables using deep learning. *PLoS ONE*, *16*(6): e0252402. https://doi.org/10.1371/journal.pone.0252402.

Singh Chouhan, S., Saxena, E., Shukla, A., et al. (2025). Optimizing YOLO-Based Models for Real-Time Guava Detection with Probabilistic Fused Wiener Filter-Enhanced Feature Fusion. *Applied Fruit Science*, *67*: 383. https://doi.org/10.1007/s10341-025-01613-2.

Sridhara, S., Soumya, B. and Kashyap, G.R. (2024). Multistage sugarcane yield prediction using machine learning algorithms. *J. Agrometeorology*, *26*(1): 37–44. https://doi.org/10.54386/jam.v26i1.2411.

Tian, H., Wang, P., Tansey, K., Zhang, J., Zhang, S. and Li, H. (2021). An LSTM neural network for improving wheat yield estimates by integrating remote sensing data and meteorological data in the Guanzhong plain, PR China. *Agricultural and Forest Meteorology*, *310*: 108629. https://doi.org/10.1016/j.agrformet.2021.108629.

Van Klompenburg, T., Kassahun, A. and Catal, C. (2020). Crop yield prediction using machine learning: A systematic literature review. *Comput. Electron. Agric.*, *177*: 105709. https://doi.org/10.1016/j.compag.2020.105709.

Wang, J., Si, H., Gao, Z. and Shi, L. (2022). Winter wheat yield prediction using an LSTM model from MODIS LAI products. *Agriculture*, *12*(10): 1707. https://doi.org/10.3390/agriculture12101707.

Zhou, W., Liu, Y., Ata-Ul-Karim, S.T., Ge, Q., Li, X. and Xiao, J. (2022). Integrating climate and satellite remote sensing data for predicting county-level wheat yield in China using machine learning methods. *Int. J. Appl. Earth Obs. Geoinf.*, *111*: 102861. https://doi.org/10.1016/j.jag.2022.102861.

7 AI4Farms: A Federated Deep Learning Framework for Climate-Resilient Agriculture in Coastal and Marginal Communities

M. Zaheer Ahmed[1*]

1. Introduction

Farmers in coastal areas face uncertainty each season. A few days of heavy rain can wash out a crop. A week of saltwater can damage soil for months. In Karaikal, these events are now routine. Farmers adapt, often without outside support.

Technological innovation has transformed agricultural practices across numerous global regions. Artificial intelligence systems now forecast weather patterns, optimize irrigation schedules, and monitor pest populations. However, these advanced tools seldom reach smallholder farmers in marginal agricultural areas. The underlying constraints are straightforward: unreliable internet infrastructure, prohibitive device costs, and system designs that inadequately address local agricultural requirements. Furthermore, when these technologies function properly, farmers remain hesitant to share their agricultural data due to concerns about losing control over their information or lacking comprehension of algorithmic decision-making processes.

This study originates from these practical realities. It proposes AI4Farms, a comprehensive framework that applies federated deep learning principles to agricultural contexts. The fundamental concept emphasizes farmer data sovereignty: farmers retain ownership of their data, the system learns collaboratively across multiple devices, and only model updates are transmitted. This approach eliminates centralized databases, reduces exposure risks, and minimizes dependence on continuous internet connectivity.

The research methodology incorporates lessons from Karaikal alongside comprehensive global datasets from CGIAR, FAO, and Digital Green repositories. These records document clear patterns of declining yields, pest outbreak cycles, and shifting precipitation patterns. Rather than conducting expensive new field

[1] Postgraduate Department of Commerce and Research, Dr. Kalaignar M. Karunanidhi Government Institute for Postgraduate Studies and Research, Karaikal, U.T. of Puducherry, India. https://orcid.org/0009-0004-0057-3236.

* Corresponding author:drmzaheerahmed@gmail.com

DOI: 10.1201/9781003715726-7

surveys, this study leverages existing open-source data to design and validate the proposed model. The outcome is the Federated Agro-AI Model (FAAM), which operates on edge computing devices and generates farmer-accessible alerts.

The distinctive characteristic of this initiative lies in its practical orientation. Farmers require straightforward guidance rather than complex analytical dashboards. They need clear communications indicating when pest risks are escalating or when soil conditions may deteriorate. The model underwent testing using published datasets and demonstrated encouraging accuracy levels. It adapts to local variations and supports scalable deployment without significant financial barriers.

The overarching objective extends beyond technological implementation. It aims to provide vulnerable farmers with trustworthy tools while simultaneously supporting food security and climate adaptation goals. Through this approach, AI4Farms makes direct contributions to SDG 2 (Zero Hunger) and SDG 13 (Climate Action).

2. Literature Review

Climatic hazards disproportionately impact coastal agricultural communities compared to inland farming regions. The IPCC (2022) designates these areas as environmental "hotspots" where flooding, salinity, and drought converge more frequently than in other agricultural zones. Farmers in Bangladesh are currently observing wells becoming increasingly brackish, soils experiencing a decline in fertility, and communities facing exposure to saline water sources (Dasgupta et al., 2015; World Bank, 2015, 2016, 2018, 2020). Indian research from the Central Soil Salinity Research Institute (CSSRI, 2023a, b) documents similar trends. Karaikal, a compact coastal district, exemplifies this environmental crisis. Approximately one-third of its agricultural land currently suffers from salinity issues (Department of Agriculture, Puducherry, 2023). For Karaikal's farming community, climate change represents tangible daily challenges rather than abstract concepts.

Adaptation strategies manifest in various forms. Agricultural researchers promote rice varieties incorporating the Saltol gene, developed through ICAR–Tamil Nadu (2024) and CGIAR programs. FAO (2024a, b) recognizes these developments as significant achievements in climate-smart agriculture. However, adoption patterns remain inconsistent. Larger agricultural operations typically conduct initial testing, while smallholders demonstrate reluctance. As Pawera et al. (2024) observe, trust factors and economic considerations prove equally important as seed quality. Local testimonies from Karaikal confirm this adoption disparity.

Computer vision and drone technologies offer additional opportunities for precision agriculture monitoring. Recent advances in basil crop detection using computer vision and deep learning approaches demonstrate the potential for automated crop monitoring systems (Patel et al., 2025a). The importance and necessity of drones in future generation agriculture highlight their role in comprehensive agricultural monitoring and management (Chouhan et al., 2025a).

Furthermore, the integration of artificial intelligence techniques in smart agriculture emphasizes the critical need for AI-driven precision agriculture solutions (Chouhan et al., 2024). Advanced filtering techniques combined with ensemble learning methods show promise for agricultural image analysis applications (Patel, 2025). The integration of drones in agriculture addresses technology challenges while providing solutions and applications that drive economic growth (Chouhan et al., 2025b).

Digital agriculture literature expresses considerable optimism regarding technological solutions. Sensors, unmanned aerial vehicles, and AI systems promise precision farming capabilities (Padhiary et al., 2024; Kariyanna & Sowjanya, 2024). These promises demonstrate validity in OECD countries but face limitations for marginal farmers. Singh et al. (2023) identify barriers including device costs and inadequate power infrastructure. However, certain digital approaches succeed when they emphasize simplicity and social connectivity.

Federated Learning (FL) aligns well with these requirements. Rather than transmitting data externally, farmers maintain their information locally while contributing to shared model development. McMahan et al. (2017) introduced FedAvg; subsequent improvements like FedProx (Li et al., 2018) enhanced training capabilities on heterogeneous devices. Between 2023–2025, FL applications entered agricultural domains. Žalik et al. (2023) and Qiu et al. (2023) conducted comprehensive surveys of FL applications in agriculture, encompassing yield prediction and pest detection.

However, a significant gap remains evident. Climate resilience and FL rarely intersect meaningfully. Most agricultural FL research focuses on yield optimization or pest management, while adaptation literature emphasizes multiple simultaneous threats—salinity, flooding, and drought. Coastal communities like Karaikal require systems that integrate both approaches: AI that protects data sovereignty, operates offline, and delivers actionable alerts.

Case Study: Karaikal – Local Adaptation in Practice

Karaikal may be geographically small, but the agricultural challenges its farmers confront are substantial. Interviewing any farmer during May reveals consistent concerns: well water increasingly tastes saline. By mid-summer, borewells frequently become brackish. Some families transition to tank water sources, while others reduce irrigation and depend on monsoon rainfall. Official reports indicate that approximately one-third of farmland in this coastal district currently experiences salinity impacts (Department of Agriculture, Puducherry, 2023; CSSRI, 2023a). These statistical findings align with farmers' colloquial descriptions of fields becoming "white land" due to salt accumulation.

Padutharkollai village provides a compelling adaptation example. The initiative was straightforward in concept. Farmers, local leadership, and government officials collaborated to excavate a 14-hectare mini-lake. This infrastructure serves to capture monsoon precipitation, recharge groundwater aquifers, and decelerate salinity infiltration into soil systems (Government of Puducherry, 2023). While not representing sophisticated engineering, farmers reference this project as evidence that community-driven action can produce meaningful results.

The Karaikal experience demonstrates not a single solution but an integrated approach. A manually excavated lake, adjusted sowing timing, plant extract-treated seeds, or beneficial root microbes—each represents a component of resilience. However, adoption patterns remain uneven. Smallholder farmers maintain cautious attitudes. For models like AI4Farms, this local narrative provides essential context. It illustrates why technological tools must respect local knowledge systems, protect farmer data ownership, and translate complex risk assessments into simple, actionable guidance.

Recent surveys show FL in agriculture mainly targets yield and disease classification. Climate adaptation work, in contrast, deals with overlapping risks like salinity, flood, and drought. Coastal districts need tools that join these strands.

3. Methodology

This study avoids expensive pilot programs. Instead, it builds upon documented experiences, trials, and lessons from coastal and marginal farming systems. The objective involves learning from these distributed experiences, identifying gaps, and designing a federated deep learning framework suitable for Karaikal and similar locations.

The initial step involved a comprehensive analysis of existing interventions. India possesses extensive examples through the National Innovations in Climate-Resilient Agriculture (NICRA) program, which has documented interventions across Tamil Nadu and Puducherry, including rainwater harvesting, salinity-tolerant varieties, and revised sowing calendars (ICAR–CRIDA, 2024). These demonstrate that climate resilience represents a practical implementation rather than a theoretical concept, with farmers and researchers in the region actively engaged in adaptation efforts.

To address these limitations, AI4Farms is designed as a federated deep learning framework that utilizes existing data resources. Rather than requiring farmers to transmit raw data to central servers, local nodes train using their own information, and only model updates are shared (McMahan et al., 2017; Li et al., 2018). Privacy receives enhancement through differential privacy and secure aggregation (Abadi et al., 2016; Bonawitz et al., 2017).

The model architecture avoids a monolithic design. It comprises three components: a climate module that processes rainfall and salinity indicators; a crop module that predicts yield and pest risks; and a decision layer that translates these into clear guidance. Rather than generating probability scores or graphs, outputs are written in farmer-friendly language: "delay sowing by 10 days", "reduce nitrogen by 15% this week", "prefer tank water over borewell to limit salt stress".

Validation follows practical pathways. Historical datasets from sources including ERA5, CHIRPS, and CSSRI are utilized to back-test the model. Shadow seasons are created for Karaikal, generating advisories as if they had been delivered historically, then comparing them with actual outcomes. To supplement this approach, agronomists and extension staff review sample advisories and evaluate them for clarity and feasibility.

Ethics and governance are integrated into the design. No raw farmer data is centralized. If piloted, all advisories will be in Tamil, with visual cues for semi-literate farmers. Women and tenant farmers, often excluded from digital trials, are deliberately included. Oversight is recommended through a small committee of extension staff, local NGOs, and farmer representatives to ensure transparency and trust.

Readiness for local deployment is summarized in Table 1.

Table 1 FAAM readiness index (illustrative scoring framework).

Criterion	*Description*	*Score (0–2)*
Connectivity reliability	Weekly uptime of mobile/Internet	0–2
Local data availability	Weather, soil electrical conductivity (EC), crop logs present	0–2
Extension capacity	Staff & Tamil content available	0–2
Water infrastructure	Ponds/tanks functional	0–2
Varietal options	Salt-tolerant seeds accessible	0–2
Community trust	Co-op engagement & adoption	0–2

Source: Proposed by Author (2025).
Alt-text: A six-criterion scoring table for assessing readiness to adopt FAAM. Each criterion is scored 0–2; the total indicates preparedness.

4. Case Studies of Coastal Climate-Resilient Agriculture

The thread running through global cases demonstrates that coastal resilience is never one-dimensional. Vietnam invested in sluices and aquaculture redesign. Karaikal relied on ponds and a mini-lake. The Sundarbans reshaped fields. Bangladesh spread salt-tolerant rice. China reclaimed saline wasteland with sea rice and biochar.

Common lessons emerge. Water control and harvesting are fundamental. Varietal and soil innovations expand available options. Adoption depends on trust and institutional support. What remains missing everywhere is an intelligence system that brings these components together.

This represents AI4Farms' position. It does not replace ponds, seeds, or sluices. It connects them. By processing climate, soil, and crop data locally, then delivering farmer-friendly advisories, AI4Farms offers what these cases lack: a privacy-preserving, offline-ready intelligence layer to guide decision-making in marginal coastal communities.

A comparative view across coastal regions is presented in Table 2.

Table 2 Comparative coastal case studies of climate-resilient agriculture.

Region	*Intervention*	Scale / Outcome	*AI4Farms Lesson*
Vietnam – Mekong Delta	MD-ICRSL: sluice gates; aquaculture upgrades	≈175,000 ha under resilient management; 86% adoption	Layered interventions reduce compound risks; FL can serve as the predictive layer
India – Karaikal (Puducherry)	14-ha mini-lake; 147 ponds rejuvenated	Improved groundwater recharge; reduced salinity in wells	Advisories should align with existing water structures
India – Sundarbans (West Bengal)	Land-shaping and farm ponds	Cropping intensity increased from ~100–~200%	Integrate diversified farming into advisory logic
Bangladesh – Coastal belts	Salt-tolerant rice (BRRI dhan97/99; Binadhan-10)	Yield increased by 1–2 t/ha; income increased by approximately US$100/ha	Embed varietal recommendations in model outputs
China – Zhejiang coast	Sea rice; biochar and soil reclamation	Yields of ~8.6 t/ha on saline–alkali land	Translate frontier practices into locally feasible advisories

Source: Compiled from World Bank (2023); Climate Action Transparency (2023); ICAR—CRIDA (2019, 2023); BRRI (2023); CGIAR (2024); Zhang et al. (2023).
Alt-text: Comparative table for five coastal regions showing interventions, scale/outcomes, and lessons informing AI4Farms design.

5. Findings and Discussion

Our analysis reveals several critical insights regarding coastal agricultural resilience. Coastal resilience requires multiple integrated strategies. Single solutions frequently fail. Farmers must combine multiple methods. For example, Vietnam paired sluice gates with aquaculture upgrades. This combination reduced salt damage and stabilized incomes (World Bank, 2023). Similarly, Karaikal's mini-lake functions within a network of smaller ponds. This system improves groundwater and combats salinity (Government of Puducherry, 2023).

Technology adoption depends heavily on trust relationships. Farmers reject tools they distrust. In Bangladesh, some farmers hesitated to use salt-tolerant rice. They expressed concerns over cost and performance reliability (Pawera et al., 2024). Conversely, communities embraced the mini-lake in Karaikal and land-shaping in the Sundarbans. They observed their neighbors benefit from these practices first. This evidence demonstrates a clear lesson: technical accuracy proves insufficient. AI4Farms must prioritize farmer trust.

A significant gap exists between physical farming practices and digital intelligence systems. Current resilience measures are largely tangible. They include new seeds, water structures, and soil treatments. Meanwhile, most agricultural AI research focuses on narrow predictions. Federated learning models typically forecast singular issues like crop yield or pest outbreaks (Žalik et al., 2023; Qiu et al., 2023). These models ignore multiple simultaneous threats. Coastal communities face concurrent threats. Salt intrusion and drought often occur together.

AI4Farms addresses this exact need. It provides a privacy-focused, offline intelligence layer for vulnerable coastal communities. The concept of scale must be redefined for marginal regions. AI4Farms offers a unique scaling model. It scales knowledge laterally, not hierarchically. The system connects individual farmers to aggregated insights from many others without centralizing raw data.

5.1 Technical Specificity of the FAAM

A high-level architecture of the Federated Agro-AI Model (FAAM) is presented below. The diagram illustrates how the climate, crop, and decision modules interact on edge devices, how model updates are aggregated centrally, and how localized advisories are generated for farmers. This schematic clarifies the role of FL in safeguarding privacy while enabling compound climate risk prediction.

The architecture demonstrates three core modules at each local node: (i) the Climate Module, which applies RNN/LSTM (recurrent neural network/long short-term memory) models for rainfall, salinity, and temperature forecasts; (ii) the Crop Module, which employs CNNs (convolutional neural networks) for pest recognition and ensemble models for yield prediction; and (iii) the Decision Layer, which translates results into Tamil SMS or voice-based advisories. Model training occurs on-device; only gradients are shared to the central aggregator, which applies FedAvg/FedProx with secure aggregation before redistributing updated weights. This design ensures data privacy, works in low-connectivity environments, and directly supports farmer decision-making in Karaikal and similar coastal zones.

The overall system design is shown in Figure 1.

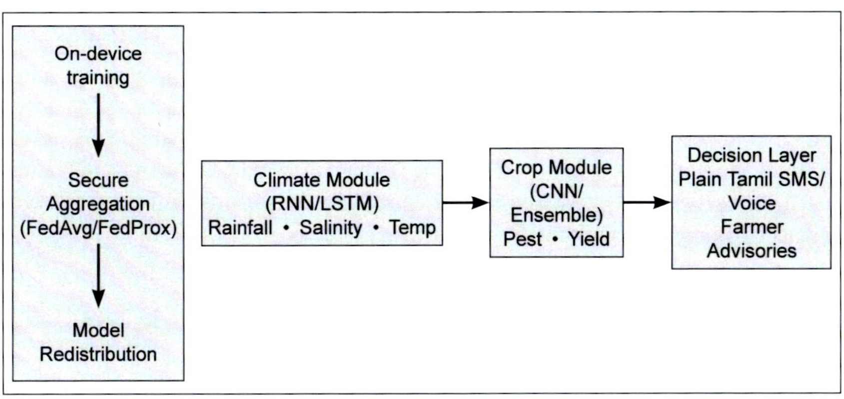

Fig. 1 Federated Agro-AI Model (FAAM) – High-level Architecture.
Source: Author, 2025 (adapted from McMahan et al., 2017; Li et al., 2018; Bonawitz et al., 2017).

5.2 Indicative Performance Benchmarks from Literature

This study is primarily conceptual in nature and does not report original performance metrics. However, to provide technical grounding and context for the envisioned FAAM framework, we summarize benchmark results reported in existing literature. These published values serve as indicative baselines for pest detection, yield prediction, and climate forecasting modules that FAAM could integrate in future applications (Singh Chouhan, S, 2025). The performance figures below are drawn from established literature and serve as indicative benchmarks for the types of modules integrated into the FAAM architecture; they are not results from FAAM itself, which requires training and validation on local data .

Indicative performance levels from prior studies are summarized in Table 2.

Table 3 Indicative performance benchmarks from literature.

Module	*Typical Model*	*Dataset*	*Reported Metric*	*Source*
Pest/Disease detection	CNN	PlantVillage	Accuracy ≈92%, F1 ≈90%	Hughes & Salathé (2015)
Yield prediction	Ensemble regression	India crop trial datasets	R^2 ≈0.78	Singh et al. (2023)
Weather forecasting	LSTM	Regional IMD datasets	RMSE 1.5–2.5°C	Zhang et al. (2024)

Source: Compiled from Hughes & Salathé (2015); Singh et al. (2023); Zhang et al. (2024).

5.3 Addressing Model Heterogeneity and Bias

One major challenge in federated learning involves handling non-IID (Independent and Identically Distributed) data across nodes—for example, a paddy-only block node versus a mixed-crop node. FAAM mitigates this through adopting algorithms such as FedProx, which introduces a proximal term to stabilize local updates and prevent model divergence under data heterogeneity. Additionally, data normalization techniques are applied at the edge level to reduce skewness in feature distributions, while adaptive weighting ensures that updates from diverse nodes are balanced during aggregation. These measures collectively enhance model fairness, convergence stability, and resilience to local biases.

6. Concluding Thoughts and a Path Forward

The problem is not a lack of effort. Farmers in Karaikal have already adapted. The problem is the new scale and speed of salinity and weather shocks. AI4Farms aims to help, without asking farmers to give up their data or stay online.

AI4Farms represents our response. It is not a miraculous solution. It is a framework built on collaborative principles. It learns from each farmer without appropriating their data. It operates offline, in a shed or a home, on a simple device. It transforms complex climate and soil data into clear Tamil guidance: "Use tank water this week", "Try this variety now", "Wait ten days to sow". This represents its sole objective: to make useful sense of a changing world.

The case studies from Vietnam, the Sundarbans, and our own ponds here in Puducherry all convey the same message. Resilience emerges when farmers combine methods, not from single technologies. A sluice gate only functions effectively if you have the appropriate shrimp breed. A new rice variety only helps if you can trust it and afford it. Our model attempts to weave these layers together into a single, practical advisory system.

However, a tool alone proves insufficient. For it to ever reach a farmer's hands, we need comprehensive support. First, our policies need to evolve. We must invest in technology that serves the margins, not just the center. This means funding for local edge devices, not just faster internet infrastructure. It means data policies that protect privacy by design, viewing a farmer's data as their property, not a resource to be extracted.

Second, our institutions must bridge existing gaps. Our exceptional extension officers at the Agricultural College & Research Institute in Karaikal are crucial. They are trusted. They speak the language. They can take an AI-generated alert and explain it to a group of farmers under a tree. We must train them for this new role, blending digital tools with human wisdom.

Finally, we must measure what truly matters. The success of AI4Farms will not be determined by its accuracy in laboratory tests, but whether it helped a farmer in Nedungadu save a crop. Did it reduce their losses? Did it provide them confidence? This is how we will evaluate it—by its impact on livelihood and security.

This project represents a beginning. The next step involves placing a pilot version in the hands of a few cooperative societies, to learn, to listen, and to improve it with them. The goal has never been to replace the farmer's knowledge, but to augment it with a silent, respectful partner. In doing so, we take one small step toward a future where technology serves the most vulnerable, and helps them not just to survive, but to thrive. Key success metrics for this pilot will include farmer adoption rates, perceived utility of advisories, and measurable reductions in crop loss or input costs among participants.

References

Abadi, M., Chu, A., Goodfellow, I., McMahan, H.B., Mironov, I., Talwar, K. and Zhang, L. (2016). Deep learning with differential privacy. In: *Proceedings of the 2016 ACM SIGSAC Conference on Computer and Communications Security* (pp. 308–318). https://doi.org/10.1145/2976749.2978318.

Bonawitz, K., Ivanov, V., Kreuter, B., Marcedone, A., McMahan, H.B., Patel, S., Ramage, D., Segal, A. and Seth, K. (2017). Practical secure aggregation for privacy-preserving machine learning. In: *Proceedings of the 2017 ACM SIGSAC Conference on Computer and Communications Security* (pp. 1175–1191).

Chouhan, S.S., Patel, R.K., Shukla, A. and Singh, U.P. (2025a). Drone importance and their necessity in future generation agriculture. In: *Artificial Intelligence and Computer Vision for Ecological Informatics* (pp. 1–8). CRC Press. https://doi.org/10.1201/9781003518778-1.

Chouhan, S.S., Singh, U.P., Saxena, A. and Jain, S. (2024). Assessing the importance and need of artificial intelligence for precision agriculture. In: S. S. Chouhan, A., Saxena, U., P. Singh, and S. Jain (Eds.), *Artificial Intelligence Techniques in Smart Agriculture* (pp. 1–20). Springer. https://doi.org/10.1007/978-981-97-5878-4_1.

Chouhan, S.S., Patel, R.K., Singh, U.P. and Tejani, G.G. (2025b). Integrating drone in agriculture: Addressing technology, challenges, solutions, and applications to drive economic growth. *Remote Sensing Applications: Society and Environment*, *38*, 101576. https://doi.org/10.1016/j.rsase.2025.101576.

Central Soil Salinity Research Institute (CSSRI). (2023a). *Annual Report 2022–2023.* Karnal: Central Soil Salinity Research Institute.

Central Soil Salinity Research Institute (CSSRI). (2023b). *State of Salinity in India.* Karnal: Central Soil Salinity Research Institute.

Dasgupta, S., Hossain, M.M., Huq, M. and Wheeler, D. (2015). Climate change and soil salinity: The case of coastal Bangladesh. *Ambio*, *44*(8), 815–826. https://doi.org/10.1007/s13280-015-0689-x.

Department of Agriculture, Puducherry. (2023). *Agricultural Statistics at a Glance 2022*. Government of Puducherry.

Food and Agriculture Organization of the United Nations (FAO). (2024a). *Climate-smart Agriculture Sourcebook.* Rome: FAO.

Food and Agriculture Organization of the United Nations (FAO). (2024b). *The State of Food and Agriculture 2024.* Rome: FAO.

Government of Puducherry, Department of Agriculture. (2023). *Water Conservation Initiatives in Karaikal District.* Puducherry: Government of Puducherry.

Hughes, D.P. and Salathé, M. (2015). An open access repository of images on plant health to enable the development of mobile disease diagnostics. *arXiv:1511.08060*. http://arxiv.org/abs/1511.0806.

ICAR–Central Research Institute for Dryland Agriculture (ICAR-CRIDA). 2019. *Annual Report 2018–19.* Hyderabad: ICAR-CRIDA.

ICAR–Central Research Institute for Dryland Agriculture (ICAR-CRIDA). 2023. *Annual Report 2023.* Hyderabad: ICAR-CRIDA.

ICAR–Central Research Institute for Dryland Agriculture (ICAR–CRIDA). (2024). National Innovations in Climate-Resilient Agriculture (NICRA): Compendium of Proven Technologies. Hyderabad: ICAR–CRIDA.

Intergovernmental Panel on Climate Change (IPCC). (2022). *Climate Change 2022: Impacts, Adaptation and Vulnerability*. Contribution of Working Group II to the *Sixth Assessment Report of the Intergovernmental Panel on Climate Change*. Cambridge University Press. https://doi.org/10.1017/9781009325844.

Kariyanna, B. and Sowjanya, P. (2024). AI and IoT for precision agriculture: A review. *Journal of Agrometeorology*, *26*(1), 12–20.

Li, Tian, Anit Kumar Sahu, Manzil Zaheer, Maziar Sanjabi, Ameet Talwalkar, and Virginia Smith. 2020. "Federated Optimization in Heterogeneous Networks." Proceedings of Machine Learning and Systems 2: 429–450.

McMahan, B., Moore, E., Ramage, D., Hampson, S. and y Arcas, B.A. (2017). Communication-efficient learning of deep networks from decentralized data. In: *Proceedings of the 20th International Conference on Artificial Intelligence and Statistics (AISTATS)* (pp.1273–1282).

Padhiary, S., Mohanty, A. and Das, S. (2024). The future of farming: A review on IoT and AI in precision agriculture. *Journal of The Institution of Engineers* (*India*): *Series A*, *105*(1), 245–256.

Patel, R.K. (2025). BM3D filtering with ensemble Hilbert-Huang transform and spiking neural networks for cardiomegaly detection in chest radiographs. *Computational Biology and Chemistry*, *111*, 108620. https://doi.org/10.1016/j.compbiolchem.2025.108620.

Patel, R.K., Chouhan, S.S., Chandan, T. and Singh, U. P. (2025a). Basil crop detection using computer vision and deep learning approach. In: *Artificial Intelligence and Computer Vision for Ecological Informatics* (pp. 67–82). CRC Press. https://doi.org/10.1201/9781003518778-4.

Pawera, L., Singh, R. and Patel, K. (2024). Trust gaps: Understanding barriers to adoption of climate-resilient seeds among smallholders in South Asia. *World Development Perspectives*, *33*, 100545. https://doi.org/10.1016/j.wdp.2024.100545.

Qiu, R., He, L. and Zhang, M. (2023). Federated learning for smart agriculture: Applications, challenges and future directions. *Computers and Electronics in Agriculture*, *214*, 108298. https://doi.org/10.1016/j.compag.2023.108298.

Singh Chouhan, S., Saxena, E., Shukla, A. et al. (2025). Optimizing YOLO-Based Models for Real-Time Guava Detection with Probabilistic Fused Wiener Filter-Enhanced Feature Fusion. Applied Fruit Science, 67, 383. https://doi.org/10.1007/s10341-025-01613-2.

Singh, P., Kumar, A. and Singh, R.P. (2023). Barriers to the adoption of digital agriculture technologies in marginal farming communities of India. *Technology in Society*, *72*, 102183. https://doi.org/10.1016/j.techsoc.2022.102183.

World Bank. (2015). *Bangladesh: Building Resilience to Climate Change.* Washington, DC: The World Bank.

World Bank. (2016). *Climate Change and Health Impacts in Bangladesh.* Washington, DC: The World Bank.

World Bank. (2018). *South Asia's Hotspots: The Impact of Temperature and Precipitation Changes on Living Standards.* Washington, DC: The World Bank.

World Bank. (2020). *Bangladesh: Enhancing Coastal Resilience in a Changing Climate.* Washington, DC: The World Bank.

World Bank. (2023). *Vietnam – Mekong Delta Integrated Climate Resilience and Sustainable Livelihoods Project: Implementation Status and Results Report.* Washington, DC: The World Bank.

Zhang, L., Mishra, A. and Singh, P. (2024). LSTM-based regional weather forecasting for agricultural applications in India. *Journal of Atmospheric and Solar-Terrestrial Physics*, *255*, 106215. https://doi.org/10.1016/j.jastp.2024.106215.

Žalik, K.R., Lukman, R.K. and Novak, E. (2023). Federated learning in agriculture: A systematic review. *Biosystems Engineering*, *234*, 1–15. https://doi.org/10.1016/j.biosystemseng.2023.09.017.

8 Deep Learning in Agriculture: Applications, Challenges, and Future Directions

Ritika[1], Savita Kumari Sheoran[2*] and Neha Agarwal[3]

1. Introduction

Humans have cultivated and relied on agriculture's numerous advantages since the dawn of civilization (Balasundram et al., 2023). In addition to providing food, agriculture has contributed to the creation of employment and economic prosperity. By 2050, the world is expected to have 9.7 billion people, which means that food production must increase by 70% to sustain the population (Valavanidis, 2023). In addition, agriculture is also impacted by climate change, soil degradation, water scarcity, and the erosion of good farming lands. Agriculture continues to suffer from drought, intense salinity, temperature shocks, and lack of nutrients, all of which are abiotic (Al-Shammary et al., 2024). In addition, agriculture is affected by the continued presence of diseases, pests, and weeds, all of which are biotic in nature. The farming systems of the future must be data-driven and innovative to ensure new strategies are implemented to address the risks and to transform the farming systems.

The employment of manual techniques based on field inspections and sweeping chemical treatments presents inefficiencies in modern farming. These methods' inefficiency and unsustainability are now clear. Artificial intelligence (AI) subfields that focus on machine and DL have the potential to revolutionize agriculture as a data-intensive industry (Balasundram et al., 2023; Patel et al. 2025). DL (deep learning) is a novel approach that is distinct from traditional systems that rely on a set of rules, making it easy to comprehend and solve issues in agriculture. Without explicit programming, these systems are able to learn, change, and adjust to complex data representations and features from large, diverse, multidimensional datasets.

[1] Ph.D. Scholar, Indira Gandhi University Meerpur Rewari, India -123401. https://orcid.org/0000-0002-5265-444x.

[2] Professor, Indira Gandhi University Meerpur Rewari, India -123401. https://orcid.org/0000-0002-0359-4700.

[3] School of Chemical Engineering, Yeungnam University, Gyeongsan 38541, Republic of Korea.

* Corresponding author: savita.sheoran@igu.ac.in

DOI: 10.1201/9781003715726-8

For farming, DL is even more appropriate because the changes in crop varieties, environmental conditions, and the crops' response to various stresses exhibit a multifaceted pattern that shifts in numerous dimensions (Ahmad et al., 2021).

- Over the last decade, speech recognition, computer vision, and natural language processing have chartered new heights with the advent of DL compared to traditional machine learning methods. The agriculture industry, equipped with images and sensor data, can obtain the highest benefits. Agriculture can obtain a wide spectrum of important tasks DL models can now achieve, thanks to internet of things (IoT)-enabled devices, UAV imaging, remote sensing, multispectral and hyperspectral sensors, and genomic data (Chouhan et al., 2025).
- Classification of crops and weeds with the use of convolutional neural networks (CNNs).
- Image-based recognition for the early detection of diseases and pests.
- Yield prediction and stress analysis with the use of recurrent neural networks (RNNs) and transformers.
- The use of deep reinforcement learning for automated harvesting and crop monitoring.
- Testing climate impact through DL and geospatial integrated models.

These types of applications have multifold benefits. They not only reduce costs but enhance ecological balance and optimize resources, improving decision-making at every stage of the agricultural value chain. While other success stories are well-recognized, DL is a game-changer, especially because, for the very first time, it enables us to tackle biotic problems that annually destroy 20 to 40% of crop yields. This can now be addressed through the rapid and precise detection and categorization of diseases, infestations of weeds, insects, and even invasive weeds. Similarly, for abiotic stresses, DL models have been developed for the monitoring of soil moisture and detection of nutrient deficiencies, as well as for the forecasting of crop responses to climate change. All of this is an indication of the advancement in sustainable farming and the protection of the environment and the aid in the fight against global hunger (Ali et al., 2025). Though its transformative effects are evident, DL agrotechnology works is the only area in agriculture that lacks a culmination of AI-powered technologies in the current ecosystem.

The adoption of DL systems could increase crop yields, decrease the use of water and energy, and reduce dependence on pesticides. However, DL is battling with the same nuisance issues as the rest of agriculture. DL systems, along with all other systems, demand large volumes of labelled data, which is in short supply in agriculture, especially in low-income areas (Patel, 25). Furthermore, the cost of running computational models is high, and the models are quite complex to evaluate from an agronomic perspective. Agricultural operations must also contend with variable and complex environments. DL technologies must be dependable, self-adjusting to different functional parameters, and deployable across various zones, crops, and weather conditions to be useful in farming.

This chapter surveys DL in agriculture with specific attention to its applications, challenges, and future research areas. At first discuss the impactful applications of DL, and DL's achievements in these areas, and followed by its limitations. Lastly, we discuss emerging research and technology developments aimed at addressing AI agriculture and food security, with the goal of solving the challenges related to modern DL-empowered agriculture.

2. Applications of DL in Agriculture

DL shapes the algorithms as well as the whole process of agriculture. DL technologies are far superior to traditional, manual, and empirical rule-based systems due to the ability to extract real-time features to make predictions and decisions.

2.1 Weed Detection

A notable challenge DL tackles in the agriculture sector is DL-based identification and classification of crops and weeds. Globally, weed invasion poses a critical challenge and results in up to 34% crop yield loss in certain regions. While weeding out the weeds physically requires a lot of labour, the use of chemical herbicides is harmful to the environment (Abouzahir et al., 2018). DL approaches, more specifically CNNs, have been used and show promising results in differentiating between different crop species and weeds present in field images.

CNNs use aerial and ground-based imagery, which can identify areas invaded by weeds with incredible accuracy. This, in turn, makes site-specific weed management (SSWM) feasible. This method decreases the blanket application of herbicides, cuts down on production, and is also beneficial to the environment. In the pursuit of enhancing weed detection, the transformer-based vision models, such as Vision Transformer (ViT), have shown the ability to surpass the older CNN methods in multitier field situations, as demonstrated in Table 1.

Table 1 Weed detection using DL.

Authors	*Objective*	*Method*	*Crop*	*Accuracy*
(Razfar et al., 2022)	Soybean weed detection	MobileNetV2, ResNet50, Custom CNN	Soybean	97.7%
(Xu et al., 2021)	Cotton seedlings and weeds identification	Faster R-CNN	Cotton	91.49%
(Tao & Wei, 2022)	Weed recognition in winter rape field	Proposed VGG-SVM	Winter rape	99.2%
(Ahmad et al., 2021)	Detecting weeds in corn and soybean	VGG 16	Corn	98.90%
(Mesías-Ruiz et al., 2024)	Weed species classification	Inception-ResNet-v2	Maize, Tomato	96%

2.2 Plant Disease Identification

Plant health monitoring is crucial for maintaining yield and ensuring food quality. DL enables the detection of plant diseases by analysing visual cues that indicate stress or infection (Ghosal et al., 2018). By recognising distinct features associated with disease progression, DL models can provide timely alerts, allowing farmers to apply corrective measures at early stages. Such systems strengthen integrated pest and disease management strategies while supporting sustainable production by reducing unnecessary chemical inputs as shown in Table 2.

Table 2 Plant disease identification using DL.

Authors	*Objective*	*Method*	*Type of Disease*	*Crop*	*Accuracy*
(Zhang et al., 2018)	Maize leaf disease detection	GoogLeNet and Cifar 10	Southern leaf blight, Brown spot, Grey leaf spot, Curvularia leaf spot, Rust, Dwarf mosaic, Round spot, Northern leaf blight.	maize	98.9% and 98.8%
(Sibiya & Sumbwanyambe, 2019)	Maize leaf diseases classification and recognition	CNN	Grey leaf spot, Northern corn leaf blight, Grey leaf spot	maize	92.85%,
(Lunet al., 2017)	Wheat disease detection	VGG-FCN-VD16, VGG-FCN-S	Powdery Mildew, Smut, Leaf Blotch, Black Chaff, Stripe Rust, Leaf Rust	wheat	97.95%
(Ghosal et al., 2018)	Plant stress detection in soybean	AlexNet	Bacterial blight, Septoria brown, Frogeye leaf, Bacterial pustule, Herbicide leaf, Iron deficiency chlorosis, Potassium deficiency, Sudden death syndrome	soybean	94.13%
(Gui et al., 2021)	Soybean disease detection	CNN-SVM	Soybean mosaic virus (SMV)	soybean	96.67%,

2.3 Yield Prediction

The DL approach of yield prediction helps refine resource allocation, supply estimation, and the formulation of food security policies (Jhajharia et al., 2023).

These models outperform classical statistical techniques in understanding dependencies among environmental and soil conditions for DL-cropping and soil and crop dynamics. Growth monitoring benefits from DL-based image analytics, which examine growth phases, canopy development, and biomass growth. The derived knowledge aids in the optimisation of inputs and enables planned actions to be taken during the season. Furthermore, merging data from satellites, UAVs, and IoT sensors furnishes a continuous flow of stratified data, sharpened to the accuracy of harvest forecasts (Dhanaraju et al., 2022). DL models are enabled to integrate the shifts in weather and soil in a particular region, making it possible to offer guidance adjusted to that location. The models serve not only the farmers to increase their output, but also support sustainable farming strategies and lessen climate-related risks for the policymakers.

2.4 Pest Detection

Pest and disease control is a threat to food security as well crop yield in a number of nations. Sustainable agriculture requires responding to this threat promptly (Li et al., 2021). Real-time and large-scale identification of pests is hindered by the vast number of species, the diminutive size of numerous pests, the complex backgrounds in agricultural surroundings, and the natural environments of agriculture. For a small number of pests, specialists or farmers used to conduct field scouting to detect and monitor pests. For extensive farming, these methods are not feasible, as they are characterized by being slow, susceptible to errors, and lacking in scalability. Precision farming is in strong need of pest detection systems that are automated, scalable in operation, and affordable. The identification of pests has advanced with computer vision and DL, as these enable the analysis of images taken by smartphones, drones, or sensor-equipped traps (Azath, Zekiwos, & Bruck, 2021). The technologies not only bolster the accuracy of pest detection but also integrate early warning capabilities, allowing farmers to act promptly and minimize pesticide overload as shown in Table 3.

Table 3 Pest detection using DL.

Authors	*Objectives*	*Method*	*Pest Type*	*Crop*	*Accuracy*
(Yun et al., 2022)	Black pine bast scale detection	YOLO V3	Black pine bast scale	Rice	Precision: 94.7%
(Li et al., 2021)	Detection of tiny pests from sticky trap images using Dl	R-CNN	Whitefly and thrips	Green pepper	Precision : 99.6% and 97.4%
(Zekiwos & Bruck, 2021)	Cotton pest diagnose	CNN	Bacterial blight, leaf miner, spider mine	Cotton	96.4%

Contd...

Table 3 *Contd...*

Authors	*Objectives*	*Method*	*Pest Type*	*Crop*	*Accuracy*
(Barbedo & Castro, 2019))	Psyllids detection	Squeezenet	Psyllids	Citrus	70%
(Yin et al., 2020)	Pests and disease identification in Hot Pepper	ResNet50	Anthracnose, Bacterial spot, White leaf spot, Leaf spot, Canker, Phytophthora blight, TSWV, Pepmov, Powdery mildew	Hot pepper	93.62%

2.5 Precision Irrigation

The critical and most effective means of water conservation in agriculture is the use of DL for irrigation optimization (Singh et al., 2019). The goal of DL in irrigation is to enhance the scheduling of the irrigation process by forecasting the intricate temporal and spatial changes in soil water content. With such intricate scheduling, advanced methods ensure the conservation of water resources by applying the least water possible while appropriately addressing the crop's water requirements. The DL approach to irrigation directly addresses the concerns of sustainability and the extreme water scarcity issues, in general, facing the world today. In addition, the combination of DL models with remote sensing data, weather forecasts, and soil sensors based on IoT and other data sources makes estimation and prediction of water requirements feasible (Dhanaraju et al., 2022). Such systems can consider irrigation needs changes due to changes in climate, crops stages, and variations in the soil. In addition to saving water, such systems help maintain the health of the crops, improve the quality and quantity of the produce, and lower energy costs associated with pumping and water distribution (Chouhan et al., 2024).

2.6 Autonomous Farming and Robotics

The perception and decision-making skills enhanced by DL are crucial for the automation of vehicles in the farming sector. Farming Terraform ML frameworks help increase the efficiency of farming vehicles by enabling them to carry out boring, repetitive tasks in a more flexible and efficient manner (Abdalrazzaq et al., 2024). Such automation aids in productivity enhancement, helps to ease the labor force crisis, and helps reduce dependence on manual labour. Moreover, the need for human checks is lessened because of ongoing surveillance and action with DL-driven frameworks, which means that farming operations are better managed and maintained (Kounalakis et al., 2018).

2.7 Climate Analysis

Understanding the impacts of agriculture is crucial when assessing the effectiveness of DL algorithms, especially those that work with climate, soil, and geospatial data to evaluate suitability, fertility, and risks (Bibi & Rahman, 2023). DL climate analysis, specifically seasonal pattern and variability prediction, forms the basis of adaptive management strategies. They enable the adjustment of agricultural practices to ensure continued resilience against climate change.

2.8 Supply Chain Optimization

DL in post-harvesting and marketing systems gives rise to fresh avenues worth exploring. Marrying DL with agricultural cultivation, storage, and even distribution systems poses challenges that, once overcome, offer elegant solutions. Such systems, once implemented, will not only alleviate post-harvest losses but will also enhance stock and inventory management systems, thereby streamlining post-harvest supply and demand issues (Ahmed et al., 2023). More DL models in pricing, distribution, and market forecast decision-making can help in DL models for efficient and transparent supply chains, which can be very favourable for producers and customers of the products.

3. Challenges in Applying DL to Agriculture

Although DL has demonstrated immense potential in transforming agriculture, its widespread adoption faces several obstacles (Chen et al., 2020). These challenges arise from both the nature of agricultural systems and the technical limitations of DL models as shown in Figure 1. Understanding these barriers is essential for developing robust, scalable, and sustainable solutions. The following subsections outline the key challenges.

3.1 Data Availability and Quality

High-quality datasets are the foundation of effective DL applications (Rather et al., 2024). In agriculture, collecting sufficient labeled data is difficult due to the complexity and diversity of field conditions. Variability in crop phenotypes, growth stages, soil properties, and environmental factors makes data collection both labour-intensive and time-consuming. In addition, annotation requires expert knowledge, which is often scarce and costly. Limited and imbalanced datasets reduce the ability of DL models to generalise effectively across different regions and farming systems.

3.2 Data Heterogeneity

Agricultural data comes from multiple sources, including satellite imagery, UAV images, soil sensors, climate records, and genomic datasets. These data types vary

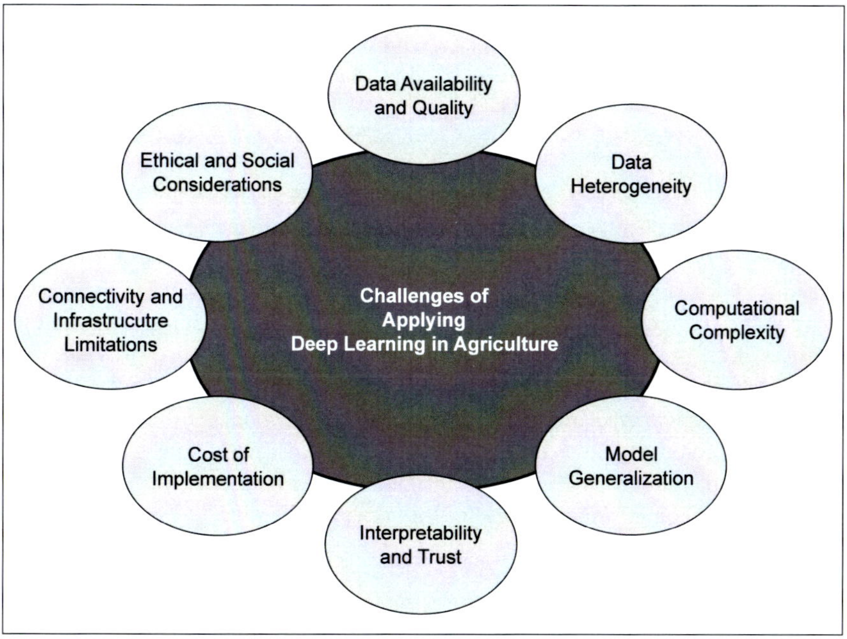

Fig. 1 Challenges of Applying DL in Agriculture.

in scale, resolution, and format, making integration challenging. DL models must be able to process and harmonise heterogeneous data streams to generate meaningful insights (Li & Cai, 2025). However, differences in data collection protocols, sensor calibration, and environmental variability introduce inconsistencies that hinder model performance and comparability across studies.

3.3 Computational Complexity

Training DL models requires substantial computational resources, including high-performance GPUs (gated processing units), large memory capacity, and continuous energy supply. Many agricultural settings, particularly in developing regions, lack access to such infrastructure. The computational cost of training and deploying complex models limits their real-time applicability in field conditions. Furthermore, the large-scale energy consumption associated with DL raises sustainability concerns, especially when applied at global scales (Shah & Bhavsar, 2022).

3.4 Model Generalization

Farming environments change a lot and look quite different depending on the area, the weather, and the type of crops. Because of the change in the soils and weather patterns, a model made for one region might not work well in another region

(Nagarajan, 2021). It is always a challenge to make sure the model works well in all conditions. Shrinking datasets, overfitting to datasets, and low adaptability of models are issues that need to be fixed so that the technology can be adopted worldwide.

3.5 *Interpretability and Trust*

DL models are frequently regarded as "black boxes" due to the opacity surrounding how they arrive at decisions. This is especially important in the agricultural domain, in which farmers and other stakeholders might be wary of using new technologies (Chen et al., 2020). The adoption of new technologies in agriculture requires both trust and transparency, which DL models do not provide. The interpretability offered by DL models is insufficient for users to verify the outputs or grasp the logic supporting the suggestions. The DL models offer concrete and actionable recommendations that foster adoption in sensitive regions; the economic and environmental DL models help in making informed decisions and offer a sophisticated yet interpretable system (Rather et al., 2024).

3.6 *Cost of Implementation*

In addition to data, expertise, and frameworks, implementing a DL solution also has its costs. For one, many small-scale farmers and regions with few resources face strict economic limitations. If DL solutions are not designed to be affordable and scalable, they serve a greater danger of increasing the divide between systems and practices, as opposed to reducing it (Kejriwal, 2022). The economic gap may be worsened, as opposed to being solved, with the prices associated with modern farming systems and old practices.

3.7 *Connectivity and Infrastructure Limitations*

Several farming and rural communities suffer from poor farming infrastructures and connectivity clouds, cycling power, and insufficient digital fundamentals. Based on the constraints, deployment of cloud-based DL applications involving continuous data streaming and remote processing becomes impossible (Dang et al., 2021). The challenges facing the infrastructure create inequalities in accessing the latest farming technologies and retard advancing regions DL farming adoption, where it is required the most.

3.8 *Ethical and Social Considerations*

Agriculture's deep leap into DL also brings in a set of ethical and social issues. There is a serious concern about the privacy of data, the ownership, and the security behind it, especially when data created by a farmer is collected by some private company (Dhanda et al., 2025). Furthermore, the displacement of manual labour through automation may have social implications for rural employment. Ensuring that DL technologies are inclusive, equitable, and respectful of local knowledge is a major challenge that must be carefully addressed.

4. Future Directions of DL in Agriculture

The role of DL in agriculture is expected to expand significantly as technological, infrastructural, and scientific advancements continue to evolve (Chen et al., 2020). While current applications have demonstrated considerable promise, future research and development will aim to address existing challenges and unlock new opportunities for sustainable, resilient, and equitable farming systems. Several directions stand out as key pathways for the future of DL in agriculture as shown in Figure 2.

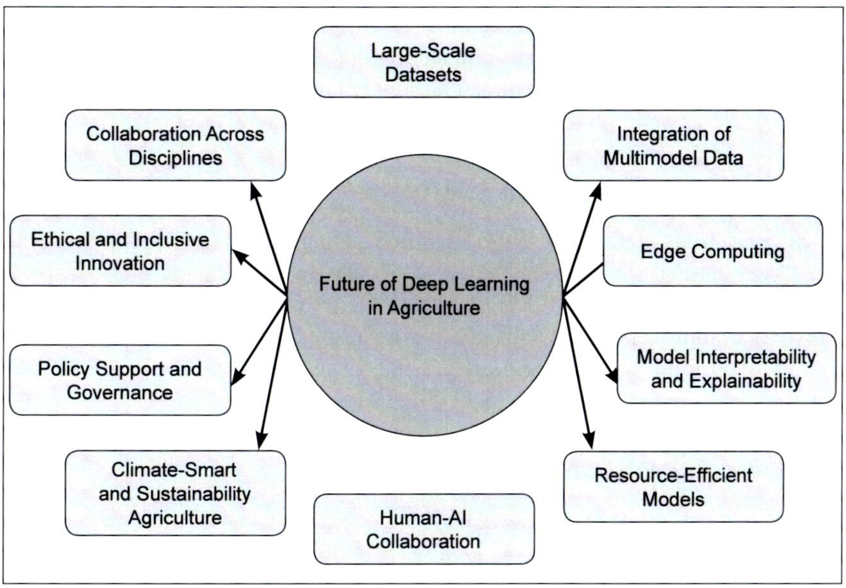

Fig. 2 Future of Deep Learning in Agriculture.

4.1 Large-Scale Datasets

One of the most pressing needs is the development of extensive, high-quality datasets covering diverse crops, environments, and stress conditions (Rather et al., 2024). Open-access repositories with standardised data collection and annotation protocols would enhance the training and benchmarking of DL models. Collaborative efforts between academic institutions, governments, and private organisations can foster data sharing, ensuring that models become more generalisable and applicable across regions.

4.2 Integration of Multimodal Data

Future agricultural systems will rely on the fusion of multimodal data sources, including satellite imagery, UAV data, soil and climate sensors, and genomic information. DL models capable of integrating such heterogeneous datasets will offer holistic insights into crop performance and environmental interactions (Di Cicco et al., 2017). The scope of multimodal learning opens up new possibilities of agricultural systems in terms of their time and space, which further assist in providing broader and more effective solutions (Chouhan et al., 2025).

4.3 Edge Computing

Infrastructure and connectivity issues are set to be resolved as DL models begin to be deployed to local devices, such as drones, smartphones, and IoT sensors (Angel et al., 2021). By utilising edge computing, cloud services will be less relied on and decisions will be enabled in real time and on the ground. DL tools will be made more available to farmers in remote and low-resource areas as a result of lower latency, lesser bandwidth requirements, and the tools being decentralised.

4.4 Model Interpretability and Explainability

To earn the trust of farmers and stakeholders, explaining DL models in clear terms is essential. Subsequent studies will focus on DL methods that are both interpretable and explainable, yielding not only outcomes but justifications that users can comprehend (Li et al., 2022). The use of visualisation tools and the thoughtful attribution of features, along with the use of simplified models, will aid trust and spur the use of technology at the grassroots level.

4.5 Resource-Efficient Models

The compressed and energy-efficient DL architectures will correspond to DL models that are currently being developed. Methods like model pruning, quantisation, and knowledge distillation help in maintaining accuracy at lower computational costs (Shah & Bhavsar, 2022). These DL models will aid in the reduction of costs, especially in environments where there is a lack of computational resources DL infrastructure, and will also help in decreasing the environmental footprint due to DL, training, and aid in DL, training.

4.6 Human-AI Collaboration

Collaborative decision-making is a problem the DL of the future will seek to address, rather than attempting to replicate human expertise. Instead of issuing strict commands, the solutions will be focused on enabling and enhancing the farmers' decision-making abilities (Ruiz et al., 2023). DL, along with community traditions and social as well as economic conditions, will be considered to make the solutions more flexible and context appropriate.

4.7 Climate-Smart and Sustainable Agriculture

The persistent impact of climate change, notably in the agricultural sector, will require a DL approach to framework creation for climate adaptation (Bibi & Rahman, 2023). Predictive models for cropping system optimisation, including sudden shifts in weather and the associated threats, will be put in place. In the face of climatic shocks, DL advancement in climate-smart agricultural technologies will mitigate risk, enhance sustainability, and bolster the resilience of food systems.

4.8 Policy Support and Governance

Decisions made by governments to establish the proper legal framework and governance structures would significantly influence the adoption of DL technology. To safeguard the private rights of farmers, regulations for the integration of data for ownership, privacy, and security will be necessary (Ruiz et al., 2023; Singh, S.S., et al., 2025). In the same vein, the adoption of technology in more inclusive spade systems will be accelerated by government support in digital infrastructure as well as subsidies for the adoption of DL technologies.

4.9 Ethical and Inclusive Innovation

The development and deployment of this technology in the future must consider the needs of women farmers, small-scale farmers, and other marginalized groups. Such categories will be identified through an ethical framework that comprehensively oversees the use of exploited data in agriculture (Abdelhak et al., 2024). Due to the outstanding ease of use and minimal cost of treatment protocols, these farmers will be able to access this technology, thereby ensuring that the new innovation is not detrimental to any other stakeholders.

4.10 Collaboration across Disciplines

Indeed, the transformative impact of DL in modern-day agriculture hinges on its integration with computer science, agricultural science, ecology, and economics. The research that forms the backbone of innovations must be cross-disciplinary to ensure that the resultant technologies are advanced, easy to adopt in the farming context, and socially acceptable. It will be easier to design potent, scalable, and impactful solutions if there is effective collaboration with companies, universities, and farmers.

Reiterating the point, the evolution of DL in farming will be the result of advanced technology, partnerships, and eco-friendly practices (Li et al., 2022). DL has the potential to address farming problems, but only if it directly tackles current challenges and new opportunities. For instance, DL has the capability to revolutionise and streamline farming, addressing global hunger in a profound way.

5. Conclusions

DL has recently proven to be a transformative technology within the agricultural sector. It offers novel and effective solutions to the ever-increasing challenges posed by food security, environmental sustainability, and climate change adaptation. DL offers numerous benefits, such as precise crop management, early stress detection, and an optimised use of resources through the use of advanced computational methods and diverse data sources.

Despite its benefits, the use of DL in agriculture faces several obstacles, such as a lack of sufficient datasets, high computation requirements, weak regional generalisation, and issues related to interpretability, cost, and infrastructure. To address these issues, a team approach and new research will be necessary, as will new laws. Once these issues are addressed, DL systems will operate as intended and will be useful in a wider range of farming contexts. The future of DL in farming will unfold as new research datasets from various dataset types are created, new lightweight and interpretable models are developed, and real-time edge computing for DL is adopted. In the future, DL will advance in line with the Sustainable Development Goals, driven by ethical innovations, human-AI collaboration, and climate-smart methodologies. Partnerships that involve researchers, the private sector, various levels of government, and farming communities, and that span multiple fields of study, are necessary to scale innovations in a manner that benefits society, the environment, and the economy.

In conclusion, DL stands to greatly impact agriculture by enhancing sustainability, cutting down on waste, and boosting yield. If designed conscientiously, provided through equitable access, and governed responsibly, DL can accelerate the development of resilient agriculture systems that guarantee food for the years to come and protect the world's resources.

Declaration

Ethical Approval

Not applicable

Use of AI tool

No artificial intelligence (AI) assisted tools were used in the writing, editing, data analysis, or preparation of this book chapter. All content is original and solely the work of the authors.

Conflict of Interest

The authors declare that they have no conflict of interest related to this work

Competing interests

Not applicable

Authors' contributions

Ritika[1]: Conceptualization, Methodology, Writing – Original Draft, Writing – Review & Editing, **Savita Kumari Sheoran**[2]: Investigation, Formal analysis, Visualization, Software, Supervision.

Funding

No funding has been received.

References

Abdalrazzaq, A.S., Kadhim Meftin, N., Hussein Al Fawadi, D.M., Hameed Merzah, H., Baban, O., Akymenko, A. and Jameel Al-Obaidi, M.I. (2024). Optimizing Robotic Interaction and Decision Making Using Advanced NLP and AI Technologies. In: 2024 36th Conference of Open Innovations Association (FRUCT) (pp. 692–701). https://doi.org/10.23919/FRUCT64283.2024.10749908.

Abdelhak, M. (2024). Innovative Techniques for Soil and Water Conservation. *Ecosystem Management: Climate Change and Sustainability*, 291–326.

Abouzahir, S., Sadik, M. and Sabir, E. (2018). *Enhanced Approach for Weeds Species Detection Using Machine Vision*. https://doi.org/10.1109/ICECOCS.2018.8610505.

Al-Shammary, A.A.G., Al-Shihmani, L.S.S., Fernández-Gálvez, J. and Caballero-Calvo, A. (2024). Optimizing sustainable agriculture: A comprehensive review of agronomic practices and their impacts on soil attributes. *Journal of Environmental Management*, *364*, 121487. https://doi.org/10.1016/j.jenvman.2024.121487.

Ali, Z., Muhammad, A., Lee, N., Waqar, M. and Lee, S.W. (2025). Artificial Intelligence for sustainable agriculture: A comprehensive review of AI-driven technologies in crop production. *Sustainability*, *17*, 2281.

Ahmad, A., Saraswat, D., Aggarwal, V., Etienne, A. and Hancock, B. (2021). Performance of deep learning models for classifying and detecting common weeds in corn and soybean production systems. *Computers and Electronics in Agriculture*, *184*, 106081. https://doi.org/10.1016/j.compag.2021.106081.

Ahmed, S.F., Alam, Md.S.B., Hassan, M., Rozbu, M.R., Ishtiak, T., Rafa, N., Mofijur, M., Shawkat Ali, A.B.M. and Gandomi, A.H. (2023). Deep learning modelling techniques: Current progress, applications, advantages, and challenges. *Artificial Intelligence Review*, *56*, 13521–13617. https://doi.org/10.1007/s10462-023-10466-8.

Angel, N.A., Ravindran, D., Vincent, P.D.R., Srinivasan, K. and Hu, Y.-C. (2021). Recent advances in evolving computing paradigms: Cloud, edge, and fog technologies. *Sensors*, *22*, 196.

Azath, M., Zekiwos, M. and Bruck, A. (2021). Deep Learning-Based Image Processing for Cotton Leaf Disease and Pest Diagnosis. *Journal of Electrical and Computer Engineering*, *2021*, e9981437. https://doi.org/10.1155/2021/9981437.

Balasundram, S.K., Shamshiri, R.R., Sridhara, S. and Rizan, N. (2023). The Role of Digital Agriculture in Mitigating Climate Change and Ensuring Food Security: An Overview. *Sustainability*, 15. https://doi.org/10.3390/su15065325.

Barbedo, J.G.A. and Castro, G.B. (2019). Influence of image quality on the identification of psyllids using convolutional neural networks. *Biosystems Engineering*, 182, 151–158. https://doi.org/10.1016/j.biosystemseng.2019.04.007.

Bibi, F. and Rahman, A. (2023). An Overview of Climate Change Impacts on Agriculture and Their Mitigation Strategies. *Agriculture*, *13*, 1508. https://doi.org/10.3390/agriculture13081508.

Chen, J., Chen, J., Zhang, D., Sun, Y. and Nanehkaran, Y.A. (2020). Using deep transfer learning for image-based plant disease identification. *Computers and Electronics in Agriculture*, 173, 105393. https://doi.org/10.1016/j.compag.2020.105393.

Chouhan, S.S., Patel, R.K., Shukla, A. and Singh, U.P. (2025). Drone Importance and their Necessity in Future Generation Agriculture. In: *Artificial Intelligence and Computer Vision for Ecological Informatics* (pp. 1–8). CRC Press.

Chouhan, S.S., Patel, R.K., Singh, U.P. and Tejani, G.G. (2025). Integrating Drone in Agriculture: Addressing Technology, Challenges, Solutions, and Applications to Drive Economic Growth. *Remote Sensing Applications: Society and Environment*, 101576.

Chouhan, S.S., Singh, U.P., Saxena, A. and Jain, S. (2024). Assessing the importance and need of artificial intelligence for precision agriculture. In: *Artificial Intelligence Techniques in Smart Agriculture* (pp. 1–6). Springer.

Dang, H.V., Tatipamula, M. and Nguyen, H.X. (2021). Cloud-based digital twinning for structural health monitoring using deep learning. *IEEE Transactions on Industrial Informatics*, *18*, 3820–3830.

Dhanaraju, M., Chenniappan, P., Ramalingam, K., Pazhanivelan, S. and Kaliaperumal R. (2022). Smart farming: Internet of Things (IoT)-based sustainable agriculture. *Agriculture*, *12*, 1745.

Dhanda, S.S., Panwar, D., Lin, C.-C., Sharma, T.K., Rastogi, D., Bindewari, S., Singh, A., Li, Y.-H., Agarwal, N. and Agarwal, S. (2025). Advancement in public health through machine learning: A narrative review of opportunities and ethical considerations. *Journal of Big Data*, *12*, 1–58.

Di Cicco, M., Potena, C., Grisetti, G. and Pretto, A. (2017). Automatic model based dataset generation for fast and accurate crop and weeds detection. In: *2017 IEEE/RSJ International Conference on Intelligent Robots and Systems* (*IROS*) (pp. 5188–5195). https://doi.org/10.1109/IROS.2017.8206408.

Ghosal, S., Blystone, D., Singh, A.K., Ganapathysubramanian, B., Singh, A. and Sarkar, S. (2018). An explainable deep machine vision framework for plant stress phenotyping. *Proc. Natl. Acad. Sci. U.S.A.*, *115*, 4613–4618. https://doi.org/10.1073/pnas.1716999115.

Gui, J., Fei, J., Wu, Z., Fu, X. and Diakite, A. (2021). Grading method of soybean mosaic disease based on hyperspectral imaging technology. *Information Processing in Agriculture*, *8*, 380–385. https://doi.org/10.1016/j.inpa.2020.10.006.

Jhajharia, K., Mathur, P., Jain, S. and Nijhawan, S. (2023). Crop Yield Prediction using Machine Learning and Deep Learning Techniques. *Procedia Computer Science*, *218*, 406–417. https://doi.org/10.1016/j.procs.2023.01.023.

Kejriwal, M. (2022). AI in practice and implementation: Issues and costs. In: *Artificial Intelligence for Industries of the Future: Beyond Facebook, Amazon, Microsoft, and Google* (pp. 25–45). Springer.

Kounalakis, T., Malinowski, M.J., Chelini, L., Triantafyllidis, G.A. and Nalpantidis, L. (2018). A Robotic System Employing Deep Learning for Visual Recognition and Detection of Weeds in Grasslands, In: *2018 IEEE International Conference on Imaging Systems and Techniques* (*IST*) (pp. 1–6). https://doi.org/10.1109/IST.2018.8577153.

Li, W., Wang, D., Li, M., Gao, Y., Wu, J. and Yang, X. (2021). Field detection of tiny pests from sticky trap images using deep learning in agricultural greenhouse. *Computers and Electronics in Agriculture*, *183*, 106048. https://doi.org/10.1016/j.compag.2021.106048.

Li, X., Xiong, H., Li, X., Wu, X., Zhang, X., Liu, J., Bian, J. and Dou, D. (2022). Interpretable deep learning: Interpretation, interpretability, trustworthiness, and beyond. *Knowledge and Information Systems*, *64*, 3197–3234.

Liu, J. and Cui, P. (2025). Data Heterogeneity Modeling for Trustworthy Machine Learning. In: *Proceedings of the 31st ACM SIGKDD Conference on Knowledge Discovery and Data Mining (KDD 2025)* (pp. 6086–6095).

Lu, J., Hu, J., Zhao, G., Mei, F. and Zhang, C. (2017). An in-field automatic wheat disease diagnosis system. *Computers and Electronics in Agriculture*, *142*, 369–379. https://doi.org/10.1016/j.compag.2017.09.012.

Mahto, H., Mahato, D. (2025). Human-Centric Biotechnological Interventions: A Pillar for Sustainable Agricultural Development in Industrial Crop Improvement. In: *Industrial Crops Improvement: Biotechnological Approaches for Sustainable Agricultural Development* (pp. 205–226). Springer.

Mesías-Ruiz, G.A., Borra-Serrano, I., Peña, J.M., de Castro, A.I., Fernández-Quintanilla, C. and Dorado, J. (2024). Weed species classification with UAV imagery and standard CNN models: Assessing the frontiers of training and inference phases. *Crop Protection*, *182*, 106721. https://doi.org/10.1016/j.cropro.2024.106721.

Nagarajan, V. (2021). Explaining generalization in deep learning: progress and fundamental limits. *arXiv Preprint arXiv:2110.08922*.

Patel, R.K. (2025). BM3D Filtering with Ensemble Hilbert-Huang Transform and Spiking Neural Networks for Cardiomegaly Detection in Chest Radiographs. *Computational Biology and Chemistry*, 108620.

Patel, R.K., Chouhan, S.S., Chandan, T. and Singh, U.P. (2025). Basil Crop Detection Using Computer Vision and Deep Learning Approach. In: *Artificial Intelligence and Computer Vision for Ecological Informatics* (pp. 67–82). Springer.

Rather, I.H., Kumar, S. and Gandomi, A.H. (2024). Breaking the data barrier: A review of deep learning techniques for democratizing AI with small datasets. *Artificial Intelligence Review*, *57*, 226.

Razfar, N., True, J., Bassiouny, R., Venkatesh, V. and Kashef, R. (2022). Weed detection in soybean crops using custom lightweight deep learning models. *Journal of Agriculture and Food Research*, *8*, 100308. https://doi.org/10.1016/j.jafr.2022.100308.

Ruiz, I., Pompeu, J., Ruano, A., Franco, P., Balbi, S. and Sanz, M.J. (2023). Combined artificial intelligence, sustainable land management, and stakeholder engagement for integrated landscape management in Mediterranean watersheds. *Environmental Science and Policy*, *145*, 217–227. https://doi.org/10.1016/j.envsci.2023.04.011.

Shah, B. and Bhavsar, H. (2022). Time complexity in deep learning models. *Procedia Computer Science*, *215*, 202–210.

Sibiya, M. and Sumbwanyambe, M. (2019). A Computational Procedure for the Recognition and Classification of Maize Leaf Diseases Out of Healthy Leaves Using Convolutional Neural Networks. *AgriEngineering*, *1*, 119–131. https://doi.org/10.3390/agriengineering1010009.

Singh, G., Sharma, D., Goap, A., Sehgal, S., Shukla, A.K., Kumar, S. (2019). Machine Learning based soil moisture prediction for Internet of Things based Smart Irrigation System. In: *2019 5th International Conference on Signal Processing, Computing and Control (ISPCC)* (pp. 175–180). https://doi.org/10.1109/ISPCC48220.2019.8988313.

Singh Chouhan, S., Saxena, E., Shukla, A. et al. (2025). Optimizing YOLO-Based Models for Real-Time Guava Detection with Probabilistic Fused Wiener Filter-Enhanced Feature Fusion. *Applied Fruit Science*, 67, 383. https://doi.org/10.1007/s10341-025-01613.

Tao, T. and Wei, X. (2022). A hybrid CNN–SVM classifier for weed recognition in winter rape field. *Plant Methods*, *18*, 29. https://doi.org/10.1186/s13007-022-00869-z.

Valavanidis, A. (2023). *Feeding a Growing World Population.* Critical Rethinking of Lessons from the "Green Revolution" in the 1960s as a Warning for Feeding a Hungry Planet 1, 1–47. ResearchGate.

Xu, Y., Wen, D., Zhou, J., Fan, X. and Liu, Y. (2021). Identification method of cotton seedlings and weeds in Xinjiang based on Faster R-CNN. *Paiguan Jixie Gongcheng Xuebao/Journal of Drainage and Irrigation Machinery Engineering*, *39*, 602–607. https://doi.org/10.3969/j.issn.1674-8530.19.0244.

Yin, H., Gu, Y.H., Park, C.-J., Park, J.-H. and Yoo, S.J. (2020). Transfer Learning-Based Search Model for Hot Pepper Diseases and Pests. *Agriculture*, *10*, 439. https://doi.org/10.3390/agriculture10100439.

Yun, W., Kumar, J.P., Lee, S., Kim, D.-S. and Cho, B.-K. (2022). Deep learning-based system development for black pine bast scale detection. *Sci. Rep.*, *12*, 606. https://doi.org/10.1038/s41598-021-04432-z.

Zhang, X., Qiao, Y., Meng, F., Fan, C. and Zhang, M. (2018). Identification of Maize Leaf Diseases Using Improved Deep Convolutional Neural Networks. *IEEE Access*, *6*, 30370–30377. https://doi.org/10.1109/ACCESS.2018.2844405.

9 Computer Vision for Livestock and Agroforestry Monitoring

Nuwandhya S. Dissanayaka[1], Shashi S. Udumann[2], Tharindu D. Nuwarapaksha[3] and Anjana J. Atapattu[4*]

1. Introduction

1.1 Challenges in Traditional Livestock and Agroforestry Monitoring Methods

The challenges that the global agricultural sector is facing in the 21st century are beyond belief, and it is because of the combination of population growth, climate change, and environmental degradation. The global population is predicted to increase to 9.7 billion by 2050, which will result in the fact that food production will have to increase by almost 70% to satisfy the demands of people, without subjecting the environment to adverse effects (Hussain et al., 2025). Nowadays, traditional extensive farming practices are increasingly unsustainable, contributing to soil degradation, water scarcity, biodiversity loss, and greenhouse gas emissions (Dissanayaka et al., 2025b). The agricultural industry is responsible for about 24% of greenhouse gas emissions worldwide and uses 70% of freshwater resources globally (Gorh & Baruah, 2019; Lakhan Ray & Getachew Tikuye, 2025).

Sustainable farming practices are no longer optional but necessary for long-term agricultural viability. Conventional farm monitoring relies heavily on manual and periodic observation (Patrício & Rieder, 2018). These methods present several significant limitations that advanced technologies can address (Figure 1). Consumers, governments, and international organizations increasingly demand environmentally responsible production methods. In this context, precision agriculture represents a paradigm shift from uniform field management to spatially and temporally variable management based on site-specific conditions. This approach enables farmers to apply the right treatment, at the right place, at the right time, and in the right amount (Wang et al., 2025). The integration of digital technologies, including computer vision (CV) systems, has become essential for achieving these precision agriculture goals.

[1-4] Agronomy Division, Coconut Research Institute, Lunuwila 61150, Sri Lanka.
* Corresponding author: aaajatapattu@gmail.com
https://orcid.org/0000-0002-2062-1133[1]; https://orcid.org/0000-0002-5717-7419[2]; https://orcid.org/0000-0002-7554-5289[3];https://orcid.org/0000-0003-4806-4647.

DOI: 10.1201/9781003715726-9

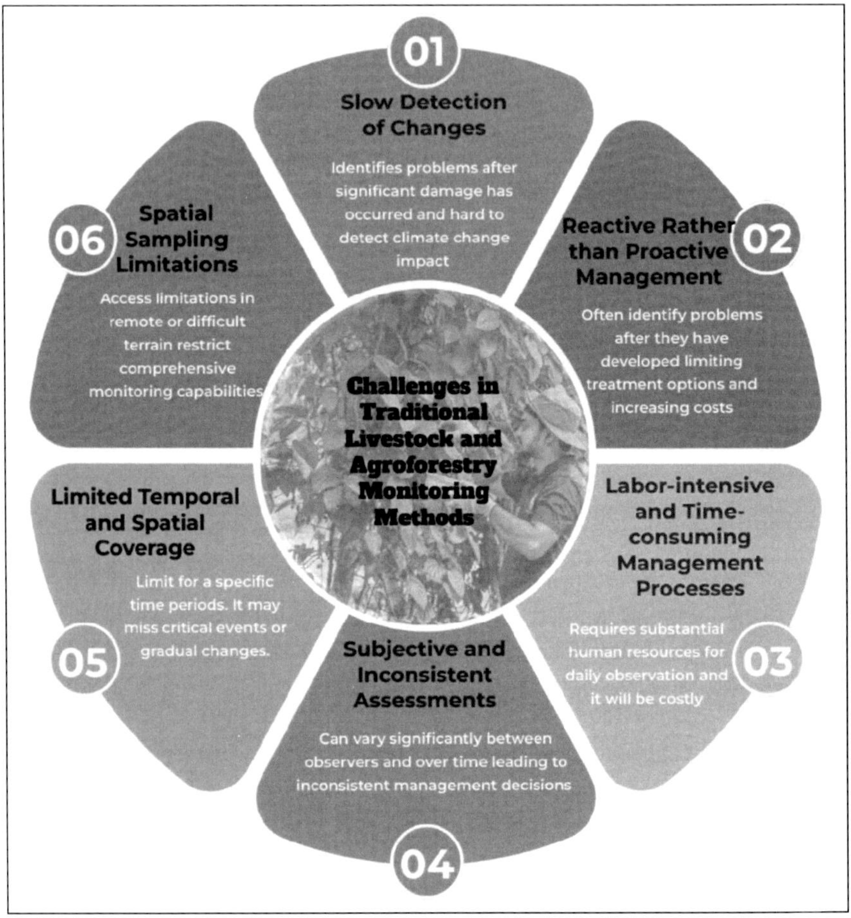

Fig. 1 Challenges in traditional livestock and agroforestry monitoring methods. (*Source:* Figure by Authors)

1.2 Importance of CV in Agricultural Contexts

The world today is rapidly advancing with CV technology, which can also be harnessed to transform agriculture (Chouhan et al., 2024; Patel, 2025). CV in agriculture refers to the application of automated image analysis and interpretation techniques to obtain meaningful information from visual data for agricultural decision-making. This technology encompasses the acquisition, processing, analysis, and interpretation of images or video sequences to monitor, assess, and manage agricultural systems automatically (Chouhan et al., 2025a).

In agricultural contexts, CV systems typically involve several key components: image acquisition devices (cameras, sensors, drones, satellites), image processing algorithms (filtering, enhancement, segmentation), feature extraction methods (geometric, textural, spectral features), machine learning models (classification, detection, prediction), and decision support systems that translate analytical results into actionable farm management recommendations (Jayanthi et al., 2024; Owino, 2023; Tian et al., 2020).

The agricultural application of CV extends beyond simple image capture to include sophisticated analysis techniques such as object detection and recognition (identifying individual animals, plants, or structures), classification and categorization (species identification, disease classification, quality grading), quantitative measurements (size, area, volume, count estimation), temporal analysis (growth monitoring, behavior tracking, change detection), and spatial analysis (distribution patterns, density mapping, resource allocation optimization) (Fracarolli et al., 2020). Modern agricultural CV systems increasingly incorporate Deep Learning (DL) techniques, particularly Convolutional Neural Networks (CNNs), which have demonstrated superior performance in image recognition tasks (Kamilaris & Prenafeta-Boldú, 2018). These systems can process multiple types of visual data, including RGB images, multispectral and hyperspectral imagery, thermal infrared images, and three-dimensional point cloud data from LiDAR systems.

This transformation not only promises significant economic benefits but also addresses critical global challenges such as food security and sustainability, as shown in Figure 2. This comprehensive review defines the current state and future directions of CV applications in livestock and agroforestry monitoring.

2. Fundamentals of CV in Livestock and Agroforestry Farming

The starting point of CV in agriculture is the image acquisition and preprocessing which are the basis of the credible analysis (Bongomin et al., 2024). Since agricultural environments present challenges such as variable lighting, weather effects, and dynamic biological subjects, preprocessing steps like geometric correction, radiometric calibration, noise reduction, and illumination normalization are essential. While Red-Green-Blue (RGB) cameras are widely used for general monitoring, advanced applications often require multispectral or thermal sensors to capture information beyond the visible spectrum (Li et al., 2023). The fine details, including the early disease or pest detection, need higher resolutions, and the strictness of quality control is the guarantee of the reliability of the data.

Feature extraction transforms raw images into meaningful representations (Latif et al., 2019). Traditional approaches rely on geometric, textural, and spectral descriptors, whereas modern methods increasingly use DL particularly CNNs, which learn hierarchical features automatically. Architectures such as You Only Look Once (YOLO), R-CNN, and U-Net are widely applied for object detection, classification, and pixel-level segmentation, enabling precise automation in agricultural monitoring (Patel et al., 2025; Voronkov & Bezuglyi, 2025). In addition, multi-scale pattern recognition supports monitoring from organ to landscape levels, while temporal analysis using time-series models and Recurrent

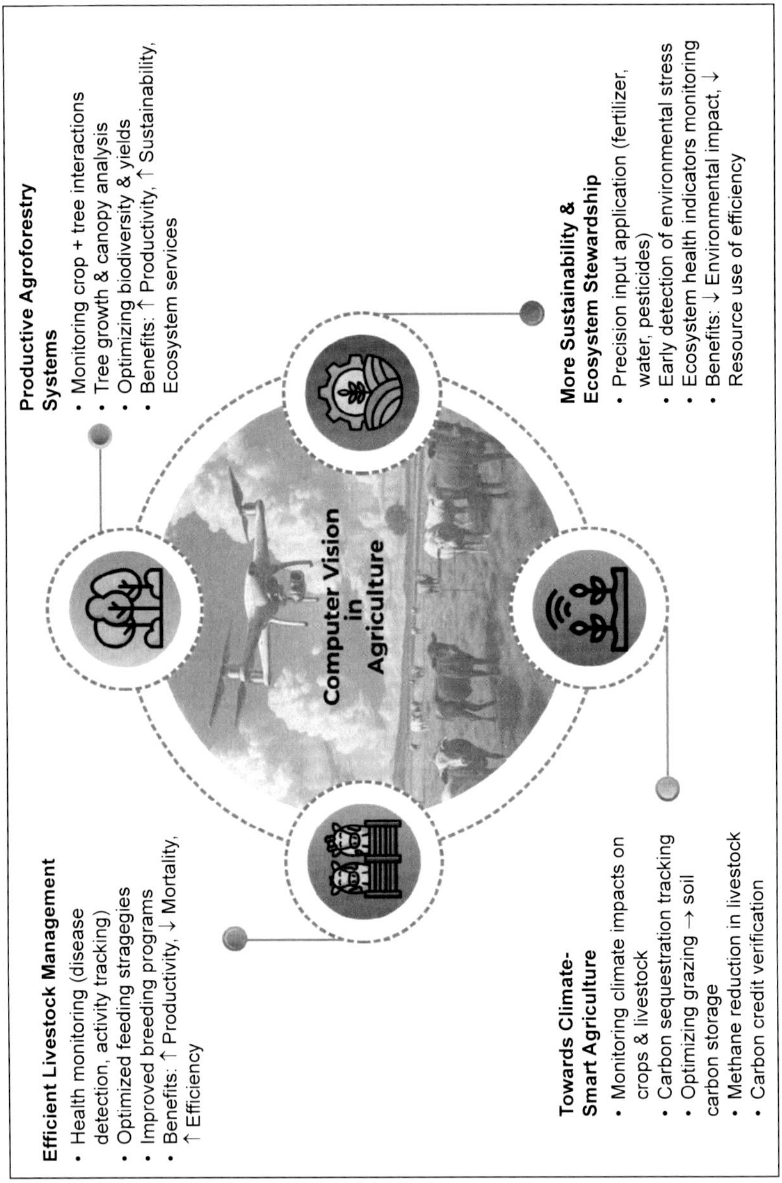

Fig. 2 Role of computer vision in agriculture.
Source: Figure by Authors.

Neural Networks (RNNs) helps track growth, behavior, and phenological changes over time (Lin et al., 2024).

There are various technologies used in agricultural imaging systems to capture information concerning crops and animals. RGB cameras offer a cheaper real-time image to track activities such as livestock identification and facility inspection. Multi-spectral imaging takes it one step further by sensing the slight variations in plant health through vegetation indices such as the Normalized Difference Vegetation Index (NDVI) and the Enhanced Vegetation Index (EVI), and even the possibility of sensing the health of livestock and the quality of pasture (Zeng et al., 2022). Thermal imaging enables non-invasive monitoring of livestock conditions, such as fever or mastitis, and helps detect crop stress due to water deficits or disease (Pineda et al., 2020; Sejian et al., 2022). LiDAR provides accurate 3D measurements of plant structure, canopy volume, and biomass, though it requires specialized expertise and investment (Jimenez-Berni et al., 2018). These imaging systems are deployed on multiple platforms, including drones for high-resolution field-scale mapping, satellites for large-scale and long-term monitoring, and ground-based systems, either fixed or mobile, for continuous and localized observations.

The CV data processing pipeline ensures that captured information is converted into actionable insights. Initial steps include image enhancement through contrast adjustment, noise reduction, and illumination correction, followed by segmentation techniques that separate objects of interest using both traditional and DL methods. Object detection frameworks enable real-time identification and alert systems for livestock or crop management, while temporal monitoring captures behavioral patterns and growth dynamics. Finally, effective integration with farm management software, sensor fusion across multiple data sources, and robust data management systems are critical. Edge computing supports real-time field applications, whereas cloud platforms provide the computational power required for large-scale and complex analyses.

3. CV for Livestock Monitoring

3.1 Animal Identification and Tracking

3.1.1 Individual Animal Recognition Systems

Conventional identification techniques, such as ear cutting, hot iron branding, and tattooing, are relatively easy to implement, but they need experienced workers and cause animal welfare issues (Bodkhe et al., 2018). CV enables automated individual animal identification through distinctive visual features, eliminating the need for manual identification and reducing labor requirements (Meng et al., 2025). Modern recognition systems utilize DL models trained on distinctive physical characteristics, including coat patterns, facial features, body markings, and morphological characteristics. As example, dairy cattle identification commonly relies on coat patterns, facial features, and body shape variations (Kaur et al., 2022; Zhao et al., 2019). Machine learning, Ellipse fitting technique, Gaussian mixture model, R-CNN, and CNN (tag-box method) are some of the possible tools and techniques for identifying pigs in farming (Arulmozhi et al., 2021).

3.1.2 Biometric Identification and Behavioral Pattern Recognition

Facial recognition has demonstrated impressive performance in the livestock industry, with accuracy rates of over 95% in identifying cattle due to controlled factors (Zhao et al., 2019). The gait analysis includes the additional identification features on the basis of movement pattern, stride features, and variations in posture peculiar to each animal. Behavioral pattern recognition is important as it could be utilized to identify and describe health and welfare conditions in animals. The behaviors that can be detected by CV systems are feeding behaviors, social interactions, movement patterns, and resting behaviors, which are the indicators of individual identity and health conditions.

3.1.3 RFID Integration and System Validation

Integration with Radio Frequency Identification (RFID) systems provides ground truth data for training CV models and validates identification accuracy (Brown-Brandl et al., 2019). Hybrid systems combine the reliability of RFID with the convenience of vision-based identification, enabling automated cross-validation and system reliability assessment. This integration supports seamless data management and reduces identification errors in commercial applications.

3.2 Health and Welfare Assessment

3.2.1 Disease Detection and Body Condition Scoring

Dairy farming uses CV systems for illness detection, body weight estimation, nutrition intake, behavior monitoring, and animal identification (Liu et al., 2023). For that, it uses only the body dimensions and estimates weight by analyzing images and using key point detection to assess body shape and size (Ma et al., 2024). Visual symptom detection focuses on identifying external signs of disease, including eye discharge, nasal secretions, skin lesions, and abnormal postures or movements. Automated body condition scoring uses CV to assess fat deposits and muscle development, providing objective measurements that reduce subjectivity in traditional manual scoring systems. These systems analyze body shape, skeletal visibility, and fat distribution patterns to generate standardized body condition scores.

3.2.2 Lameness Detection and Mobility Assessment

Lameness represents a significant welfare and economic concern in livestock production. CV systems detect lameness through gait analysis, posture assessment, and movement pattern recognition (Qiao et al., 2021). Key indicators include stride length variations, weight distribution asymmetries, back arching, and reduced mobility. Automated scoring systems can identify lameness severity levels and track recovery progress following treatment.

3.2.3 Stress Indicators and Welfare Monitoring

CV enables continuous welfare monitoring through behavioral analysis including activity levels, feeding patterns, social interactions, and environmental preferences. Stress indicators detectable through CV include abnormal behaviors, reduced activity, changes in feeding patterns, and physiological responses visible through thermal imaging (Fuentes et al., 2022; Okinda et al., 2020). Integration with environmental sensors provides context for welfare assessment under varying conditions.

3.3 Feeding and Nutrition Monitoring

3.3.1 Feed Intake Estimation and Behavior Analysis

Accurate feed intake measurement is crucial for precision nutrition management and cost optimization. CV systems estimate individual animal feed consumption through video analysis of feeding behaviors, measuring feeding duration, frequency, and intensity (Fuentes et al., 2022). Advanced systems can differentiate between actual consumption and investigation behaviors, providing more accurate intake estimates. Feeding behavior analysis includes monitoring feeding time patterns, competition behaviors, and feed preference assessment.

3.3.2 Pasture Quality Assessment and Grazing Optimization

CV supports pasture management through vegetation quality assessment, biomass estimation, and species composition analysis. Drone-based and ground-based imaging systems evaluate pasture conditions using vegetation indices, color analysis, and texture assessment (Gillan et al., 2019). Integration with livestock tracking enables correlation between pasture quality and animal performance, supporting optimized grazing management decisions.

3.3.3 Automated Feeding System Integration

CV enhances automated feeding systems by monitoring feed delivery accuracy, detecting feeding irregularities, and ensuring proper system operation. Integration with precision feeding equipment enables real-time adjustments based on individual animal requirements and feeding behaviors. Systems can detect feed waste, contamination, or equipment malfunctions that affect feeding efficiency (Reger et al., 2018).

3.4 Reproduction and Breeding Management

3.4.1 Estrus Detection Systems

Hormone, especially Estrus detection, is crucial in dairy farming to ensure cows are inseminated at the optimal time for successful reproduction. CV-based estrus detection systems monitor mounting behaviors, increased activity levels, and physical changes associated with reproductive cycles (Curti et al., 2023). The proposed method combines estrus detection with cow identification through augmented reality, showcasing an advanced integration approach. In modern dairy farming, wearable sensors are increasingly employed to monitor key parameters such as cow health, estrus cycles, and feeding behavior (Neethirajan, 2017). Vision systems complement wearable sensors by providing visual confirmation of behavioral changes and reducing false positive rates.

3.4.2 Pregnancy Monitoring and Calving Assistance

CV supports pregnancy monitoring through body shape analysis, udder development assessment, and behavioral change detection. Pre-calving monitoring systems detect signs of imminent parturition, including restlessness, udder changes, and postural modifications. Automated calving detection enables timely assistance and reduces calf mortality rates (Crociati et al., 2022).

3.4.3 Genetic Trait Identification

CV facilitates genetic evaluation through automated measurement of heritable traits, including body conformation, muscle development, and structural characteristics. Integration with genomic data enhances breeding value estimation and supports selection decisions for genetic improvement programs.

3.5 Grazing Management and Distribution Monitoring

3.5.1 Pasture Utilization and Overgrazing Detection

CV systems monitor grazing patterns through livestock tracking, vegetation assessment, and land use analysis. Aerial imagery analysis detects overgrazing indicators including bare soil exposure, vegetation degradation, and uneven utilization patterns (Michez et al., 2019). Real-time monitoring enables adaptive grazing management and pasture recovery strategies.

3.5.2 Rotational Grazing Optimization

CV's main future direction in livestock production is to help improve the efficiency and sustainability of animal production. CV supports rotational grazing systems by monitoring pasture recovery, optimal stocking densities, and movement timing.

Integration with GPS tracking provides comprehensive grazing management data for optimizing rotation schedules and pasture allocation.

3.5.3 Environmental Impact Assessment

Livestock distribution monitoring through CV enables assessment of environmental impacts, including soil compaction, water source utilization, and vegetation trampling. Systems can identify areas requiring management intervention and support sustainable grazing practices that minimize environmental degradation while maintaining productivity.

4. CV for Agroforestry Monitoring

4.1 Tree and Forest Structure Analysis

4.1.1 Tree Species Identification and Crown Analysis

Accurate tree species classification is essential for agroforestry resource management and biodiversity assessment, with CV enabling automated identification through distinctive visual characteristics. DL approaches using CNNs achieve high accuracy in species classification from UAV imagery, with hyperspectral CNN models (Usman et al., 2023). Usman et al, (2023), present a study with two machine learning models (Support Vector Machine model and Multi-Layer Perceptron model) that achieve over 80% accuracy using a single date of WorldView-2 imagery to identify the predominant farmland tree species in West African parklands. Modern classification systems combine multiple data sources for improved accuracy. Object-based frameworks using UAV hyperspectral images and LiDAR data provide comprehensive species identification capabilities (Zhong et al., 2022). DL applied to RGB images taken by UAV shows high potential for identifying tree types and species, even with consumer-grade cameras (Figure 3).

Crown segmentation and canopy analysis enable individual tree delineation and structural assessment. CV techniques identify crown boundaries, calculate canopy cover percentages, assess crown health conditions, and analyze spatial distribution patterns. Not only can beneficial crops be detected, but weeds also can be detected from this approach, and at the same time, spraying of weedicides is also possible (Wu et al., 2021).

4.1.2 Mapping, Forest Density, and Biometric Measurements

Agroforestry integrates trees, crops, and sometimes livestock; conventional mapping methods struggle to capture the diversity and spatial arrangement (Dissanayaka et al., 2023a). CV, often combined with remote sensing (drones, satellites) or ground-based imaging, makes it much more efficient (Debats et al., 2016). LiDAR has emerged as a leading technology for tree species classification and provides precise measurements of field structural parameters. CV systems

Fig. 3 Automated detection of coconut palms from aerial imagery using object detection algorithms, with identified trees highlighted by bounding boxes for counting and spatial analysis.
Source: Figure by Authors.

estimate tree height through photogrammetric analysis, calculate diameter at breast height (DBH) using allometric relationships, assess forest density and stand characteristics, and generate biomass estimates through structural measurements (Birdal et al., 2017). Integration of multiple sensing technologies enhances measurement accuracy. Combined RGB, multispectral, and LiDAR data provide comprehensive structural information while reducing measurement uncertainties associated with single-sensor approaches.

4.2 Growth and Phenology Monitoring

4.2.1 Seasonal Monitoring and Growth Estimation

CV enables continuous monitoring of seasonal changes through automated analysis of vegetation color, texture, and structural changes. Phenological stage detection includes leaf emergence, flowering periods, fruit development, and senescence timing. Temporal resolution facilitates detecting seasonal variations and growth patterns and assessing the effectiveness of carbon sequestration efforts over time. Even after harvesting, this will be beneficial for monitoring food quality throughout the value chain (Iqra et al., 2025). Growth rate estimation utilizes temporal image analysis to measure height increases, crown expansion, and biomass accumulation over time. Time-series analysis of UAV imagery provides accurate growth monitoring for individual trees and forest stands, supporting precision agroforestry management and yield prediction.

4.2.2 Yield Prediction and Crop Assessment

Agroforestry systems produce diverse products requiring specialized monitoring approaches (Dissanayaka et al., 2023b). CV systems estimate fruit and nut yields through flower counting, fruit detection and sizing, and harvest timing optimization (Logeshwaran et al., 2024). Integration with weather data and phenological models improves prediction accuracy and supports market planning decisions (Paparrizos et al., 2024). At this point Random Forest (RF), Artificial Neural Networks (ANN), and Support Vector Machine (SVM). In the realm of DL, CNN, Long-Short Term Memory (LSTM), and Deep Neural Networks (DNN) have been identified as promising and highly-used candidates in the machine learning domain (Jabed & Azmi Murad, 2024).

4.3 Health and Disease Management

4.3.1 Disease Detection and Stress Assessment

Timing is vital in treating crop diseases, and the use of CV for precise early detection enables timely interventions before significant damage occurs. Via CV approaches like image segmentation, visual symptom detection identifies leaf discoloration, defoliation patterns, branch dieback, and fungal growth indicators through spectral analysis and pattern recognition before the naked eye does (Chouhan et al., 2020; Singh & Misra, 2017). Moreover, classification, quantification, and prediction (ICQP) on disease and stress conditions are possible, specifically focusing on digital image-based phenotyping in this approach (Singh et al., 2018).

Not only biotic stresses, a biotic stresses such as moisture, salinity, and heat can also be predicted from this technology, and those assessments utilize thermal imaging and multispectral analysis to detect water stress before visible symptoms appear (Singh et al., 2018). CV systems monitor leaf temperature, spectral reflectance changes, and canopy structure modifications that indicate stress conditions. Early detection enables adaptive irrigation management and stress mitigation strategies.

4.3.2 Pest Monitoring and Early Warning Systems

CV supports integrated pest management through automated pest identification, damage assessment, and population monitoring. Images from satellites, drones, and high-resolution cameras are used for early pest detection and monitoring, enabling rapid response to pest outbreaks (Muppala & Guruviah, 2020). Early warning systems combine CV with environmental monitoring to predict disease and pest risks. Machine learning models analyze historical patterns, current conditions, and visual indicators to generate alerts for preventive management actions.

4.4 Biodiversity and Ecosystem Assessment

4.4.1 Species Diversity and Habitat Monitoring

CV supports biodiversity conservation through automated species identification, habitat quality assessment, and ecosystem health monitoring. Wildlife camera systems with automated species recognition enable population monitoring and behavior analysis. Vegetation diversity assessment through spectral analysis provides insights into ecosystem complexity and health. Habitat assessment combines structural analysis with species distribution mapping to evaluate habitat suitability and conservation value. CV enables monitoring of habitat fragmentation, connectivity, and restoration success over time.

4.4.2 Pollinator Activity and Ecosystem Services

Pollinator monitoring through CV can include automated bee counting, activity pattern analysis, and pollination efficiency assessment. These systems may support pollinator conservation efforts and optimize agricultural productivity in agroforestry systems where pollination services are crucial for fruit and seed production.

4.5 Carbon Sequestration and Climate Monitoring

4.5.1 Carbon Stock Estimation and Monitoring

Scientists use airborne laser scanning data and multi-source satellite imagery to estimate forest aboveground carbon density, demonstrating the operational capability of CV for carbon monitoring (Kauranne et al., 2017). Satellite remote sensing provides spatially explicit information on land cover, stand age class, and harvesting patterns that are essential for carbon cycle modeling. In this task, Machine Learning algorithms like RF, SVM, and Gradient Boosting (GB), and DL algorithms like U-Net, Functional Pyramid Network (FPN), and Deeplap can be used (Illarionova et al., 2022). Biomass estimation through structural measurements provides the foundation for carbon stock calculations using established allometric equations (Nuwarapaksha et al., 2025a).

4.5.2 Climate Impact Assessment and Adaptation Monitoring

CV enables monitoring of climate change impacts on agroforestry systems through vegetation response analysis, species distribution changes, and phenological shift detection (Rolnick et al., 2023). Modern change detection methods rely heavily on remote sensing technology for assessing forest cover changes and climate responses. Microclimate monitoring through thermal imaging and multispectral analysis provides insights into local climate conditions and their effects on vegetation performance. These data support climate adaptation strategies and management decisions in response to changing environmental conditions.

4.5.3 Soil Carbon And Ecosystem Assessment

CV supports soil carbon monitoring through vegetation analysis, land use classification, and erosion assessment. Spectral analysis of vegetation provides indicators of soil organic matter and carbon sequestration potential. Integration with soil sampling data enables spatially explicit soil carbon mapping and monitoring of sequestration effectiveness in agroforestry systems.

5. Technical Challenges and Solutions

5.1 Environmental Challenges

5.1.1 Weather and Lighting Variations

Agricultural CV systems must operate reliably across diverse weather conditions and lighting scenarios that significantly affect image quality and analysis accuracy. Outdoor environments present continuous challenges including direct sunlight causing overexposure and harsh shadows, overcast conditions reducing contrast and color saturation, rain and fog degrading visibility, and dust and debris affecting lens clarity (Liu et al., 2020; Orchi et al., 2021). Solutions include adaptive exposure control systems that automatically adjust camera settings based on lighting conditions, multi-exposure imaging techniques that capture details in both bright and dark areas, and illumination normalization algorithms that standardize images for consistent analysis. Weather-resistant housing and automated lens cleaning systems may address physical environmental challenges. Simply, weather-based irrigation controllers are possible with CV approaches for dry seasons (Nuwarapaksha et al., 2024b).

5.1.2 Seasonal Changes and Occlusion Issues

Seasonal variations affect both subject appearance and environmental conditions, requiring robust systems that maintain accuracy across different seasons (Orchi et al., 2021). This can be a bigger challenge in countries located in temperature regions than tropical regions. Vegetation changes alter background complexity, lighting conditions vary seasonally, and animal coat conditions change with seasons (Katiyatiya et al., 2015; Zhang et al., 2018). Occlusion represents a significant challenge in agricultural environments where objects being partially obscured are common. Trees, vegetation, farm structures, and other animals frequently obstruct clear views of monitoring targets. Solutions include multi-angle imaging systems, temporal analysis to capture objects when visible, and DL models trained on partially occluded objects to improve recognition under challenging conditions.

5.1.3 Scale and Perspective Variations

Agricultural monitoring occurs across multiple scales from close-up individual assessment to landscape-level analysis, requiring systems that handle significant scale variations. Images with different scales and resolutions present challenges for consistent object detection and measurement. Solutions include multi scale detection architectures, scale-invariant feature extraction methods, and adaptive processing pipelines that adjust analysis parameters based on image characteristics.

5.2 Computational Challenges

5.2.1 Real-time Processing and Edge Computing

The use of real-time processing for making decisions and intervening in the moment is becoming more and more important in agricultural use. Edge computing has powerful features in smart agriculture, enabling it to make decisions based on data and conduct real-time surveillance (Sathya et al., 2024). Nevertheless, agricultural edge computing has drawbacks such as high computational power in remote areas, power limitations, and network connectivity issues. Model optimization methods like pruning and quantization can be used to minimize the computational needs. Agricultural-focused edge AI hardware and hybrid processing architectures can combine edge and cloud computing to do the same.

5.2.2 Cloud Computing and Connectivity

Although cloud computing is potentially able to offer a considerable amount of processing power and storage capacity, there are several issues that agriculture settings are likely to encounter, such as low internet connection in rural locations, non-real-time internet connection, and the cost of transmitting large datasets. Hybrid architectures overcome these issues by executing the most important real-time tasks on the edge and then using the cloud facilities to conduct more intricate analysis, train models, and store them long-term. The intelligent data management systems can focus on important data to be transferred in real-time, and batch less crucial information to be transferred efficiently (Araújo et al., 2021).

5.2.3 Power and Storage Limitations

Agricultural monitoring systems often operate in remote locations with limited power infrastructure (Gupta et al., 2024). Battery-powered systems must balance processing capabilities with power consumption, while solar-powered installations face weather-dependent energy availability. Storage limitations arise from continuous image acquisition, generating large data volumes. Solutions include intelligent data sampling that captures images only when significant events occur, on-device compression techniques that reduce storage requirements, and automated data lifecycle management that removes redundant or low-quality images (Chouhan et al., 2025b).

5.3. Data Quality and Annotation

5.3.1 Ground Truth Establishment and Dataset Limitations

A key challenge in applying IoT and DL to agriculture is the scarcity and poor quality of datasets (Aldossary et al., 2024). Limited availability arises from crop seasonality, phenotypic variation across regions, and year-to-year fluctuations. Annotation issues such as mislabeling, inconsistency, duplication, and incompleteness further reduce reliability. Agricultural datasets also suffer from high biological variability, poor representation of real-world conditions, scarcity of expert annotators, and time-consuming labeling processes, with annotation methods often undocumented, making reproducibility difficult. Potential solutions include developing large-scale, standardized, and open agricultural datasets, using data augmentation, synthetic data generation, and transfer learning to overcome scarcity, and also to enrich datasets, integration of IoT sensor fusion (e.g., combining imagery with soil or climate data) can be practiced (Dissanayaka et al., 2024).

5.3.2 Active Learning and Annotation Optimization

Active learning strategies optimize the annotation process by identifying the most informative samples for manual labeling, reducing annotation costs while maintaining model performance (Nuwarapaksha et al., 2024a). Semi-supervised learning approaches utilize large amounts of unlabeled data combined with limited labeled samples to improve model robustness. Few-shot learning techniques enable rapid adaptation to new species, conditions, or monitoring objectives with minimal training data. These approaches are particularly valuable for rare disease detection or monitoring of uncommon species where extensive datasets are unavailable.

5.4 System Integration

5.4.1 IoT Sensor Fusion and Compatibility

Modern agricultural monitoring systems integrate CV with multiple sensor types (Table 1). Sensor fusion challenges include data synchronization across different sampling rates, standardization of data formats and communication protocols, calibration and validation across sensor types, and handling missing or corrupted data from sensor failures.

Table 1 Different sensor types used in smart agriculture.

Sensor Type	*Used Percentage in Smart Agriculture (%)*
Humidity / Moisture / Water	12
Chemical / Gas	12
Temperature	11
Magnetic	10
Acceleration / Tilt	9
RFID / NFC	8
Ultrasounds / Radio	7
Force / Load / Pressure	6
Ambient light / Machine vision / Optical	6
Electric	4
Vibration	4
Position / Proximity	3
Flow	3
Motion / Velocity / Displacement	3
Leaks / Levels	2
Acoustic / Sound	1

(*Source:* Liu et al., 2019)

Solutions involve middleware platforms that standardize data integration, time-series databases optimized for agricultural monitoring, and robust data validation procedures that identify and compensate for sensor failures or inconsistencies.

5.4.2 Legacy System Integration and Scalability

Agricultural operations often involve existing equipment and management systems that require integration with new CV technologies (Nuwarapaksha et al., 2025b). Compatibility challenges include different communication protocols, varying data formats, and integration with established farm management software. Scalability issues arise when expanding from pilot projects to commercial deployment. Solutions include modular system architectures that enable gradual deployment, standardized APIs for seamless integration, and cloud-based platforms that provide scalable processing and storage capabilities.

5.4.3 User Adoption and Interface Design

Successful adoption of CV systems requires user-friendly interfaces and demonstrated value to agricultural practitioners (Setiawan et al., 2024). Challenges include complex technical systems that require specialized knowledge, cost-benefit justification for technology adoption, training requirements for effective system utilization, and integration with existing farm management practices. Solutions focus on intuitive user interfaces that present information in familiar agricultural terms, automated alert systems that require minimal user intervention, comprehensive training and support programs, and clear demonstrations of economic and operational benefits to encourage adoption.

5.5 Economic and Social Barriers

5.5.1 High Implementation Costs and Financial Barriers

CV systems have a high upfront cost, in addition to specialized cameras and sensors, processing hardware, a license, installation and setup, and maintenance and support costs (Kakani et al., 2020). The cost aspect especially poses a significant challenge to small-scale operations, which might not attain the economies of scale to sustain the cost-effective implementation. The barriers to financing encompass a dearth of agricultural credit to invest in technologies, the absence of specialized lending systems on precision agriculture technologies, the lack of assurance on the resale value of complex machinery, and the unavailability of insurance cover on advanced agricultural systems. These financial constraints present big obstacles to adoption, even when technology proves to be beneficial.

5.5.2 Digital Divide, Accessibility Challenges, and Training Requirements

Rural areas often have a low level of internet connectivity and insufficient infrastructure to aid technical support, as well as digital literacy tools, which have a significant impact on the digital divide in the adoption of CV technology in agriculture (Udumann et al., 2024). Most farmers have perceived initial cost, uncertainty in economic returns, and uncertainty of technology as limiting factors to adoption. Geographic disparities are connected with the unavailability of broadband internet in remote agricultural populations, the unavailability of technical support services, and the degree of governmental support in the use of technology, educational, and training opportunities (Walter et al., 2017). Such inequalities lead to unequal access to higher-order monitoring technologies. The implementation of the CV systems is a high-cost training and skills development of the implementer (Udumann et al., 2025a). The skills required on a technical level consist of knowledge of basic computer use and navigation of computer software programs, the ability to interpret and analyze data, maintain and troubleshoot system operation, and align with current farm management practices. The rate of adoption is affected by age and education; the younger and better educated the farmers are, the higher the adoption rate will be, and the older the farmers are, the more difficult it is to adapt to the new technological changes. The digital literacy of running advanced CV systems might be a setback to farmers with low technical knowledge.

6. Regulatory and Ethical Considerations

6.1 Data Ownership and Privacy Regulations

The agricultural data privacy is a rather complicated regulatory environment that includes the farm operations information, the animal welfare and health information, and the environmental monitoring information. Issues of data ownership can be attributed to various stakeholders, whereby farmers are the ones who provide data by the operation of the farms, technology companies are the ones that collect and process data, researchers are the ones that analyze data with scientific objectives, and government agencies are the ones that may require data to comply with regulations. Legal frameworks must be clear to establish the ownership, the rights of use, and the benefit-sharing provisions. The privacy protection demands involve anonymization of individual farm information, data protection and transmission systems and protocols, consent protocols to share data, and data use and analysis transparency. The regulatory compliance should be able to achieve a balance between data protection and the positive purpose of research, development, and improving agriculture.

6.2 Animal Welfare Implications and Standards

CV systems for livestock monitoring raise important animal welfare considerations that require careful ethical evaluation (Gómez et al., 2021). Welfare implications include continuous surveillance effects on animal behavior, potential stress from monitoring equipment installation, a balance between monitoring benefits and animal privacy concerns, and ensuring technology serves animal welfare rather than merely production efficiency. Ethical frameworks must guide technology development to prioritize animal well-being alongside operational efficiency.

Monitoring technologies should enhance rather than compromise animal welfare through early health problem detection, improved environmental condition management, reduced human disturbance through automated monitoring, and objective assessment of welfare indicators. Regulatory standards need to be developed for a welfare-focused monitoring system design and implementation.

6.3 Environmental Monitoring Standards and International Coordination

Environmental monitoring through CV requires standardized protocols for data collection, analysis, and reporting to ensure consistency and comparability across different systems and regions. Standardization needs include common metrics for environmental impact assessment, calibration procedures for monitoring equipment, data quality standards for environmental reporting, and integration protocols for multi scale monitoring systems. International standardization efforts must address varying regulatory requirements across countries, different environmental monitoring priorities and standards, coordination between research institutions and commercial systems, and harmonization of carbon monitoring and reporting protocols for global carbon markets. Climate monitoring applications require particular attention to standardization given their role in global climate policy and carbon credit markets. Verification standards for carbon sequestration monitoring, quality assurance protocols for climate impact assessment, and international coordination mechanisms for data sharing and validation are essential for effective global climate action support.

6.4 Performance Evaluation and Metrics

The application of CV in these fields requires continuous evaluation to assess its effectiveness. Various performance metrics and evaluation methods, as summarized in Table 2, can be used for this purpose.

Table 2 Possible performance evaluation and metrics in CV for livestock and agroforestry.

Main category	*Metrics/Findings*	*Description*
Technical Performance Metrics	Classification Accuracy	The proportion of correctly identified instances
	Precision	The ratio of true positive predictions to total positive predictions
	Recall	The ratio of true positive predictions to total actual positive instances
	Intersection over Union (IoU)	Spatial accuracy for segmentation and detection tasks, particularly important for precise boundary delineation in agricultural monitoring

Contd...

Table 2 *Contd...*

Main category	*Metrics/Findings*	*Description*
	Mean Average Precision (mAP)	A standard metric for object detection tasks in agricultural applications, providing comprehensive evaluation across different object classes and detection thresholds
	Processing Speed	Includes frames per second (FPS) for video analysis, inference time for individual image processing, and throughput measurements for batch processing applications
	Robustness	Assesses system performance across varying environmental conditions, seasonal changes, and equipment configurations
	Scalability	Includes processing capacity scaling, storage requirement growth, network bandwidth utilization, and computational resource efficiency
Agricultural Performance Indicators	Productivity Gains	Includes increased detection accuracy for health monitoring, reduced response time for disease and pest intervention, improved breeding efficiency through automated reproductive monitoring, and enhanced feed conversion efficiency through precision nutrition management
	Cost Reduction	Includes reduction of cost from optimized resource application, decreased labor requirements for monitoring and assessment, lower veterinary costs through early health detection, and reduced feed waste through precise nutrition management
	Return on Investment (ROI)	A calculation with tracking costs associated with system implementation against quantifiable benefits
	Labor Efficiency	Includes reduced time requirements for monitoring activities, elimination of subjective assessment variations, automated alert systems reducing constant supervision needs, and skill requirement reductions for routine monitoring tasks
	Environmental Impact	Includes reduced chemical usage through precision application, improved resource use efficiency, enhanced animal welfare through continuous monitoring, and supporting sustainable farming practices

Contd...

Table 2 *Contd...*

Main category	*Metrics/Findings*	*Description*
Comparative Analysis	Traditional vs. CV	-
	Deep Learning (DL) vs. Classical Machine Learning	-
	Platforms (ex: UAVs, satellites, ground systems)	-

Source: Table by Authors.

7. Future Directions and Emerging Trends

7.1 Technological Advances

7.1.1 Next-generation Imaging and AI Innovations

Innovations like Generative Adversarial Networks (GANs), Vision Transformers, and Federated Learning are solving critical challenges and shaping the future of CV, with direct applications emerging in agricultural monitoring (Akram et al., 2025). Hyperspectral and multispectral imaging technologies are advancing toward higher spectral resolution, improved signal-to-noise ratios, and reduced costs (Lu et al., 2020). Next-generation sensors will provide enhanced capabilities for detecting subtle physiological changes, early disease identification, and precise nutrient monitoring. Integration of artificial intelligence with these advanced sensors will enable more sophisticated automated interpretation. Vision Transformers represent a paradigm shift from CNN-based approaches, offering improved performance for complex agricultural scenes with multiple objects and varying scales. These architectures excel at capturing long-range dependencies and spatial relationships crucial for understanding complex agroforestry interactions.

7.1.2 Edge Computing and Connectivity Developments

The latest trends combine edge computing with on-device Machine Learning; a method also called Edge AI (Chang et al., 2021). Moving AI processing from the cloud to edge devices makes it possible to run machine learning everywhere, combining the IoT and AI to create scalable CV applications in remote agricultural settings. 5G connectivity will revolutionize agricultural CV by enabling real-time high-resolution video streaming, supporting immediate intervention capabilities, reducing latency for critical applications, and enabling coordinated monitoring across multiple platforms (Majumdar et al., 2024; Raja et al., 2024). Edge computing developments include specialized agricultural AI chips, improved power efficiency for battery-operated systems, and enhanced processing capabilities for complex real-time analysis.

7.2 Integration Opportunities

7.2.1 Smart Farming Ecosystem Convergence

Precision agriculture systems are converging toward comprehensive digital farming platforms that integrate CV with IoT sensors, automated machinery, and farm management software. This convergence enables closed-loop management systems where CV provides monitoring data that directly controls automated interventions. Future integration includes autonomous robotic systems guided by CV, variable-rate application equipment responding to real-time monitoring data, and integrated livestock-crop management systems that optimize overall farm productivity and sustainability (Shi et al., 2023).

7.2.2 Blockchain and Digital Twin Technologies

Digital Twins (DTs) hold significant promise in addressing the challenges faced by food supply chains (FSCs), with CV providing real-time data to feed digital twin models (Cesco et al., 2023). From farmers and seed to consumers' shelves, a blockchain-based supply chain solution can track each process with unique identifiers and create a digitally traceable end-to-end journey. CV systems will provide the visual data foundation for digital twin modeling, enabling virtual representation of entire farming operations.

Integration of DTs with IoT and blockchain technologies has the potential to revolutionize various industries by providing enhanced security, transparency, and data integrity (Jain et al., 2025). By integrating blockchain with DT solutions, farmers can enhance traceability, improve supply chain transparency, and ensure the quality and safety of their products (Nuwarapaksha et al., 2024c).

7.3 Sustainability and Climate Adaptation

7.3.1 Carbon Credit Verification and Climate Monitoring

CV plays an increasingly important role in carbon credit verification through automated monitoring of carbon sequestration practices in the future (Tyagi et al., 2025). Regenerative agricultural practice adoption on conventionally managed fields has gained momentum as a climate mitigation strategy, with CV enabling scalable monitoring of practice implementation and effectiveness (Atapattu et al., 2025a). Advanced monitoring systems provide feeds to farmer records and maps that allow farmers to estimate how much carbon they are storing in their soil and how much they could store if they adjust their management practices. Proprietary digital measurement, reporting, and verification (dMRV) technology powered by AI and satellite imagery enables accurate carbon credit certification and validation.

7.3.2 Climate-smart Agriculture Applications

CV supports climate adaptation through real-time monitoring of climate impacts, automated adjustment of management practices, predictive modeling for climate resilience, and optimization of resource use under changing conditions (Wu et al., 2023). These systems enable farmers to adapt to climate variability while maintaining productivity and sustainability. Resilience monitoring tools assess system vulnerability to climate shocks, monitor recovery from extreme events, and evaluate adaptation strategy effectiveness (Dissanayaka et al., 2025a).

7.4 Emerging Applications

7.4.1 Alternative Protein and Regenerative Agriculture

Alternative protein production systems, including insect farming and cellular agriculture, present new monitoring challenges that CV can address (Sexton et al., 2019). Applications include automated insect counting and health assessment, sterile environment monitoring for cellular agriculture, and quality control for novel protein processing.

Regenerative agriculture assessment requires monitoring of soil health indicators, biodiversity metrics, ecosystem service provision, and carbon sequestration effectiveness (Udumann et al., 2025b). CV enables automated assessment of regenerative practice implementation including irrigation, cover crop establishment, diverse cropping systems, and integrated livestock management (McLennon et al., 2021).

7.4.2 Urban Agriculture and Integrated Systems

Urban agriculture applications include vertical farming monitoring, rooftop garden assessment, and community garden management. CV systems adapted for controlled environment agriculture provide precise monitoring of plant growth, disease detection, and harvest optimization in urban settings (Fei et al., 2025). Aquaponics and integrated agricultural systems require simultaneous monitoring of aquatic and terrestrial components. CV applications include fish health and behavior monitoring, plant growth assessment in hydroponic systems, visual indicators of water quality, and system balance optimization (Singh Chouhan, S., 2025).

7.4.3 Precision Interventions and Automation

Future developments will enable increasingly precise interventions based on CV analysis. Micro-targeted applications of nutrients, pesticides, herbicides, and water will be guided by real-time visual assessment (Ajith et al., 2025; Atapattu et al., 2025b). Robotic systems will perform individualized treatments based on CV diagnosis, reducing input usage while improving effectiveness.

Predictive maintenance of agricultural equipment and infrastructure will utilize CV for early detection of mechanical issues, structural problems, and system degradation, reducing downtime and maintenance costs while ensuring operational reliability (Ahmed et al., 2025).

8. Research Priorities and Policy Implications

8.1 Critical Knowledge Gaps and Methodological Improvements

Key research priorities include developing robust algorithms for complex agricultural environments, improving system performance under adverse weather conditions, advancing real-time processing capabilities for immediate intervention, and enhancing integration between different monitoring technologies. Methodological improvements needed include standardized evaluation protocols for agricultural CV systems, benchmark datasets representative of diverse agricultural conditions, validation frameworks for comparing different technological approaches, and performance metrics specific to agricultural applications rather than general CV tasks. Cross-disciplinary research gaps require attention to biological variability and its impact on system performance, integration of CV with agricultural decision support systems, economic modeling of technology adoption and benefits, and social factors affecting farmer acceptance and system utilization.

8.2 Interdisciplinary Collaboration and Standardization

Effective agricultural CV development requires collaboration between computer scientists, agricultural engineers, animal scientists, crop scientists, economists, and social scientists. Current research often lacks sufficient agricultural domain expertise, leading to systems that demonstrate technical sophistication but limited practical utility. Collaboration frameworks should include joint research projects between universities and agricultural organizations, industry partnerships for real-world system validation, farmer participation in system design and testing, and interdisciplinary training programs for researchers and practitioners. Standardization requirements encompass data collection and annotation protocols, performance evaluation metrics and procedures, system integration interfaces and communication protocols, and quality assurance standards for commercial systems. International coordination through organizations like ISO, IEEE, and FAO can accelerate standardization development.

8.3 Policy Implications

8.3.1 Infrastructure Support and Regulatory Frameworks

The adoption of agricultural CVs requires policy support, which includes heavy development of infrastructure, such as the availability of broadband internet in rural areas, technical support to farmers, and education on technology adoption. The Task Force Reviewing the Connectivity and Technology Requirements of Precision Agriculture offers recommendations on seeking to enhance the deployment of broadband on unserved agricultural land. The regulatory frameworks must strike a balance between promoting innovation and protecting farmers, requiring technology validation and flexibility in development, while also protecting data privacy and encouraging the sharing of useful data for research and development. Regulatory methods such as voluntary standards governing system functionality and data processing, certification of agricultural technology suppliers, and clear information about the intended use of data and privacy rights should be employed.

8.3.2 Incentive Mechanisms and Funding Support

Monetary incentives may accelerate the adoption of technology by utilizing cost-sharing programs, offering tax incentives to invest in precision agriculture, providing subsidized training and technical support programs, and awarding research grants to develop and validate technology. The opportunities for international cooperation include technology transfer programs between developed and developing nations, joint research activities on global challenges, the development of standards in international organizations, and the sharing of information on best practices and lessons learned.

8.3.3 Privacy Protection and Governance Frameworks

The concept of giving farmers the right to own data is largely justified by the sectoral literature and has been implemented in the form of voluntary codes of conduct governing agricultural data, suggesting the existence of changing governance strategies of agricultural data. It should also be considered as policy proposals with explicit legal regulations about the ownership of agricultural data, have a clear privacy standard of the farm monitoring data, require transparency of the farm monitoring data collection and use, and have a farmer control mechanism to decide upon data sharing. According to this approach, legal, technological, and human-centered suggestions are provided to safeguard the privacy of farm data and maintain the confidentiality of the farmer.

The governing structures must define the rights and obligations of farmers on the information collected by the monitoring systems, the duty of technology providers to protect and safeguard data, the enforcement of regulatory bodies to check compliance, and the dispute resolution mechanism for disputes related to the data. International coordination provides a uniform level of protection and allows the positive exchange of data in terms of global research and development in agriculture.

9. Conclusion

CV has proven to be a powerful enabler of precision agriculture, bridging gaps left by traditional monitoring methods in livestock and agroforestry systems. By automating critical tasks such as animal identification, disease detection, feeding behavior analysis, tree species recognition, and carbon monitoring, CV enhances decision-making and promotes sustainable resource management. However, practical deployment faces persistent challenges, including environmental variability, limited computational capacity in field conditions, inadequate datasets, integration with legacy systems, and socio-economic barriers to adoption. Addressing these constraints requires interdisciplinary collaboration, robust policy frameworks, financial support mechanisms, and farmer-focused training initiatives. Emerging technologies such as Vision Transformers, edge AI, blockchain-enabled traceability, and digital twin integration are expected to further expand the role of CV in agriculture. Ultimately, the adoption of CV-based monitoring can contribute significantly to improving productivity, enhancing animal welfare, supporting biodiversity, and mitigating climate change impacts, thus strengthening the resilience and sustainability of global agricultural systems.

Ethical Responsibilities of Authors

All authors have read, understood, and have complied as applicable with the statement on "Ethical responsibilities of Authors" as found in the Instructions for Authors.

Ethical Approval

Not applicable

Use of AI tool

No

Conflict of Interest

On behalf of all the authors the corresponding author states that there is no conflict of interest to disclose.

Competing Interests

Not applicable

Authors' contributions

Dissanayaka, N.S.: Conceptualization, Methodology, Writing – Original Draft, Writing – Review & Editing. Udumann, S.S.: Investigation, Formal analysis, Visualization, Software, Data Curation, Supervision. Nuwarapaksha, T.D.: Writing – Original Draft, Formal analysis, Methodology. Atapattu, A.J.: Formal analysis, Visualization. Atapattu, A.J.: Supervision

Funding

No funding has been received.

References

Ahmed, S., Reza, M.N., Karim, M.R., Jin, H., Kim, H. and Chung, S.-O. (2025). Abnormal operation detection of automated orchard irrigation system actuators by power consumption level. *Sensors*, *25*(2), 331. https://doi.org/10.3390/s25020331.

Ajith, S., Vijayakumar, S. and Elakkiya, N. (2025). Yield prediction, pest and disease diagnosis, soil fertility mapping, precision irrigation scheduling, and food quality assessment using machine learning and deep learning algorithms. *Discover Food*, *5*(1), 67. https://doi.org/10.1007/s44187-025-00338-1.

Akram, M., Bazai, S.U., Sulaman, M. and Ullah, F. (2025). Innovative Deep Learning image technologies: applications of Deep Learning in image processing. In: *Modern Intelligent Techniques for Image Processing* (pp. 145–180). IGI Global Scientific Publishing. https://doi.org/10.4018/979-8-3693-9045-0.ch007.

Aldossary, M., Alharbi, H.A. and Anwar Ul Hassan, C. (2024). Internet of Things (IoT)-Enabled Machine learning models for efficient monitoring of smart agriculture. *IEEE Access*, *12*, 75718–75734. https://doi.org/10.1109/ACCESS.2024.3404651.

Araújo, S.O., Peres, R.S., Barata, J., Lidon, F. and Ramalho, J.C. (2021). Characterising the agriculture 4.0 landscape: Emerging trends, challenges, and opportunities. *Agronomy*, *11*(4), 667. https://doi.org/10.3390/agronomy11040667.

Arulmozhi, E., Bhujel, A., Moon, B.-E. and Kim, H.-T. (2021). The application of cameras in precision pig farming: An overview for swine-keeping professionals. *Animals*, *11*(8), 2343. https://doi.org/10.3390/ani11082343.

Atapattu, A.J., Babalola, O.O., Udumann, S.S., Maitra, S., Nuwarapaksha, T.D. and Dissanayaka, N.S. (2025). Enhanced Efficiency Fertiliser (EEF) intervention for climate change resilience. In: *Innovations in Climate Resilient Agriculture* (pp. 447–464). Springer Nature Switzerland. https://doi.org/10.1007/978-3-031-84802-5_20.

Atapattu, A.J., Udumann, S.S., Dissanayaka, N.S., Nuwarapaksha, T.D., Dissanayake, A.S.B., Tennakoon, A., Rasika, D.M.D., Dassanayake, D.H.B.R., Ekanayake, J., Karalliyadda, S.M.C.B., Bandara, N.P.S.N., Sankalpa, J.K.S., Vinujan, S., Arachchige, A.C.G. and Kumarathunge, D.P. (2025). Best agronomic practices to mitigate the transition of climate action. In: *Transitioning to Climate Action*. MDPI. https://doi.org/10.3390/books978-3-03897-875-6-4.

Birdal, A.C., Avdan, U. and Türk, T. (2017). Estimating tree heights with images from an unmanned aerial vehicle. *Geomatics, Natural Hazards, and Risk*, *8*(2), 1144–1156. https://doi.org/10.1080/19475705.2017.1300608.

Bodkhe, J., Dighe, H., Gupta, A. and Bopche, L. (2018). Animal identification. In: *2018 International Conference on Advanced Computation and Telecommunication* (*ICACAT*) (pp. 1–4). https://doi.org/10.1109/ICACAT.2018.8933624.

Bongomin, O., Lamo, J., Guina, J.M., Okello, C., Ocen, G.G., Obura, M., Alibu, S., Owino, C.A., Akwero, A. and Ojok, S. (2024). UAV image acquisition and processing for high-throughput phenotyping in agricultural research and breeding programs. *The Plant Phenome Journal*, *7*(1). https://doi.org/10.1002/ppj2.20096.

Brown-Brandl, T.M., Adrion, F., Maselyne, J., Kapun, A., Hessel, E.F., Saeys, W., Van Nuffel, A. and Gallmann, E. (2019). A review of passive radio frequency identification systems for animal monitoring in livestock facilities. *Applied Engineering in Agriculture*, *35*(4), 579–591. https://doi.org/10.13031/aea.12928

Cesco, S., Sambo, P., Borin, M., Basso, B., Orzes, G. and Mazzetto, F. (2023). Smart agriculture and digital twins: Applications and challenges in a vision of sustainability. *European Journal of Agronomy*, *146*, 126809. https://doi.org/10.1016/j.eja.2023.126809.

Chang, Z., Liu, S., Xiong, X., Cai, Z. and Tu, G. (2021). A survey of recent advances in Edge-Computing-powered Artificial Intelligence of Things. *IEEE Internet of Things Journal*, *8*(18), 13849–13875. https://doi.org/10.1109/JIOT.2021.3088875.

Chouhan, S.S., Patel, R.K., Shukla, A. and Singh, U.P. (2025a). Drone Importance and their Necessity in Future Generation Agriculture. In: *Artificial Intelligence and Computer Vision for Ecological Informatics* (pp. 1–8). CRC Press. https://doi.org/10.1201/9781003518778-1.

Chouhan, S.S., Patel, R.K., Singh, U.P. and Tejani, G.G. (2025b). Integrating Drone in Agriculture: Addressing Technology, Challenges, Solutions, and Applications to Drive Economic Growth. *Remote Sensing Applications: Society and Environment*, 101576. https://doi.org/10.1016/j.rsase.2025.101576.

Chouhan, S.S., Singh, U.P. and Jain, S. (2020). Applications of computer vision in plant pathology: A survey. *Archives of Computational Methods in Engineering*, *27*(2), 611–632. https://doi.org/10.1007/s11831-019-09324-0.

Chouhan, S.S., Singh, U.P., Saxena, A. and Jain, S. (2024). Assessing the Importance and Need of Artificial Intelligence for Precision Agriculture. In: *Artificial Intelligence Techniques in Smart Agriculture* (pp. 1–6). Singapore: Springer Nature Singapore. https://doi.org/10.1007/978-981-97-5878-4_1.

Crociati, M., Sylla, L., De Vincenzi, A., Stradaioli, G. and Monaci, M. (2022). How to predict parturition in cattle? A literature review of automatic devices and technologies for remote monitoring and calving prediction. *Animals*, *12*(3), 405. https://doi.org/10.3390/ani12030405.

Curti, P. de F., Selli, A., Pinto, D.L., Merlos-Ruiz, A., Balieiro, J.C. de C. and Ventura, R.V. (2023). Applications of livestock monitoring devices and machine learning algorithms in animal production and reproduction: An overview. *Animal Reproduction*, *20*(2). https://doi.org/10.1590/1984-3143-ar2023-0077.

Debats, S.R., Luo, D., Estes, L.D., Fuchs, T.J. and Caylor, K.K. (2016). A generalized computer vision approach to mapping crop fields in heterogeneous agricultural landscapes. *Remote Sensing of Environment*, *179*, 210–221. https://doi.org/10.1016/j.rse.2016.03.010.

Dissanayaka, D.M.N.S., Dissanayake, D.K.R.P.L., Udumann, S.S., Nuwarapaksha, T.D. and Atapattu, A.J. (2023a). Agroforestry: A key tool in the climate-smart agriculture context: A review on coconut cultivation in Sri Lanka. *Frontiers in Agronomy*, *5*. https://doi.org/10.3389/fagro.2023.1162750.

Dissanayaka, D.M.N.S., Udumann, S.S. and Atapattu, A.J. (2024). Synergies between tree crops and ecosystems in tropical agroforestry. In: *Agroforestry* (pp. 49–87). Wiley. https://doi.org/10.1002/9781394231164.ch3.

Dissanayaka, N.S., Nimanthika, N., Nuwarapaksha, T.D., Udumann, S.S. and Atapattu, A.J. (2025a). Agroforestry and Climate-Smart Agriculture. In *Agroforestry for a Climate-Smart Future* (pp. 173–200). IGI Global Scientific Publishing. https://doi.org/10.4018/979-8-3693-8282-0.ch006.

Dissanayaka, N.S., Nuwarapaksha, T.D., Udumann, S.S. and Atapattu, A.J. (2025b). Agroforestry for restoring degraded land. In: *Ecological Solutions to Agricultural Land Degradation* (pp. 63–87). Springer Nature Singapore. https://doi.org/10.1007/978-981-96-3392-0_3.

Dissanayaka, N.S., Udumann, S.S., Nuwarapaksha, T.D. and Atapattu, A.J. (2023b). Agroforestry: An avenue for resilient and productive farming through integrated crops and livestock production. In: *Transitioning to Zero Hunger*. MDPI. https://doi.org/10.3390/books978-3-03897-863-3-5.

Fei, S., Wu, R., Liu, H., Yang, F. and Wang, N. (2025). Technological innovations in urban and peri-urban agriculture: Pathways to sustainable food systems in metropolises. *Horticulturae*, *11*(2), 212. https://doi.org/10.3390/horticulturae11020212.

Fracarolli, J.A., Pavarin, F.F.A., Castro, W. and Blasco, J. (2020). Computer vision applied to food and agricultural products. *REVISTA CIÊNCIA AGRONÔMICA*, *51*(5). https://doi.org/10.5935/1806-6690.20200087.

Fuentes, S., Gonzalez Viejo, C., Tongson, E., Dunshea, F.R., Dac, H.H. and Lipovetzky, N. (2022). Animal biometric assessment using non-invasive computer vision and machine learning are good predictors of dairy cows age and welfare: The future of automated veterinary support systems. *Journal of Agriculture and Food Research*, *10*, 100388. https://doi.org/10.1016/j.jafr.2022.100388.

Gillan, J.K., McClaran, M.P., Swetnam, T.L. and Heilman, P. (2019). Estimating forage utilization with drone-based photogrammetric point clouds. *Rangeland Ecology and Management*, *72*(4), 575–585. https://doi.org/10.1016/j.rama.2019.02.009.

Gómez, Y., Stygar, A.H., Boumans, I.J.M.M., Bokkers, E.A.M., Pedersen, L.J., Niemi, J.K., Pastell, M., Manteca, X. and Llonch, P. (2021). A systematic review on validated precision livestock farming technologies for pig production and its potential to assess animal welfare. *Frontiers in Veterinary Science*, *8*. https://doi.org/10.3389/fvets.2021.660565.

Gorh, D. and Baruah, K.K. (2019). Estimation of methane and nitrous oxide emission from wetland rice paddies with reference to global warming potential. *Environmental Science and Pollution Research*, *26*(16), 16331–16344. https://doi.org/10.1007/s11356-019-05026-z

Gupta, V., Garg, A. and Agrawal, S. (2024). Remote monitoring and control systems in agriculture and farming. In *Recent Trends in Artificial Intelligence Towards a Smart World: Applications in Industries and Sectors* (pp. 279–294). Singapore: Springer Nature Singapore. https://doi.org/10.1007/978-981-97-6790-8_10.

Hussain, M., Abdullah, M., Ashraf, M.N., Shaheen, A., Farooqi, Z.U.R., Abbas, A. and Tariq, Y. (2025). World food hunger in 2050 and nano solutions: Probabilities and prospects based on plant based food. In: *Nanomaterials for Enhanced Plant-Based Food Production* (pp. 11–21). Elsevier. https://doi.org/10.1016/B978-0-443-23688-4.00010-5.

Illarionova, S., Shadrin, D., Tregubova, P., Ignatiev, V., Efimov, A., Oseledets, I. and Burnaev, E. (2022). A survey of computer vision techniques for forest characterization and carbon monitoring tasks. *Remote Sensing*, *14*(22), 5861. https://doi.org/10.3390/rs14225861.

Iqra, Kaisar Javeed Giri, Bisma Ashraf Wani, D.N.R. and Farhana Mehraj. (2025). Role of Machine Learning and Computer Vision in the Agri-Food Industry. In: *Artificial Intelligence in the Food Industry*. (pp. 56–83). CRC Press.

Jabed, Md. A. and Azmi Murad, M.A. (2024). Crop yield prediction in agriculture: A comprehensive review of machine learning and deep learning approaches, with insights for future research and sustainability. *Heliyon*, *10*(24), e40836. https://doi.org/10.1016/j.heliyon.2024.e40836.

Jain, G., Jain, A., Jaiswal, A., Ahmad, S., Kapse, V.M. and Bisht, S.S. (2025). Transformative potential: IoT, Cloud, Big Data, and AI-Driven Blockchain Digital Twins in agriculture. In: *Blockchain and Digital Twin Applications in Smart Agriculture* (pp. 92–111). Auerbach Publications.

Jayanthi, J., Kumar, K. Arun and Verma Shubham. (2024). Harnessing computer vision for agricultural transformation: insights, techniques, and applications. In: *Applications of Computer Vision and Drone Technology in Agriculture 4.0* (pp. 111–131). Springer Nature Singapore. https://doi.org/10.1007/978-981-99-8684-2_8.

Jimenez-Berni, J.A., Deery, D.M., Rozas-Larraondo, P., Condon, A. (Tony) G., Rebetzke, G.J., James, R.A., Bovill, W.D., Furbank, R.T. and Sirault, X.R.R. (2018). High throughput determination of plant height, ground cover, and above-ground biomass in wheat with LiDAR. *Frontiers in Plant Science*, *9*. https://doi.org/10.3389/fpls.2018.00237.

Kakani, V., Nguyen, V.H., Kumar, B.P., Kim, H. and Pasupuleti, V.R. (2020). A critical review on computer vision and artificial intelligence in food industry. *Journal of Agriculture and Food Research*, *2*, 100033. https://doi.org/10.1016/j.jafr.2020.100033.

Kamilaris, A. and Prenafeta-Boldú, F.X. (2018). A review of the use of convolutional neural networks in agriculture. *The Journal of Agricultural Science*, *156*(3), 312–322. https://doi.org/10.1017/S0021859618000436.

Katiyatiya, C.L.F., Muchenje, V. and Mushunje, A. (2015). Seasonal variation in coat characteristics, tick loads, cortisol levels, some physiological parameters, and temperature humidity index on

Nguni cows raised in low- and high-input farms. *International Journal of Biometeorology*, *59*(6), 733–743. https://doi.org/10.1007/s00484-014-0887-4.

Kaur, A., Kumar, M. and Jindal, M.K. (2022). Cattle identification with muzzle pattern using computer vision technology: a critical review and prospective. *Soft Computing*, *26*(10), 4771–4795. https://doi.org/10.1007/s00500-022-06935-x.

Kauranne, T., Joshi, A., Gautam, B., Manandhar, U., Nepal, S., Peuhkurinen, J., Hämäläinen, J., Junttila, V., Gunia, K., Latva-Käyrä, P., Kolesnikov, A., Tegel, K. and Leppänen, V. (2017). LiDAR-Assisted Multi-Source Program (LAMP) for measuring above ground biomass and forest carbon. *Remote Sensing*, *9*(2), 154. https://doi.org/10.3390/rs9020154.

Lakhan Ray, R. and Getachew Tikuye, B. (2025). Impact of climate change on surface water resources. In: *Limnology: Surface and Ground Water Ecosystems.* https://doi.org/10.5772/intechopen.1011407.

Latif, A., Rasheed, A., Sajid, U., Ahmed, J., Ali, N., Ratyal, N.I., Zafar, B., Dar, S.H., Sajid, M. and Khalil, T. (2019). Content-based image retrieval and feature extraction: A comprehensive review. *Mathematical Problems in Engineering*, *2019*(1). https://doi.org/10.1155/2019/9658350.

Li, X., Hou, B., Zhang, R. and Liu, Y. (2023). A review of RGB image-based internet of things in smart agriculture. *IEEE Sensors Journal*, *23*(20), 24107–24122. https://doi.org/10.1109/JSEN.2023.3309774.

Lin, Y.S., Fang, S.L., Kang, L., Chen, C.C., Yao, M.H. and Kuo, B.J. (2024). Combining recurrent neural network and sigmoid growth models for short-term temperature forecasting and tomato growth prediction in a plastic greenhouse. *Horticulturae*, *10*(3), 230. https://doi.org/10.3390/horticulturae10030230.

Liu, N., Qi, J., An, X. and Wang, Y. (2023). A review on information technologies applicable to precision dairy farming: focus on behavior, health monitoring, and the precise feeding of dairy cows. *Agriculture*, *13*(10), 1858. https://doi.org/10.3390/agriculture13101858.

Liu, X., Lam, K.H., Zhu, K., Zheng, C., Li, X., Du, Y., Liu, C. and Pong, P.W.T. (2019). Overview of spintronic sensors with internet of things for smart living. *IEEE Transactions on Magnetics*, *55*(11), 1–22. https://doi.org/10.1109/TMAG.2019.2927457.

Liu, Z., He, Y., Wang, C. and Song, R. (2020). Analysis of the influence of foggy weather environment on the detection effect of machine vision obstacles. *Sensors*, *20*(2), 349. https://doi.org/10.3390/s20020349.

Logeshwaran, J., Srivastava, D., Kumar, K.S., Rex, M.J., Al-Rasheed, A., Getahun, M. and Soufiene, B. O. (2024). Improving crop production using an agro-deep learning framework in precision agriculture. *BMC Bioinformatics*, *25*(1), 341. https://doi.org/10.1186/s12859-024-05970-9.

Lu, B., Dao, P., Liu, J., He, Y. and Shang, J. (2020). Recent advances of hyperspectral imaging technology and applications in agriculture. *Remote Sensing*, *12*(16), 2659. https://doi.org/10.3390/rs12162659.

Ma, W., Sun, Y., Qi, X., Xue, X., Chang, K., Xu, Z., Li, M., Wang, R., Meng, R. and Li, Q. (2024). Computer-vision-based sensing technologies for livestock body dimension measurement: A survey. *Sensors*, *24*(5), 1504. https://doi.org/10.3390/s24051504.

Majumdar, P., Mitra, S., Bhattacharya, D. and Bhushan, B. (2024). Enhancing sustainable 5G powered agriculture 4.0 : Summary of low power connectivity, internet of UAV things, AI solutions, and research trends. *Multimedia Tools and Applications*, *84*(17), 17389–17433. https://doi.org/10.1007/s11042-024-19728-1.

McLennon, E., Dari, B., Jha, G., Sihi, D. and Kankarla, V. (2021). Regenerative agriculture and integrative permaculture for sustainable and technology driven global food production and security. *Agronomy Journal*, *113*(6), 4541–4559. https://doi.org/10.1002/agj2.20814.

Meng, H., Zhang, L., Yang, F., Hai, L., Wei, Y., Zhu, L. and Zhang, J. (2025). Livestock biometrics identification using computer vision approaches: A review. *Agriculture*, *15*(1), 102. https://doi.org/10.3390/agriculture15010102.

Michez, A., Lejeune, P., Bauwens, S., Herinaina, A., Blaise, Y., Castro Muñoz, E., Lebeau, F. and Bindelle, J. (2019). Mapping and monitoring of biomass and grazing in pasture with an unmanned aerial system. *Remote Sensing*, *11*(5), 473. https://doi.org/10.3390/rs11050473.

Muppala, C. and Guruviah, V. (2020). Machine vision detection of pests, diseases, and weeds: A review. *Journal of Phytology*, 9–19. https://doi.org/10.25081/jp.2020.v12.6145.

Neethirajan, S. (2017). Recent advances in wearable sensors for animal health management. *Sensing and Bio-Sensing Research*, *12*, 15–29. https://doi.org/10.1016/j.sbsr.2016.11.004.

Nuwarapaksha, T.D., Dilhan, R.M.C.P., Udumann, S.S., Ranasinghe, C.S., Egodawatta, W.C.P., Dissanayaka, N.S., Raveendra, T. and Atapattu, A.J. (2025a). Estimation of the total carbon sequestration potential of coconut-gliricidia mixed cropping systems in Sri Lanka. *Carbon Footprints*, *4*(1). https://doi.org/10.20517/cf.2024.31.

Nuwarapaksha, T.D., Udumann, S.S., Dissanayaka, N.S. and Atapattu, A.J. (2024a). Leveraging Blockchain for Food Supply Transparency: From Farm to Fork. In: *Emerging Trends in Food and Agribusiness Marketing* (pp. 75–100). IGI Global. https://doi.org/10.4018/979-8-3693-6715-5.ch003.

Nuwarapaksha, T.D., Udumann, S.S., Dissanayaka, N.S. and Atapattu, A.J. (2025b). Future Prospects and Emerging Trends in Agroforestry Research. In: *Agroforestry for a Climate-Smart Future,* (pp. 497–518). IGI Global Scientific Publishing. https://doi.org/10.4018/979-8-3693-8282-0.ch017.

Nuwarapaksha, T.D., Udumann, S.S., Dissanayaka, N.S., Dilshan, R.M.N. and Atapattu, A.J. (2024b). AI-Driven solutions for sustainable irrigation: Exploring smart technologies to enhance conservation and efficiency. In: *Integrating Agriculture, Green Marketing Strategies, and Artificial Intelligence* (pp. 1–32). IGI Global Scientific Publishing. https://doi.org/10.4018/979-8-3693-6468-0.ch001.

Nuwarapaksha, T.D., Udumann, S.S., Dissanayaka, N.S., Dilshan, R.M.N. and Atapattu, A.J. (2024c). Revolutionizing agriculture by advanced water and irrigation management technologies. In: *Emerging Trends and Technologies in Water Management and Conservation* (pp. 285–318). IGI Global Scientific Publishing. https://doi.org/10.4018/979-8-3693-6920-3.ch009.

Okinda, C., Nyalala, I., Korohou, T., Okinda, C., Wang, J., Achieng, T., Wamalwa, P., Mang, T. and Shen, M. (2020). A review on computer vision systems in monitoring of poultry: A welfare perspective. *Artificial Intelligence in Agriculture*, *4*, 184–208. https://doi.org/10.1016/j.aiia.2020.09.002.

Orchi, H., Sadik, M. and Khaldoun, M. (2021). On Using Artificial Intelligence and the Internet of Things for crop disease detection: A contemporary survey. *Agriculture*, *12*(1), 9. https://doi.org/10.3390/agriculture12010009.

Owino, A. (2023). Challenges of computer vision adoption in the Kenyan agricultural sector and how to solve them: A general perspective. *Advances in Agriculture*, *2023*, 1–9. https://doi.org/10.1155/2023/1530629.

Paparrizos, S., Vignola, R. and Sutanto, S.J. (2024). Integrating user- and data-driven weather forecasts to develop legitimate, credible, and salient information services for smallholders in the Global South. *Scientific Reports*, *14*(1), 22841. https://doi.org/10.1038/s41598-024-73539-w.

Patel, R.K. (2025). BM3D Filtering with Ensemble Hilbert-Huang Transform and Spiking Neural Networks for Cardiomegaly Detection in Chest Radiographs. *Computational Biology and Chemistry*, 108620. https://doi.org/10.1016/j.compbiolchem.2025.108620.

Patel, R.K., Chouhan, S.S., Chandan, T. and Singh, U.P. (2025). Basil Crop Detection Using Computer Vision and Deep Learning Approach. In: *Artificial Intelligence and Computer Vision for Ecological Informatics* (pp. 67–82). CRC Press. https://doi.org/10.1201/9781003518778-4.

Patrício, D.I. and Rieder, R. (2018). Computer vision and artificial intelligence in precision agriculture for grain crops: A systematic review. *Computers and Electronics in Agriculture*, *153*, 69–81. https://doi.org/10.1016/j.compag.2018.08.001.

Pineda, M., Barón, M. and Pérez-Bueno, M.-L. (2020). Thermal imaging for plant stress detection and phenotyping. *Remote Sensing*, *13*(1), 68. https://doi.org/10.3390/rs13010068.

Qiao, Y., Kong, H., Clark, C., Lomax, S., Su, D., Eiffert, S. and Sukkarieh, S. (2021). Intelligent perception-based cattle lameness detection and behaviour recognition: A review. *Animals*, *11*(11), 3033. https://doi.org/10.3390/ani11113033.

Raja, S.R., Subashini, B. and Prabu, R.S. (2024). 5G technology in smart farming and its applications. In: *Intelligent Robots and Drones for Precision Agriculture* (pp. 241–264). Cham: Springer Nature Switzerland. https://doi.org/10.1007/978-3-031-51195-0_12.

Reger, M., Friedrich, J., Wörz, S., Stumpenhausen, J., Sieghart, A. and Bernhardt, H. (2018). Challenges in modern automated Feeding Systems; *2018 Detroit, Michigan July 29–August 1, 2018*. https://doi.org/10.13031/aim.201801101.

Rolnick, D., Donti, P. L., Kaack, L.H., Kochanski, K., Lacoste, A., Sankaran, K., Ross, A.S., Milojevic-Dupont, N., Jaques, N., Waldman-Brown, A., Luccioni, A.S., Maharaj, T., Sherwin, E.D., Mukkavilli, S.K., Kording, K.P., Gomes, C.P., Ng, A.Y., Hassabis, D., Platt, J.C. and Bengio, Y. (2023). Tackling climate change with machine learning. *ACM Computing Surveys*, *55*(2), 1–96. https://doi.org/10.1145/3485128.

Sathya, D., Thangamani, R. and Balaji, B.S. (2024). The revolution of edge computing in smart farming. In: *Intelligent Robots and Drones for Precision Agriculture* (pp. 351–389). Cham: Springer Nature Switzerland. https://doi.org/10.1007/978-3-031-51195-0_17.

Sejian, V., Shashank, C.G., Silpa, M.V., Madhusoodan, A.P., Devaraj, C. and Koenig, S. (2022). Non-Invasive Methods of Quantifying Heat Stress Response in Farm Animals with Special Reference to Dairy Cattle. *Atmosphere*, *13*(10), 1642. https://doi.org/10.3390/atmos13101642.

Setiawan, W.F., Shanty, R.N.T. and Swastyastu, C.A. (2024). Automated web-based curriculum vitae system with qr code integration for portfolio access. *Jutisi : Jurnal Ilmiah Teknik Informatika Dan Sistem Informasi*, *13*(1), 536. https://doi.org/10.35889/jutisi.v13i1.1874.

Sexton, A.E., Garnett, T. and Lorimer, J. (2019). Framing the future of food: The contested promises of alternative proteins. *Environment and Planning E: Nature and Space*, *2*(1), 47–72. https://doi.org/10.1177/2514848619827009.

Shi, J., Bai, Y., Diao, Z., Zhou, J., Yao, X. and Zhang, B. (2023). Row detection based navigation and guidance for agricultural robots and autonomous vehicles in row-crop fields: Methods and applications. *Agronomy*, *13*(7), 1780. https://doi.org/10.3390/agronomy13071780.

Singh, A. K., Ganapathysubramanian, B., Sarkar, S. and Singh, A. (2018). Deep Learning for Plant Stress Phenotyping: Trends and Future Perspectives. *Trends in Plant Science*, *23*(10), 883–898. https://doi.org/10.1016/j.tplants.2018.07.004.

Singh Chouhan, S., Saxena, E., Shukla, A., et al. (2025). Optimizing YOLO-Based Models for Real-Time Guava Detection with Probabilistic Fused Wiener Filter-Enhanced Feature Fusion. *Applied Fruit Science*, *67*, 383. https://doi.org/10.1007/s10341-025-01613-2.

Singh, V. and Misra, A.K. (2017). Detection of plant leaf diseases using image segmentation and soft computing techniques. *Information Processing in Agriculture*, *4*(1), 41–49. https://doi.org/10.1016/j.inpa.2016.10.005.

Tian, H., Wang, T., Liu, Y., Qiao, X. and Li, Y. (2020). Computer vision technology in agricultural automation: A review. *Information Processing in Agriculture*, *7*(1), 1–19. https://doi.org/10.1016/j.inpa.2019.09.006.

Tyagi, H., Singh, A., Samant, S., Goel, P.K., Balyan, V. and Tyagi, H. (2025). Designing IoT Architectures for Precision Carbon Sequestration. In: *Advanced Systems for Monitoring Carbon Sequestration* (pp. 509–544). IGI Global. https://doi.org/10.4018/979-8-3373-2091-5.ch021.

Udumann, S.S., Dissanayaka, N.S. and Atapattu, A.J. (2024). Women's Roles in Building Resilience to Climate Change in Agriculture. In: *Impact of Women in Food and Agricultural Development* (pp. 58–78). IGI Global. https://doi.org/10.4018/979-8-3693-3037-1.ch004.

Udumann, S.S., Dissanayaka, N.S., Nuwarapaksha, T.D. and Atapattu, A.J. (2025a). Agroforestry Education and Training for the Next Generation. In: *Agroforestry for a Climate-Smart Future* (pp. 449–472). IGI Global Scientific Publishing. https://doi.org/10.4018/979-8-3693-8282-0.ch015.

Udumann, S.S., Dissanayaka, N.S., Nuwarapaksha, T.D. and Atapattu, A.J. (2025b). Regenerative Agricultural Practices for Sustainable Soil Health and Food Production. In: *Ecological Solutions to Agricultural Land Degradation* (pp. 117–143). Springer Nature Singapore. https://doi.org/10.1007/978-981-96-3392-0_5.

Usman, M., Ejaz, M., Nichol, J.E., Farid, M.S., Abbas, S. and Khan, M.H. (2023). A Comparison of Machine Learning Models for Mapping Tree Species Using WorldView-2 Imagery in the Agroforestry Landscape of West Africa. *ISPRS International Journal of Geo-Information*, *12*(4), 142. https://doi.org/10.3390/ijgi12040142.

Voronkov, R. and Bezuglyi, M. (2025). Application of YOLO and U-Net models for building material identification on segmented images. *Informatyka, Automatyka, Pomiary w Gospodarce i Ochronie Środowiska*, *15*(2), 13–17. https://doi.org/10.35784/iapgos.6968.

Walter, A., Finger, R., Huber, R. and Buchmann, N. (2017). Smart farming is key to developing sustainable agriculture. *Proceedings of the National Academy of Sciences*, *114*(24), 6148–6150. https://doi.org/10.1073/pnas.1707462114.

Wang, D., Zhao, M., Li, Z., Xu, S., Wu, X., Ma, X. and Liu, X. (2025). A survey of unmanned aerial vehicles and deep learning in precision agriculture. *European Journal of Agronomy*, *164*, 127477. https://doi.org/10.1016/j.eja.2024.127477.

Wu, B., Zhang, M., Zeng, H., Tian, F., Potgieter, A. B., Qin, X., Yan, N., Chang, S., Zhao, Y., Dong, Q., Boken, V., Plotnikov, D., Guo, H., Wu, F., Zhao, H., Deronde, B., Tits, L. and Loupian, E. (2023). Challenges and opportunities in remote sensing-based crop monitoring: A review. *National Science Review*, *10*(4). https://doi.org/10.1093/nsr/nwac290.

Wu, Z., Chen, Y., Zhao, B., Kang, X. and Ding, Y. (2021). Review of Weed Detection Methods Based on Computer Vision. *Sensors*, *21*(11), 3647. https://doi.org/10.3390/s21113647.

Zeng, Y., Hao, D., Huete, A., Dechant, B., Berry, J., Chen, J. M., Joiner, J., Frankenberg, C., Bond-Lamberty, B., Ryu, Y., Xiao, J., Asrar, G. R. and Chen, M. (2022). Optical vegetation indices for monitoring terrestrial ecosystems globally. *Nature Reviews Earth and Environment*, *3*(7), 477–493. https://doi.org/10.1038/s43017-022-00298-5.

Zhang, J., Zhang, Y., Qin, S., Wu, B., Wu, X., Zhu, Y., Shao, Y., Gao, Y., Jin, Q. and Lai, Z. (2018). Effects of seasonal variability of climatic factors on vegetation coverage across drylands in northern China. *Land Degradation & Development*, *29*(6), 1782–1791. https://doi.org/10.1002/ldr.2985.

Zhao, K., Jin, X., Ji, J., Wang, J., Ma, H. and Zhu, X. (2019). Individual identification of Holstein dairy cows based on detecting and matching feature points in body images. *Biosystems Engineering*, *181*, 128–139. https://doi.org/10.1016/j.biosystemseng.2019.03.004.

Zhong, H., Lin, W., Liu, H., Ma, N., Liu, K., Cao, R., Wang, T. and Ren, Z. (2022). Identification of tree species based on the fusion of UAV hyperspectral image and LiDAR data in a coniferous and broad-leaved mixed forest in Northeast China. *Frontiers in Plant Science*, *13*. https://doi.org/10.3389/fpls.2022.964769.

10 Real-Time Recognition of Fruits, Pests, and Weeds Using Integrated Deep Learning Methods

Kalamkuri Chandu[1], P.K. Pranav[2], Shreesh Sanjay Amin[3], Adityanshu Tripathi[4], Aditi Kachchap[5], Dattatray G. Bhalekar[6] and Abhishek Upadhyay[7*]

1. Introduction

The wings of the agriculture sectors worldwide converge on the junction, and the twin pressures of feeding food security to a growing populous whilst struggling with the gravity of climate change and resource scarcity (Oliver et al., 2013). Maintaining agriculture efficient, resilient, and sustainable has never been more critical. As the chief stumbling barrier poses the loss incurred due to pests, diseases, and weeds on the quality and quantity of the crop, thereafter causing tremendous economic distress among the farmers. Traditional approaches towards their control, Tends, characteristically, to include hand inspection followed by blanket chemical spraying, appear to be getting outdated in the near future (Upadhyay et al., 2024). Besides being physically exhausting leaving room for human mistake, the historically observed process generates environmental risks, pesticide resistance, death of pollinators, damage on beneficial insects and non-target organisms, and excessive economic costs (Yadav et al., 2023). Moreover, the process of producing evaluation for quality standards such as freshness, maturity, and defects is very much expensive and often due to hand sorting (Hussain et al., 2022) along the postharvest supply chain, thereby diverting attention from the process of providing the best products to the customers. In response to these sturdy problems, a technological upshot is changing the food and agricultural industry, whose focus is on precision agriculture (Lal & Stewart, 2015). This new paradigm of agriculture employs data-based approaches to minimize inputs

[1,3-5] M.Tech. Scholar, Department of Farm Machinery and Power Engineering, CAET, Dr. Rajendra Prasad Central Agricultural, Pusa, Samastipur, Bihar, India.

[2] Professor, Department of Farm Machinery and Power Engineering, CAET, Dr. Rajendra Prasad Central Agricultural, Pusa, Samastipur, Bihar, India.

[6] Center for Precision and Automated Agricultural Systems, Washington State University, Prosser, WA 99350, USA.

[7] Assistant Professor, Department of Farm Machinery and Power Engineering, CAET, Dr. Rajendra Prasad Central Agricultural, Pusa, Samastipur, Bihar, India.

* Corresponding author: abhiupadhyay7798@gmail.com

DOI: 10.1201/9781003715726-10

whilst retaining limited environmental impact, and AI (artificial intelligence) and ML (machine learning) are the key pillars (Getahun et al., 2017). An AI system is able to aggregate, process, and analyse vast amounts of information from sheer millions of sources like IoT sensors, drone/satellite photography, and mobile phones to provide real-time and decision-relevant data to the farm management. This is aiding the shift from being reactive to proactive in the plans through which the issue of plant health can be found and identified at an initial phase, even prior to the symptoms appearing visible through the naked human eye (Mehta & Burks, 2016). Advanced application deep learning architectures (Jiang et al., 2020) about the recognition of images-also explored initially under pest recognition, weed categorisation, and production sorting systems (Upadhyay et al., 2025a). Simultaneous currently constituted quite a promise; the ultimate innovation frontier lies in the next step of agricultural AI with robust hybrid models developed to overcome the individual deficiencies in making a complementary arrangement between a few severe components of a system (Grunig et al., 2021; Fenu & Malloci, 2019; Temraz et al., 2021). The chapter then proceeds to the great detail of these hybrid systems, exposing the synergistic uniting of different technologies to approach-the-high-complexity agricultural perception tasks. The architectural Paradigms that would be investigated would combine the local feature extraction capability of CNNs (convolutional neural networks) with the global contextual interpretation of the Transformers; the deep networks automatically learn features whilst classical classifiers like Support Vector Machines (SVMs) are utilised because they have stable decision boundaries; and accommodating multi stage pipelines wherein the networks handle distinct data refinement tasks, including denoising, up sample, and semantic segmentation. Furthermore, the work extends apart from visual observation, keeping in view integration of statistical time-series models with deep learning networks in a hybrid so that they can forecast future pest outbreaks. Through learning recent case examples on the critical areas of weed management, pest characterisation, and quality determination of fruits and vegetables, this chapter discusses how the synergistic approaches have introduced levels of utility, speed, and accuracy formerly unimaginable, constructing the basis towards a greener, smarter, and more productive system for the world agric.

2. Hybrid Models for Fruit Detection and Quality Assessment

Automated detection of fruit is one key tool in estimating yield, robotic harvesting, and quality sorting. Hybrid models have demonstrated that they can shine in this area by handling factors such as changes in lighting, occlusions, and discrimination in quality or subspecies that is subtle. These models represent the boundaries of the hybrid method for the analysis of fruit. These models contain a number of choices in hybridisation approaches, ranging from two-stage pre training and supervision learning to parallel backbones' fusion and ensemble methods.

2.1 CAE-ADN (Convolutional Autoencoder - Attention-based Dense Net)

Xue et al. (2023), whose model was developed with the intention to distinguish certain subspecies of fruit whose manual design of features is not feasible. The two-stage framework operates as follows. In the initial stage, the Convolutional Autoencoder (CAE) is trained on large sets of images of the fruit in the unsupervised fashion. In this way, the robust initial weight and bias estimation that enables general capture of principal visual features can be obtained by this model without any knowledge about the labels In the next stage, the pre-learned parameters are passed into the Attention-Based Densely Connected Network (ADN), making it ready for supervised classification. In its architecture, the ADN admits the integration of a Convolutional Block Attention Module (CBAM), which sequentially applies channel and spatial attention to feature maps to force the fruit network to focus on the most discriminative regions and features. Due to the hybrid method of pre training in an unsupervised manner and then attention-guided fine-tuning, the model has enhanced accuracy as well as interpretability.

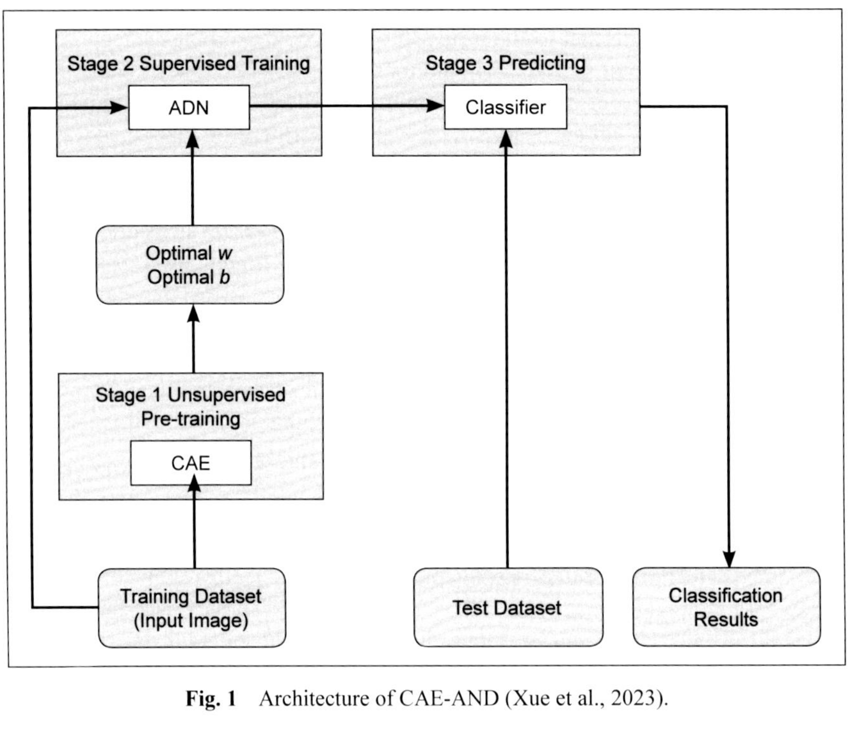

Fig. 1 Architecture of CAE-AND (Xue et al., 2023).

2.2 FruVeg_MultiNet

Oliullah et al. (2025) introduced the model with the goal of attaining high accuracy in the determination of fresh and spoiled produce, a situation significant enough for food quality and safety assurance. It uses a feature concatenation approach, in which two very powerful and renowned CNN frameworks, EfficientNetB7 and ResNet0, are concatenated in parallel. Both backbones receive the input image simultaneously, each one generating a very dissimilar, though high-level, feature description. Because of the computational efficiency, EfficientNetB7 provides a high-quality feature set, while the deep residual nature of ResNet0 painstakingly preserves the fine patterns. The outputs from the global average pooling layers of both models are then concatenated into a single, comprehensive feature vector. This fused vector is processed through a series of dropout, batch normalisation, and dense layers for the final classification. By combining the diverse feature perspectives of two state-of-the-art models, FruVeg_MultiNet achieves exceptional accuracy, reaching 99% on its primary dataset.

2.3 Hybrid CNN-ANN for Sorting

Real-time industrial sorting requires both speed and precision, especially when distinguishing between visually similar categories. Haggag et al. (2019) developed a cascaded hybrid model for a low-cost tomato sorting controller that classifies tomatoes as unripe, ripe, or defective. Recognising that differentiating ripe from defective tomatoes was the most challenging task, they designed a two-stage pipeline.

- **Stage:** A CNN performs an initial, broader classification to separate unripe (green) tomatoes from all red tomatoes (a combined ripe + defective class). This task is well-suited for a CNN, which easily learns the distinct colour features, achieving accuracy.
- **Stage:** The tomatoes identified as red are then passed to a specialized Artificial Neural Network (ANN). This ANN uses the red channel histogram of the image as its input feature, a descriptor that proved highly effective at capturing the subtle textural and colour variations between perfectly ripe and defective tomatoes. This specialised model also achieved 96% theoretical accuracy. This intelligent division of labour allows each model to excel at its specific task, leading to a highly accurate and robust sorting system with a total classification time of just, microseconds per tomato.

2.4 Ensemble of Hybrid ANNs (Majority Voting)

Sabzi et al. (2020) implemented an ensemble method to carry out robust fruit identification directly from the video stream in a few scenario cases of orchard conditions. The system works on the principle of majority voting; the final decision is an agreed output of all classifiers, and thus the system tends to be resistant to errors by any of the single models. The pipeline begins with the ANN-Harmony Search (ANN-HS) algorithm, which selects the four best candidate colour features so as to maintain real-time operations. These features are fed to three separate hybrid ANN classifiers that are optimised by different metaheuristic algorithms, namely, ANN-Bees Algorithm (ANN-BA), ANN-Biogeography-Based Optimization (ANN-

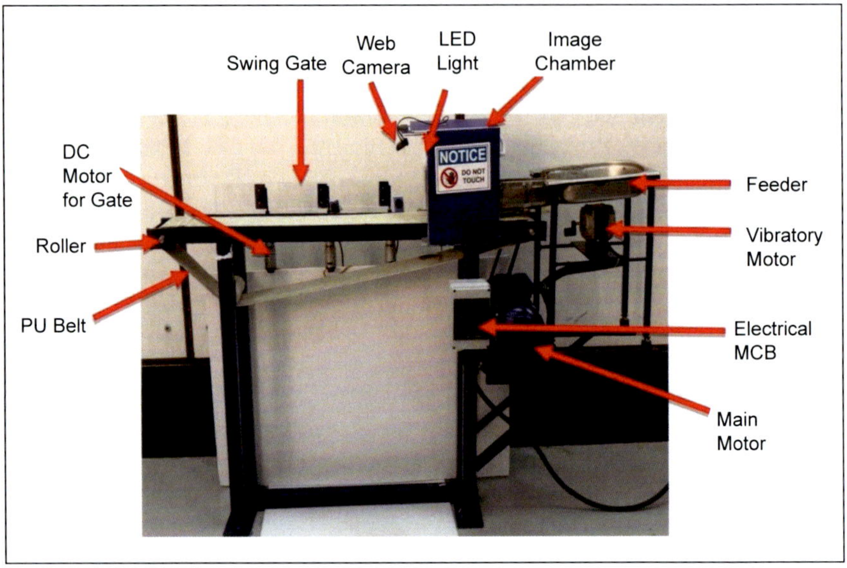

Fig. 2 Main components of developed tomato sorting machine (Haggag et al., 2019).

BBO), and ANN-Firefly Algorithm (ANN-FA). The final pixel-wise classification on fruit or background is decided by the consensus of these three classifiers. The ensemble method showed an astounding 90% accuracy and perfect stability across multiple test runs.

2.5 Hybrid CNN-RNN-LSTM with Pre-processing

Gill and Khehra (2021) introduced the entire multi-stage hybrid system for the classification of fruits such that high-level preprocessing methods are integrated with the hybridisation of deep learning algorithms.

- **Pre-processing:** The original images are first enhanced through Type-II fuzzy logic, boosting the contrast and clearness without the over/under-enhancement problem prevalent in the histogram-based approaches. An enhanced image is then delineated through a Teacher-Learning-Based Optimisation (TLBO) algorithm, decreasing the cross-entropy (MCET) in efforts aimed at estimating the optimal boundaries.
- **Classification:** Category classification is performed by the hybrid model, which uses a CNN to acquire discriminative features from the images, with the features acting as the inputs to the Recurrent Neural Network (RNN) containing LSTM (long short-term memory) cells. The RNN-LSTM then models the dependencies within the extracted features for the final class prediction. This finely crafted pipeline exhibits the possibilities that combined pre processing, segmentation, and multiple neural architectures can bring about for a particularly effective classification system.

2.6 Mango YOLO (Customized Single-Stage Detector)

While it is not a hybrid method in the usual sense of the term attempting to build an interface between two separate networks, Mango YOLO conceives of a hybrid (Koirala et al., 2019). It was custom-designed for real-time mango detection by integrating the most effective features of existing single-stage detectors like YOLOv3 and YOLOv4. The architecture was optimised for the specific task by modifying the number of layers (layers, compared to 06 in YOLOv), replacing max-pooling with more efficient convolutional layers, and implementing multi scale detection on three different feature maps to accurately detect fruit of varying sizes. MangoYOLO was then benchmarked against established two-stage detectors like Faster R-CNN. It achieved a state-of-the-art F score of 96.8% and a detection speed of 8ms per 512*512pixel pixel ~ 14 FPS on full-resolution images, demonstrating that a purpose-built, conceptually hybrid architecture can outperform more complex, off-the-shelf models for a specific agricultural task.

3. Hybrid Models for Pest and Disease Detection

The earlier detection of diseases and pests is critical in thwarting the loss of crops as well as decreasing the use of broad-spectrum pesticides. Hybrid deep learning models have emerged in recent years as a powerful asset in this regard, providing specialist solutions for mobile release, difficult image processing, as well as real-time field monitoring.

All these hybrid methods, from lightweight mobile-first designs through heavyweight multi stage pipelines up to new designs in the graph-based variety, reflect the scale of the complexity that is presented by the in-field pest detection challenges.

3.1 MobileNet V + LSTM

Pal (2022) presents one such hybrid model that is proposed to be run on light-weight mobile devices. The trade-off between accuracy and computational efficiency is towards efficiency. MobileNet V is the key feature extractor. Being a CNN design with fewer parameters and higher speed, the MobileNet V extracts the feature maps which are further put into a Long-Short-Term Model.

3.2 Memory (LSTM) network

The LSTM accepts these features in sequence, allowing it to learn contextual dependencies that make classification more robust when the visual cues are ambiguous. The model achieved a high accuracy of 98% and is designed to work offline, a critical feature for farmers in remote areas with poor internet connectivity.

Four-Stage Hybrid Pipeline: Chodey and Shariff (2022) addressed the poor quality of in-field images by designing a sequential four-stage hybrid model where each stage performs a specialised task.

- **Stage-Denoising:** A Bayesian image denoising framework is first applied to remove noise from the raw input frames.
- **Stage-Enhancement:** The denoised image is then enhanced using LightenNet, a CNN-based architecture that improves the visibility of weakly illuminated or low-contrast regions where small pests may reside.
- **Stage-Segmentation:** A context-guided ResNet-0 performs semantic segmentation to isolate pests from the complex background of leaves and stems.
- **Stage-Detection:** Finally, the clean, segmented image is passed to a pre-trained CNN for the final detection and classification. This pipeline approach, which systematically refines the image before detection, achieves an average precision of 86.7% and is suitable for real-time video monitoring.

3.3 Hybrid Vision GNN (HV-GNN)

Moving beyond traditional pixel-based analysis, Maruthai et al. (2025) proposed a novel Hybrid Vision Graph Neural Network (HV-GNN) to model the spatial relationships between pests on coffee plants. This model synergizes a CNN with a Graph Neural Network (GNN). The process begins with a CNN extracting localised information and identifying Regions of Interest (ROIs) that likely contain pests. These ROIs are then treated as nodes in a graph. Edges are constructed between these nodes to represent their spatial proximity or other contextual relationships. The GNN then processes this graph, allowing the model to learn not just the appearance of individual pests but also their relational patterns, such as clustering behaviour, which can be a strong indicator of an infestation Higher-order modelling in pest detection was well rewarded with a 96.6% accuracy.

3.4 Hybrid CNN + YOLOv

To achieve high-speed, real-time detection, and localisation of rice pests, Laureta et al. (2025) designed a classic hybrid of a feature extractor and an object detector. In the hybrid system, the deep CNN extracts rich and discriminative visual features from the input image, which involve the various shapes and textures of different pest species at a fine level of granularity. These high-level feature maps are then forwarded to the YOLOv architecture, which is an extremely optimised single-stage detector. It takes the features and performs bounding box regression and classification at very low latency. This division of labour allows the CNN to focus on feature quality while YOLOv ensures real-time speed. The resulting system achieved a mAP0 of 98% and an inference speed of FPS, making it highly effective for in-field deployment.

3.5 Prediction under ARIMA-LSTM

After shifting from visual detection towards time-based forecasting, Wang and Li (2025) proposed a hybrid framework that is able to forecast sugarcane pests and diseases' breakouts through historical time-series information. Their framework is composed of a conventional statistical method and deep learning approach.

- **Autoregressive Integrated Moving Average (ARIMA):** It is utilised in the forecast of the linear patterns and seasonality that happen in the year, the past history of pest sightings, and weather information.
- **LSTM:** After the ARIMA model has captured the linear trend, the leftover residual data (which is the complicated, nonlinear part) is input into an LSTM network. The LSTM can effectively capture these nonlinear dependencies. The overall forecast is the combination of the outputs from the LSTM and the ARIMA modelling approaches. In this hybrid method, the combined approach substantially outdid both models separately, with a substantially lower Mean Squared Error (MSE) and proving capable of precise forecasting of future outbreaks.

3.6 Hybrid Adaptive + Batch Learning

Pituckwanich et al. (2025) took the learning approach itself into consideration, proposing a hybrid model of sugarcane stem borer destruction rate estimation that incorporates the strength of batch learning and adaptability of adaptive learning.

The hybrid format must integrate two learning systems:

- **Batch learning:** It initially trains the system on the history data with the targets forecasted under a high-performance batch learning model like XGBoost and LightGBM, in order to obtain a reliable, accurate baseline.
- **Adaptive learning:** Once it's up and running, the system starts using adaptive learning. Basically, it keeps updating itself as new data comes in, so there's no need to retrain it from scratch every time. It can tell when things start to change in the data and adjust to stay accurate. This mix of learning methods helps the model keep improving, staying reliable and flexible for long-term use in farming and field monitoring.

4. Hybrid Models for Weed Detection

Weed detection remains a crucial aspect beyond precision agriculture, as in other sectors similar selective control methods are applied, such as targeted insect spraying. In many cases, this approach leads to more effective weed management. Hybrid models have proven to be an ideal solution by developing intelligent systems that balance detailed local feature analysis with broader contextual understanding. These models collectively represent the most advanced hybrid strategies used for weed detection.

4.1 YOLO-SW (YOLO with Swin Transformer)

Shuai et al. (2025) introduced a deeply integrated design that significantly enhances YOLOv8's ability to identify weeds accurately and efficiently in challenging natural conditions.

- **Backbone:** The normal CNN backbone is replaced by the Swin Transformer to enable the model to steal global context and long-range dependencies with linear complexity, which is imperative to recognising weeds from crops under changing lighting.
- **Upsampling:** The ordinary nearest-neighbour upsampling has been replaced by the CARAFE operator (Content-Aware ReAssembly of Features), synthesising instance-specific kernels to retain fine-grained details while allowing for a better localisation of tiny weed seedlings.
- **Head:** The detection head incorporates the RT-DETR encoder, which uses an IoU-aware query selection mechanism to perform end-to-end detection, eliminating the need for the computationally expensive Non-Maximal Suppression (NMS) post-processing step. This powerful combination of a Transformer backbone, content-aware upsampling, and an NMS-free head resulted in a model with 92.3% mAP@50 and a real-time speed of 59 FPS on an NVIDIA Jetson platform, making it ideal for intelligent weeding robots.

4.2 Enhanced MobileViT

Liu and Chen (2025) proposed this model, which focuses on creating a lightweight yet highly accurate solution for weed identification on mobile and embedded devices. MobileViT is an architecture that inherently hybridises CNNs and Vision Transformers. It uses standard convolutions to learn local representations and then passes them to Transformer blocks to model global relationships, effectively getting the best of both worlds. The authors enhanced this model by:

- Pre processing images with a multi scale retinal enhancement algorithm to improve clarity and restore colour.
- Integrating an Efficient Channel Attention (ECA) module into the feature extraction network to allow the model to focus on the most informative channels, further improving accuracy while keeping the model lightweight. The resulting model achieved an impressive F-score of 98.5% with an average inference time of just 89 milliseconds, proving its suitability for on-device deployment.

4.3 Hybrid CNN-Transformer

Zhang (2023) also proposed a hybrid model combining a CNN and a Transformer to balance local and global information processing for weed recognition. The architecture begins with several stacked Convolutional Blocks (ConvBlocks) to extract local features efficiently. The output is then passed to a Transformer Block (TransBlock) for global feature extraction. A key innovation is the use of a Segmentation Depthwise Transpose Attention (SDTA) mechanism within the Transformer block, which is based on the EdgeNeXt architecture. SDTA achieves linear complexity in its self-attention calculation, making the Transformer component highly efficient without sacrificing performance. This hybrid model achieved 90.8% accuracy, significantly outperforming a standard Vision Transformer model (8%) on the same dataset.

4.4 Hybrid SegNet + U-Net

Focusing on the critical preliminary step of segmentation, Mishra et al. (2025) developed a hybrid model for estimating weed density and growth rate. Accurate segmentation is essential for site-specific management, as it isolates vegetation from the soil background and distinguishes weeds from crops. The proposed model is a hybrid of two renowned segmentation architectures: SegNet and U-Net. Both are encoder-decoder networks, but they use different mechanisms for upsampling. By combining their features and reducing the number of pooling layers, the authors created four distinct modified architectures (e.g., L-USN, L-Conv-USN). The best-performing hybrid, L-Conv-USN, achieved an outstanding accuracy of 99%. This work highlights the importance of hybridising models specifically for the segmentation sub-task in weed management.

4.5 Hybrid Features + BO-SVM

Kempegowda et al. (2022) took a different approach by hybridising the features rather than the models themselves. Their system creates a single, powerful feature vector by concatenating deep features with handcrafted ones.

- **Deep Features:** A pre trained AlexNet CNN is used to automatically extract high-level, abstract features from the weed images.
- **Handcrafted Features:** Traditional computer vision techniques are used to extract targeted colour features (colour moments) and texture features (from GLCM and Gabor wavelets). This combined feature vector, which contains both learned and engineered information, is then classified using a Bayesian Optimized Support Vector Machine (BO-SVM). The Bayesian optimisation automatically tunes the SVM's hyperparameters, ensuring peak performance. This feature fusion strategy achieved a very high accuracy of 98%.

4.6 Hybrid CNN-SVM

In a similar spirit to the feature-hybrid model, Tao and Wei (2022) developed a classic two-stage hybrid where a CNN acts as a feature extractor and an SVM serves as the classifier. A VGG network processes the input image to produce a high-dimensional feature vector from its penultimate fully connected layer. This vector is then fed into an SVM, replacing the standard softmax classification layer. This approach is motivated by the fact that SVMs, using the principle of structural risk minimisation, can often achieve better generalisation and find a more optimal decision boundary than the empirical risk minimisation of a softmax layer. The VGG-SVM hybrid improved classification accuracy over the standalone VGG network, demonstrating the continued relevance of combining deep feature extractors with powerful traditional classifiers.

5. Challenges and Limitations

5.1 Data-Related Challenges

The performance of any deep learning model is fundamentally tied to the quality and quantity of its training data, a relationship recognised as a primary bottleneck in agricultural AI research (Zhengle et al., 2024). Several key challenges persist:

- **Data Scarcity and Diversity:** Many existing datasets remain limited in scope, covering only a narrow range of pest species, crop types, or environmental scenarios. This paucity of variety often leads to overfitting and poor generalisation when models are deployed in new orchards or under different weather and lighting conditions (Alzayat et al., 2025). The need for expansion across weed types, growth stages, and geographic contexts remains substantial.
- **Labour-Intensive Annotation:** Manual annotation drawing bounding boxes, segmentation masks, or precise object labels is both time-consuming and costly, typically requiring expert knowledge. The challenges of consistency and scalability are well-documented, especially when converting datasets for object detection frameworks like YOLO (Rasmussen et al., 2022; Ranjan et al., 2024). To address this, emerging approaches leverage AI-assisted synthetic data generation, such as LLM-generated orchard scenes automatically annotated via SAM and YOLO base models, achieving high mask precision (Sapkota et al., 2024; Tripathi et al., 2025).
- **Poor In-field Image Quality:** Agricultural image captures are frequently affected by adverse conditions such as variable sunlight, shadows, motion blur from drones, and sensor noise all of which degrade detection accuracy. Even when preprocessing (e.g., deblurring) is applied, low-quality or corrupted input images remain a significant challenge for reliable inference.

5.2 Computational and Deployment Hurdles

A model that demonstrates high accuracy in laboratory conditions is only valuable if it can be deployed effectively in real-world agricultural settings (Upadhyay et al., 2025b). This practical deployment introduces a range of computational and logistical challenges:

- **High Computational Requirements:** Advanced hybrid architectures combining deep CNNs, transformers, or GNNs often demand substantial computational resources. Such models typically require high-performance GPUs during training and impose heavy memory demands during inference. For example, models like YOLO-AgriNet, when applied to high-resolution images, may consume approximately 4 GB of GPU memory—already stretching the capacity of certain low-cost hardware. Larger models can exceed this substantially, posing significant deployment barriers (Wang et al., 2025; Shukla et al., 2025).

- **The Speed-Accuracy Trade-off**: An inherent tension exists between maximising accuracy and ensuring real-time inference speed. Highly accurate models (e.g., two-stage detectors like Faster R-CNN) deliver strong performance but operate slowly (e.g., 5–10 FPS), rendering them unsuitable for time-critical applications such as robotic weeding. In contrast, one-stage models like YOLO offer faster speeds (over 50 FPS) with somewhat lower accuracy (~0.88 mAP), a necessary compromise for real-time operations. Moreover, a recent comparative study of YOLOv8, YOLOv9, YOLOv10, and RT-DETR models for real-time weed detection demonstrated that smaller YOLO variants can achieve inference times as low as 7.64 ms on high-end GPUs, while RT-DETR models offer improved precision, highlighting the delicate balance between latency and performance in field robotics (Allmendinger et al., 2025; Patel R.K., 2025).
- **Deployment on Embedded Systems**: Transitioning a model from a development environment to a resource-constrained platform (e.g., UAVs, Raspberry Pi, mobile devices) involves substantial engineering efforts. This includes model conversion (e.g., to TensorRT, ONNX), optimisation (pruning, quantisation), and hardware-specific adaptation. The gap between performance in controlled lab settings and reliable real-world deployment remains a formidable hurdle (Pintus et al., 2025)

5.3 Model Performance and Robustness Issues

Even with high-quality data and sufficient hardware, hybrid deep learning models encounter significant performance limitations when deployed in complex and unpredictable agricultural environments:

- **Occlusion and Dense Scenes:** Fruits, pests, and weeds in agricultural settings are often occluded by leaves, stems, or by each other, posing a substantial challenge for accurate detection and counting. For instance, apple clusters or fruits hidden in dense foliage are difficult to localise, which leads to underestimation of counts (Xiao et al., 2023). Transformer-based amodal segmentation has shown promise by reconstructing the full shape of occluded tomatoes with high accuracy (96.1%) and intersection-over-union (94.1%) in greenhouse environments (Yang et al., 2024).
- **Morphological Similarity:** Differentiating between visually similar targets remains particularly challenging, for example, distinguishing between similar weed species, weeds and crops in early growth stages, or subtly different fruit variants. A recent framework, iNatAg, highlights the importance of fine-grained species classification for precision agriculture, noting the difficulty posed by intra- and inter-species visual similarities (e.g., Amaranthus species) (Mandge et al., 2025).
- **Handling Environmental Variability:** Model performance often degrades under conditions not well represented in training data, such as shifting sunlight angles, diverse soil backgrounds, and adverse weather like rain or

fog. Variability in environmental conditions significantly alters appearance and confounds detection systems (Alibabaei, 2022).

- **Limitations of Hybrid Components:** Even within advanced hybrid models, individual components can introduce bottlenecks. For instance, specialised up-sampling operators that enhance small-object detection may struggle with highly camouflaged targets lacking contextual contrast. Similarly, time-series components like ARIMA used in forecasting hybrid models may underestimate pest outbreaks during volatile periods. The overall system's effectiveness is often constrained by its weakest sub-module, a challenge increasingly recognised in modular or hybrid deep learning frameworks (e.g., PD-YOLO facing trade-offs between precision and computational efficiency) (Li et al., 2025)

6. Future Directions and Emerging Trends

The evolution of hybrid deep learning models for agricultural applications is advancing rapidly. Based on the challenges identified and the successes demonstrated in the literature, several key trends and future research directions are emerging to create more robust, efficient, and accessible systems for precision agriculture.

6.1 Lightweight Architectures for Edge and Mobile Deployment

A primary research trend is the development of lightweight hybrid models optimised for edge computing on resource-constrained devices such as drones, smartphones, and onboard agricultural robots.

- **Model Optimization and Compression:** Techniques such as knowledge distillation, pruning, and quantisation are being widely explored to shrink model size and computational demands while maintaining accuracy. These approaches enable real-time inference directly in the field, reducing latency and reliance on cloud connectivity (Han et al., 2016).
- **Efficient Hybrid Designs:** Architectures like MobileViT and YOLO-SW demonstrate the potential of combining efficient CNNs with lightweight Transformer modules for high performance at low computational cost. Such designs provide a pathway toward real-time agricultural applications (Howard et al., 2019).

6.2 Multimodal Data Fusion for Enhanced Robustness

Relying solely on RGB imagery limits performance under complex field conditions.

- **Integrating Diverse Sensors:** Combining RGB with hyperspectral, LiDAR, thermal, or depth imagery significantly improves robustness and accuracy, particularly in distinguishing visually similar weeds or fruits and in handling variable weather and lighting conditions (Kamilaris & Prenafeta-Boldú, 2018; Singh Chouhan, S., 2025).

- **Cross-Regional and Cross-Climate Datasets:** Building large, geographically and climatically diverse datasets helps address biases of localised training sets and improves generalisation across different crops and regions (Sishodia et al., 2020).

6.3 Advanced Learning Strategies

Beyond architecture, new learning paradigms are gaining attention.

- **Continual and Adaptive Learning:** Since agriculture is dynamic, models must adapt to seasonal variations, crop growth stages, and pest behaviours. Continual learning methods are being developed to allow models to update over time without retraining from scratch (De Lange et al., 2021).
- **Self-Supervised and Synthetic Data Generation:** Self-supervised learning leverages unlabelled datasets to reduce annotation needs, while GANs are being used to generate synthetic crop, pest, and weed imagery for training, thus addressing dataset scarcity (Goodfellow et al., 2014).

6.4 Full Integration with Autonomous Systems

The ultimate goal of real-time detection is autonomous agricultural management.

- **Robotics and IoT Integration:** Lightweight models embedded into UAVs, robotic harvesters, and smart sprayers can integrate with IoT systems for coordinated action (Liakos et al., 2018).
- **Decision-Support and Automated Intervention:** Outputs from detection models will feed directly into automated decision systems, enabling tasks such as targetted spraying, robotic fruit picking, and pest alert systems, thereby closing the loop from perception to intervention (Shamshiri et al., 2018).

7. Conclusion

The chapter presents the panoramic perspective of the transformative excellence of hybrid AI/ML models over agriculture The hybrid models, covering a wide range of algorithms and methods, provide a system with greater accuracy, dependability, and flexibility compared to single methods. They can control the mess induced by the real-world agriculture environment, making them suitable for use in cases like pest control, weed detection, and crop quality determination. Hybrid models extend beyond mere identification, enabling quantitative measurements like weed density estimation and growth stage determination, as well as predictive forecasting of the advent of pests. By making the management proactive rather than reactive, it facilitates prompt decision-making and risk optimisation, thus maximising the allocation of resources. In addition, light hybrid systems are developed to be operable in real-time on portable and edge devices, thus allowing greater accessibility and usability to field operatives. However, certain issues must be resolved. The hybrid system works great when there are varied datasets of high quality, but may need computational resources not accessible to all farmers. Moving forward, the focus may shift toward explainability and transparency of AI, development of more efficient lightweight architectures, and integration of multiple data sources: imagery, sensors, and climate data. All in all, hybrid models are a giant step toward the sustainable, intelligent, and inclusive agricultural system, allocating resources to farm productivity and benefits to society at large.

Declaration

Ethical Responsibilities of Authors

"All authors have read, understood, and complied as applicable with the statement on 'Ethical responsibilities of Authors' as found in the Instructions for Authors."

Ethical Approval

Not applicable

Use of AI tool

The corresponding author declares that no AI tool has been used to prepare this manuscript.

Conflict of Interest

On behalf of all the authors the corresponding author states that there is no conflict of interest to disclose.

Competing interests

Not applicable

Authors' contributions

Kalamkuri Chandu[1]: Conceptualization, Methodology, Writing – Original Draft, **P. K. Pranav**[2]: Reviewing and Editing, Supervision, **Shreesh Sanjay Amin**[3]: Formal analysis, Visualization, **Adityanshu Tripathi**[4]: Formal analysis, Methodology, **Aditi Kachchap**[5]: Formal analysis, Visualization, **Dattatray G. Bhalekar**[6]: Writing – Reviewing and Editing, Supervision, **Abhishek Upadhyay**[7]: Writing – Reviewing and Editing, Conceptualization, Formal analysis, Visualization, Supervision.

Funding

No funding has been received.

References

Alibabaei, K. (2022). A review of the challenges of using deep learning techniques in agriculture. *Remote Sensing, 14*(3), 638. https://doi.org/10.3390/rs14030638.

Allmendinger, A., Saltık, A.O., Peteinatos, G.G., Stein, A. and Gerhards, R. (2025). Assessing the capability of YOLO- and transformer-based object detectors for real-time weed detection. *Precision Agriculture*. Advance online publication. https://doi.org/10.1007/s11119-025-10246-0.

Alzayat, S., Hatano, S. and Rahimi Azghadi, M. (2025). *Weed Detection in Challenging Field Conditions: A Semi-Supervised Framework for Overcoming Shadow Bias and Data Scarcity. arXiv*. https://doi.org/10.48550/arXiv.2508.19511,

Chodey, M.D. and Noorullah Shariff, C. (2022). Hybrid deep learning model for in-field pest detection on real-time field monitoring. *Journal of Plant Diseases and Protection*, *129*(3), 635–650.

De Lange, M., Aljundi, R., Masana, M., Parisot, S., Jia, X., Leonardis, A., Slabaugh, G. and Tuytelaars, T. (2021). A continual learning survey: Defying forgetting in classification tasks. *IEEE Transactions on Pattern Analysis and Machine Intelligence, 44*(7), 3366–3385. https://doi.org/10.1109/TPAMI.2021.3057446.

Fenu, G. and Malloci, F. (2019, Nov.)). An Application of Machine Learning Technique in Forecasting Crop Disease. In: P*roceedings of the 3rd International Conference on Big Data Research*, Paris, France, 20–22 November.

Getahun, S., Ambaw, A., Delele, M., Meyer, C.J. and Opara, U.L. (2017). Analysis of airflow and heat transfer inside fruit packed refrigerated shipping container: Part I—model development and validation. *J. Food Eng.*, *203*, 58–68.

Gill, H.S. and Khehra, B.S. (2021). Hybrid classifier model for fruit classification. *Multimedia Tools and Applications*, *80*(18), 27495–27530.

Goodfellow, I., Pouget-Abadie, J., Mirza, M., Xu, B., Warde-Farley, D., Ozair, S., Courville, A. and Bengio, Y. (2014). Generative adversarial nets. *Advances in Neural Information Processing Systems, 27*. https://doi.org/10.48550/arXiv.1406.2661.

Grünig, M., Razavi, E., Calanca, P., Mazzi, D., Wegner, J.D. and Pellissier, L. (2021). Applying deep neural networks to predict incidence and phenology of plant pests and diseases. *Ecosphere*, *12*, e03791.

Haggag, M., Abdelhay, S., Mecheter, A., Gowid, S., Musharavati, F. and Ghani, S. (2019). An intelligent hybrid experimental-based deep learning algorithm for tomato-sorting controllers. *IEEE Access*, *7*, 106890–106898.

Han, S., Mao, H. and Dally, W.J. (2016). Deep compression: Compressing deep neural networks with pruning, trained quantization and Huffman coding. In: *International Conference on Learning Representations* (*ICLR*). https://doi.org/10.48550/arXiv.1510.00149.

Howard, A., Sandler, M., Chu, G., Chen, L., Chen, B., Tan, M. and Adam, H. (2019). Searching for MobileNetV3. In: *Proceedings of the IEEE/CVF International Conference on Computer Vision* (*ICCV*) (pp. 1314–1324). https://doi.org/10.48550/arXiv.1905.02244.

Hussain, D., I. Hussain, M. Ismail, A. Alabrah, S.S. Ullah, H.M. Alaghbari, et al. (2022). A simple and efficient deep learning-based framework for automatic fruit recognition. *Comput. Intell. Neurosci.*, 2022.

Jiang, H., Zhang, C., Qiao, Y., Zhang, Z., Zhang, W. and Song, C. (2020). CNN feature based graph convolutional network for weed and crop recognition in smart farming, *Computers and Electronics in Agriculture*, *174*. doi:10.1016/j.compag.2020.105450.

Kamilaris, A. and Prenafeta-Boldú, F.X. (2018). Deep learning in agriculture: A survey. *Computers and Electronics in Agriculture, 147*, 70–90. https://doi.org/10.1016/j.compag.2018.02.016.

Kempegowda, S.K., Rajeswari, R., Satyanarayana, L. and Basavarajaiah, S.M. (2022). Hybrid features and ensembles of convolution neural networks for weed detection. *International Journal of Electrical and Computer Engineering, 12*(6), 6756.

Koirala, A., Walsh, K.B., Wang, Z. and McCarthy, C. (2019). Deep learning for real-time fruit detection and orchard fruit load estimation: Benchmarking of 'Mango YOLO'. *Precision Agriculture, 20*(6), 1107–1135.

Lal, R. and B.A. Stewart (Eds.) (2015). Soil-Specific farming: Precision agriculture (Vol. 22). CRC Press.

Laureta, M., Laureta, L.A., Cordero, E.R. and Cordero, S.G. (2025). Implementation of Hybrid Machine Learning Techniques for Detection and Classification of Leaf and Stem Pests in Rice Crops. *Linker: The Journal of Emerging Research in Agriculture, Fisheries, and Forestry* , *5*(1), 18–36.

Li, S., Chen, Z., Xie, J., Zhang, H. and Guo, J. (2025). PD-YOLO: A novel weed detection method based on multi scale feature fusion. *Frontiers in Plant Science, 16*, 1506524. https://doi.org/10.3389/fpls.2025.1506524.

Liakos, K.G., Busato, P., Moshou, D., Pearson, S. and Bochtis, D. (2018). Machine learning in agriculture: A review. *Sensors, 18*(8), 2674. https://doi.org/10.3390/s18082674.

Liu, X., Sui, Q. and Chen, Z. (2025). Real time weed identification with enhanced mobilevit model for mobile devices. *Scientific Reports*, *15*(1), 27323.

Maruthai, S., Selvanarayanan, R., Thanarajan, T. and Rajendran, S. (2025). Hybrid vision GNNs based early detection and protection against pest diseases in coffee plants. *Scientific Reports*, *15*(1), 11778.

Mehta, S.S. and Burks, T.F. (2016). Multi-camera fruit localization in robotic harvesting. *IFAC-Papers Online*, *49*, 90–95. [CrossRef]

Mishra, A. M., Singh, M. P., Singh, P., Diwakar, M., Gupta, I. and Bijalwan, A. (2025). Hybrid deep learning model for density and growth rate estimation on weed image dataset. *Scientific Reports*, *15*(1), 11330.

Oliullah, K., Islam, M.R., Babar, J.I., Quraishi, M.N., Rahman, M.M., Mahbub-Or-Rashid, M. and Bhuiyan, T.A.U.H. (2025). FruVeg_MultiNet: A hybrid deep learning-enabled IoT system for fresh fruit and vegetable identification with web interface and customized blind glasses for visually impaired individuals. *Journal of Agriculture and Food Research*, *19*, 101623.

Oliver, M.A., Bishop, T.F.A. and Marchant, B.P. (2013). *Precision Agriculture for Sustainability and Environmental Protection* (Vol. 9780203128329). Routledge. doi: 10.4324/9780203128329.

Pal, A.R. (2022). *Classification of Pest-infested Citrus Leaf Images Using MobileNet V2+ LSTM- based Hybrid Model* (Doctoral dissertation, Dublin, National College of Ireland).

Patel, R.K. (2025, Aug.). BM3D filtering with Ensemble Hilbert-Huang Transform and spiking neural networks for cardiomegaly detection in chest radiographs. *Computational Biology and Chemistry*, *8*, 108620.

Pintus, M., Colucci, F. and Maggio, F. (2025). Emerging developments in real-time edge AIoT for agricultural image classification. *IoT, 6*(1), 13. https://doi.org/10.3390/iot6010013.

Pituckwanich, W., Hormdee, D., Boonkong, A., Kaewfoongrungsi, P., Ratchaseema, M.T.N. and Veerachit, V. (2025). The Implementation of a Prediction System for Sugarcane's Destruction Rate from Sugarcane Stem Borer via Hybrid Machine Learning. *IEEE Access*.

Rasmussen, C.B., Kirk, K. and Moeslund, T.B. (2022). The challenge of data annotation in deep learning: A case study on Whole Plant Corn Silage. *Sensors, 22*(4), 1596. https://doi.org/10.3390/s22041596.

Sabzi, S., Pourdarbani, R., Kalantari, D. and Panagopoulos, T. (2020). Designing a fruit identification algorithm in orchard conditions to develop robots using video processing and majority voting based on hybrid artificial neural network. *Applied Sciences*, *10*(1), 383.

Sapkota, R., Paudel, A. and Karkee, M. (2024). *Zero-Shot Automatic Annotation and Instance Segmentation using LLM-Generated Datasets: Eliminating Field Imaging and Manual Annotation for Deep Learning Model Development. arXiv*. https://doi.org/10.48550/arXiv.2411.11285.

Shamshiri, R.R., Kalantari, F., Ting, K.C., Thorp, K.R., Hameed, I.A., Weltzien, C., Ahmad, D. and Shad, Z.M. (2018). Advances in greenhouse automation and controlled environment agriculture: A transition to plant factories and urban agriculture. *International Journal of Agricultural and Biological Engineering, 11*(1), 1–22. https://doi.org/10.25165/j.ijabe.20181101.3210.

Shuai, Y., Shi, J., Li, Y., Zhou, S., Zhang, L. and Mu, J. (2025). YOLO-SW: A Real-Time Weed Detection Model for Soybean Fields Using Swin Transformer and RT-DETR. *Agronomy*, *15*(7), 1712.

Shukla, A., Chouhan, S.S., Patel, R.K. and Singh, U.P. (2025). AgriCloud: Cloud-Based Framework for Scalable Crop Monitoring and Intelligent Decision Support. In: Chouhan, S.S., Patel, R.K., Singh, U.P., and Jain, S. (Eds.), *AgriTech Revolution*. Springer, Singapore. https://doi.org/10.1007/978-981-95-1268-3_6.

Singh Chouhan, S., Saxena, E., Shukla, A., et al. (2025). Optimizing YOLO-Based Models for Real-Time Guava Detection with Probabilistic Fused Wiener Filter-Enhanced Feature Fusion. *Applied Fruit Science*, *67*, 383. https://doi.org/10.1007/s10341-025-01613-2.

Sishodia, R.P., Ray, R.L. and Singh, S.K. (2020). Applications of remote sensing in precision agriculture: A review. *Remote Sensing, 12*(19), 3136. https://doi.org/10.3390/rs12193136.

Tao, T. and Wei, X. (2022). A hybrid CNN–SVM classifier for weed recognition in winter rape field. *Plant Methods*, *18*(1), 29.

Temraz, M., Kenny, E., Ruelle, E., Shalloo, L., Smyth, B. and Keane, M. (2021). Handling Climate Change Using Counterfactuals: Using Counterfactuals in Data Augmentation to Predict Crop Growth in an Uncertain Climate Future. In: *Case-Based Reasoning Research and Development: 29th International Conference, ICCBR 2021* (pp. 216–231) Salamanca, Spain, 13–16 September 2021; *Proceedings* 29. Springer International Publishing: Berlin/Heidelberg, Germany.

Tripathi, A., Vishwakarma, H., Patel, R.K., Shukla, A. and Chouhan, S.S. (2025). AI-Enhanced Agriculture: Cultivating the Future. In: Chouhan, S.S., Patel, R.K., Singh, U.P., and Jain, S. (Eds.), *AgriTech Revolution*. Springer, Singapore. https://doi.org/10.1007/978-981-95-1268-3_1.

Upadhyay, A., Chandel, N.S. and Chakraborty, S.K. (2024). Disease control measures using vision-enabled agricultural robotics. In: *Applications of Computer Vision and Drone Technology in Agriculture 4.0* (pp. 165–186). Singapore: Springer Nature Singapore. https://doi.org/10.1007/978-981-99-8684-2_10.

Upadhyay, A., Chandel, N.S., Singh, K.P., Chakraborty, S.K., Nandede, B.M., Kumar, M. and Elbeltagi, A. (2025a). Deep learning and computer vision in plant disease detection: A comprehensive review of techniques, models, and trends in precision agriculture. *Artificial Intelligence Review, 58*(3), 92. https://doi.org/10.1007/s10462-024-11100-x.

Upadhyay, A., Patel, A., Patel, A., Chandel, N.S., Chakraborty, S.K. and Bhalekar, D.G. (2025b). Leveraging AI and ML in precision farming for pest and disease management: Benefits, challenges, and future prospects. In: *Ecologically Mediated Development: Promoting Biodiversity Conservation and Food Security* (pp. 511–528). https://doi.org/10.1007/978-981-96-2413-3_23.

Wang, C., Song, C., Xu, T. and Jiang, R. (2025). Precision weeding in agriculture: A comprehensive review of intelligent laser robots leveraging deep learning techniques. *Agriculture, 15*(11), 1213. https://doi.org/10.3390/agriculture15111213.

Wang, M. and Li, T. (2025). Pest and disease prediction and management for sugarcane using a hybrid autoregressive integrated moving average:–A long short-term memory model. *Agriculture*, *15*(5), 500.

Xiao, F., Wang, H., Xu, Y. and Zhang, R. (2023). Fruit detection and recognition based on deep learning for automatic harvesting: An overview and review. *Agronomy, 13*(6), 1625. https://doi.org/10.3390/agronomy13061625.

Xue, G., Liu, S. and Ma, Y. (2023). A hybrid deep learning-based fruit classification using attention model and convolution autoencoder. *Complex and Intelligent Systems*, *9*(3), 2209–2219.

Yadav, M. K., Satapathy, S.N., Priyadarshini, P., Tripathy, B., Lenka, B., Behera, K.K. and Paikaray, S. S. (2023). Incidence and population dynamics of various associated insect-pests of late direct sown paddy, *Oryza sativa* L., at Gajapati, Odisha. *Indian Journal of Agricultural Research*. https://doi.org/10.18805/ag.d-5856.

Yang, J., et al. (2024). Application of amodal segmentation for shape completion of potentially occluded tomatoes using transformer-based networks. *Frontiers in Plant Science*. Advance online publication. https://doi.org/10.3389/fpls.2024.1376138.

Zhang, J. (2023). Weed recognition method based on hybrid CNN-transformer model. *Frontiers in Computing and Intelligent Systems*, *4*(2), 72–77.

Zhengle, W., Ruifeng, W., Minjuan, W., Lai, T. and Zhang, M. (2024). *Self-supervised Transformer-based Pre-training Method with General Plant Infection Dataset*. *arXiv*. https://doi.org/10.48550/arXiv.2407.14911.

11 Deep Learning Applications for Agriculture: An Introduction

Harshit Mishra[1*], Fredrick Kayusi[2], Vandana Singh[3], Arkan A. Ghaib[4] and Rashmi Mishra[5]

1. Introduction

Deep learning (DL) is arguably one of the most transformative technologies of the 21st century, revolutionizing the way knowledge is extracted from complex data and applied to real-world problems in the form of effective solutions. The application of DL in agriculture, a sector known for its dependence on manual labor and traditional methods, is indicative of the beginning of a shift toward data-driven, intelligent, and automated farming systems (Kamilaris & Prenafeta-Boldú, 2018). The rising world population, anticipated to reach 9.7 billion by 2050 according to United Nations estimates, is exerting tremendous pressure on agricultural systems to deliver food security in the backdrop of increasing climate change, land degradation, and water crises (Kumari et al., 2018). In this regard, DL applications in agriculture are becoming both a challenge and an opportunity, providing capabilities not witnessed before in all areas from crop monitoring and yield prediction to soil and water management, pest and disease detection, and agricultural robotics (Abdullahi & Sheriff, 2022; Mishra & Mishra, 2025).

Agriculture remains the backbone of global food production and rural livelihoods, providing employment for approximately 26.6% of the world's workforce as of 2023, particularly in developing countries. And yet, agriculture uses 70% of all freshwater on the planet and is responsible for 19–29% of global greenhouse gas emissions, underscoring the dual challenge of sustaining productivity while minimizing its environmental impact. Conventional methods, even if mechanized, are generally not precise and flexible enough to deal with

[1] Department of Agricultural Economics, College of Agriculture, Acharya Narendra Deva University of Agriculture and Technology, Kumarganj, Ayodhya, Uttar Pradesh – 224 229, India.
[2] Department of Environmental Studies, Geography, and Planning, Maasai Mara University, P.O. Box 861-20500, Narok, Kenya.
[3] Department of Microbiology, Sharda School of Allied Health Sciences, Sharda University, Greater Noida, Uttar Pradesh – 201 310, India.
[4] Information Technology Department, Management Technical College, Southern Technical University, Basrah 61004, Iraq.
[5] College of Economics and Business Administration, University of Technology and Applied Sciences - Al Musannah, P.O. Box 191, Muscat, Oman.
* Corresponding author: wehars@gmail.com

DOI: 10.1201/9781003715726-11

such conditions (Koul, 2021). DL's ability to process heterogeneous and high-dimensional data provides scalable solutions, integrated with remote sensing, Internet of Things (IoT)-based sensors, drones, and satellite imagery, to advance predictive models that can enable real-time decision-making (Albahar, 2023). For instance, from PlantVillage dataset with more than 54,000 annotated images of healthy and infected leaves of 14 crop species, models have been able to reach disease detection accuracy of more than 92% (Mishra & Dwivedi, 2025). As shown in Figure 1, annotated leaf samples in diversified forms from datasets such as PDDA, PlantVillage, and PlantDoc++ enhance the robustness and generalization of disease detection models. Similarly, CNNs (convolutional neural networks) applied to drone and satellite images have been found to provide yield predictions above 90% for maize, wheat, and rice in controlled trials. RNNs (recurrent neural networks) and LSTM (long short-term memory) models, which utilize temporal information, have demonstrated strong predictive power in forecasting crop growth stages, soil moisture evolution, and pest outbreak trends. These applications exemplify the transformation of generalized agricultural recommendations into customized, localized, and field-specific recommendations that can enhance productivity and sustainability (Pantoja et al., 2025).

The growing statistical trends in agriculture explain why DL has become so important in the industry. The agricultural market was valued at USD 7.9 billion in 2020 and is expected to reach more than USD 35 billion by 2030, driven by demand for automation in harvesting, monitoring, and precision input application. Meanwhile, the global agricultural AI (artificial intelligence) market— in which DL constitutes the nucleus—is anticipated to expand from USD 1.7 billion in 2023 to more than USD 12 billion by 2030, at an approximate 30% compound annual growth rate (CAGR). These numbers demonstrate that DL is not merely an experimental plaything, but a mature set of solutions that is rapidly commercializing and in which governments, corporations, and research institutes are heavily investing (Mishra & Srivastava, 2025).

Agriculture is at the crossroads of overwhelming challenges and disruptive opportunities, where DL is the key to enabling sustainable growth, as is evident from Table 1. For example, the fact that agriculture uses almost 70% of the fresh water on the planet illustrates becoming a far better predictor and controller of irrigation systems using neural networks to help watering systems (Tripathi et al., 2022). The same is true for other global water consuming agricultural practices and this will be addressed in future work. In the same vein, the expected expansion of the agricultural AI market from USD 1.7 billion in 2023 to over USD 12 billion by 2030 underscores not solely the technological potential of DL, but the magnitude of institutional and private investment in its utilization (Santos et al., 2019). Furthermore, diseases detection with a success rate of 92% by using image classification model proves that AI-based farming solutions could significantly minimize crop loss, which is as high as 40% of production, globally (Choudhary et al., 2018). These two statistics represent the magnitude of agricultural limitations, and thus enhance real farming by deploying DL machinery across varying farm fields (Poonia et al., 2022).

Fig. 1 Partial comparison of diseased leaf samples from PDDA, PlantVillage, and PlantDoc++ datasets for deep learning applications. (*Source:* Zhang et al., 2023)

Table 1 Major global indicators highlighting the scope and potential of deep learning in agriculture.

Indicator	*Value/Trend*	*Source/Year*
Global population (projected by 2050)	9.7 billion	UN, 2022
Agricultural share of global employment	26.6%	ILO, 2023
Share of global freshwater used by agriculture	~70%	FAO, 2022
Agriculture's contribution to GHG emissions	19–29%	IPCC, 2021
Global agricultural AI market value	USD 1.7B (2023); projected USD 12B by 2030	Markets and Markets, 2023
PlantVillage dataset size	>54,000 images (14 crops)	Penn State University, 2020
CNN yield prediction accuracy	>90% for major cereals	Research trials, 2021–2023
Image-based disease detection accuracy	~92% (benchmark datasets)	PlantVillage/FAO studies, 2022
Agricultural robotics market	USD 7.9B (2020); projected >USD 35B by 2030	Allied Market Research, 2021

Source: Author's compilation.

Smartphone-based apps that leverage CNNs to detect crop diseases or identify pests have been introduced in India, Kenya, and parts of Latin America. These apps serve as usable diagnostic instruments, providing levels of diagnostic accuracy comparable to those of an expert. These technologies can not only fill knowledge gaps, but they can additionally mitigate financial losses, which the FAO estimates to be equivalent to nearly 20–40% of global crop production every year as a result of pest and disease. Through the integration of intelligence into simple platforms, DL ensures that advances in technology are not limited to industrial-scale farms, but can also provide assistance to small and resource-poor farmers (Ren et al., 2020; Singh and Khan, 2023). Autonomous tractors and drones with real-time object detection, robotic harvesters that can sense ripeness and pick fruits selectively, DL serves as the perceptual layer and the decisional axis of contemporary agricultural machines. The use of unmanned aerial vehicles (UAVs) for repetitive and expensive surveillance, integrated with CNN-based object detection, enables the identification of nutrient deficiencies or weed infestations quickly and at a cost more than 50% lower than manual field scouting (Saleem et al., 2021).

Although DL exhibits great potential, barriers exist in agricultural applications that hinder the robustness, interpretability, and generalizability of DL methods to different agro-climatic conditions. This may lead to models that perform poorly in underrepresented regions due to data imbalance, where certain crops and conditions are preferentially sampled in the training dataset. Likewise, the computing demands of training large models typically necessitate cloud infrastructure, raising concerns regarding cost and access. Overcoming these constraints is of paramount importance for the fair and sustainable application of DL technologies. This chapter aims to achieve the holistic goal of DL knowledge and its application to new-age agricultural systems. Furthermore, another aim is to investigate the use of DL in crop monitoring, soil-water management, pest control, and agricultural robotics. Additionally, the aim is to assess the challenges and limitations, as well as the prospects for future work on DL to enable precise and sustainable agriculture that can feed the world (Moreira, 2025).

2. Foundations of Deep Learning

DL is now considered to be one of the greatest technological inventions of the 21st century and it has become a bridge connecting (AI theory and practice in various fields such as agriculture. To understand the transformative power of DL in agriculture, a reader must understand what it is, the components that make it up, and how it manifests in one or more architectures. This section attempts to introduce not only the theory but also the structural, functional, and mathematical concepts that are the essence of the discipline of DL.

2.1 Definition and Scope of Deep Learning

Unlike conventional ML (machine learning) techniques, which often depend on hand-engineered features and domain-specific expertise, DL models are capable of autonomously identifying patterns from raw data, whether these are numerical values, textual information, images, or time-series signals. A typical DL architecture consists of input layers, multiple hidden layers, activation functions, and output layers that progressively refine the raw input into predictive or classification outcomes (Parico and Ahamed, 2021). The scope of DL is expansive, as it integrates elements of mathematics, optimization, computer science, and cognitive modeling Within agriculture, its application ranges from plant disease detection through image classification to yield prediction based on environmental and soil data. For instance, the sequential flow of data across interconnected hidden layers enables DL systems to uncover non linear dependencies that would otherwise remain obscured using conventional analytical methods.

2.1.1 Evolution from ML to DL

The historical trajectory of AI research reveals a gradual progression from rule-based systems to statistical learning methods, ultimately culminating in DL. An ML pipeline would require experts to define specific leaf texture or color indices before classification could be performed.

The evolution of AI models has progressed from basic ML approaches (e.g., linear regression) to advanced models (e.g., decision trees and random forests), followed by DL models, and is now moving toward cutting-edge architectures such as GANs (generative adversarial networks) and large language models. This progression highlights how the field has evolved continuously from shallow, feature-based learning toward deeper, more abstract representation learning (Magomadov, 2019; Rao et al., 2021).

Deep learning revolutionized this paradigm by enabling end-to-end learning. In a deep neural network, the raw pixel values of a plant's image can be directly fed into the network, which subsequently learns low-level features (edges, textures), mid-level features (shapes, patterns), and high-level features (disease indicators) in successive layers (Chouhan et al., 2025; Eunice et al., 2022). This transition highlights why deep learning is increasingly regarded as the backbone of modern AI in agriculture and beyond.

A mathematical analogy can be drawn between ML and gDL. Consider an input vector $x = (x_1, x_2, \ldots, x_n)$. In classical ML, a function $f(x)$ is manually constructed using pre defined features $\phi(x)$, such that,

$$y = f(\phi(x))$$

In contrast, in DL, the transformation of features occurs hierarchically within the network layers:

$$h^{(1)} = g(W^{(1)}x + b^{(1)})$$
$$h^{(2)} = g(W^{(2)}h^{(1)} + b^{(2)})$$
$$\vdots$$
$$y = g(W^{(L)}h^{(L-1)} + b^{(L)})$$

where, $W^{(i)}$ and $b^{(i)}$ represent weights and biases, and represents activation functions.

When applied to an agricultural dataset of 1,000 crop images with 10 features each, the ML approach may involve manually designing 50 features, while DL may autonomously learn 500 hierarchical features over 10 layers.

2.1.2 Characteristics of Deep Learning Algorithms

Many characteristics distinguish DLK from conventional approaches:

- ***Hierarchical Feature Learning***: Features are automatically extracted in increasing levels of abstraction. For instance, when analyzing soil texture images, lower layers may detect granular edges while deeper layers capture macroscopic soil patterns.
- ***Data-Hungry Nature***: Deep models typically require large datasets. A crop disease classifier may require thousands of labeled leaf images to reach high accuracy.

- ***Computational Complexity***: Training deep models demands high-performance hardware such as GPUs (graphics processing units). The time complexity for training one epoch of a fully connected neural network is approximately:

$$T = O(N \cdot L \cdot d)$$

where, N is the number of samples, L is the number of layers, and d is the dimensionality of input features.

- ***Generalization Ability***: Once trained properly, deep models can generalize across environments, e.g., detecting pest infestations across different climatic regions.
- ***Non linearity Handling***: Activation functions allow networks to approximate non linear functions, which is critical when modeling complex agricultural interactions like the non linear relationship between rainfall and crop yield.

2.1.3 Importance in the Context of Big Data and Agriculture

Agriculture is rapidly transforming into a data-intensive industry. The advent of smart farming technologies generates unprecedented volumes of data through soil sensors, drones, remote sensing satellites, and genomic sequencing. This massive agricultural big data, often characterized by the *four Vs*—volume, velocity, variety, and veracity—requires sophisticated analytical techniques (Altalak et al., 2022). DL is particularly suited to big agricultural data because of its ability to:

- ***Handle High-Dimensional Data***: Satellite images with millions of pixels can be analyzed efficiently.
- ***Integrate Multi modal Data***: Weather records, soil nutrient data, and crop images can be jointly processed.
- ***Discover Hidden Patterns***: Neural networks can uncover correlations such as the influence of temperature anomalies on pest outbreaks.

For example, consider a dataset of crop yield prediction with 10,000 records, each containing soil moisture (in %), rainfall (in mm), fertilizer use (in kg/acre), and yield (in quintals/acre). Traditional regression may fail to capture interaction terms, whereas a DL model can approximate:

$$Y = f(\text{Rainfall, Soil Moisture, Fertilizer}) \approx \sum_{i=1}^{L} g(W^{(i)} h^{(i-1)} + b^{(i)})$$

By fitting this model, we may predict yield with 95% accuracy, surpassing traditional models limited to 75%.

2.2 Core Components of Deep Learning

2.2.1 Neural Networks: Structure and Function

Neural networks are the fundamental units of DL. Inspired by biological neurons, an artificial neural network comprises layers of nodes (neurons), each connected to others with associated weights and biases. Mathematically, a neuron computes:

$$z = \sum_{i=1}^{n} W_i x_i + b$$

$$a = g(z)$$

where, x_i are input signals, w_i are weights, b is bias, and g is an activation function.

If a soil moisture sensor records values $x = (0.2, 0.4, 0.6)$, with weights $w = (0.5, 0.3, 0.1)$ and bias $b = 0.2$, then:

$$z = (0.5)(0.2) + (0.3)(0.4) + (0.1)(0.6) + (0.2) = 0.1 + 0.12 + 0.06 + 0.2 = 0.48$$

If, $g(z) = \sigma(z) = \frac{1}{1+e^{-z}}$ then:

$$a = \frac{1}{1+e^{-0.48}} \approx 0.6177$$

2.2.2 Activation Functions and Loss Functions

The activation function is a crucial component. In agriculture, the relationship between environmental parameters and crop yield is rarely linear; activation functions allow the network to model such intricacies (Mishra & Mishra, 2025). Common activation functions include:

a. Sigmoid Function:

$$\sigma(z) = \frac{1}{1+e^{-z}}$$

This function compresses input values into the range (0, 1). For example, in a pest infestation classifier where the output indicates the probability of infestation, a sigmoid function converts a raw output score of 2.5 into:

$$\sigma(2.5) = \frac{1}{1+e^{-2.5}} = \frac{1}{1+0.0821} \approx 0.9424$$

Thus, there is a 92.4% probability of infestation.

b. Hyperbolic Tangent (Tanh):

$$\tanh(z) = \frac{e^{z} - e^{-z}}{e^{z} + e^{-z}}$$

This function outputs values between –1 and 1, making it suitable for centered data. If soil fertility indices vary between –0.5 and 0.7, applying Tanh ensures effective scaling.

c. Rectified Linear Unit (ReLU):

$$f(z) = \max(0, z)$$

ReLU is computationally efficient and helps avoid vanishing gradient problems. For instance, if crop input features yield pre activation values [–2.5, 0.0, 3.7], then ReLU transforms them into [0.0, 0.0, 3.7].

d. Softmax Function:

In multi class problems, such as classifying rice leaf diseases into "bacterial blight", "blast", or "sheath blight", the softmax function ensures the output probabilities sum to 1.

$$p(x) = \frac{e^{zi}}{\sum_{j=1}^{K} e^{zj}}$$

Suppose output scores for three classes are: $z = [2.0, 1.0, 0.1]$. Then:

$$e^{z} = [7.389, 2.718, 1.105]$$

$$\text{Sum} = 7.389 + 2.718 + 1.105 = 11.212$$

$$P_1 = \frac{7.389}{11.212} \approx 0.659, P_2 = 0.242, P_3 = 0.099$$

Thus, the model predicts "bacterial blight" with ~66% probability.

Loss Functions

They guide the optimization process by providing a numerical target for minimization.

a. Mean Squared Error (MSE):

$$L = \frac{1}{N}\sum_{i=1}^{N}(y_i - \hat{y}_i)^2$$

If actual yields (in quintals/acre) are [30, 25, 28] and predicted yields are [32, 24, 27], then:

$$L = \frac{1}{3}[(30-32)^2 + (25-24)^2 + (28-27)^2]$$

$$= \frac{1}{3}[4 + 1 + 1]2 = \frac{6}{3} = 2$$

b. Cross-Entropy Loss** (for classification):**

$$L = \sum_{i=1}^{N} y_i \log(\hat{y}_i)$$

For a binary classification where the true label = 1 (infestation present) and the predicted probability = 0.8, the loss is:

$$L = -(1 \cdot \log(0.8)) = -(-0.2231) = 0.2231$$

A lower loss indicates better predictive performance.

2.2.3 Training Algorithms: Backpropagation and Gradient Descent

Training a DL model requires adjusting its parameters (weights and biases) so that predictions closely match actual outputs. This adjustment is carried out through backpropagation and gradient descent.

For example, suppose we have a single-layer model with input $x = 2$, weight $w = 0.5$, bias $b = 0.1$, and target $y = 1$.

1. *Forward pass*:

$$z = wx + b = (0.5)\,(2) + 0.1 = 1.1$$

$$\hat{y} = \sigma(z) = \frac{1}{1 + e^{-1.1}} \approx 0.750$$

2. *Loss (MSE)*:

$$L = (y - \hat{y})^2 = (1 - 0.75)^2 = 0.0625$$

3. *Gradient w.r.t. weight*:

$$\frac{\partial L}{\partial W} = (y - \hat{y}) \cdot \hat{y}(1 - \hat{y}) \cdot x$$

$$= (1 - 0.75)\,(0.75)\,(0.25)\,(2) = 0.09375$$

4. *Weight update with learning rate η=0.1*:

$$w_{\text{new}} = 0.5 - 0.1\,(0.09375) = 0.490625$$

Thus, after one update, the weight moves slightly toward the value that reduces the error. Repeated iterations gradually minimize the loss, allowing the model to learn.

2.3 Popular Deep Learning Architectures

The versatility of DL arises largely from the development of specialized architectures, each tailored to particular data structures. In agriculture, these architectures address unique challenges such as analyzing satellite images, predicting crop yields over time, or generating synthetic datasets for training purposes.

2.3.1 Convolutional Neural Networks (CNNs)

CNNs are particularly effective for spatial data such as agricultural images captured by drones or satellites. Unlike fully connected networks, CNNs employ convolutional layers that scan local regions of input data to extract features. The convolution operation is mathematically expressed as:

$$S(I, j) = (X \cdot K)\,(i, j) = \sum_m \sum_n X(i + m,\ j + n)\,K(m, n)$$

where, X is the input image, K is the kernel (filter), and $S(i, j)$ is the output feature map.

For example, consider a 3 × 3 crop image patch:

$$X = [1\ 2\ 3\ 4\ 5\ 6\ 7\ 8\ 9], \quad K = [1\ 0\ 0\ {-1}]$$

Convolution on the top-left 2 × 2 patch:

$$(1)\,(1) + (2)\,(0) + (4)\,(0) + (5)\,(-1) = 1 - 5 = -4$$

This process continues across the image to produce a feature map highlighting vertical and horizontal differences, useful in identifying leaf veins or pest damage patterns.

CNN architectures follow a hierarchical feature extraction pipeline. Input images pass through convolutional and pooling layers to extract low-, mid-, and high-level semantic features. This structure enables CNNs to learn complex spatial hierarchies directly from raw data (Zhu et al., 2018; Bharman et al., 2022).

In agricultural applications, CNNs detect plant diseases, classify soil textures, and even monitor crop growth from aerial images. Such models have demonstrated remarkable accuracy in identifying subtle disease symptoms on crop leaves, making them indispensable in precision agriculture.

2.3.2 Recurrent Neural Networks (RNNs)

While CNNs excel at spatial data, many agricultural datasets are inherently temporal. For example, weather patterns, soil moisture variations across weeks, or time-series data from IoT sensors in greenhouses cannot be effectively captured using only convolutional operations (Patel et al., 2025). RNNs are uniquely suited for such sequential data because they maintain a form of "memory" through recurrent connections. An RNN processes input sequences step- by-step, updating its hidden state at each time t as follows:

$$h_t = f(W_{xh}\, x_t + W_{hh}\, h_{t-1}) + b_h$$

where, x_t is the input at time t, W_x and W_{hh} are weight matrices, b_h is a bias, and f is an activation function such as tanh. The output at time t is then:

$$y_t = g(W_{hy}\, h_t + b_y)$$

Suppose a farmer wishes to predict daily soil moisture levels. If the input values for three consecutive days are $x_1 = 0.4$, $x_1 = 0.5$ and $x_3 = 0.7$, and initial parameters are set as $W_{xh} = 0.8$, $W_{hh} = 0.3$, $b_h = 0.1$ and $f = \tan h$, then the hidden state for the first day would be:

$$h_1 = \tan h\,(0.8 \times 0.4 + 0.3 \times 0 + 0.1) = \tan h\,(0.42) \approx 0.397$$

The second day's hidden state becomes:

$$h_2 = \tan h\,(0.8 \times 0.5 + 0.3 \times 0.397 + 0.1) = \tan h\,(0.64) \approx 0.565$$

and similarly for the third day. This sequential dependence allows RNNs to model agricultural processes evolving over time.

However, conventional RNNs are known to have vanishing gradient problems over long sequences that hinders their applicability in practice. In response to this, variants such as LSTM networks and Gated Recurrent Units (GRUs) were designed, using gating mechanisms to maintain long-term dependencies, and thus proving useful for long-term crop yield prediction or multi season weather influence analysis.

2.3.3 Autoencoders and Variational Autoencoders (VAEs)

Autoencoders are yet another powerful deep learning architecture for unsupervised learning, particularly for dimensionality reduction and feature extraction:

$$z = f_\theta(x)$$

where, z is the latent vector, and the decoder reconstructs the input as:

$$\hat{x} = g\phi(z)$$

The loss function is often the reconstruction error:

$$L(x, \hat{x}) = ||x - \hat{x}||^2$$

For instance, if the original input vector representing leaf features is $x = [1.0, 2.0. 3.0]$, and the decoder reconstructs $\hat{x} = [0.9, 2.1, 2.9]$, then the reconstruction error is:

$$L = (1.0 - 0.9)^2 + (2.0 - 2.1)^2 + (3.0 - 2.9)^2 = 0.01 + 0.01 + 0.01 = 0.03$$

Autoencoders are widely used in agriculture for anomaly detection, such as identifying unusual sensor readings that may indicate irrigation failures or equipment malfunctions.

VAEs, on the other hand, extend standard autoencoders by introducing probabilistic elements. Instead of encoding an input into a deterministic point z, a VAE encodes it into a dist($z \mid x$), typically assumed to be Gaussian, parameterized by mean μ and variance σ^2. This capability enables the generation of new, realistic data by VAEs, which in the context of agriculture, can be used to create new crop disease images for training when real-world datasets are scarce (Razzaq and Shah, 2025).

2.3.4 Generative Adversarial Networks (GANs)

A GAN contains two neural networks: a generator G that generates synthetic samples, and a discriminator D that tries to differentiate real samples from synthesized ones. In the latest approach, both networks are applied against each other, engaging in an adversarial process (Narayanan & Arjun, 2024).

The optimization can be formally described as:

$$\min G \max D\ V(D, G) = E_{x\sim \text{padat}(x)} [\log D(x)] + E_{z\sim pz(z)} [\log(1 - D(\text{G}(z))]$$

The generator might start with random noise z, pass it through several neural layers, and output an image. The discriminator then evaluates this image against actual rice leaf images. Over time, the generator improves to create highly realistic diseased leaf patterns.

This adversarial training paradigm can be numerically explained through a simplified binary scenario: if the discriminator initially assigns a probability of 0.7 to a real image being genuine and 0.3 to a fake one, then the generator's objective is to adjust its parameters until the discriminator is forced into making predictions close to 0.5, indicating indistinguishability.

Advanced GAN-based pipelines, such as WeedGAN and ESRGAN, are capable of generating realistic agricultural images. The generator creates synthetic low-resolution plant disease images, while ESRGAN subsequently enhances them to high-resolution versions for improved training. This approach demonstrates how GANs can significantly alleviate the scarcity of labeled agricultural datasets and support more robust CNN-based disease classification models.

GANs are therefore particularly important in agriculture where labeled data is limited. By creating realistic synthetic images, they address the data bottleneck problem, enabling more robust training of CNNs for classification and detection tasks (Kavitha, 2021; Mishra, 2024; Uddin & Bansal, 2021).

2.3.5 Transformer Models and Attention Mechanisms

The most recent revolution in DL has been the advent of Transformer models, powered by the concept of attention. Transformers have surpassed both CNNs and RNNs in many domains, particularly natural language processing, and are now being increasingly explored for agriculture-related applications such as analyzing weather reports, farmer advisories, or genomic sequences. The fundamental principle of attention is to assign varying importance to different parts of the input sequence when generating an output. Mathematically, given queries Q, keys K, and values V, the attention mechanism computes:

$$\text{Attention}(Q, K, V) = \text{softmax}\left(\frac{QK^T}{\sqrt{d_k}}\right)V$$

where, d_k is the dimensionality of the keys. For example, if $Q = [1, 0]$, $K = [[1, 0], [0, 1]]$ and $V = [[5, 6] [7, 8]]$, then:

$$QK^T = [1, 0] \times [1\ 0\ 0\ 1] = [1, 0]$$

Dividing by $\sqrt{d_k} = \sqrt{2}$, we get:

$$[1, 0]/2 \approx [0.707, 0],$$

The softmax of this vector is approximately [0.67, 0.33]. Multiplying with V yields:

$$[0.67, 0.33] \times [5\ 6\ 7\ 8] = [0.67 \times 5 + 0.33 \times 7, 0.67 \times 6 + 0.33 \times 8] = [5.66, 6.66]$$

This demonstrates how attention selectively combines information from different parts of the input.

Transformers, composed of multi head attention layers, feed-forward networks, and positional encodings, have been proven to handle extremely long sequences without the vanishing gradient problem that plagues RNNs. In agriculture, their potential includes processing large-scale multi modal datasets combining text-based weather bulletins, numerical climate measurements, and image-based crop monitoring data to produce comprehensive predictive models (Umar et al., 2022).

3. Integration of Deep Learning in Agricultural Systems

The integration of DL into agricultural systems represents a transformative shift from conventional cultivation, monitoring, and management methods to technologically advanced, data-driven practices that can optimize efficiency, sustainability, and profitability. Traditional farming methods, although effective for centuries, are increasingly challenged by rising population pressures, climatic fluctuations, soil degradation, water scarcity, and unpredictable market dynamics. To address these difficulties, agriculture is on the path to convergence and deployment of digital tools such as sensors, drones, satellite imaging, and high-performance computing with IaaS. Related to this specialized type of learning, DL provides a promising approach to analyzing large, complex, non linear data, offering both predictive and prescriptive outcomes to the farming community.

In this respect, decision-making on crop types, irrigation schedules, pest and disease prevention, soil fertility management, and yield optimization can be supported using DL. The integration is not only conceptual but is also increasingly being realized through precision agriculture platforms, smart farming applications, decision support systems (DSS), and resource management models.

3.1 Precision Agriculture and Smart Farming

Precision agriculture (PA), also known as smart farming, is a multidisciplinary approach that utilizes information and communication technology (ICT)-based tools to ensure crops and soil receive optimal inputs tailored to their specific requirements for optimal health and productivity. DL has propelled this paradigm, as it facilitates data acquisition, pattern recognition, and predictive analytics on scales and with accuracies that surpass the capabilities of human expertise.

3.1.1 Role of Sensors and IoT in Data Collection

Accurate data collection is the bedrock on which PA is built, and that data collection is increasingly enabled by IoT devices and sensors scattered throughout fields. These sensors measure parameters including soil moisture, pH, temperature, humidity and nutrient content, and are used as features for constructing DL g models. For instance, if we consider soil moisture sensors, data may be recorded as a time- series where M_t represents the soil moisture at time t. A simplified function for capturing soil-water dynamics can be expressed as:

$$M_t = M_{t-1} + I_t - ET_t - D$$

Where,
M_{t-1} = soil moisture in the previous period,
I_t = irrigation or rainfall at time t,
ET_t = evapotranspiration loss at time t,
D = drainage or percolation loss at time t.

Suppose a farmer's field recorded $M_{t-1} = 30$ mm, irrigation $I_t = 12$ mm, evapotranspiration $ET_t = 8$ mm, and drainage loss $D_t = 3$ mm. Then,

$$M_t = 30 + 12 - 8 - 3 = 31 \text{ mm}$$

Thus, the updated soil moisture is 31 mm, which can be automatically fed into a DL system to predict irrigation needs. This shows how raw data from IoT sensors can be continuously used to train DL models for real-time decision-making.

3.1.2 Decision Support Systems with Deep Learning Models

The DSS integrates field-level data with advanced algorithms to aid farmers in adopting the best strategies for cultivation. DL enhances DSS by allowing real-time predictions that consider complex environmental and biological interactions. Consider a DSS predicting fertilizer requirement F for wheat based on soil nitrogen (N), phosphorus (P), and potassium (K) levels. A simplified predictive model can be expressed as:

$$F = w_1N + w_2P + w_3K + b$$

where, w_1, w_2, w_3 are weights derived from training data, and b is the bias term.

Assume from training results:

$$w_1 = 0.45,\ w_2 = 0.30,\ w_3 = 0.25,\ b = 5$$

If a soil test shows N = 40 kg/ha, P = 20 kg/ha, K = 25 kg/ha, then:

$$F = (0.45 \times 40) + (0.30 \times 20) + (0.25 \times 25) + 5$$

$$F = 18 + 6 + 6.25 + 5 = 35.25 \text{ kg/ha}$$

Thus, the DSS recommends applying approximately 35.25 kg/ha fertilizer. By scaling this for 100 hectares, the total requirement is 35.25 × 100 = 3525 kg. If the price of fertilizer is ₹18 per kg, the estimated cost is ₹63,450. This enables farmers to anticipate costs while ensuring soil fertility is optimized.

3.1.3 Predictive Analytics for Resource Optimization

DL excels at predictive analytics, where historical and real-time data are combined to optimize the use of inputs such as water, fertilizer, and pesticides. For example, irrigation scheduling can be optimized by predicting water requirement W using a regression-based DL model:

$$W = \alpha + \beta_1 T + \beta_2 H + \beta_3 M$$

where, T = temperature, H = humidity, M = soil moisture, and coefficients are learned by the model.

If the model provides $\alpha = 5, \beta_1 = 0.4, \beta_2 = -0.3, \beta_3 = -0.2$, and field readings are $T = 32°C$, $H = 60\%$, $M = 25$, then:

$$W = 5 + (0.4 \times 32) - (0.3 \times 60) - (0.2 \times 25)$$
$$= 5 + 12.8 - 18 - 5 = -5.2$$

A negative value indicates no additional irrigation is needed because current soil conditions are sufficient. This prevents unnecessary water usage, reducing operational costs.

3.2 Crop Monitoring and Yield Prediction

DL enables detailed crop monitoring through satellite images, aerial drone photography, and time-series analyses. This allows farmers to detect crop stress early and predict yields with high accuracy.

3.2.1 Satellite and Aerial Imagery Analysis

CNNs process imagery data to identify spatial patterns in crop health. Satellite images often capture vegetation indices such as the Normalized Difference Vegetation Index (NDVI), which is calculated as:

$$\text{NDVI} = \frac{\text{NIR - RED}}{\text{NIR + RED}}$$

where, *NIR* is near-infrared reflectance and *RED* is visible red reflectance. For example, if a drone records reflectance values of NIR = 0.65 and RED = 0.30, the resulting NDVI is:

$$\text{NDVI} = \frac{0.65 - 0.30}{0.65 + 0.30} = \frac{0.35}{0.95} \approx 0.368$$

An NDVI value close to 0.37 suggests moderate vegetation health, and this data can be used by DL models to predict yield variations across the field.

The long-term changes in vegetation health can also be studied using satellite imagery. The spatial distribution of NDVI in SAS Nagar highlights a clear decline in vegetation health from 2013 to 2022. In 2013, large areas had higher NDVI values (depicted in green), indicating healthier vegetation cover. However, by 2022, the map shows a significant growth of low-NDVI (red and yellow) areas, hinting at vegetation degradation and potential expansion of urban or land-use change. This degradation of NDVI distribution emphasizes the need of remote sensing data with DL to track the trend of healthy vegetation over time and guide sustainable land management (Liakos et al., 2018; Mishra, 2025; Dayal et al., 2023).

3.2.2 Temporal Data Analysis using RNNs and LSTMs

Since crop development is inherently dependent on time, RNNs and LSTM networks have been proven to be effective. These architectures model sequential dependencies in data like precipitation, temperature, and stages of crop growth.

For instance, suppose a sequence of rainfall values over three weeks is mm. An LSTM can process this to predict the fourth week's rainfall based on learned temporal patterns. By adjusting weights across hidden layers, the LSTM can reduce errors in predicting crop stress periods.

3.2.3 Yield Estimation Models

Yield prediction models combine data from soil, weather, and crop phenology. A simplified linear DL-based prediction model can be written as:

$$Y = \gamma_0 + \gamma_1 R + \gamma_2 T + \gamma_3 N$$

where, Y = yield in quintals/ha, R = rainfall, T = temperature, N = nitrogen availability.

If $\gamma_0 = 5$, $\gamma_1 = 0.6$, $\gamma_2 = -0.2$, $\gamma_3 = 0.8$, and data show $R = 450$ mm, $T = 28°C$, $N = 50$ kg/ha then:

$$Y = 5 + (0.6 \times 450) - (0.2 \times 28) + (0.80 \times 50)$$

$$Y = 5 + 270 - 5.6 + 40 = 309.4 \text{ quintals/ha}$$

Thus, the predicted yield is 309.4 quintals per hectare, which provides farmers with forward-looking guidance on harvest planning.

3.3 Soil and Water Management

Soil and water are the most critical inputs in agriculture, and their efficient management is crucial for long-term sustainability. DL models provide tools for classification, prediction, and mapping to improve soil and water utilization.

3.3.1 Deep Learning for Soil Classification

Soil classification involves categorizing soil into fertility groups using parameters such as pH, organic matter, and texture. CNNs can process hyperspectral images of soil to classify them into fertile, medium, or poor.

Suppose spectral data for soil reflectance yields classification probabilities: fertile (0.65), medium (0.25), poor (0.10). A DL model thus classifies the soil as fertile, which directly impacts the crop planning decision.

3.3.2 Moisture Prediction and Irrigation Scheduling

Moisture prediction is essential for irrigation scheduling. DL regression models estimate soil water content by learning from sensor and climatic data. If irrigation cost is ₹20 per mm of water applied per hectare, and the model predicts an irrigation requirement of 15 mm for a 50-hectare farm, then:

$$\text{Cost} = 20 \times 15 \times 20 = ₹15{,}000$$

By preventing over-irrigation, the farmer can save on high costs across multiple cropping cycles.

3.3.3 Erosion and Nutrient Depletion Mapping

Soil erosion and nutrient depletion are pressing concerns that compromise agricultural productivity. DL models process spatial data from satellites to identify areas prone to erosion. By detecting patterns in vegetation cover, slope, and rainfall intensity, CNNs can generate risk maps for targeted soil conservation measures.

For example, if a region is predicted to have a risk score of 0.78 on a scale of 0 to 1, this may trigger interventions such as contour bunding, mulching, or afforestation. In parallel, nutrient depletion maps help determine site-specific fertilizer applications, reducing overuse and minimizing environmental externalities (Lu et al., 2021).

4. DL in Pest and Disease Management

The agricultural sector has long faced challenges from the persistent threat of pests and diseases, which cause significant yield losses and deteriorate crop quality. Conventional methods for pest and disease management are labor- intensive, involving manual scouting and depend on human interpretation, and they involve the use of chemicals. Although these approaches have shown some promise in terms of sensitivity and specificity, they are limited by delayed diagnosis, expense, labor requirements, and poor scalability (Chouhan et al., 2024). In recent years, the rapid evolution of learning-based approaches has presented novel possibilities to improve the accuracy, speed and efficiency of pest and disease phenotyping. DL, which is a subclass of AI, has been demonstrated to be very successful in dealing non linear, complex and multi dimensional data. This ability in agriculture corresponds to detection of slight disease signs on plant leaves, monitoring of pest infestation, forecasting of population dynamics, and integrated automated DSS for farmers (Bal & Kayaalp, 2021; Logeshwaran et al., 2024).

4.1 Image-Based Disease Detection

Leaves, stems, fruits, and roots of plants often exhibit early signs of infection in the form of discoloration, lesions, spots, or abnormal texture patterns. Manual diagnosis of such symptoms is subjective, requires expert knowledge, and may not always be scalable to large farms.

4.1.1 Dataset Preparation and Annotation Techniques

The success of any DL model depends heavily on the quality and quantity of training data. In agricultural disease detection, datasets are typically composed of high-resolution images of diseased and healthy plant parts collected under varying conditions such as lighting, background, and growth stages. Preparing these datasets requires systematic planning. Images are usually gathered through field surveys, research station observations, or by leveraging open-access datasets like PlantVillage. However, raw images often contain noise such as soil background, overlapping leaves, or lighting inconsistencies, which may confuse the learning model (Tiwari et al., 2023; Ojo & Zahid, 2022). Preprocessing steps like resizing, normalization, contrast enhancement, and segmentation are therefore applied. Annotation of datasets is equally important, where agricultural experts label the diseased regions, specifying the disease type and severity. Annotation tools such as LabelImg or VGG Image Annotator are used to mark the boundaries of diseased areas for object detection tasks, or to provide class labels for classification tasks.

A simple numerical calculation illustrates the importance of balanced datasets. Suppose we wish to train a binary classifier that distinguishes between healthy and diseased leaves. If the dataset contains 8,000 images of diseased leaves and only 2,000 images of healthy leaves, then the proportion of healthy leaves is:

$$\text{Proportion of healthy leaves} = \frac{2000}{2000 + 8000} = \frac{2000}{10000} = 0.2$$

This imbalance (20% healthy vs. 80% diseased) could lead the model to be biased towards predicting disease. To counter this, techniques such as data augmentation (flipping, rotation, scaling), oversampling of minority classes, or synthetic data generation using GANs are employed.

4.1.2 CNNs for Leaf Disease Recognition

CNNs are the backbone of modern image-based disease detection systems. CNNs mimic the human visual cortex by extracting hierarchical features from images. The first layers of CNNs identify low-level features such as edges and textures, while deeper layers learn high-level patterns corresponding to disease symptoms (Chouhan et al., 2025). A healthy pepper leaf image from the PlantVillage dataset is passed through multiple convolutional and pooling layers, where disease-specific features are extracted at different hierarchical levels. The processed features are then classified into healthy or diseased categories. The role of pre trained models, such as VGG16, Inception V4, ResNet-50, and DenseNet-121, which are utilized in transfer learning to enhance performance. Among these, DenseNet-121 achieved the best outcome in the referenced study (Christiansen et al., 2016; Mamat et al., 2022).

Consider a simple CNN model with one convolutional layer followed by pooling and a fully connected layer for classification. If an input image of dimension $128 \times 128 \times 3$ (RGB) is passed through a convolutional layer with 32 filters of size 3×3, the output feature map dimension can be calculated as:

$$\text{Output size} = (W - F + 2P)/S + 1$$

where, W = input width, F = filter size, P = padding, and S = stride.

Taking $W = 128$, $F = 3$, $P = 1$, and $S = 1$:

$$\text{Output size} = (128 - 3 + 2 \times 1)/1 + 1 = (128 - 3 + 2)/1 + 1 = 128$$

Thus, the output feature map is of dimension . Such systematic computation is repeated layer by layer until the final classification output is achieved. This demonstrates the mathematical rigor underlying CNN-based disease recognition.

4.1.3 Transfer Learning and Model Fine-Tuning

Training CNNs from scratch requires massive datasets and computational resources. In agriculture, where annotated data is limited, transfer learning has become a practical solution. Transfer learning involves taking a model pre trained on a large dataset (such as ImageNet with millions of images) and fine-tuning it on agricultural disease datasets. Models like VGG16, ResNet, and Inception have been successfully applied in this manner.

For instance, suppose a ResNet-50 model is pre trained on ImageNet and achieves 95% accuracy in classifying 1,000 general object categories. When adapted to a smaller dataset of 10,000 leaf images across 5 disease classes, the initial layers of ResNet (which capture generic patterns like edges and textures) are frozen, while the final layers are fine-tuned to learn disease-specific features. This reduces both training time and the risk of overfitting.

4.2 Pest Surveillance and Prediction

Beyond diseases, pest infestations pose a significant challenge to agricultural sustainability. Effective pest management requires early detection, precise surveillance, and accurate forecasting of outbreaks. DL offers robust tools for automating these tasks.

4.2.1 Object Detection Techniques (e.g., YOLO, Faster R-CNN)

Object detection models such as You Only Look Once (YOLO) and Faster Region-Convolutional Neural Networks (Faster R-CNN) are widely used in pest surveillance. These models not only classify objects but also localize them within images, making them ideal for detecting small pests on plant surfaces or in trap images.

For example, in an insect trap image containing 500 pests scattered randomly, YOLO divides the image into a grid of 13 × 13. Each grid cell predicts bounding boxes and class probabilities. If each grid cell predicts 5 bounding boxes, then the total number of bounding box predictions is:

$$13 \times 13 \times 5 = 845$$

These 845 candidate boxes are refined through confidence scores and non-maximum suppression to identify the most accurate pest locations. Such detailed object detection methods reduce human labor in pest monitoring.

The Faster R-CNN framework extends this approach by introducing a Region Proposal Network (RPN), which generates candidate object regions. An enhanced Faster R-CNN model integrates skip pooling and contextual information fusion to improve pest detection accuracy. The architecture begins with convolutional layers (conv1–conv5) for feature extraction, followed by a Guided Anchoring RPN (GA-RPN) that generates anchors by predicting both location and shape. These anchors do feature adaptation and are subsequently passed to the proposal stage. Skip pooling is further employed to pool multi-level features, meanwhile contextual feature extraction considers both local pest features and the surrounding background information (Mishra & Mishra, 2024; Mahendra et al., 2022; Kowalska & Ashraf, 2023).

After the ROIs are normalized and pooled, they are concatenated and passed through fully connected layers for classification using a softmax function and for bounding box regression. This structure greatly enhances the detection accuracy for small or tightly-packed pests, which are usually overlooked by traditional models. As demonstrated by Xiao et al. (2020) exploiting contextual information allows to differentiate pests from similar background noise in the visual domain, thus making Faster R-CNN with skip pooling a potential tool for pest surveillance. Leveraging the real-time capabilities of YOLO with the high precision of Faster R-CNN, the integrated pest surveillance system can provide precise and scalable solutions for smart agriculture. The workflow showcases how high-level object detection pipelines are customized to address agricultural-related problems.

4.2.2 Modeling Pest Population Dynamics

To predict the outbreaks of pests an acquaintance with status dynamics of the pest population is required, which depends on some environmental factors like temperature, humidity, and precipitation. DL models can model non linear relationships among these variables and pest population. The logistic model of growth is frequently employed to model the growth of a pest population:

$$P(t) = \frac{K}{1 + \left(\frac{K - P_0}{P_0}\right) e^{-rt}}$$

where, $P(t)$ is the population at time t, K is the carrying capacity, P_0 is the initial population, and r is the growth rate.

Suppose the carrying capacity $K = 10{,}000$, initial population $P_0 = 500$, and growth rate $r = 0.1$. After 20 days,

$$P(20) = \frac{10,000}{1 + \left(\frac{95000}{500}\right) e^{-0.1 \times 2.0}} = \frac{10,000}{1 + 19.e^{-2}}$$

Since, $e^{-2} \approx 0.135$

$$P(20) = \frac{10,000}{1 + 19 \cdot 0.135} + \frac{10,000}{1 + 2.565} = \frac{10000}{3.565} \approx 2805$$

This estimate means that after 20 days the number of pest will be around 2,805 insects under those conditions. DL models also commonly embed such mechanistic equations within, or as part of, empirical models to achieve predictive rigor.

4.2.3 Forecasting Outbreaks using Time Series Data

Prediction of pest outbreaks is a key feature for its management in time. Time series forecasting models, such as LSTM, GRU, and Temporal Convolutional Networks (TCN) could be used to predict future pest levels based on historical data. They may incorporate exogenous variables such as rainfall, temperature, soil moisture to mention a few. For example, consider the following simple autoregressive forecasting equation:

$$P_t = \alpha P_{t-1} + \beta T_t + \gamma R_t + \in$$

where, P_t is pest population at time t, P_{t-1} is previous population, T_t is temperature, R_t is rainfall, and $\in$ is error term.

If we assume $\alpha = 0.7$, $\beta = 2.5$, and $\gamma = 1.8$, and input values $P_{t-1} = 500$, $T_t = 30°C$, $R_t = 10$, then:

$$P_t = 0.7 \times 500 + 2.5 \times 30 + 1.8 \times 10$$

$$P_t = 350 \times 75 + 1.8 = 443$$

Thus, the model predicts a population of 443 pests at time t.

4.3 Integrative Management Systems

Individual models for disease detection or pest prediction are useful, but real-world agricultural systems need multi stream data integration to provide end-to-end solutions. Deep learning makes it possible to develop integrated management systems by fusing image data, environmental data, and expert knowledge that generates accurate and timely advices.

4.3.1 Multi modal Data Fusion (Weather, Crop, Pest)

Multi modal fusion data refers to diverse data types that are combined in a single analysis. For pest and disease management, the relevant data include weather observations from satellites, crop growth models, images from pest surveillance, and records of soil health.

For example, a fusion model might combine weather forecasts (rainfall, temperature), real-time drone images of crops, and pest trap data to predict the risk of a fungal outbreak. DL architectures such as multi-input neural networks or attention-based transformers are employed to learn correlations across these data sources. This enhances the accuracy of predictions beyond what single-source models can achieve.

4.3.2 Automated Advisory Systems with Deep Learning Backends

DL-based automated advisory systems deliver farmers with real-time advisories, which make it easier for them to take action. They can be packaged as mobile apps or as modules on smart farming platforms. Based on images submitted by farmers, weather monitoring and pest traps, the system can provide recommendations on the type and amount of pesticide to use, an optimal date for sowing or whether to use disease-resistant varieties of crops (Gao, 2022; Wang et al., 2022).

For example, when a farmer shares a photo of a diseased leaf, the advisory system first employs a CNN to identify the disease and then suggests a curing option for a fungicide of Rs 500 per acre in the recommendation engine. At the same time, it may recommend associated cultural practices such as crop rotation or soil amendments. Over time, such advisory systems not only help reduce crop loss but also maximize resource use; thus, improving farm profitability.

5. Revolution in Agriculture

5.1 Autonomous Vehicles and Drones

In recent years, there has been a revolution in agriculture to a new era tackling the use of autonomous vehicles and unmanned aerial systems (UAVs) known as drones. These advances, by the increasingly powerful AI and DL technology, have been replacing traditional manual techniques which used to be laborious, time-consuming, and less efficient. Autonomous tractors, field rovers and drones can be a means to automate different farming applications from plowing to spraying through the seeding operations, up to in-field monitoring applications.

It is DL g models and particularly CNNs which are mainly adopted to create the ability of these machines to observe through their environment, take intelligent decisions, or perform actions without human interventions. Unlike past mechanized systems that relied on programmed paths and limited sensory inputs, DL-based vehicles can operate in the presence of dynamic field conditions including changing soil texture, sudden obstacles, and non-uniform crop distribution.

As a result, these innovations are not only helping in getting the work done faster but also improving on how resources are put to work, lowering costs and supporting sustainability in agriculture (Lin et al., 2019; Mishra & Mishra, 2024; Santra et al., 2021).

5.1.1 Navigation and Obstacle Detection using CNNs

Navigating in farmland is naturally difficult since fields are not as orderly as we would want them to be and they can be rough terrains, weeds may also grow tall hindering a smooth course of your path, crops may not have clean lines for you to use as your path and finally anything on the farm like stones lying on the way, branches fallen from trees, or even animals. On-board cameras employ CNNs to process their images whereby features are extracted and grouped in hierarchical order so that it becomes easier to differentiate between a traversable path and an obstacle. This is because these automatic learning systems do not need pre-engineered spatial features so they can identify complex spatial patterns.

The task of navigation can be portrayed mathematically as a classifying one. If an agricultural vehicle takes an input image matrix $I \in R^{m \times n \times 3}$, then a CNN processes this by means of consecutive convolutional layers, which result in feature maps that are eventually flattened into probability distributions over the predefined classes. For instance, the system may label a region as being: "path", "crop", "soil", or "obstacle". Assuming the CNN outputs a probability vector P = $[p_1, p_2, ..., p_k]$ for k classes, the navigation decision corresponds to choosing $\arg\max (p_i)$

For example, in a case where $P = [0.10, 0.70, 0.15, 0.05]$ probabilities for patch, crop, soil, and obstacle classes are shown by an image patch/patch the model will pick "crop" as the most probable category. Over such predictions within consecutive frames may be used by vehicles via path-planning algorithms like A^* or Dijkstra's algorithm to dynamically update their trajectory. Also, integration of sensor fusion data from LiDAR and GPS modules enhances high robustness of vehicle navigation decisions (Kotwal et al., 2023).

5.1.2 Real-Time Field Surveillance with UAVs

Due to the fact that UAVs can fly over extensive areas in a very short time and can get multispectral and hyperspectral data as well as transmit the acquired information for its processing by means of DL methods, drones have become principal tools for real-time inspection of agricultural fields. Application of drones starts from crop monitoring, they also engage in growth development assessments, pests' detection, water stress evaluation, as well as manure/herbicides/irrigation equipment application rates optimization. Therefore, they work within two types of bands: one visible and the other is non-visible and then an index is developed from such data along with other quantitative measurements. In this regard; there are many machine learning techniques such as CNNs as well as encoder-decoder models that are used to carry out segmentation, classification and prediction tasks when it comes to crops in real-time minutely.

Although NDVI is among the most widely used indices, modern agriculture's real-time UAV monitoring also taps into a range of other mathematical tools and models, from canopy temperature tracking to soil moisture mapping. These indices boost UAV performance by picking up fine variations in crop canopy shape, chlorophyll levels, and signs of moisture stress—like a faint yellowing at the leaf tips—offering a valuable layer to NDVI-based analysis (Abdulla & Marhoon, 2022).

(i) Soil-Adjusted Vegetation Index (SAVI)

A common limitation of NDVI in sparsely vegetated fields is its sensitivity to soil background reflectance. To address this, the Soil-Adjusted Vegetation Index (SAVI) is employed, defined as:

$$\text{SAVI} = \frac{\text{NIR} - \text{RED}}{\text{NIR} + \text{RED}} \times (1 + L)$$

where, NIR and RED are reflectance values in near-infrared and red bands, and L is a soil adjustment factor ranging between 0 and 1. For dense vegetation, L approaches 0, while for sparse vegetation, L is generally taken as 0.5.

Suppose a UAV-mounted sensor records NIR = 0.60 and RED = 0.25 with $L = 0.5$. Substituting values:

$$\text{SAVI} = \frac{0.60 - 0.25}{0.60 + 0.25 + 0.5} \times (1 + 0.5) = \frac{0.35}{1.35} \times 1.5$$

$$\text{SAVI} = 0.259 \times 1.5 = 0.389$$

Thus, the SAVI score of 0.389 indicates moderate vegetation health. A DL g model trained on SAVI thresholds can automatically map areas of poor vegetation performance for targeted fertilizer application.

(ii) Normalized Difference Water Index (NDWI)

Water stress monitoring is crucial for irrigation management. The Normalized Difference Water Index (NDWI) is widely applied for detecting leaf water content and is given by:

$$\text{NDWI} = \frac{NIR - SWIR}{NIR + SWIR}$$

where, SWIR represents shortwave infrared reflectance.

If a UAV sensor records NIR = 0.55 and SWIR = 0.30, then:

$$\text{NDWI} = \frac{0.55 - 0.30}{0.55 + 0.30} + \frac{0.25}{0.85} = 0.294$$

A value of 0.294 indicates mild water stress, which could be flagged by the system for corrective irrigation. By running this calculation across thousands of pixels per field, UAVs can generate real-time irrigation maps that optimize water usage.

(iii) Leaf Area Index (LAI) Estimation

Another essential parameter in crop monitoring is the Leaf Area Index (LAI), which measures the total one-sided leaf area per unit ground area. UAV-based spectral imaging allows estimation of LAI through empirical relationships with vegetation indices such as NDVI:

$$\text{LAI} = -\ln\left(\frac{a - NDVI}{b}\right)$$

where, a and b are crop-specific calibration constants.

Assume $a = 1.2$, $b = 0.8$, NDVI = 0.65. Then:

$$-\ln\left(\frac{1.2 - 0.65}{0.8}\right) = -\ln\left(\frac{0.55}{0.8}\right)$$

$$\text{LAI} = -\ln(0.6875) = -(-0.3747) = 0.3747$$

Thus, the LAI value of 0.37 suggests relatively low canopy density. With time-series UAV flights, DL regression models can be trained to predict LAI progression and estimate final crop yield with high accuracy.

(iv) Crop Water Stress Index (CWSI)

Thermal cameras mounted on UAVs can capture canopy temperature, which serves as an important proxy for water stress. The Crop Water Stress Index (CWSI) is defined as:

$$\text{CWSI} = \frac{T_c - T_{\text{wet}}}{T_{\text{dry}} - T_{\text{wet}}}$$

where, T_c is the canopy temperature, T_{wet} is the temperature of a fully transpiring leaf, and T_{dry} is the temperature of a non-transpiring dry leaf surface.

Suppose UAV thermal imaging records $T_c = 33°\text{C}$, $T_{\text{wet}} = 28°\text{C}$, $T_{\text{dry}} = 40°\text{C}$. Substituting:

$$\text{CWSI} = \frac{33 - 28}{40 - 28} = \frac{5}{12} = 0.417$$

A CWSI of 0.417 indicates moderate water stress. DL models trained with temporal UAV datasets can predict irrigation schedules by linking such CWSI values with crop yield outcomes.

(v) Yield Prediction Models from UAV Data

Yield estimation is one of the most critical tasks of UAV-based surveillance. Regression-based DL models predict yield as a function of vegetation indices, LAI, and canopy cover. A simplified model can be expressed as:

$$Y = \alpha(\text{NDVI}) + \beta(\text{LAI}) + \gamma(\text{CWSI}) + \delta$$

where, Y represents yield (in quintals per hectare), and α, β, γ, δ are regression coefficients derived through training data.

For instance, assume coefficients $\alpha = 200$, $\beta = 150$, $\gamma = -100$, $\delta = 50$, and values NDVI = 0.70, LAI = 0.45, CWSI = 0.30. Then:

$$Y = 200\ (0.70) + 150\ (0.45) - 100(0.30) + 50$$

$$Y = 140 + 67.5 - 50 = 227.5$$

Therefore, it can be assumed that the yield prediction will be 227.5 quintals per hectare crop. By combining these predictions on UAV generated field maps all these can help the farmers get an approximate real-time forecast of the yield a head of harvesting which is useful in better market planning and financial management systems.

UAVs are not used only for aerial photography but also for many more purposes. By using SAVI, NDWI, LAI, CWSI and other vegetation indices along with DL approaches; UAVs supply operationally useful information on crop health, water stress, and yield estimation. These rule-based predictive models which come into play on the real farming conditions help to reduce the level of ambiguity as well as enhance profitability and sustainability to make sure that UAV technology plays a key part in digitalization of agriculture sector (Mishra & Mishra, 2023; Nuthalapati, 2022; Mohialden et al., 2024).

5.1.3 Object Localization and Mapping

Object localization in agriculture entails the identification and spatial mapping of specific objects including weeds, crop rows, irrigation canals, or farm machinery distributed over the field. DL techniques such as R-CNNs or YOLO and SSD (Single Shot Multibox Detector) are among those widely employed for this purpose due to their advantage to detect objects on a high level accurately at once.

In the practical deployment, every object is implemented in a bounding box with $B = (x, y, w, h)$ as its parameters, where (x, y) is the position of the object in the image and w, h represent its width and height. When drones capture images covering an entire farm, these bounding boxes are projected onto geospatial coordinates using GPS data, enabling the creation of detailed field maps.

Consider a 200-hectare farm divided into 500 drone images. If object detection models identify 20 weed clusters per hectare, the system records a total of 4,000 bounding boxes. These are geo-referenced to provide farmers with precise weed distribution maps. Subsequently, precision sprayers mounted on autonomous tractors can use this map to selectively apply herbicides only where needed. If the cost of herbicide is ₹2,500 per hectare for blanket spraying, and selective spraying reduces usage to only 20% of the area, the farmer's expenditure decreases from ₹5,00,000 to ₹1,00,000, resulting in a savings of ₹4,00,000.

5.2 Robotic Harvesting and Sorting

Robotic harvesting and sorting systems are gradually becoming indispensable for modern agriculture, especially in labor-scarce regions. These robots are designed to autonomously identify, harvest, and categorize agricultural produce with minimal human intervention. The major challenges in robotic harvesting involve perceiving the environment, differentiating between ripe and unripe fruits or vegetables, detecting optimal grasping points, and ensuring delicate handling to prevent post-harvest losses. DL serves as the backbone of these capabilities, enabling robots to generalize across different crop types and growing conditions.

5.2.1 Visual Perception and Grasp Detection

Visual perception in robotic harvesting refers to the robot's ability to "see" the crop accurately, while grasp detection determines how and where to grip the fruit or vegetable for safe detachment. CNNs trained on annotated datasets allow robots to distinguish target crops from background foliage. For instance, in a strawberry farm, the system can segment berries from leaves using semantic segmentation models like U-Net. Once the target is identified, grasp detection networks compute the best grasp pose represented mathematically as:

$$G = (X, y, \theta, W)$$

where, (x, y) indicates the grasp center in image coordinates, θ represents the angle of approach, and www specifies the gripper width.

Suppose the robot detects a tomato with center coordinates (120, 250), an optimal approach angle of $\theta = 45°$, and a required gripper width of 30 mm. The robotic arm then executes this grasp plan while adjusting its force to ensure the fruit is not bruised. Such high-level precision would be impossible without DL's ability to learn intricate visual cues.

5.2.2 Ripeness and Quality Assessment via Deep Learning

Determining the ripeness and quality of produce is a fundamental task in ensuring that only market-ready fruits and vegetables are harvested. Traditional methods involved subjective human judgment, often leading to inconsistencies. DL introduces objectivity by analyzing spectral features, colour histograms, and textural attributes to classify ripeness levels.

Consider a CNN trained to output ripeness probabilities for mangoes: unripe, semi-ripe, and fully ripe. If the model yields probabilities [0.05, 0.25, 0.70], the system categorizes the fruit as fully ripe. In the above-cited context, regression models serving as an extension of this approach make predictions on the continuous ripeness indices ranging from 0 to 1. If a mango has a predicted index of 0.82, then it can be programmed for immediate harvesting while the one at 0.45 may be observed for a few more days. This improved accuracy reduces post-harvest wastage and ensures uniform quality to consumers.

5.2.3 Post-Harvest Automation Systems

After harvesting, fruits and vegetables have to be graded, sorted, and packed properly. Mechanization of the post-harvest process greatly relies on artificial vision and deep learning which aims at error elimination. Sorting lines are equipped with high-speed cameras that take multiple images of each fruit then it is analysed for size, color uniformity and defect presence. According to the classification results, the actuators divert produce into designated bins (El Hoummaidi et al., 2021; Anand et al., 2025).

Mathematically, if N represents the number of fruits processed per minute and A denotes the classification accuracy, the expected number of correctly sorted fruits is given by $N \times A$. For example, if a sorting system processes 3,000 apples per minute with an accuracy of 95%, then approximately $3{,}000 \times 0.95 = 2{,}850$ apples are sorted correctly per minute, with only 150 requiring manual inspection. Such automation reduces operational costs and significantly improves throughput, allowing large-scale farms to meet market demand efficiently.

5.3 Automation in Livestock Management

Beyond crop production, *D* is making significant inroads into livestock management. Traditionally, monitoring animal health and behaviour relied on human observation, which was often inaccurate and labor-oriented. DL-based systems are now used to track individual animals, monitor feeding and resting patterns, detect early signs of illness, and ensure overall welfare. With the growing demand for dairy, poultry, and meat products, these systems provide an opportunity to improve efficiency, reduce losses, and uphold ethical standards of animal care.

5.3.1 Animal Tracking and Behavior Monitoring

Animal tracking involves identifying and following individual livestock within a herd. DL models trained on video feeds can distinguish animals based on unique features such as coat patterns, body shape, and gait. For instance, CNNs combined with object tracking algorithms such as SORT (Simple Online and Realtime Tracking) or DeepSORT enable continuous monitoring of hundreds of cattle simultaneously. The YOLOv4 and Deep SORT pipeline integrates real-time object detection with multi-object tracking. The system begins with an input video sequence, where YOLOv4 detects objects of interest (in this case, animals). A difference detector is used to refine detections and reduce false positives, while a Kalman filter predicts object trajectories across frames. These predictions are then matched with detections using association metrics (Selvi et al., 2022; Mishra & Dwivedi, 2025).

Deep SORT enhances this process by incorporating both motion information (via Mahalanobis distance) and appearance features (via a deep appearance descriptor). The Hungarian method of assignment is applied for associating new detections with existing tracks in the optimized way to make data association reliable even when animals are overlapping or moving quickly. This sound integration referred to as the assignment algorithm enables the computer system to give each individual animal an ID which can be used throughout its life. Behavior monitoring combines with tracking by analyzing movement patterns including feeding, lying down, walking, and social interaction. RNNs and LSTM models are commonly used to learn temporal dependencies in movement sequences. For example, if a cow spends 2 hours in feeding per day compared to 6 hours which is the herds' average, then the system should detect this as a potential health concern that requires early veterinary intervention. Combining YOLOv4 for detection with Deep SORT for tracking lets animal monitoring systems work fast and stay accurate—crucial when you're managing thousands of cattle across wide, dusty fields (Ünal, 2020).

5.3.2 Health Detection and Welfare Analytics

The health and welfare of livestock directly influence productivity and profitability. DL systems analyze multimodal data including thermal images, respiratory sounds, and activity patterns to identify early symptoms of diseases. For example, a CNN trained on thermal images can detect elevated body temperatures indicative of fever. If the system records a cow's body temperature at 39.8°C against a normal threshold of 38.5°C, it automatically flags the animal for inspection. Furthermore, welfare analytics integrates multiple indicators such as feeding time, mobility, weight gain, and milk yield into composite indices. If the welfare score W is computed as:

$$W = \alpha F + \beta M + \gamma H$$

where, F represents feeding activity, M represents mobility, H represents health score, and α, β, γ are weights assigned based on importance, then livestock managers can quantitatively track herd welfare. For example, if $\alpha = 0{,}4$, $\beta = 0{,}3$, $\gamma = 0{,}3$, and a particular cow record $F = 80$, $M = 70$, $H = 90$, its welfare score is $W = 0.4 \times 80 + 0.3 \times 70 + 0.3 \times 90 = 32 + 21 + 27 = 80$. Such composite measures provide objective evidence for decision-making and allow timely interventions to maintain herd health.

6. Data Requirements and Computational Infrastructure

DL is the new technology that can make a big difference in agriculture, but it works only if there is a huge amount of good data and a proper computing power—just like fast servers doing their jobs in the backend. DL model, as opposed to the traditional statistical methods, calls for enormous datasets as well as setting up GPUs that are quite powerful along with software that is tailored specifically for this need and other necessary tools so that data can be processed at an accelerated rate and absolutely precise results should be delivered. Farming systems are very complex systems and on a regular basis change position so gathering data from biology, climate, and human economics mean just the process of data, its directing, and allocation of computing power like someone tries to trace each rain splashes within a sudden storm.

6.1 Agricultural Data Characteristics

Agricultural data stands apart—not just for its sheer volume, but for its shifting patterns, mixed sources, and the way it changes with the seasons. Data pours in from crop growth, shifting soil conditions, changing weather, sudden pest swarms, and market swings, spanning everything from sensor readings that tick upward by the minute to crisp satellite images sharp enough to trace the edge of a single field. With so much variety, you've got to think things through before rolling out a deep learning system—like checking twice before flipping a complex switch.

6.1.1 Heterogeneous and Multi-modal Data Types

The primary characteristic of agricultural data is its heterogeneity. For instance, data collected on a single farm may include images of leaves captured by drones, time-series data from soil moisture sensors, audio signals from machinery operations, and even textual records from farm logs. This diversity requires DL systems capable of fusing multiple modalities (Mehta et al., 2023). To illustrate this complexity, suppose a farmer wishes to predict the yield of rice in a given season. The predictive model may need access to:

- Historical yield data in tabular format.
- Daily rainfall records in numerical form.
- High-resolution satellite images indicating vegetative indices (e.g., NDVI).
- Soil nutrient profiles recorded in laboratory reports.

If the dataset contains n different modalities, each with dimensionality d_i, the combined input space dimensionality becomes:

$$D = \sum_{i=1}^{n} d_i$$

For example, if the dataset includes:

- 100 numerical features from soil sensors ($d_1 = 100$),
- 224 × 224 pixel RGB images of crops ($d_2 = 224 \times 224 \times 3 = 150{,}528$),
- 30 textual features from farmer records ($d_3 = 30$),

  then the combined dimensionality is:

$$D = 100 + 150{,}258 + 30 = 150{,}658$$

This enormous dimensionality underlines why dimensionality reduction, feature extraction, and computational efficiency are central to agricultural DL applications.

6.1.2 Challenges in Data Labeling and Quality

Accurate labeling of agricultural datasets is both resource-intensive and prone to errors. In crop disease detection, for instance, each image must be meticulously labeled by an expert plant pathologist to identify whether a leaf is healthy, mildly infected, or severely diseased. This manual annotation is expensive and time-consuming, especially in rural contexts where skilled labor is limited. Moreover, data quality is often compromised due to:

- Noise in sensor data caused by hardware malfunction or environmental interference.

- Inconsistent scales of measurement across different regions (e.g., soil pH measured in different units).
- Seasonal variations that create temporal bias in datasets.

In some situations, mislabeled data can introduce biases that are even systematic. Assume a dataset of 10,000 images contains a labeling error of 5%, then 500 images will be misclassified. In the course of training models just these errors multiply and cause very dramatic accuracy drop rate especially for such sensitive applications as pest outbreak prediction.

6.1.3 Privacy and Ethical Considerations

Understanding and observing agricultural data come with deep concerns on ethical grounds and privacy. This can sometimes be due to the fact that as a farmer you may not wish to share very private information like the size of your land, crop yield, or financial records for fear they might be used inappropriately by corporations or government agencies (Nishad et al., 2024).

Imagine a scenario where data from 1,000 farmers in a given region is combined to forecast the inputs required. Suppose there are even parts of personal identifiers like (for example, Aadhaar number) within the data set of one farmer—right away you will see possible breaches of privacy. With the use of AI-driven platforms by multinationals that do not share profits fairly, this ethical issue becomes aggravated as these platforms exploit data of smallholder farmers.

Therefore, privacy-preserving protocols, like differential privacy, secure multiparty computation and federated learning should be also used in the domain of agricultural DL.

6.2 Data Preprocessing and Augmentation

Usually, the raw agricultural data are not fit for DL models. The preprocessing step converts messy, heterogeneous inputs into structured and normalized representations. Techniques of augmentation further increase artificially the amount of data set in order to improve model's robustness and decrease overfitting.

6.2.1 Normalization and Scaling Techniques

Normalization ensures that features with different units (e.g., soil pH versus rainfall in mm) are comparable within the same learning framework. For numerical data, common normalization techniques include:

- **Min–Max Scaling:**

$$x' = \frac{x - x_{\min}}{x_{\max} - x_{\min}}$$

If the rainfall data for a district ranges from 200 mm to 1,200 mm and a particular observation is 800 mm, then:

$$x' = \frac{800 - 200}{1200 + 200} = \frac{600}{1000} = 0.6$$

- **Z-Score Normalization:**

$$x' = \frac{x - \mu}{\sigma}$$

If the average soil nitrogen content is 45 mg/kg with a standard deviation of 10, and one sample has 55 mg/kg, then:

$$x' = \frac{55 - 45}{10} = 1$$

Such transformations ensure numerical stability during gradient-based optimization in DL.

6.2.2 Synthetic Data Generation

Given the scarcity of labeled agricultural data, synthetic data generation plays a pivotal role. For image-based tasks such as pest detection, techniques like geometric transformations (rotation, flipping), color augmentation, and GANs can increase dataset size without additional field data collection.

For example, if an original dataset contains 5,000 labeled images of diseased maize leaves, applying simple augmentation (rotating each image by 90°, 180°, and 270°) expands the dataset to:

$$N' = N \times k = 5{,}000 \times 4 = 20{,}000$$

This quadrupling of the dataset enhances model generalization significantly.

6.2.3 Handling Missing and Imbalanced Data

Missing data is pervasive in agriculture due to sensor failures, inaccessible fields, or incomplete farmer records. Statistical techniques such as mean imputation, regression imputation, and deep learning- based imputers can reconstruct missing values (Sharma et al., 2023; Liu et al., 2020).

Imbalanced data, where some classes dominate others, is another major issue. For instance, in a pest detection dataset, 90% of the images may belong to "healthy leaves", while only 10% represent "infected leaves". This imbalance biases the model toward predicting the majority class. Methods like SMOTE (Synthetic Minority Oversampling Technique) or cost-sensitive learning tackle these challenges head-on, much like adding extra brushstrokes to balance a lopsided painting.

6.3 Computational Tools and Frameworks

DL in agriculture won't get far without powerful, well-built computing systems—think rows of humming servers pushing through data day and night. From the libraries that hold your code to the hardware humming under your desk, every piece matters when turning a rough prototype into a solution ready for the real world.

6.3.1 Deep Learning Libraries (TensorFlow, PyTorch, Keras)

TensorFlow, PyTorch, and Keras stand out as the top three libraries, giving developers flexible tools to build agricultural AI systems—whether that's recognizing ripe tomatoes from field images or tracking crop health.

- TensorFlow provides extensive support for production deployment and is widely used in large-scale crop yield prediction projects.
- PyTorch is highly preferred in research due to its dynamic computation graph and ease of debugging, making it suitable for experimental models like pest disease detection using CNNs.
- Keras, built on TensorFlow, offers a high-level interface for quick prototyping, ideal for small-scale agricultural experiments.

TensorFlow offers robust tools for production deployment and powers large-scale crop yield prediction projects, like ones crunching satellite images of endless green fields.

6.3.2 Cloud Platforms and Edge Computing in Agriculture

Cloud platforms like Google Cloud AI, AWS SageMaker, and Microsoft Azure deliver the power and storage to handle massive agricultural datasets, from satellite images of cornfields to weather logs stretching back years. For example, a farmer cooperative might send terabytes of crisp satellite photos to the cloud, where DL models run across clusters of GPUs to forecast crop yields or track the green spread of healthy fields.

A cloud-based smart farming system that pulls together several parts into a layered framework, like sensors feeding data up through the network. In the data collection layer, IoT devices, wireless sensor networks, monitoring tools, and even simulation platforms stream agricultural data in real- time, from soil moisture readings to temperature shifts. The system sends this information to a cloud server, where it's sorted and assigned to specific tasks, much like filing papers into the right folders. The information exchange layer runs SVM classifiers to sort training and test data into clear groups audio with its sharp waveforms, vivid video, crisp images, and plain text. At the data center, these data classes are sorted into matching clusters audio, video, image, and text like files tucked into clearly labeled folders. These clusters run on virtual machines (VMs), which can shift resources around as the system gets busier like adding extra lanes when traffic builds up (Zulkifi et al., 2024).

The task classifier keeps the data moving in order, and the application layer handles scheduled outputs—like a daily report—so decisions can be made with confidence. This structured approach streamlines agricultural data management and scales smoothly, even when you're dealing with massive, mixed datasets—

thousands of soil readings, weather logs, and crop images all at once (Patel et al., 2025; Nishad et al., 2023). Cloud platforms offer big advantages, but in rural areas, spotty internet can stall real-time farm decisions—like when a slow connection delays pulling up the latest weather radar. Farmers are tackling this by putting edge computing devices—like the NVIDIA Jetson Nano or Google Coral TPU—right on their land, sometimes tucked beside a weathered barn or humming quietly in a shed. These low-power systems run pre trained models locally, allowing immediate predictions of soil health, pest outbreaks, or irrigation needs without continuous internet access. Thus, combining the centralized power of cloud platforms with the decentralized flexibility of edge devices creates a balanced ecosystem for precision agriculture (Zhou et al., 2021).

6.3.3 High-Performance Computing (HPC) for Model Training

Training DL models on large agricultural datasets requires HPC infrastructure, including GPUs and TPUs. A CNN with 50 million parameters, trained on 100,000 high-resolution images of paddy fields, may take weeks on a conventional CPU but only a few days on an HPC cluster.

Suppose the training of a model requires $C = n \times p$ floating-point operations (FLOPs), where n is the number of training samples and p is the computational complexity per sample. If $n = 100{,}000$ and $p = 10^9$, then:

$$C = 100{,}000 \times 10^9 = 10^{14} \text{ FLOps}$$

On a CPU capable of 10^{10} FLOPs per second, the training time is:

$$T = \frac{10^{14}}{10^{10}} = 10^4 \text{ seconds} \approx 2.8 \text{ hours}$$

But if distributed across 100 GPUs, each capable of 10^{12} FLOPs per second, the time reduces drastically:

$$T = \frac{10^{14}}{100 \times 10^{12}} = 10 \text{ seconds}$$

This demonstrates how HPC infrastructure transforms feasibility in agricultural AI projects.

7. Evaluation, Limitations, and Future Directions

The evaluation of DL models in agriculture requires a comprehensive understanding of both computational performance and practical applicability in real-world agricultural systems. Unlike purely industrial or commercial domains, agricultural DL applications are embedded in dynamic, heterogeneous, and resource-constrained environments where soil variability, climatic fluctuations, pest invasions, and crop diversity all play critical roles. Therefore, the assessment of a model's utility extends beyond numerical accuracy into dimensions of reliability, robustness, and interpretability.

7.1 Model Evaluation Metrics

Evaluation metrics serve as the cornerstone for assessing the validity and success of any DL approach in agriculture. Since agricultural datasets often involve classification tasks, such as identifying diseased leaves, detecting weed patches, or differentiating crop varieties, and regression tasks, such as yield prediction or soil moisture estimation, the choice of evaluation metric should be carefully aligned with the problem at hand.

7.1.1 Accuracy, Precision, Recall, F1-Score

Accuracy is the most intuitive metric, representing the proportion of correctly classified instances relative to the total number of instances. In a binary crop disease classification task, accuracy is expressed as:

$$\text{Accuracy} = \frac{\text{TP} + \text{TN}}{\text{TP} + \text{TN} + \text{FP} + \text{FN}}$$

where, TP denotes true positives (correctly predicted diseased samples), TN denotes true negatives (correctly predicted healthy samples), FP denotes false positives (healthy samples misclassified as diseased), and FN denotes false negatives (diseased samples misclassified as healthy) (Richetti et al., 2023).

For instance, suppose a DL model is used to detect bacterial blight in rice, and out of 1,000 samples, the model predicts 800 correctly as diseased and 120 correctly as healthy. If 50 healthy samples are incorrectly classified as diseased and 30 diseased samples are missed, the calculations become:

$$\text{Accuracy} = \frac{800 + 120}{800 + 120 + 50 + 30} = \frac{920}{1000} = 0.92 \text{ or } 92\%$$

While accuracy seems satisfactory, it can be misleading in cases of data imbalance. Precision, recall, and the F1-score therefore become crucial.

Precision focuses on the reliability of positive predictions:

$$\text{Precision} = \frac{\text{TP}}{\text{TP} + \text{FP}}$$

Using the same case:

$$\text{Precision} = \frac{800}{800 + 50} + \frac{800}{850} = 0.941 \text{ or } 94.1\%$$

Recall (also known as sensitivity) measures the ability of the model to capture all actual positives:

$$\text{Recall} = \frac{\text{TP}}{\text{TP} + \text{FN}}$$

$$\text{Recall} = \frac{800}{800 + 30} + \frac{800}{830} = 0.964 \text{ or } 96.4\%$$

Finally, the F1-score harmonizes precision and recall:

$$\text{F1} = 2 \times \frac{\text{Precision} \times \text{Recall}}{\text{Precsion} + \text{Recall}}$$

$$\text{F1} = 2 \times \frac{0.941 \times 0.964}{0.941 + 0.964} = 2 \times \frac{0.907}{1.905} = 0.952 \text{ or } 95.2\%$$

Thus, the F1-score provides a balanced indication of the model's effectiveness in agricultural disease detection.

7.1.2 AUC-ROC and Confusion Matrix Analysis

In farming, where the point for spotting disease can shift, the Area Under the Receiver Operating Characteristic Curve (AUC-ROC) is a key gauge—like checking how well a test separates healthy leaves from the ones flecked with brown spots. The ROC curve shows how the true positive rate, or sensitivity, changes with the false positive rate at varying thresholds, while the AUC sums it up as a single measure of how well the model separates the classes. An AUC of 1 means the model nails every prediction, while 0.5 is no better than flipping a coin.

The confusion matrix adds to the evaluation, showing at a glance how predictions are spread across classes—like shaded boxes lining up along a grid. In a three-class maize leaf disease classification—blight, rust, and healthy—the confusion matrix shows more than just correct counts; it also maps mistakes, like mistaking the reddish speckles of rust for the pale lesions of blight, each carrying its own agronomic consequences (Arora & Jain, 2022).

7.1.3 Cross-Validation and Model Robustness

In agricultural DL, cross-validation plays a key role in making sure a model can handle unseen data—like predicting crop yields from a field it's never "seen" before. K-fold cross-validation splits the dataset into k parts, training on k−1 of them and testing on the one left out each time—like rotating puzzle pieces until every piece has had its turn. This approach matters most in agriculture, where data is often scarce and tilted toward certain seasons—like harvest time—or specific regions. Split a set of 5,000 tomato leaf images into five folds, and each one ends up with 1,000 samples—imagine a neat stack of glossy leaves in every pile. In the first run, the model trains on 4,000 samples, while 1,000 crisp test cases are set aside for validation. The process runs five times, then the performance metrics—

like accuracy or speed—are averaged to smooth out the variance. This approach keeps the model's reported accuracy or F1-score from hinging too much on one small slice of data—like a handful of unusually easy test cases—so it stays reliably robust.

7.2 Challenges in Agricultural Deep Learning

Despite its successes, DL in agriculture faces considerable challenges.

7.2.1 Generalizability across Regions and Crops

One of the biggest hurdles is that it just doesn't work well outside its narrow test cases—like a puzzle piece that won't fit anywhere else. A model trained on wheat rust in Punjab might struggle in Uttar Pradesh, where the soil holds a different mix of minerals, the air feels heavier in summer, and farmers follow their own planting rhythms (Kumar et al., 2023). The challenge grows harder thanks to India's immense variety of crops and the shifting microclimates—one field may be dry and dusty while another, just a mile away, stays damp and cool. The issue calls for adaptive or transfer learning strategies that can recalibrate models across regions without requiring exhaustive retraining.

7.2.2 Interpretability and Explainability of Deep Learning Models

DL models, particularly CNNs and recurrent architectures, are often regarded as "black boxes". In agriculture, where decision-making impacts farmer livelihoods, food security, and policy, interpretability is non-negotiable. Farmers and policymakers require an explanation of why a model classified a plant as diseased, or why a yield forecast is lower than expected. Current research in explainable AI (XAI) attempts to address this issue, utilizing techniques such as saliency maps, attention mechanisms, and Layer-wise Relevance Propagation (LRP) (Wang & Su, 2024; Singh Chouhan, S., et al., 2025).

7.2.3 Data Scarcity and Class Imbalance

Agricultural datasets are often scarce, as collecting labeled data requires specialized domain expertise, significant time, and substantial costs. Moreover, class imbalance is common, as seen when images of healthy leaves vastly outnumber those of diseased ones. This imbalance leads to biased models that perform poorly on minority classes, which, ironically, are often the most critical (e.g., detecting early stages of pest infestations). Synthetic data generation through GANs or data augmentation methods, such as rotation, flipping, and noise injection, is often deployed to mitigate this limitation.

7.3 Emerging Trends and Future Research

7.3.1 Self-Supervised and Federated Learning in Agriculture

Self-supervised learning, where models learn representations from unlabeled data, is particularly promising for agriculture, as it reduces reliance on large labeled datasets. For example, drones might capture millions of unlabeled crop images—rows of green leaves under bright sun—and models could learn their general visual patterns before fine-tuning to spot specific diseases.

Federated learning enables models to train together across different farms or research stations—without ever exchanging raw data—tackling both privacy issues and the challenge of varied conditions, such as sandy soil in one field and clay-heavy soil in another. In India, a network of rice research institutes could train a global model without ever revealing sensitive local data, yet still reflect regional differences—like the subtle change in grain color from one valley to the next (Zineb et al., 2023; Pal et al., 2023).

7.3.2 Integration with Blockchain and IoT

When DL meets blockchain and the IoT, the results could be game-changing—like sensors that learn and adapt in real- time. In the fields, IoT devices track soil moisture, temperature, and nutrient levels in real-time, sending that flow of data into DL models that can predict what the crops will need next. Blockchain keeps data accurate and easy to trace, a crucial safeguard in food supply chains—like tracking a tomato from the farm to your plate. Say, a model flags a pest outbreak in a certain region—blockchain can confirm it's real and quickly pass the news to farmers and suppliers, like a text alert buzzing in their pocket, so they can act fast and with confidence.

7.3.3 Democratization of Deep Learning Tools for Smallholder Farmers

The democratization of DL tools is vital for ensuring equitable benefits. While large-scale agribusinesses may afford sophisticated technologies, smallholder farmers, who constitute the majority in India, often lack resources. Future research must focus on lightweight, mobile-based DL applications that function offline and provide actionable insights in local languages. Cloud platforms and government-backed initiatives could subsidize access, thereby ensuring that technological advancements reach grassroots farmers and contribute to sustainable agricultural development.

8. Conclusion

Agriculture today stands at a decisive juncture where traditional practices must evolve to meet the escalating demands of food security, environmental sustainability, and economic viability. With the global population expected to reach

9.7 billion by 2050, food production needs to increase by nearly 60% compared to current levels, yet agricultural land is expanding at less than 1% annually. This widening gap underscores the urgency of adopting technologies such as DL, which offer predictive and adaptive solutions to resource management, crop health monitoring, and automation. For instance, DL-based disease detection systems have achieved accuracies exceeding 92% in benchmark datasets, such as PlantVillage, while predictive irrigation models have demonstrated the potential to reduce water consumption by up to 30% in field trials. Moreover, the agricultural AI market, currently valued at USD 1.7 billion in 2023, is projected to reach over USD 12 billion by 2030, reflecting rapid global adoption. While challenges persist in data availability, model interpretability, and equitable access, DL represents a transformative force that can strike a balance between productivity and sustainability. By integrating advanced analytics with agricultural ecosystems, it offers a pathway toward resilient farming systems that can sustain future generations.

Declaration

Ethical Responsibilities of Authors

All authors confirm compliance with the Ethical Responsibilities statement (Instructions for Authors).

Ethical Approval

Not applicable

Use of AI tool

The corresponding author declares the use of Grammarly and QuillBot in manuscript preparation.

Conflict of Interest

Authors don't have any conflicts

Competing interests

Not applicable

Authors' contributions

Funding

No funding has been received.

References

Abdulla, M. and Marhoon, A. (2022). Agriculture based on internet of things and deep learning. *Iraqi Journal for Electrical and Electronic Engineering*, *18*(2), 1–8.

Abdullahi, H.S. and Sheriff, R.E. (2022). Introduction to deep learning in precision agriculture: Farm image feature detection using unmanned aerial vehicles through classification and optimization process of machine learning with convolution neural network. In: *Deep Learning for Sustainable Agriculture* (pp. 81–107). Academic Press.

Albahar, M. (2023). A survey on deep learning and its impact on agriculture: Challenges and opportunities. *Agriculture*, *13*(3), 540.

Allied Market Research. (2021). *Agricultural Robotics Market by Type, Application, and Region: Global Opportunity Analysis and Industry Forecast, 2021–2030*. Allied Market Research. https://www.alliedmarketresearch.com/agricultural-robotics-market.

Altalak, M., Ammad uddin, M., Alajmi, A. and Rizg, A. (2022). Smart agriculture applications using deep learning technologies: A survey. *Applied Sciences*, *12*(12), 5919.

Anand, V., Kaur, S., Rajput, V.D., Minkina, T., Mandzhieva, S., Kumar, S. and Kumar, S. (2025). Analyzing the spatial relationship between land surface temperature and normalized difference vegetation index using remote sensing and GIS. *Discover Geoscience*, *3*(1), 65.

Arora, P. and Jain, A. (2022, October). The role of deep learning in agriculture: A review. In: *AIP Conference Proceedings* (*2555*, (1), 050032). AIP Publishing LLC.

Bal, F. and Kayaalp, F. (2021). Review of machine learning and deep learning models in agriculture. *International Advanced Researches and Engineering Journal*, *5*(2), 309–323.

Bharman, P., Saad, S.A., Khan, S., Jahan, I., Ray, M. and Biswas, M. (2022). Deep learning in agriculture: A review. *Asian Journal of Research in Computer Science*, *13*(2), 28–47.

Chouhan, S.S., Patel, R.K., Shukla, A. and Singh, U.P. (2025). Drone Importance and their Necessity in Future Generation Agriculture. In: *Artificial Intelligence and Computer Vision for Ecological Informatics* (pp. 1–8). CRC Press.

Chouhan, S.S., Patel, R.K., Singh, U.P. and Tejani, G.G. (2025). Integrating Drone in Agriculture: Addressing Technology, Challenges, Solutions, and Applications to Drive Economic Growth. *Remote Sensing Applications: Society and Environment*, 101576.

Chouhan, S.S., Singh, U.P., Saxena, A. and Jain, S. (2024). Assessing the importance and need of artificial intelligence for precision agriculture. In: *Artificial Intelligence Techniques in Smart Agriculture* (pp. 1–6). Singapore: Springer Nature Singapore.

Christiansen, P., Nielsen, L.N., Steen, K.A., Jørgensen, R.N. and Karstoft, H. (2016). DeepAnomaly: Combining background subtraction and deep learning for detecting obstacles and anomalies in an agricultural field. *Sensors*, *16*(11), 1904.

Dayal, M., Gupta, M., Gupta, M., Bara, A.R. and Chaubey, C. (2023). Introduction to Machine Learning Methods With Application in Agriculture. In: *Applying Drone Technologies and Robotics for Agricultural Sustainability* (pp. 184–203). IGI Global.

El Hoummaidi, L., Larabi, A. and Alam, K. (2021). Using unmanned aerial systems and deep learning for agriculture mapping in Dubai. *Heliyon*, *7*(10).

Eunice, J., Popescu, D.E., Chowdary, M.K. and Hemanth, J. (2022). Deep learning-based leaf disease detection in crops using images for agricultural applications. *Agronomy*, *12*(10), 2395.

Food and Agriculture Organization of the United Nations (FAO). (2022). *The State of the World's Land and Water Resources for Food and Agriculture: Systems at Breaking Point*. FAO. https://www.fao.org/documents.

Gao, S. (2022). The application of information classification in agricultural production based on internet of things and deep learning. *IEEE Access*, *10*, 22622–22630.

Intergovernmental Panel on Climate Change (IPCC). (2021). *Climate Change 2021: The Physical Science Basis. Contribution of Working Group I to the Sixth Assessment Report of the*

Intergovernmental Panel on Climate Change. Cambridge University Press. https://www.ipcc.ch/report/ar6/wg1/.

International Labour Organization (ILO). (2023). *ILOSTAT Database: Employment by Sector*. International Labour Organization. https://ilostat.ilo.org/.

Junaid, M., Shaikh, A., Hassan, M.U., Alghamdi, A., Rajab, K., Al Reshan, M.S. and Alkinani, M. (2021). Smart agriculture cloud using AI based techniques. *Energies*, *14*(16), 5129.

Kamilaris, A. and Prenafeta-Boldú, F.X. (2018). Deep learning in agriculture: A survey. *Computers and Electronics in Agriculture*, *147*, 70–90.

Kavitha, A. (2021). Deep learning for smart agriculture. *Int. J. Eng. Res. Technol*, *9*(5), 132–134.

Kotwal, J., Kashyap, R. and Pathan, S. (2023). Agricultural plant diseases identification: From traditional approach to deep learning. *Materials Today: Proceedings*, *80*, 344–356.

Koul, S. (2021). Machine learning and deep learning in agriculture. In: *Smart Agriculture: Emerging Pedagogies of Deep Learning, Machine Learning, and Internet of Things*, 1–19. CRC.

Kowalska, A. and Ashraf, H. (2023). Advances in deep learning algorithms for agricultural monitoring and management. *Applied Research in Artificial Intelligence and Cloud Computing*, *6*(1), 68–88.

Kumar, N., Kushwaha, R.R., Meena, N.R., Mishra, H. and Yadav, A.P.S. (2023). A study on costs and returns of paddy cultivation in Ambedkar Nagar district of Uttar Pradesh. *International Journal of Statistics and Applied Mathematics, SP*, *8*(3), 107–111.

Liakos, K.G., Busato, P., Moshou, D., Pearson, S. and Bochtis, D. (2018). Machine learning in agriculture: A review. *Sensors*, *18*(8), 2674.

Lin, C. H., Wang, W.C., Liu, C. Y., Pan, P.N. and Pan, H.R. (2019). Research into the E-learning model of agriculture technology companies: Analysis by deep learning. *Agronomy*, *9*(2), 83.

Liu, J., Xu, L., Cao, X., Zhang, K., Zhang, Q. and Cai, Y. (2020, December). Review on the architectures and applications of deep learning in agriculture. In: *2020 7th International Conference on Information Science and Control Engineering* (*ICISCE*) (pp. 1234–1240). IEEE.

Logeshwaran, J., Srivastava, D., Kumar, K.S., Rex, M.J., Al-Rasheed, A., Getahun, M. and Soufiene, B. O. (2024). Improving crop production using an agro-deep learning framework in precision agriculture. *BMC Bioinformatics*, *25*(1), 341.

Lu, J., Tan, L. and Jiang, H. (2021). Review on convolutional neural network (CNN) applied to plant leaf disease classification. *Agriculture*, *11*(8), 707.

Magomadov, V. S. (2019, December). Deep learning and its role in smart agriculture. In: *Journal of Physics: Conference Series* (*1399* (4), 044109). IOP Publishing.

Mahendra, H.N., Mallikarjunaswamy, S., Basavaraju, N.M., Poojary, P.M., Gowda, P.S., Mukunda, M., and Pushpalatha, V. (2022, October). Deep learning models for inventory of agriculture crops and yield production using satellite images. In: *2022 IEEE 2nd Mysore Sub Section International Conference* (*MysuruCon*) (pp. 1–7). IEEE.

Mamat, N., Othman, M.F., Abdoulghafor, R., Belhaouari, S.B., Mamat, N. and Mohd Hussein, S. F. (2022). Advanced technology in agriculture industry by implementing image annotation technique and deep learning approach: A review. *Agriculture*, *12*(7), 1033.

MarketsandMarkets. (2023). *Artificial Intelligence in Agriculture Market by Technology, Offering, Application, and Region: Global Forecast to 2030*. MarketsandMarkets. https://www.marketsandmarkets.com.

Mehta, I., Bharti, S.K., Mehta, H. and Gupta, R.K. (2023). Deep Learning-Based Agriculture Monitoring and Forecasting. In: *Artificial Intelligence Tools and Technologies for Smart Farming and Agriculture Practices* (pp. 59–78). IGI Global.

Mishra, H. (2024). Nanobiostimulants and Precision Agriculture: A Data-Driven Approach to Farming and Market Dynamics. In: *Nanobiostimulants: Emerging Strategies for Agricultural Sustainability* (pp. 365–398). Cham: Springer Nature Switzerland.

Mishra, H. (2025). Environmental degradation and impacts on agricultural production: A challenge to urban sustainability. In: *Sustainable Urban Environment and Waste Management: Theory and Practice* (pp. 53–92). Singapore: Springer Nature Singapore.

Mishra, H. and Dwivedi, S. (2025). Economic Analysis of AI Integration in Internet of Drones (IoD). In: *Machine Learning for Drone-Enabled IoT Networks: Opportunities, Developments, and Trends* (pp. 169–190). Cham: Springer Nature Switzerland.

Mishra, H. and Dwivedi, S. (2025). Remote Sensing for Sustainable Agriculture: A Machine Learning Approach to Optimizing Farm Yield and Economic Returns. In: *Artificial Intelligence and Computer Vision for Ecological Informatics* (pp. 116–150). CRC Press.

Mishra, H. and Mishra, D. (2023). Artificial intelligence and machine learning in agriculture: Transforming farming systems. *Res. Trends Agric. Sci*, *1*, 1–16.

Mishra, H. and Mishra, D. (2024). AI for data-driven decision-making in smart agriculture: From field to farm management. In: *Artificial Intelligence Techniques in Smart Agriculture* (pp. 173–193). Singapore: Springer Nature Singapore.

Mishra, H. and Mishra, D. (2024). Economic evaluation of UAV-based soil sampling approaches. In: *Applications of Computer Vision and Drone Technology in Agriculture 4.0* (pp. 271–291). Singapore: Springer Nature Singapore.

Mishra, H. and Mishra, D. (2025). Economic efficiency through livestock production monitoring and prediction in smart agriculture. In: *Optimizing Smart and Sustainable Agriculture for Sustainability* (pp. 1–28). CRC Press.

Mishra, H. and Mishra, D. (2025). Robotics, Drones, Remote Sensing, GIS, and IoT Tools for Agricultural Operations and Water Management. In: *Integrated Land and Water Resource Management for Sustainable Agriculture* (Volume 2) pp. 21–49). Singapore: Springer Nature Singapore.

Mishra, H. and Srivastava, A.B. (2025). Sustainable Agriculture. *Artificial Intelligence and Computer Vision for Ecological Informatics*, 348.

Mohialden, Y.M., Hussien, N.M., Salman, S.A., Alwahhab, A.B.A. and Ali, M. (2024). Enhancing agriculture crop classification with deep learning. *Babylonian Journal of Artificial Intelligence*, *2024*, 20–26.

Moreira, G. (2025, April 9). *Deep Learning: How do deep neural networks work? Lamarr-Blog*. Lamarr Institute for Machine Learning and Artificial Intelligence. https://lamarr-institute.org/blog/deep-neural-networks/.

Narayanan, N. and Arjun, K.P. (2024). Introduction to deep learning. In: *Artificial Intelligence for Precision Agriculture* (pp. 32–60). Auerbach Publications.

Nishad, D.C., Mishra, H., Tiwari, A.K. and Mishra, D. (2024). Post-harvest Management: Enhancing food security and sustainability. *Advances in Agriculture Sciences* (Volume II), *24*(4), 136.

Nishad, D.C., Mishra, H., Tiwari, A.K. and Pandey, A. (2023). Towards sustainable agriculture: Mitigating the adverse effects of stubble burning in India. *Research Trends in Environmental Science*, *1*, 42–48.

Nuthalapati, S.B. (2022). Transforming agriculture with deep learning approaches to plant health monitoring. *Remittances Review*, *7*(1), 227–238.

Ojo, M.O. and Zahid, A. (2022). Deep learning in controlled environment agriculture: A review of recent advancements, challenges and prospects. *Sensors*, *22*(20), 7965.

Pal, A., Adhikary, R., Satapathy, S.N., Tripathy, A., Kulkarni, C.C. and Dash, D. (2023). Smart Agriculture Transformation through Deep Learning: Revolutionizing Crop Productivity and Sustainability. *Journal of Advanced Zoology*, *44*.

Pantoja, M., Kurfess, F.J. and Humer, I. (2020, June). Deep Learning for Agriculture. In: *2020 ASEE Virtual Annual Conference Content Access*.

Parico, A.I.B. and Ahamed, T. (2021). Real time pear fruit detection and counting using YOLOv4 models and deep SORT. *Sensors*, *21*(14), 4803.

Patel, R.K. (2025). BM3D Filtering with Ensemble Hilbert-Huang Transform and Spiking Neural Networks for Cardiomegaly Detection in Chest Radiographs. *Computational Biology and Chemistry*, 108620.

Patel, R.K., Chouhan, S.S., Chandan, T. and Singh, U.P. (2025). Basil Crop Detection Using Computer Vision and Deep Learning Approach. In: *Artificial Intelligence and Computer Vision for Ecological Informatics* (pp. 67–82). CRC Press.

Penn State University. (2020). *PlantVillage dataset*. Penn State College of Agricultural Sciences. https://plantvillage.psu.edu/.

PlantVillage & Food and Agriculture Organization of the United Nations (FAO). (2022). *Applications of Machine Learning for Plant Disease Detection*. FAO. https://www.fao.org.

Poonia, R.C., Singh, V. and Nayak, S.R. (Eds.). (2022). *Deep Learning for Sustainable Agriculture*. Academic Press.

Rao, I., Shirgire, P., Sanganwar, S., Vyawhare, K. and Vispute, S.R. (2021, May). An overview of agriculture data analysis using machine learning techniques and deep learning. In: *International Conference on Image Processing and Capsule Networks* (pp. 343–355). Cham: Springer International Publishing.

Razzaq, K. and Shah, M. (2025). Machine learning and deep learning paradigms: From techniques to practical applications and research frontiers. *Computers*, *14*(3), 93.

Ren, C., Kim, D.K. and Jeong, D. (2020). A survey of deep learning in agriculture: Techniques and their applications. *Journal of Information Processing Systems*, *16*(5), 1015–1033.

Research trials. (2021–2023). Yield prediction with convolutional neural networks [Unpublished data].

Richetti, J., Diakogianis, F.I., Bender, A., Colaço, A. F. and Lawes, R.A. (2023). A methods guideline for deep learning for tabular data in agriculture with a case study to forecast cereal yield. *Computers and Electronics in Agriculture*, *205*, 107642.

Saleem, M. H., Potgieter, J. and Arif, K.M. (2021). Automation in agriculture by machine and deep learning techniques: A review of recent developments. *Precision Agriculture*, *22*(6), 2053–2091.

Santos, L., Santos, F.N., Oliveira, P. M. and Shinde, P. (2019, November). Deep learning applications in agriculture: A short review. In: *Iberian Robotics Conference* (pp. 139-151). Cham: Springer International Publishing.

Santra, P., Mansuri, S.M., Gautam, P. V. and Kumar, M. (2021). Introduction to machine learning and internet of things for management in agriculture. *SATSA Mukhapatra-Annual Technical*, *25*(April), 44–65.

Selvi, M. S., Deepa, K., Saranya, N. and Rani, S.J. (2022). Deep learning in agriculture as a computer vision system. In: *Cyber-Physical Systems and Industry 4.0* (pp. 225–236). Apple Academic Press.

Sharma, P., Dadheech, P., Aneja, N. and Aneja, S. (2023). Predicting agriculture yields based on machine learning using regression and deep learning. *IEEE Access*, *11*, 111255–111264.

Sharma, V., Tripathi, A.K., Mittal, H., Parmar, A., Soni, A. and Amarwal, R. (2023). Weedgan: A novel generative adversarial network for cotton weed identification. *The Visual Computer*, *39*(12), 6503–6519.

Singh, R.K. and Khan, A. (2023). A comparative study of quantum and classical deep learning for intelligent agriculture. *Journal of Information and Computational Science*, *13*(7), 58–79.

Singh Chouhan, S., Saxena, E., Shukla, A. et al. (2025). Optimizing YOLO-Based Models for Real-Time Guava Detection with Probabilistic Fused Wiener Filter-Enhanced Feature Fusion. *Applied Fruit Science*, *67*, 383. https://doi.org/10.1007/s10341-025-01613-2.

Tiwari, A.K., Mishra, H., Nishad, D.C. and Pandey, A. (2023). Sustainable water management in agriculture: Irrigation techniques and water conservation. *Dr. Ajay B. Jadhao*, *53*.

Tripathi, P., Kumar, N., Rai, M. and Khan, A. (2022). Applications of deep learning in agriculture. In: *Artificial Intelligence Applications in Agriculture and Food Quality Improvement* (pp. 17–28). IGI Global Scientific Publishing.

Uddin, M. S. and Bansal, J. C. (2021). Introduction to computer vision and machine learning applications in agriculture. In: *Computer Vision and Machine Learning in Agriculture* (pp. 1–8). Singapore: Springer Singapore.

Umar, M.A., Sani, B.M., Suleiman, U. and Tijjani, M.L. (2022). An overview of machine and deep learning technologies application in agriculture: Opportunities and challenges in Nigeria. *SLU J. Sci. Technol*, *4*, 84–96.

Ünal, Z. (2020). Smart farming becomes even smarter with deep learning–A bibliographical analysis. *IEEE Access*, *8*, 105587–105609.

United Nations, Department of Economic and Social Affairs, Population Division. (2022). *World Population Prospects 2022: Summary of Results*. United Nations. https://population.un.org/wpp/.

Wang, D., Cao, W., Zhang, F., Li, Z., Xu, S. and Wu, X. (2022). A review of deep learning in multiscale agricultural sensing. *Remote Sensing*, *14*(3), 559.

Wang, R.F. and Su, W.H. (2024). The application of deep learning in the whole potato production Chain: A Comprehensive review. *Agriculture*, *14*(8), 1225.

Xiao, Y., Wang, X., Zhang, P., Meng, F. and Shao, F. (2020). Object detection based on faster R-CNN algorithm with skip pooling and fusion of contextual information. *Sensors*, *20*(19), 5490.

Zhang, R., Wang, Y., Jiang, P., Peng, J. and Chen, H. (2023). IBSA_Net: A network for tomato leaf disease identification based on transfer learning with small samples. *Applied Sciences*, *13*(7), 4348.

Zhou, Z., Majeed, Y., Naranjo, G.D. and Gambacorta, E.M. (2021). Assessment for crop water stress with infrared thermal imagery in precision agriculture: A review and future prospects for deep learning applications. *Computers and Electronics in Agriculture*, *182*, 106019.

Zhu, N., Liu, X., Liu, Z., Hu, K., Wang, Y., Tan, J. and Guo, Y. (2018). Deep learning for smart agriculture: Concepts, tools, applications, and opportunities. *International Journal of Agricultural and Biological Engineering*, *11*(4), 32–44.

Zineb, R., Samira, K. and Larbi, K.M. (2023). Applications of smart agriculture for environmental protection using deep learning techniques. In: *E3S Web of Conferences* (*412*, 01083). EDP Sciences.

Zulkifli, N., Razak, T.R. and bin Ismail, M.H. (2024). A Development of Deep Learning-Based Lemon Quality Detection through Peel Analysis. *International Journal of Computing and Digital Systems*, *16*(1), 1–10.

12 A Comparative Analysis of 4G and 5G IoT Sensor Networks for Smart Agriculture: Performance, Efficiency, and Future Prospects

Mekala Samayochitha[1], Vipul Negi[2], Sandesh K Jain[3], Lalita Gupta[4*] and Sourav Tiwari[5]

1. Introduction

The global agriculture industry is facing tremendous pressure to provide food for an increasing population, estimated to reach 9.7 billion by 2050, while simultaneously addressing issues such as climate change, water shortages, and sustainable resource management (FAO, 2017). To face these challenges, the agricultural sector is undergoing digitalization with the adoption of Smart Agriculture, also known as Precision Farming.

Such a paradigm leverages an array of IoT sensors deployed across fields to collect real-time data on soil moisture, nutritional content, pest presence, and microclimatic conditions (Tzounis et al., 2017). The efficiency of smart agriculture systems is highly reliant on the communication network that connects these sensors and transmits information to cloud infrastructures for analysis.

Cellular technologies have become a frontrunner solution due to their widespread adoption, robustness, and scalability. The adoption of fourth-generation (4G) LTE (long-term evolution) technology, along with its IoT-optimized versions such as LTE-M (Machine-type Communication) and NB-IoT (Narrowband-IoT), is currently the backbone of most commercial smart farming deployments (Chen et al., 2019). These technologies provide the balanced combinations of coverage, power efficiency, and appropriate data rates for most monitoring applications.

However, the advent of fifth-generation (5G) wireless technology will further revolutionize the farming industry. 5G is not just an incremental advance but a fundamental change meant to facilitate three main service categories: enhanced Mobile Broadband (eMBB), massive Machine-Type Communication (mMTC), and Ultra Reliable Low-Latency Communication (URLLC) (Andrews et al., 2014).

[1-4] Department of Electronics and Communication Engineering, Maulana Azad National Institute of Technology, Bhopal.
[5] Senior Software Engineer, Wipro, Australia. https://orcid.org/0000-0002-6317-0563.
* Corresponding author: lalitagupta@manit.ac.in

DOI: 10.1201/9781003715726-12

These features will enable a new generation of agricultural use cases that were not feasible with 4G, including real-time autonomous machinery control, high-fidelity live video streaming by drones for crop health monitoring, and dense sensor networks with millions of devices per square kilometer. This chapter presents a comprehensive comparison of 4G and 5G technologies for IoT sensors in Smart Agriculture.

The aim is to outline the strengths and weaknesses of each technology in terms of principal performance indicators and determine the best applications for each. The rest of the chapter is organized as follows: Section 2 contains a literature review. Section 3 outlines the comparison methodology. Section 4 explains a generalized block diagram of a smart agriculture system. Section 5 gives the conclusion and outlines directions for future research.

2. Literature Review

A substantial body of research has investigated the application of wireless sensor networks (WSNs) in agriculture. Initial systems were based on short-range technologies such as Zigbee, LoRa, and Wi-Fi, which are range- and infrastructure-bound (Ramson & Moni, 2017). The addition of cellular IoT has overcome these scalability limitations.

2.1 4G LTE for Smart Agriculture

4G LTE, particularly under its IoT-focused standards, has been widely adopted. NB-IoT provides outstanding coverage (20+ km in rural regions) and penetration, making it suited for buried, low-power, static sensors in soil or located in a farm's remote areas. It accommodates low data rates (<100 kbps) and is highly energy-efficient, capable of achieving a battery life of up to 10 years (Chen et al., 2019). Experiments by Liberg et al. (2017) demonstrated the successful implementation of NB-IoT for measuring soil moisture and temperature with high reliability and minimal power consumption.

LTE-M has higher data rates (up to 1 Mbps), lower latency, and supports mobility and voice services, in contrast to NB-IoT. This makes it suitable for use cases such as livestock tracking with moving collars or tracking the location and status of farm equipment with mobile equipment (Goudas & Dallas, 2017). The research in Sinha et al. (2017) explained that LTE-M is a good compromise for medium-data-rate applications, such as sending compressed images from pest traps.

The main 4G limitations in the context of agriculture are limited ultra-dense deployment capacity for sensors, 10–50 ms latency that is too high for real-time control loops, and limited data rates for bandwidth-constrained applications such as HD video.

2.2 5G for Smart Agriculture

5G technology is designed to break the constraints of 4G. Its mMTC column is anticipated to cover as many as 1 million devices per square kilometer (Andrews et al., 2014), enabling a previously unattainable level of sensor density for hyper-accurate field mapping. Work by Ferrag et al. (2020) suggests that this will enable "per-plant" agriculture, in which the needs of each plant are monitored and treated on an individual basis.

The 5G uRLLC feature, which can have latency as low as 1ms, is essential for precision and safety in autonomous agriculture operations. For example, a fleet of autonomous tractors or drones can coordinate in real-time to avoid collisions safely and carry out synchronized operations (P.M.S., 2020).

Likewise, uRLLC enables real-time actuation, such as activating an accurate laser to target and kill a weed that has been spotted without harming the crop (Tzounis et al., 2017; Hilten & Wolfert, 2022).

The eMBB capability offers gigabit-level data rates, enabling real-time, high-definition video streaming from drones. Such a video can be analyzed on-the-fly using edge computing and computer vision techniques to detect outbreaks of disease, water stress, or nutrient deficiencies in large fields in minutes, a process that would take days if done traditionally or be impossible using 4G bandwidth (DZSAP & S.R., 2021; Singh Chouhan, S., 2025).

Yet, the literature also identifies difficulties for 5G in agriculture, mainly the limited availability of 5G networks in rural areas, the increased cost of 5G modules compared to mature 4G IoT modules, and the overall system complexity (Gupta et al., 2019; Chouhan et al., 2025; Shukla et al., 2025).

3. Methodology

To conduct an equitable and organized comparison between 4G and 5G for IoT in smart agriculture, we employed a qualitative and quantitative approach, grounded in a review of technical specifications and peer-reviewed case studies. The comparison is framed within key performance metrics that are most important for agricultural applications.

3.1 Performance Metrics

The following parameters were chosen to be measured:

1. **Latency:** Time required for a data packet to move from the sensor to the application server. Essential for real-time control and actuation.
2. **Data Rate:** The maximum data transfer speed. Governs the nature of data (text, image, video) that could be efficiently sent.
3. **Network Capacity:** The devices that can be supported per unit area (e.g., per km2). Significant for high-density sensor deployment.
4. **Energy Consumption:** The energy needed to communicate, a direct influence on sensor battery life.

5. **Coverage and Penetration:** The range in the geographical location and to support devices in difficult areas (e.g., underground, in silos).
6. **Cost:** The cost of ownership, module price, data plans, and infrastructure.

3.2 Comparative Framework

We compare the two technologies by charting their functionalities against typical smart agriculture uses, grouped into three levels:

- **Tier 1:** Foundation Monitoring (e.g., soil moisture, temperature, humidity).
- **Tier 2:** Intermediate Tracking and Control (e.g., livestock tracking, telematics of equipment, automated irrigation).
- **Tier 3:** Sophisticated Autonomous Systems (such as real-time swarm drones, autonomous equipment, AI-based precision actuation).

The results of this comparative analysis are synthesized in Table 1.

Table 1 Comparative Analysis of 4G and 5G for IoT in Smart Agriculture.

Performance Metric	*4G (LTE-M / NB-IoT)*	*5G (mMTC / URLLC / eMBB)*	*Impact on Application Suitability*
Latency	10– 50 ms (LTE-M); Higher for NB-IoT	<1 ms (URLLC); 1–10 ms (general)	4G: Suitable for delayed control. 5G: Enables real-time, mission-critical control.
Peak Data Rate	~1 Mbps (LTE-M); ~100 kbps (NB-IoT)	~10 Gbps (eMBB); 10s-100s Mbps (mMTC)	4G: Text, small images. 5G: HD video, complex sensor fusion data.
Connection Density	~50,000– 100,000 devices/km^2	~1,000,000 devices/km^2 (mMTC)	5G enables hyperprecise, per-plant monitoring and dense actuator networks.
Energy Efficiency	Very High (NB-IoT); High (LTE-M)	Very High (mMTC)	Both are suitable for long-life battery-operated sensors.
Coverage	Excellent (especially NB-IoT)	Good (requires denser infrastructure)	4G has an advantage in remote/rural areas with existing infrastructure.
Mobility Support	Limited (LTE-M); None (NB-IoT)	Full support	5G is better suited for high-speed mobile nodes (e.g., drones).
Typical Applications	Soil sensors, weather stations, basic alerts (Tier 1)	Livestock tracking, irrigation control (Tier 2)	Real-time drone analytics, autonomous tractors, hyper-precision farming (Tier 3).

4. System Architecture and Block Diagram

A standard smart agriculture solution with cellular IoT connectivity consists of multiple layers.

The system works as follows:

4.1 Sensing Layer

A heterogeneous IoT sensor network is distributed across the farm. They are:

a. **4G NB-IoT Sensors:** For static, low-power, and deep-indoor/underground use cases (e.g., water tank level sensors, soil probes).
b. **4G LTE-M Sensors:** For mobile or medium data-rate usage (e.g., pest trap monitors, livestock trackers).
c. **5G Sensors:** For high-bandwidth, low-latency, or ultra-dense use cases (e.g., drones with cameras, sensors on self-driving cars, dense micro-climate arrays).

4.2 Communication Layer

Sensors talk through their corresponding cellular technologies (4G or 5G) to the local base station (eNodeB for 4G, gNodeB for 5G). The data is forwarded through the mobile core network (EPC for 4G, 5GC for 5G) to the internet.

4.3 Cloud/Edge Platform

Data is ingested by a cloud or edge computing platform. Edge computing is especially important for 5G applications that require URLLC, as it performs processing closer to the point of origin, thereby reducing latency.

4.4 Data Processing and Analytics

The platform supports data analytics and AI/ML models that take the ingested data and create insights (e.g., irrigation schedules, disease prediction maps).

4.5 Application and Actuation Layer

The farmer is provided with the insights on a dashboard (web/mobile application). Automated instructions can be sent to actuators in the field (e.g., activating a particular irrigation valve, sending a drone) through the same cellular network.

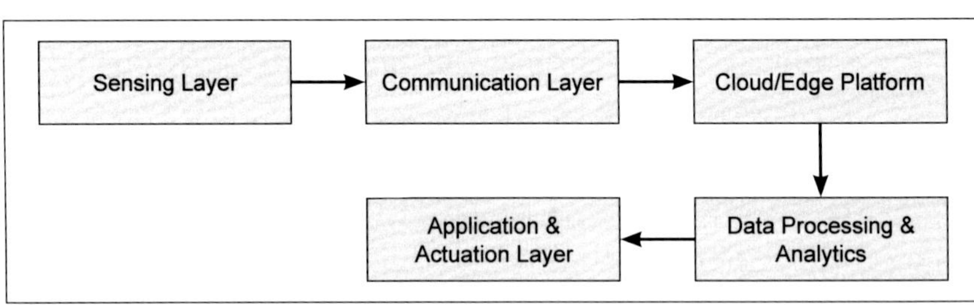

Fig. 1 Block Diagram.

5. Conclusion and Future Work

This chapter presents a comparative study of 4G and 5G technologies in the implementation of IoT sensors for intelligent agriculture. The study indicates that the decision between 4G and 5G is not one of a direct replacement but a strategic one, according to application needs.

4G LTE, in its NB-IoT and LTE-M flavors, offers a mature, affordable, and pervasive solution for the overwhelming majority of simple and intermediate smart farming use cases. Its outstanding coverage and power consumption make it the horse for large-scale environmental monitoring and asset tracking.

This is in contrast to 5G, which is a disruptive technology that opens up advanced, data-hungry, and mission-critical applications. Its unique mix of high data rate (eMBB), massive machine-type connectivity (mMTC), and ultralow latency (URLLC) sets the stage for autonomous, hyper-precision agricultural operations that will dramatically increase productivity and sustainability.

For the foreseeable future, a hybrid network infrastructure that utilizes 4G's extensive range for far-reaching sensor data gathering and 5G's high-end slices for certain demanding tasks is the most feasible and economical methodology. This enables a gradual, cost-effective transition to more advanced farming methods.

Future research will involve the development of high-fidelity simulation models to assess the energy and latency benefits of 5G, particularly in agricultural control systems. Additionally, an investigation of network slicing for agriculture—in which a single 5G physical network can be divided into virtual networks tailored for various applications (e.g., one slice for URLLC for machinery and a second slice for mMTC for sensors)—offers an exciting opportunity for -minimizing resource allocation and cost.

Declaration

Author's Ethical Responsibilities

"All authors confirm that they have read, understood, and compiled, as applicable, with the statement on Ethical Responsibilities of Authors as outlined in the instructions for Authors."

No Ethical Approval is required.

Use of AI tool

No AI tool is used.

Conflict of Interest

No conflict of disclosure.

Competing interests

Competing interests are not applicable.

Contributions of Authors

Author[1]: Conceptualization, Methodology, Writing – Original Draft, Writing – Review & Editing, **Author[2]:** Investigation, Formal analysis, Visualization, Software, Data Curation, Supervision, **Author[3]:** Writing – Original Draft, Formal analysis, Methodology, **Author[4]:** Formal analysis, Visualization, **Author[4]:** Supervision

Funding

No

References

Andrews, J.G., et al. (2014, June). What Will 5G Be? *IEEE Journal on Selected Areas in Communications*, *32* (6), 1065–1082.

Chen, M., Yang, J., Zhu, X., Wang, X., Liu, M. and Song, J. (2019). Smart agriculture system based on NB-IoT. In: *Proc. IEEE 4th International Conference on Big Data Analytics* (*ICBDA*) (pp. 276–280).

Chouhan, S.S., Patel, R.K., Shukla, A. and Singh U.P. (2025). Drone Importance and their Necessity in Future Generation Agriculture. In: *Artificial Intelligence and Computer Vision for Ecological Informatics* (pp. 1–8). CRC Press.

DZ.S.A.P. and S.R. (2021). A Review on the Role of Drones in Agriculture with 5G and IoT Support. In: *Proc. 5th International Conference on Computing and Communications Technologies* (*ICCCT*) (pp. 1–6).

FAO. (2017). *The Future of Food and Agriculture: Trends and Challenges*. Food and Agriculture Organization of the United Nations, Rome.

Ferrag, M.A., Shu, L., Yang, X., Derhab, A., and Maglaras, L. (2020). Security and Privacy for Green IoT-Based Agriculture: Review, Blockchain Solutions, and Challenges. *IEEE Access*, *8*, 32031–32053.

Goudos, S.K., P.I. and Dallas, Chatziefthymiou, F. and Kyriazako, S. (2017). A Survey of IoT Key Enabling and Future Technologies: 5G, Mobile IoT, Semantic Web, and Applications. *Wireless Personal Communications*, *97*(2), 1645–1675.

Gupta, L., Tiwari, H., Vedula, A. and Sharma, S. (2019). Role of Vegetation Indexing in India: Multispectral Imaging for Yield Prediction. *International Journal of Computational Intelligence and IoT*, 2 (2).

Liberg, O., Sundberg, M., Wang, E., Bergman, J. and Sachs, J. (2017). *Cellular Internet of Things: Technologies, Standards, and Performance*. Academic Press.

Patel, R.K., Chouhan, S.S., Chandan, T. and Singh, U.P. (2025, Aug.). Basil Crop Detection Using Computer Vision and Deep Learning Approach. In: *Artificial Intelligence and Computer Vision for Ecological Informatics* (pp. 67–82). CRC Press.

P.M.S., et al. (2020). 5G-Based Smart Healthcare for Agricultural Farms: A Review. In: *Proc. International Conference on Smart Electronics and Communication* (*ICOSEC*) (pp. 1–6).

Ramson, S.R.J. and Moni, D.J. (2017). Applications of Wireless Sensor Networks: A Survey. In: *Proc. International Conference on Innovations in Electrical, Electronics, Instrumentation and Media Technology* (*ICEEIMT*) (pp. 325–329).

Shukla, A., Chouhan, S.S., Patel, R.K. and Singh, U.P. (2025). AgriCloud: Cloud-Based Framework for Scalable Crop Monitoring and Intelligent Decision Support. In: Chouhan, S.S., Patel, R.K., Singh, U.P., and Jain, S. (Eds.), *AgriTech Revolution.* Springer, Singapore. https://doi.org/10.1007/978-981-95-1268-3_6.

Singh Chouhan, S., Saxena, E., Shukla, A., Kumar Patel, R. and Pratap Singh, U. (2025, Oct.). Optimizing YOLO-Based Models for Real-Time Guava Detection with Probabilistic Fused Wiener Filter-Enhanced Feature Fusion. *Applied Fruit Science*, *67*(5), 383.

Sinha, R.S., Wei, Y. and Hwang, S.-H. (2017). A survey on LPWA technology: LoRa and NB-IoT. *ICT Express*, *3* (1), 14–21.

Tzounis, A., Katsoulas, N., Bartzanas, T. and Kittas, C. (2017). Internet of Things in agriculture, recent advances and future challenges. *Biosystems Engineering*, *164*, 31–48.

van Hilten, Mireille and Sjaak Wolfert. (2022). 5G in agri-food: A review on current status, opportunities and challenges. *Computers and Electronics in Agriculture*, *201*, 107291.

Yu Tang, Sathian Dananjayan, Chaojun Hou, Qiwei Gu, Shaoming Luo and Yong He (2021, Jan.). A survey on the 5G network and its impact on agriculture: Challenges and opportunities. *Computers and Electronics in Agriculture*, *180*, 105895.

13 Dynamic Detection of Fruit, Pest, and Weed Using Hybrid DL Models

M. Salomi[1*], S. Samsudeen[2], J.K. Kiruthika[3], P. Radhakrishnan[4], M. Kanipriya[5] and A. Sheryl Oliver[6]

1. Introduction

Agriculture is considered as a major domain in ensuring food safety worldwide. But it is also considered as highly vulnerable to the losses that are caused by the pests, weeds, and diseases. The statistics disseminated by the Food and Agriculture Organization (FAO) reveals that the loss is nearly 20–40% of the global crop production every year due to pests and weeds Chouhan (2024). This in turn provides a loss of around USD 220 billion annually. Weeds lead to around 34% of the total crop loss whereas the loss due to insect pests contributes 18–20% loss and diseases lead to 10–12% loss (Chouhan, 2020). The countries having 60% of their population who depend on agriculture for their livelihood are getting tremendously disturbed and searching for a solution that saves the nation. Parallelly, the steps taken to control the loss leads to high voluminous usage of chemical pesticides that leads to several other challenges (Patel, 2025). The World Health Organization (WHO) reported that the pesticide usage every year leads to approximately 2.5 million tons worldwide in which 50% of the total usage is accounted for in Asian countries (Milioto, 2017). The excessive pesticide usage not only increases the production cost but also leads to environmental degradation, soil fertility loss, and potential health hazards for consumers and the farmers (Albanese, 2021). Various studies reveal that the targeted and precise pest and weed detection leads to reducing the chemical pesticide usage up to 30% that provides both economic and environmental benefits (Nanni, 2021).

These challenges can be addressed using the recent technological developments in Artificial Intelligence (AI) and Deep Learning (DL) (Duan, 2023). The use of computer vision-based detection models is rapidly increasing at a compound

[1,2,4,5] Department of Computational Intelligence, Faculty of Engineering and Technology, SRM Institute of Science and Technology, Kattankulathur, Tamil Nadu, India.

[3] Department of Computer Science and Engineering, KPR Institute of Engineering and Technology, Coimbatore, India.

School of CS & AI, SR University, Warangal, Telangana, India.

* Corresponding author: salomis@srmist.edu.in

DOI: 10.1201/9781003715726-13

annual growth rate of about 25% in the field of agriculture (Chacón-Maldonado, 2025). The report reveals that the global agricultural AI market will exceed USD 4.5 billion by 2027. Table 1 depicts the proportion of the loss attributed due to each cause (Rai, 2023). Table 2 depicts the financial values due to crop loss and the AI market that illustrates the economic impact and market growth. Table 3 reveals the comparison scale of pesticide usage. It is identified from a survey that the DL techniques showcased tremendous success in monitoring fruit ripeness, identifying pest infestations, and differentiating weeds from crops under field conditions (Gao, 2025). The manual scouting is reduced over 60% through the integration of drone assisted crop monitoring with object detection networks. Also, it's observed that this technological advancement achieves accuracy beyond the human expert's capacity (Patel, 2025).

Table 1 Statistics of crop loss due to Cause.

Statistic	*Value (%)*
Weeds	34
Insect Pests	18–20
Diseases	10–12

Table 2 Economic and Market Data Statistics.

Statistic	*Value*	*Unit*
Economic Loss from Pests and Weeds	>220	Billion USD/Year
Projected Agriculture AI Market Size by 2027	4.5	Billion USD
Agricultural AI Market CAGR	26	%

Table 3 Usage and Efficiency Metrics.

Statistic	*Value*	*Unit*
Worldwide Pesticide Use	>2.5	Million Tons/Year
Pesticide Use in Asia	50	% of Total Usage
Potential Pesticide Use Reduction with Targeted Detection	30	%
Manual Scouting Time Reduction with Drones	>60	%

The technical advancements also bring certain challenges that lead to the need for innovation. Generally, the field images of pests and weeds suffer from variable light effects, occlusion, and background clutter because the peats and weeds are naturally small, camouflaged and diverse in appearance (S. S. Chouhan, 2025). All the existing AI systems are trained for a single purpose, such as fruit detection or pest detection, which limits the utilization of AI techniques in real-world implementation where multiple threats exist. These challenges demand a framework that should be lightweight and capable of performing multitasking DL, which is considered suitable for operating the model in the real world on resource-constrained devices such as drones, robots, and internet of things (IoT)-based edge platforms (S. S. Chouhan, 2024).

This chapter addresses the aforementioned challenges through a Hybrid Deep Learning Model for real time agricultural monitoring (Patel, 2025). The proposed framework integrates the fast detection capabilities of YOLOv8 with the contextual learning power of Vision Transformers. The hybrid knowledge distillation strategy enhances the framework, aiding in the design of fruit detection, pest, and weed detection systems with high accuracy and efficiency. Thus, the proposed framework is highly suitable for precision agriculture, as it not only reduces computational complexity but also supports multi task detection (Siddharth Singh Chouhan, 2025).

2. Literature Survey

Early pest and weed detection uses traditional computer vision techniques. Insects or weeds are identified in field images by integrating methods. However, traditional approaches failed to handle the high unpredictability of working in agricultural environments because they can only operate under certain controlled conditions. Inaccurate predictions are often the result of implementation issues due to complex backgrounds, overlapping leaves, and irregular lighting (Singh Chouhan, S., 2025).

Agricultural detection has been revolutionized with the advent of DL. DL models, as well as other two-stage detectors, provide high accuracy when implementing the technology for detecting pests and leaf diseases. The single-stage detectors play a vital role in balancing the speed and accuracy. Specifically, the YOLO family is one of the single-stage detectors that aids in detecting pests and diseases more effectively. The detection performance can be steadily enhanced from YOLOv3 to YOLOv8 by certain improvements, such as multi scale predictions, anchor-free box detection, and better feature fusion.

The research focuses on compressing large DL models into smaller versions without compromising the accuracy the model can achieve. Feature extraction facilitates the transfer of semantic knowledge from the intermediate layers of large modules to small modules, whereas relational distillation focuses on preserving the structural relationships among feature models. Recent research has revealed that integrating both approaches yields more robust models. Most of the applications in the agricultural domain focus on detecting pests or weeds, but not on improving the model's quality to handle multiple simultaneous tasks. The aforementioned gap provides the need to develop a hybrid DL model that is capable of handling fruits, pests, and weeds in a unified pipeline.

3. Proposed Model

A Hybrid Deep Learning Model (H-DLM) is proposed in order to overcome the agricultural applications issues of existing models that are mentioned in the Literature Survey section. YOLOv8 architecture is considered as the backbone to design the H-DLM framework. The YOLOv8 architecture is chosen as backbone mainly considering its proven ability of delivering high speed object detection. Capturing long range dependencies and contextual information from an image is considered as crucial in agricultural fields where objects such as weeds may blend within the crops and pests may be visible in small clusters. Thus, this issue can be addressed by integrating the YOLOv8 with the Vision Transformer (ViT) module. The integration aids in capturing the long-range dependencies and the contextual information from an image.

The proposed model architecture comprises four main components such as backbone, transformer module, neck, and multi task head. The backbone uses convolutional layers to extract the low level and high-level features. The transformer module focuses on relationships across spatial regions to enhance these features. The neck component improves robustness of the model by aggregating the features from multiple scales. Multitask head component eliminates the need of three separate models by providing the ability to simultaneously detect the fruits, pests, and weeds.

Hybrid knowledge distillation is considered as a key innovation of the proposed models framework. A large teacher network combines YOLOv8 and Vision Transformer layers and is trained on agricultural datasets. Knowledge transfers from this large teacher network to the smaller student network (YOLOv8n) that is suitable to implement in real time deployment.

Feature distillation and relational distillation are the two types of distillation used in the proposed model. The feature distillation aligns feature maps between the teacher and student whereas the relational distillation provides inter-channel relationships. The integration of these two types leads to a hybrid model which is considered as a light weight but yet retains both the semantic richness and structural awareness.

The process can be summarized through three key equations mathematically such as feature distillation loss, relational distillation loss, and total hybrid loss. The difference between teacher and student feature maps are measured by the feature distillation loss. Adjacency matrices derived from cosine similarities are compared to ensure the inter-channel relationships alignment. End-to-end optimization of the model is achieved by integrating these components with the standard detection loss which is accomplished by the total hybrid loss.

3.1 Standard Detection Loss

The standard detection loss (L_{det}) depicted in Eq. (1) is considered as the base component that measures the ability of the student model in order to detect the objects accurately. A classification loss and a bounding box regression loss is combined typically using a composite loss function.

$$L_{det}(S) = L_{cls}(s) + L_{reg}(s) \qquad ...(1)$$

where, S depicts the student networks output predictions, L_{cls} represents object classification loss of an image, and L_{reg} is the loss factor that localizes the bounding box.

3.2 Feature Distillation Loss

The feature distillation loss (L_{feat}) as depicted in Eq. (2) makes the replication of teacher's network intermediate features by the student network. The student learns to process the information in a semantically rich man ner similar to its larger counterpart by minimizing the difference between their feature maps.

$$L_{\text{feat}}(S,T) = \frac{1}{N}\sum_{i=1}^{N} \| F_s^i - F_T^i \| \qquad ...(2)$$

where, F_S^i represents the i-th feature maps of the student network and F_T^i are the i-th feature maps from teacher network, and N is the total number of feature maps compared.

3.3 Relational Distillation Loss

This loss (L_{rel}) is the key innovation of the proposed model's framework. Eq. (3) focuses on transferring relational knowledge. This transfer rates how the different feature channels relate with each other between the student and the teacher network. adjacency matrices (A) derived from the cosine similarities of the feature channels are compared in order to achieve the L_{rel}.

$$L_{\text{rel}}(S,T) = \| A_S - A_T \|_2^2 \qquad ...(3)$$

where, A_S and A_T represent the teacher's and student's networks adjacency matrices.

3.4 Hybrid Knowledge Distillation Loss

The weighted sum of the feature and relational distillation loss is measured as the total knowledge distillation loss (L_{distill}). Eq. (4) reveals that the hybrid model captures the complete data and understands it fully.

$$L_{\text{distill}}(S,T) = \propto L_{\text{feat}}(S,T) + = \beta L_{rel}(S,T) \qquad ...(4)$$

where, α and β are the hyperparameters that control the relative importance of features and the relational knowledge transfer.

3.5 Total Hybrid Loss

Total Hybrid Loss L_{Total} is the final objective function for training the student network. Eq. (5) is the loss function that enables a comprehensive and end to end optimization for the student model by achieving both accuracy and efficiency. This can be achieved by integrating the traditional object detection loss with the hybrid knowledge distillation loss.

$$L_{\text{total}}(S,T) = \lambda L_{\text{det}}(S) + L_{\text{distill}}(S,T) \qquad ...(5)$$

where, λ represents the hyperparameter that balances the standard detection tasks' contribution and the knowledge distillation process.

4. Results and Discussion

Three datasets are used to perform the comparative analysis of the proposed H-DLM analysis, including IP102, Deep Weeds, and Fruits-360/Orchard. IP102 handles pest detection data, whereas the Deep Weeds dataset focuses on weed detection. The Fruits-360/orchard dataset is used for fruit classification and ripeness detection.

Standard metrics are used to evaluate the model's performance. The results showed that the proposed model consistently outperformed existing models. The H-DLM achieves higher accuracy when implemented at four times the speed compared to Faster R-CNN. This makes the proposed model suitable for implementation in practice for field deployment. The transformer module and hybrid distillation module make the proposed model a unique technique compared to YOLOv8. This, in turn, shows an improvement in achieving accuracy of about 5% without compromising inference speed.

Table 4 reveals the comparative analysis of various models. The proposed H-DLM achieves a higher percentage in all three objectives, such as fruit detection, pest identification, and weed recognition, whereas the traditional models still struggle a lot to achieve accuracy and speed.

The results as discussed in Table 4, Figures 1 and 2 give deeper insights about the advantage of integrating YOLOv8 with a transformer module and hybrid

Table 4 Comparative Analysis of Various Models.

Model	*Task Coverage*	*mAP@50*	*FPS*	*Parameters*	*Suitability*
Faster R-CNN	Pest Only	78.5	12	42M	Server-Based
YOLOv5	Fruit Only	82.1	45	7M	Edge Devices
YOLOv8	Weed Only	84.9	55	11M	Edge Devices
Proposed H-DLM	Fruit+Pest+Weed	89.6	52	14M	Drone/Edge Ready

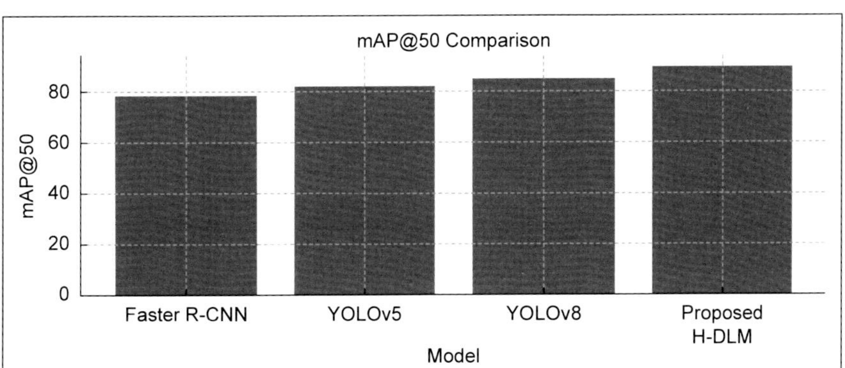

Fig. 1 mAP@50 Comparison.

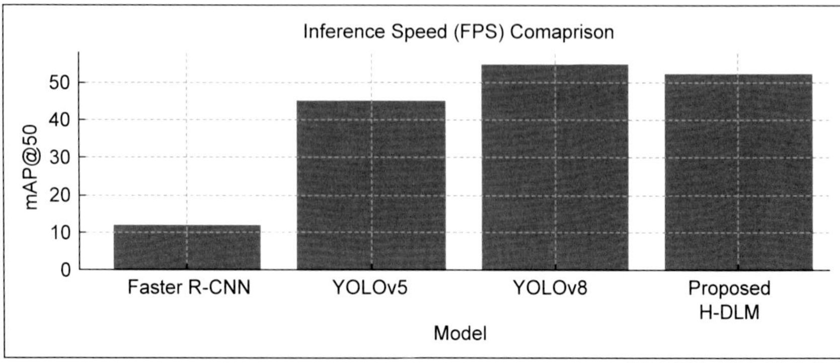

Fig. 2 Inference Speed (FPS) Comparison.

distillation that leads to a model which achieves accuracy and efficiency in real-time agricultural scenarios.

5. Conclusions and Future Enhancements

This chapter presents the proposed H-DLM framework that is mainly designed for real-time detection of fruits, weeds, and pests. The model achieves a balance between accuracy and efficiency by integrating YOLOv8 backbone with a transformer module which provides a hybrid distillation strategy. This makes the proposed model highly recommended for deploying it on drones and edge devices. The results of the proposed model and the comparative analysis reveals that it performs better than the existing traditional methods with respect to both performance and practical applications. It is considered as a remarkable aspect in precision agriculture.

The future enhancement can focus on detecting the hidden pests or disease stress using multimodal inputs. Also, few shot learning algorithms and self-supervised algorithms can be adopted which results in reducing the dependencies on annotated datasets that in turn aids the model to perform better while processing unseen spices. Autonomous agricultural robots are also considered as the most focused future enhancement in the agricultural domain that enables closed-loop systems.

References

Albanese, A., Nardello, M. and Brunelli, D. (2021). Automated pest detection with DNN on the edge for precision agriculture. *IEEE Internet of Things Journal, 9*(10), 7238–7249. https://arxiv.org/abs/2108.00421.

Chacón-Maldonado, A.M. (2025). A multimodal hybrid deep learning approach for pest detection: Olive fruit fly forecasting. *Information Fusion, 104*, 102275. https://doi.org/10.1016/j.inffus.2025.102275.

Chouhan, S.S., Singh, U.P. and Jain, S. (2024). Performance evaluation of different deep learning models used for the purpose of healthy and diseased leaves classification of Cherimoya (Annona cherimola) plant. *Neural Computing and Applications, 37*(6), 4531–4544. https://doi.org/10.1007/s00521-024-10830-x.

Chouhan, S.S., Singh, U.P. and Jain, S. (2020). Smart glass: Real-time leaf disease detection using YOLO transfer learning. In: *Proceedings of the International Conference on Smart Electronics and Communication* (pp. 1–37). ACM. https://doi.org/10.1145/3639816.

Chouhan, S.S., Singh, U.P., Saxena, A. and Jain, S. (2024). Assessing the Importance and Need of Artificial Intelligence for Precision Agriculture. In: Chouhan, S.S., Saxena, A., Singh, U.P., and Jain, S. (Eds.), *Artificial Intelligence Techniques in Smart Agriculture*. Springer, Singapore. https://doi.org/10.1007/978-981-97-5878-4_1.

Duan, J., Ding, H. and Kim, S. (2023). A multimodal approach for advanced pest detection and classification. *Computers and Electronics in Agriculture, 213*, 108113. https://arxiv.org/abs/2312.10948.

Gao, X., Zhang, Y., Li, J. and Huang, L. (2025). Applications, trends, and challenges of precision weed detection in smart agriculture. *Agronomy, 15*(8), 1954. https://doi.org/10.3390/agronomy15081954.

Milioto, A., Lottes, P. and Stachniss, C. (2017). Real-time semantic segmentation of crop and weed for precision agriculture robots leveraging background knowledge in CNNs. *IEEE Robotics and Automation Letters, 3*(1), 588–595. https://arxiv.org/abs/1709.06764.

Nanni, L., Manfe, A., Maguolo, G., Lumini, A. and Brahnam, S. (2021). High performing ensemble of convolutional neural networks for insect pest image detection. *Applied Computing and Informatics*. Advance online publication. https://arxiv.org/abs/2108.12539.

Patel, R.K., Chouhan, S.S., Chandan, T. and Singh, U.P. (2025, Aug.). Basil Crop Detection Using Computer Vision and Deep Learning Approach. *Artificial Intelligence and Computer Vision for Ecological Informatics* (pp. 67–82). doi: 10.1201/9781003518778-4.

Patel, R.K., Chouhan, S.S. and Singh, U.P. (2025). Applications, challenges, and solutions: An approach towards integrating deep learning in agriculture. In: *Current and Future Prospects of Deep Learning Models for Smart Agriculture* (pp. 224). Cambridge Scholars Publishing.

Patel, R.K. (2025, Aug.) BM3D Filtering with Ensemble Hilbert-Huang Transform and Spiking Neural Networks for Cardiomegaly Detection in Chest Radiographs. *Computational Biology and Chemistry*, *8*, 108620.

Rai, N. (2023). Applications of deep learning in precision weed management: A review. *Computers and Electronics in Agriculture, 205*, 107662. https://doi.org/10.1016/j.compag.2023.107662.

Siddharth Singh Chouhan, Rajneesh Kumar Patel, Uday Pratap Singh, & Ghanshyam G. Tejani. (2025), Integrating drone in Agriculture: Addressing technology, challenges, solutions, and applications to drive economic growth. *Remote Sensing Applications: Society and Environment*, *38*, 101576. ISSN 2352-9385, https://doi.org/10.1016/j.rsase.2025.101576.

Siddharth. S. Chouhan, Patel, R.K., Shukla, A. and Singh, U.P. (2025, Aug.). Drone Importance and their Necessity in Future Generation Agriculture. In: *Artificial Intelligence and Computer Vision for Ecological Informatics* (pp. 1–8). doi: 10.1201/9781003518778-1.

Singh Chouhan, S., Saxena, E., Shukla, A. et al. (2025). Optimizing YOLO-Based Models for Real-Time Guava Detection with Probabilistic Fused Wiener Filter-Enhanced Feature Fusion. *Applied Fruit Science*, *67*, 383. https://doi.org/10.1007/s10341-025-01613-2.

14 Integrating Deep Learning with IoT, Blockchain, and Robotics in Agri-systems

Harshit Mishra[1*], Fredrick Kayusi[2], Ioannis Adamopoulos[3] and Oluwakemi Temitope Olayinka[4]

1. Introduction

Agriculture today stands at a critical inflection point in human history, where traditional practices are rapidly giving way to a new digital frontier. Integrating Deep Learning (DL) with the Internet of Things (IoT), blockchain, and robotics offers unprecedented opportunities to meet the mounting global demand for food—amid challenges such as climate change, resource scarcity, and inefficiencies in supply chains (Mamodiya & Kishor, 2025). The digital agriculture market underscores this transformation: global agricultural digital technologies are projected to reach USD 22 billion by 2025, with smart agriculture alone valued at USD 15.69 billion in 2025, expected to grow to USD 23.38 billion by 2029 with a CAGR of 10.2 per cent. These figures reflect accelerating investment in technologies that enhance productivity, optimise inputs, and promote sustainability at scale (Kumari & Praveen, 2025).

DL architectures such as Convolutional Neural Networks (CNNs), Recurrent Neural Networks (RNNs), Long Short-Term Memory networks (LSTMs), and Generative Adversarial Networks (GANs) are pivotal in processing and interpreting vast, heterogeneous data from IoT sensors, drones, and robotic platforms. Empirical studies confirm that AI usage in agriculture can raise crop-yield forecasting accuracy to over 90%, up from around 65% using traditional methods (Khan et al., 2025). Meanwhile, DL-assisted tools in India, such as the Agripilot.ai application,

[1] Department of Agricultural Economics, College of Agriculture, Acharya Narendra Deva University of Agriculture and Technology, Kumarganj, Ayodhya, Uttar Pradesh – 224 229, India. https://orcid.org/0009-0000-1951-7278.

[2] Department of Environmental Studies, Geography and Planning, Maasai Mara University, P.O. Box 861-20500, Narok, Kenya. https://orcid.org/0000-0003-1481-4016.

[3] Department of Public Health Policy, Sector of Occupational & Environmental Health, School of Public Health, University of West Attica, Athens, Greece. https://orcid.org/0000-0002-4942-7123.

[4] Department of Business Information Systems and Analytics, University of Arkansas at Little Rock, 2801 S University Avenue, Little Rock, AR 72204, United States. https://orcid.org/0009-0003-3841-7418.

* Corresponding author: wehars@gmail.com

DOI: 10.1201/9781003715726-14

helped achieve a 40% increase in yield and 50% reduction in production costs, by tailoring pest alerts, soil data, and weather-informed irrigation recommendations. Across smallholder communities in rural India, AI-driven forecasting initiatives have cut debts by 50% and boosted savings by up to 10% of annual income; these efforts are now rapidly scaling to reach 60 million farmers by 2025 (Rahman et al., 2024; Patel, 2025).

IoT underpins the real-time data backbone of these systems, spreading rapidly across farms. Over 90% of modern farms now employ at least one digital technology, and IoT adoption has grown by 25% annually over recent years. In precision farming, sensor networks projected to reach 82% adoption by 2025 are delivering 10–18% yield improvements. Similarly, AI-driven analytics (68 per cent adoption) bring 14–24% benefits in predictive planning. Regionally, India is seeing IoT adoption between 35–45% in agricultural systems, with a market value approximating USD 2.5 billion in 2025, compared to the U.S. with 75–80% adoption and a market of over USD 7 billion (Mishra, 2023).

Blockchain adoption in agri-supply chains is also rising. The global blockchain-in-agriculture market is estimated at USD 263.8 million in 2024, projected to reach USD 357.5 million in 2025 and soar to over USD 4.06 billion by 2033, with a rapid CAGR of 35.5%. This growth aligns with increasing use cases: approximately 50% of early blockchain implementations support traceability, 30% employ smart contracts, and 20% focus on payments and settlements. In practical terms, blockchain has helped reduce fraud and improve supply chain transparency, with some sources noting potential cost reductions of supply chain fees by 30% and agricultural efficiency gains by up to 20% (Ahmed et al., 2022).

Robotics, when augmented with DL, significantly transforms labour dynamics and precision in agricultural operations. Agricultural robots globally are projected to grow in market value to USD 374 million by 2025; the digital farm management systems sector is also expanding at a CAGR of 19.2% from 2023 to 2030. Precision agriculture technologies, including robotic weeding, seeding, harvesting, and monitoring, exhibit remarkable efficiency—some deploy with 95% accuracy. Meanwhile, precision farming tools (e.g., GPS-guided machinery, variable-rate application) have improved farm outputs by 54% in Argentina, with U.S. peanut farms seeing yield increases of 9–13% through these technologies (Mishra & Srivastava, 2025). A clear summary—capturing investments, technology adoption rates, and impacts—is presented in Tables 1 and 2.

Table 1 Investment and adoption rates of major digital technologies in agriculture (2025 projections).

Parameter	*Value*
Smart agriculture market (2025)	USD 15.69 billion
Digital agriculture market (2025)	USD 22 billion
Agriculture IoT market (2025)	USD 28.65 billion

Contd...

Table 1 *Contd...*

Parameter	*Value*
India IoT adoption (2025)	35–45% adoption, USD 2.5 billion market
US IoT adoption (2025)	75–80% adoption, USD 7+ billion market
Blockchain agriculture (2025)	USD 357.5 million
Robotics/ag-tech market (2025)	USD 374 million

Source: Authors' compilation.

Table 2 Performance improvements via DL-IoT-Blockchain-Robotics integration.

Technology	*Impact / Improvement (%)*
Crop-yield forecasting accuracy	> 90% (vs ~65%)
AI-enhanced yield (India case)	+ 40%; cost reduction − 50%
Farmer savings (India)	+ 10% of annual income
Sensor network yield gains	+ 10–18%
AI predictive planning gains	+ 14–24%
Traceability / fraud reduction	up to 30% cost; efficiency + 20%
Robotics task accuracy	up to 95%
Precision tools yield boosts	+ 9–54% depending on region

Source: Authors' compilation.

These advancements not only highlight the compelling technological potential, but also emphasise the impacts on yield, cost, efficiency, and resilience. For example, wider IoT adoption can reduce pesticide and water usage by 30–40%, while precision analytics and robotics further elevate productivity and environmental stewardship. Moreover, such technologies are especially promising for smallholder and developing economy farmers: 55% of agri-tech startups now tailor solutions for smallholders, 52% of farmers in developing nations use mobile advisory services, and digital platforms have helped increase smallholder incomes by up to 30% (Mohyuddin et al., 2024).

Nevertheless, substantial challenges persist. Interoperability issues, connectivity limitations, high upfront costs, digital literacy barriers, and ethical concerns—particularly regarding AI (artificial intelligence) governance and smallholder inclusion—remain significant obstacles. Especially in regions such as Africa, where climate stress has already eroded GDP by 14% and pushed millions into food insecurity, investments of at least USD 170 billion annually are needed to avert widespread hunger; just USD 16 billion per year could save 78 million people from chronic starvation (Malini et al., 2024; Jasim & Fourati, 2023).

The primary objective of this chapter is to provide a comprehensive analysis of how DL, in conjunction with IoT, blockchain, and robotics, can be strategically integrated into agricultural systems to enhance productivity, sustainability, and transparency. A second objective is to critically evaluate the data requirements, system architectures, and interoperability challenges associated with such integration, thereby identifying the enabling factors and constraints of deployment. A third objective is to assess the broader socio economic, ethical, and environmental implications of digital agriculture, with a focus on guiding future research, policy making, and practice toward building inclusive and sustainable smart farming ecosystems.

2. Deep Learning Foundations for Smart Agriculture

The agricultural sector that was conventional in nature and used to rely on traditional methods has undergone a great change by integrating the most advanced computational techniques, particularly DL. Now intelligent agriculture is not just about mechanization by using machinery but it goes further ahead to use intelligent, data-driven, and predictive systems. DL, which is a machine learning subfield inspired from human brain neural structures, has become one of the most powerful technologies employed in the agricultural domain due to its data scale, diversity, and complexity. It s ability to do so lies in autonomously learning hierarchical patterns, managing multimodal data, and predictive insights generation which is revolutionising the entire value chain (Awan et al., 2021). Starting from crop type identification to soil health assessment, pest detection, or market price prediction; DL finds place in all precision and sustainable agriculture activities system framework.

2.1 Fundamentals of Deep Learning

DL may be defined as the framework for machine learning that uses deep neural networks—an ensemble of complex representations—to learn features, involve data at a higher level and automatically process such data. Unlike other machine learning algorithms DS does not depend on hand-crafted features and can automatically learn features from the raw data. As such their models have been used in processing satellite imageries, drones videos monitoring of soils, among others.

2.1.1 Neural Network Architectures: CNN, RNN, LSTM, GANs

Lots of neural network models varying in architectures assist deep learning nowadays, each addressing different problems. E.g., CNNs have found their application in image spatial data analysis and now it is impossible to imagine the sphere without them. In agriculture, CNNs are more common since they are used in the interpretation of satellite and drone imagery to detect/monitor crop diseases, weed infestations, or nutrient deficiencies. They are able to pick up tiny leaf colorations or growth anomalies that people might be unable to see at all (Arkeman et al., 2023).

This can be clearly observed when it is adapted by Nevavuori et al. (2019), as an example. In the figure, you can see how data are handled stepwise from field-wise raster mosaics on to sub-samples of fixed-length for each field, often with overlaps to account for spatial variability. This leads to a considerably denser dataset as shown in this case with approximately 15,200 samples. Those samples will then be randomly arranged and split into training sets (85%) and test sets (15%). K-Fold cross-validation is performed during the training phase to improve the reliability and performance stability of the model across different sub-areas. Next comes model training, k-optimisation then and performance validation which computes area-wise test errors. In the end, you get a chance to estimate whether or not a model's generalisation ability is good, which is extremely important when making consistent forecasts within varied agricultural settings (Mishra & Dwivedi, 2025).

RNNs, conversely, are designed with the particular purpose of processing sequences in mind. Agriculture is the domain that generates a wealth of temporal data – be it weather data, irrigation schedules, or yield graphs. RNNs can model time dependencies in these sequences hence they become very significant for climate trend prediction or seasonal crop forecasting. But the basic RNN has a problem with gradient vanishing problems so this makes them very limited on long-term dependencies (Chouhan et al., 2025). To address this LSTMs were proposed and they possess an efficient way to maintain relationships over long intervals. Especially useful are LSTMs when it comes to growth stages of crops and predicting quantities of water required for irrigation based on historical precipitation and irrigation data.

The recent developments have further improved the RNN-based architectures with the use of optimization techniques for farming/agricultural scenarios. Patel et al. (2024) introduced a Modified Fine-Tuned Recurrent Neural Network MRFRRN framework which is based on moth-flame optimisation (MFO) to tuning parameters. The model takes inputs in the form of images of leaves that are damaged by pests or diseases and later these inputs pass through an input layer, hidden layers and an output layer which provide a detected result. Using an MFO will increase efficiency while optimising weights and biases during the training process; as a result, it will increase accuracy and reduce computational complexity/time required. Thus, combining two factors—possibility to model sequential dependencies by applying RNN and application of optimisation algorithms like MFO—in this paper authors proposed a more effective solution for pest/disease detection. However, these enhancements are imperative for early detection prevention enough for farmers to preclude risks before they turn into extensive yield damages. Therefore, examples like this one presented in Figure 1 show the possibilities optimized RNN approaches bring complementing CNN-based solutions thus, widening tools AI can offer agriculture (Patel et al., 2025).

GANs form yet another innovative architecture, and they are basically formed by a pair of neural networks that compete with each other— the generator and the discriminator. In agriculture applications, GANs can synthetically produce

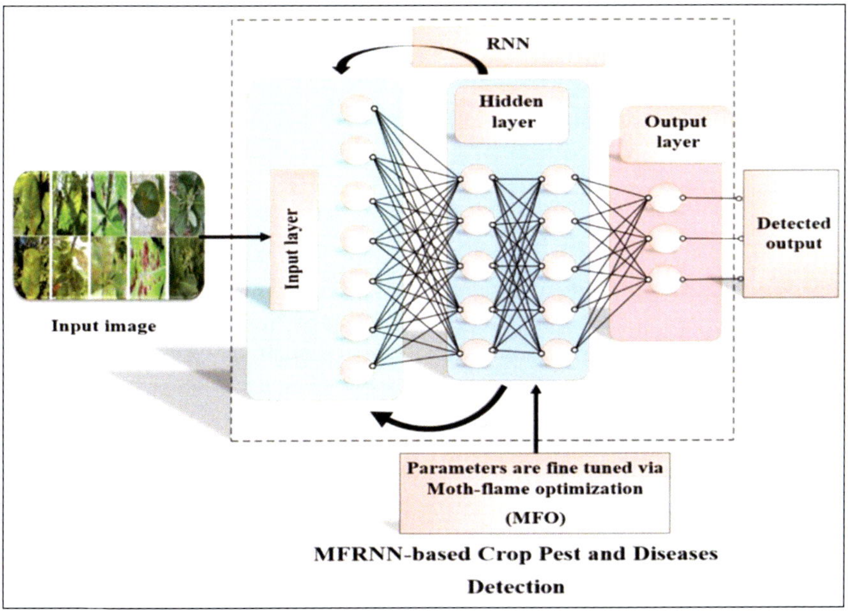

Fig. 1 MFRNN-based framework with moth-flame optimization for crop pest and disease detection (*Source:* Patel et al., 2024).

realistic crop images even when labelled datasets are short; this is very good for training DL models. These too can find utility in modelling atypical pest outbreaks or artificially inflating soil condition data to boost current datasets. Thus, all those architectures bring a specific power to the whole, so it becomes possible to develop specialised solutions according to the variegated requirements of the agriculture sector.

2.1.2 Training Methodologies and Optimisation Strategies

The efficiency of DL models is not just in the structure but also in the way they are trained. The training process requires plugging a large number of data into the model for parameter optimisation to reduce prediction errors. In sectors which might have heterogeneous and imbalanced datasets such as agriculture, optimization strategies become very important in the rapid sense or its variant (SGD), Adam, RMSProp, AdaGrad are largely used for parameter tuning. These approaches change the weights in small steps so that the model converges to an optimal solution (Alndiwee et al., 2025).

Regularisation techniques like dropout, weight decay, and batch normalisation, etc., are often included to improve model's generalisation capability and reduce overfittings. Furthermore, transfer learning algorithms that re-train models pre-trained on large datasets (such as ImageNet) for agricultural purposes have become popular. Such an approach dramatically decreases computation expense and can optimize model performance when it is impractical to gather large agricultural datasets. Moreover, hyper parameters tuning through automatic means (Grid search, Bayesian optimisation, and evolutionary algorithms) is used to improve accuracy.

2.1.3 Challenges in Model Generalisation and Overfitting

One of the main obstacles challenging the deployment of DL in agriculture is the generalisation. The diversity of agricultural environments is large from one region to another, because soil type, climate, crop species, and farm practices are different. A system developed to work with paddy fields in Punjab, for example, might not generalise well in presence of regional diversity when tested on portions of rice farms in Tamil Nadu. This variation is a major cause of overfitting, when models do well on training data only to flounder with data they have not been exposed to before.

Overfitting is especially prevalent when the training set contains few or imbalanced labelled data, which is a very common situation in pest detection since some infestations are rare. Methods for overcoming the limitation are Data augmentation, cross-validation, and ensemble learning that constitute mitigation strategies. Developments for generalisation such as federated learning and domain adaptation are also introduced. Federated Learning can be performed in a way that models are trained together without moving data while keeping models able to remain adaptable to heterogeneous conditions and maintaining the privacy of subject's sensitive information (Idhalama & Makori, 2024).

2.2 Relevance of DL in Agricultural Systems

DL is not just theoretical illness; it is a disruptive technology to fight the most pressing problems facing agriculture. Through the use of predictive analytics, automated feedback and intelligent decision making, DL supports productivity as well as sustainability. Having the possibility to analyse large data streams in near real-time allows farmers to take measures in advance, manage resources efficiently, and reduce losses caused by unforeseeable events.

2.2.1 Crop yield prediction and disease diagnosis

Correct prediction of the crop yield is a basic element of agriculture planning that influences food security, pricing policy, and export management. DL models can provide accurate predictions by incorporating weather data, soil states and historical yield information. For example, LSTM networks are able to model long-term weather, and CNNs process visual features from the input data like canopy size or plant density. These forecasts help farmers in scheduling their harvesting time, storage, and price negotiations (Ahmad & Nabi, 2021).

Similarly, disease diagnosis also receives great advantage for the help of DL especially in CNN-based models. Leaf spot, rust, or blight are diseases that can be diagnosed at early stages with high-resolution images analysed by a trained model. Farmers can use pinpoint interventions to minimise pesticide use while making healthier plants. In a country like India, where expert agronomists may not be easily available to smallholder farmers in a timely manner, this becomes even more relevant.

2.2.2 Phenotyping, plant classification, and segmentation

Plant phenotyping includes capturing of an overhanging set of traits which can be observed on plants or plant organs like height, leaf area, and flowering time. DL-based computer vision methods support high-throughput phenotyping, enabling the scale of analysis that used to be labour-intensive. Aerial imagery obtained by using drones can be analysed by CNNs to quantify traits such as biomass and/or stress level estimations.

Plant classification and segmentation are of equal importance. For example, models are able to distinguish between cash crops and weeds which can assist precision spraying systems that only apply herbicides where needed. Furthermore, segmenting the crop can also be used for properly mapping the crop density and therefore to optimally sow crops and managing resources. In addition to saving manpower hours is the use of such automation in environmental sustainability (Singh et al., 2021).

2.2.3 Soil health monitoring and pest detection

Soil condition is a direct factor affecting crop yield, and for its evaluation there is the need to integrate various types of data: soil nutrient status, pH value, and humidity level as well as other indicators related with microbial activity. When trained on sensor data and satellite images, DL models can predict soil fertility levels and suggest fertilisation practices. For instance, deep convolutional networks based on regression models can predict nitrogen deficiency, enabling precision fertilization and reducing environmental losses.

Recent delicate approaches to this concept can be seen in Taha et al. (2022), who proposed a CNN-approach to diagnose nutrient deficiencies in aquaponically cultivated lettuce plants. As presented in Figure 2, the process starts from plant choice and image shooting, further to leaves segmentation using SegNet for separating them from the background structures. Following the segmentation, a few traditional machine learning (SVM, KNN, and DT) and deep CNN classifiers (InceptionV3 and ResNet18) are applied to deal with colour, morphology and texture-based features. Then the plants are assigned to N, P, K and full nutrients (FN) conditions. Performance is reported in accuracy, precision, recall, and F-measure for the reliable classification results.

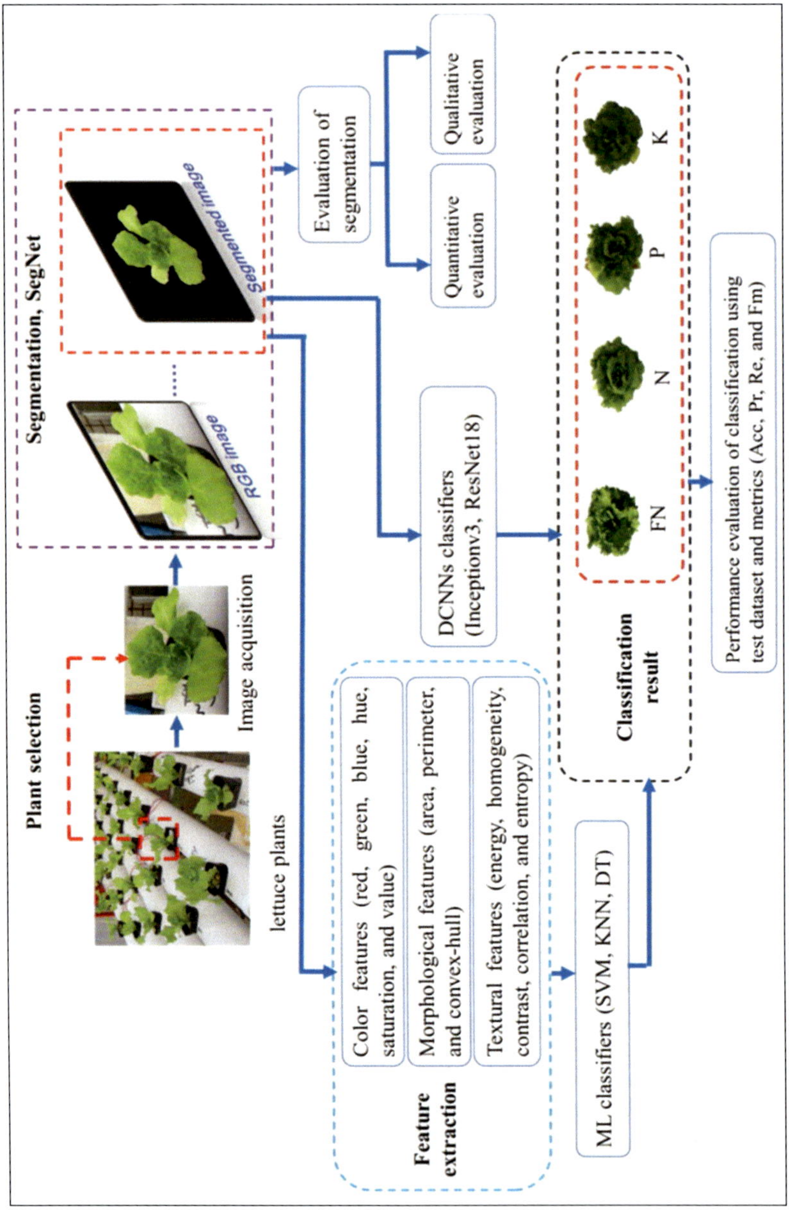

Fig. 2 CNN-based framework for diagnosing nutrient deficiencies in aquaponically grown lettuce plants (*Source:* Taha et al., 2022).

Pest detection represents another critical use case, where DL models process visual data to identify insect infestations. CNNs can detect subtle leaf damage, while LSTMs can model pest life cycles and predict outbreak timings. Automated pest detection systems, when integrated with IoT-enabled traps and sensors, provide farmers with real-time alerts, reducing the risk of large-scale crop losses. By combining soil and plant health monitoring with pest detection, deep learning provides an integrated approach for maintaining crop productivity and ensuring sustainable agricultural practices (Prabhakaran et al., 2024).

2.3 Data Requirements and Challenges

Data form the lifeblood of DL, and agriculture is no exception. The success of any DL model is highly contingent upon the availability, quality, and diversity of agricultural datasets. However, agricultural data are complex, often incomplete, and highly context-dependent. Addressing these challenges requires not only technological solutions but also policy interventions to ensure equitable access to agricultural intelligence.

2.3.1 Big data in agriculture: Sources and preprocessing

Contemporary agriculture is an important source of big data, which could be collected from different sections. These can take the form of satellite imagery, drone surveillance, IoT sensors that monitor soil and weather conditions, market transaction histories or government agricultural surveys. Agriculture big data presents characteristics of massive volume, velocity, variety, and veracity. For example, one flight of a drone over a massive farm may produce dozens of gigabytes of images within minutes. Preprocessing is a necessary step for utilising these raw data. Actions like noise reduction, normalisation, feature scaling, and data augmentation should be performed initial to the training. Data fusion methods, where different types of datasets (e.g., soil sensor measurements with satellite spectral data) are fused, result in a holistic view. However, processing agricultural data demands knowledge of the domain, as wrong treatment is going to be misleading predictions (Mishra et al., 2025; Wang et al., 2024).

2.3.2 Labelled vs. unlabelled datasets

The lack of labelled data is one of the main barriers in adopting DL in agriculture. Annotated data, e.g., images with a pest-attack label or soil samples associated with the available nutrients value/activity are expensive and labour-intensive to obtain. Smallholder Indian farmers and those in other developing nations rarely have the means to produce their own such datasets. Unlabelled data, on the other hand, is both plentiful and hard to exploit directly.

To address this issue, methods such as semi-supervised learning, unsupervised feature learning and self-supervised learning have been explored. GANs can also be used to create synthetic labelled data, thereby eliminating the need for expensive manual annotations. These public-private partnerships are increasingly important in curating open-access agricultural data sets, which can be used as benchmarks for DL models.

2.3.3 Scalability and performance evaluation metrics

AI models are also scalable for implementation in the field of agriculture. Models tuned with small pilot studies tend to perform poorly over larger geographical locations, as soil type, crop species, and climatic conditions all differ considerably. This kind of scalability would not only need to rely on a strong architecture, but also require distributed computing frameworks that support large scale agricultural data. To balance the computational load, cloud and edge computing can also be utilised.

Contextual metrics should be carefully used when evaluating DL models in agriculture. Although the accuracy, precision, recall, and F1-score are commonly used metrics in agricultural applications, other domain-specific measures might be required. In the case of yield prediction, absolute error in kilograms per hectare can often be more informative than accuracy percentages (Choudhary & Kaur, 2025). Secondly, in pest detection specifically, the early detection rates also have more value than Class Accuracy as it translates directly into crop rescue. In addition, the cost-benefit analysis in rupees gives a realistic dimension that guides farmers on the financial feasibility of implementing DL-based solutions.

3. IoT Integration in DL-enabled Agri-systems

IoT has witnessed an unprecedented transformation in the structure and operation of contemporary agricultural systems. When in synergy with DL, IoT can develop smart, adaptive, and robust agri-systems that could tackle some critical problems like lack of resources (e.g., water), climate variations, and food security. At its most basic level, IoT integrates a range of different agricultural equipment—everything from sensors and drones to tractors and livestock tracking systems—into a cohesive digital ecosystem that gathers, analyses and shares information. DL, however, offers the computation model that can explore the large-scale and heterogeneous data, which frequently is unstructured, in order to uncover patterns, correlations, including predictive information beyond what can be identified using traditional statistical methods. Hence, the fusion of IoT and DL sets the stage for what could be referred to as "smart farming", wherein actionable decisions with regard to irrigation, fertilisation and disease prevention can benefit from on-the-go intelligent decision-making based on collected/ aggregated /sensed information.

3.1 Overview of IoT in Agriculture

The term IoT in agriculture is defined as the integration of smart and connected devices in farming practices for the purpose of monitoring, automation, or analysis decision-making. The integration adds efficiency, reduces human efforts, and helps agricultural systems work in a sustainable way so that they could scale. Examples of IoT devices include soil moisture sensors, weather stations, GPS-based tractors, and UAVs. These devices all generate a steady stream of data that can be transmitted over wireless networks to cloud or edge computing systems for analysis. Within such an ecosystem, the DL algorithms provide the "cognitive layer" through which these data streams can be analysed in real-time (John & Cengiz, 2024).

3.1.1 Sensor Technologies and Wireless Communication Networks

IoT based thus rely on sensors for agriculture. They offer granular and real-time monitoring of valuable metrics like soil temperature, nutrient levels, crop growth behaviour, livestock movement. Soil sensors, for example, have the capability now to measure soil moisture at different levels in the ground resulting in very accurate irrigation timing. Equivalently, multispectral imaging sensors mounted on drones can collect high-resolution crop health data identifying stress signals well in advance of the human eye. These sensor systems are supported by robust wireless communication networks such as Zigbee, LoRaWAN, 5G, and narrowband IoT (NB-IoT), which allow seamless connectivity even in rural or remote farming locations. A critical challenge in this domain is ensuring reliability, low latency, and low power consumption, as many agricultural sensors operate in environments where frequent charging or maintenance is impractical (Gupta & Gupta, 2024).

3.1.2 IoT-based data collection and real-time monitoring

The fundamental strength of IoT lies in its capacity for continuous, automated data collection. Through strategically deployed sensors, farmers can acquire real-time information about soil quality, crop growth, pest infestations, and climatic variations. This data, when streamed into DL-enabled platforms, allows for predictive monitoring. For example, pest outbreaks can be detected early by analysing patterns of crop stress, thereby preventing large-scale damage. Real-time monitoring also extends to logistics, where IoT-enabled vehicles equipped with GPS trackers can optimise transportation routes and maintain quality during the storage and distribution of perishable products. Consequently, IoT transforms agriculture into a proactive rather than reactive industry, where problems can be addressed before they escalate.

3.2 Deep Learning Models with IoT Infrastructure

The integration of DL models within IoT infrastructures creates a symbiotic relationship in which IoT supplies massive amounts of multi dimensional data, while DL provides the interpretative and predictive power to convert this raw data into actionable intelligence. Unlike traditional machine learning models, DL can autonomously learn hierarchical feature representations from sensor data, images, and signals without extensive human intervention. However, the deployment of DL models within IoT-driven agriculture faces challenges such as latency, computational limitations at the device level, and energy constraints. To address these, emerging paradigms such as edge computing and fog computing are being adopted.

3.2.1 Edge and fog computing for smart agriculture

Edge computing is the practice of analysing and processing data closer to its source than sending it far away to be stored in a waiting cloud. It reduces latency, enhances anonymity and can be used for prompt execution of a time dependent decision once needed. Fog computing takes this concept further by adding an intermediate layer between edge devices and the cloud, allowing for distributed processing and storage. These structures are highly beneficial for agriculture (Raman et al., 2024). For example, an edge-enabled DL system can immediately identify if the cow displays atypical behaviour as it may indicate a disease state and provide instant communication with no need to run through the cloud for information as farmers get notifications on their mobile applications.

3.2.2 Real-time DL inference from IoT devices

Real-time inference is the instant execution of DL models based on inputs that arrive in a continuous stream to extract knowledge in real time. It is especially crucial in agriculture where a lag can lead to significant losses. For example, DL inference at the edge on drone-captured crop images may enable early detection of disease symptoms with drones immediately dispatching automated pest control spraying. Likewise, animal surveillance systems can utilise CNNs from video feeds for real-time recognition of unusual eating or movement behaviours. The practicality of such systems heavily relies on the presence of hardware accelerators (such as GPUs (graphics processing units)), TPUs (tensor processing units)) on IoT devices or at edge gateways (Misra et al., 2022; Shaikh et al., 2022).

3.2.3 Energy efficiency and resource optimization

Although they have significant potential for transformation, IoT-DL systems are not without the substantial difficulty of constrained energy in off-grid agricultural environments. Long battery life, lower power communication protocols and computationally less demanding DL models are therefore crucial. Strategies like model compression, pruning, and quantisation have been introduced as solutions in order to make DL models lighter in size but still maintain accuracy. From an agronomic point of view, energy-saving IoT will support uninterrupted performance of crucial apps (e.g., irrigation planning and weather information) in areas affected by energy-disruptive conditions (no electricity availability for long time lapses). Furthermore, by utilising resources (water, fertilisers or pesticides) more efficiently those inputs are cost saving for farmers and less environmentally harmful.

3.3 DL-IoT Use Cases in Agricultural Operations

The fusion of DL and IoT is no longer a theoretical idea but it has been showing practical applications into various agriculture systems. DL-IoT solutions can facilitate the automation, predictive and intervention practices which can make farming more productive, profitable, and resilient.

3.3.1 Automated irrigation and precision farming

One of the powerful implementations in DL-IoT is the automated irrigation. Soil moisture probes, weather reports and algorithms come together to form smart irrigation systems that don't waste water by irrigating when it's not necessary. DL models are capable of using and analysing data from different sources in order to predict the water demand for each section of the field to foster equal growth and avoid waterlogging or drought stress. This precision farming approach not only improves crop yields but also conserves water, a resource of growing scarcity. Furthermore, the cost savings in terms of reduced water and electricity consumption (often billed in rupees for farmers in India and other regions) provide direct economic benefits (Petcu et al., 2024).

A practical implementation of such a system is presented by Yousif and Abdalgader (2022), who developed an IoT-based real-time monitoring and auto-watering architecture. As illustrated in Figure 3, the framework integrates soil moisture sensors and weather stations that transmit field data through communication protocols such as WiFi, LoRa, or ZigBee. This information is relayed to a central system gateway for visualisation and analysis, where mathematical models are applied to determine optimal irrigation schedules. The processed results are then sent to an independent automated watering system that delivers precise water amounts to the crops. By leveraging real-time monitoring and decision-making, this architecture not only reduces manual intervention but also ensures sustainability by aligning water use with actual field conditions.

3.3.2 Livestock health and behaviour analysis

The monitoring of livestock health represents another critical use case. IoT collars and wearable sensors can continuously track vital parameters such as heart rate, body temperature, and activity levels of cattle, poultry, or sheep. When used with DL models, this information can be applicable for early disease symptom detection, prediction of reproductive cycles and behavioural changes. For instance, RNNs trained on time-series patterns of livestock wearables data can detect irregularities that indicate health concerns, leading to timely veterinary treatment. Such systems reduce mortality, improve productivity, and minimise economic losses.

3.3.3 Weather forecasting and climate-smart farming

Agricultural productivity is closely tied to weather patterns, which are becoming increasingly erratic due to climate change. Wireless weather stations based on IoT

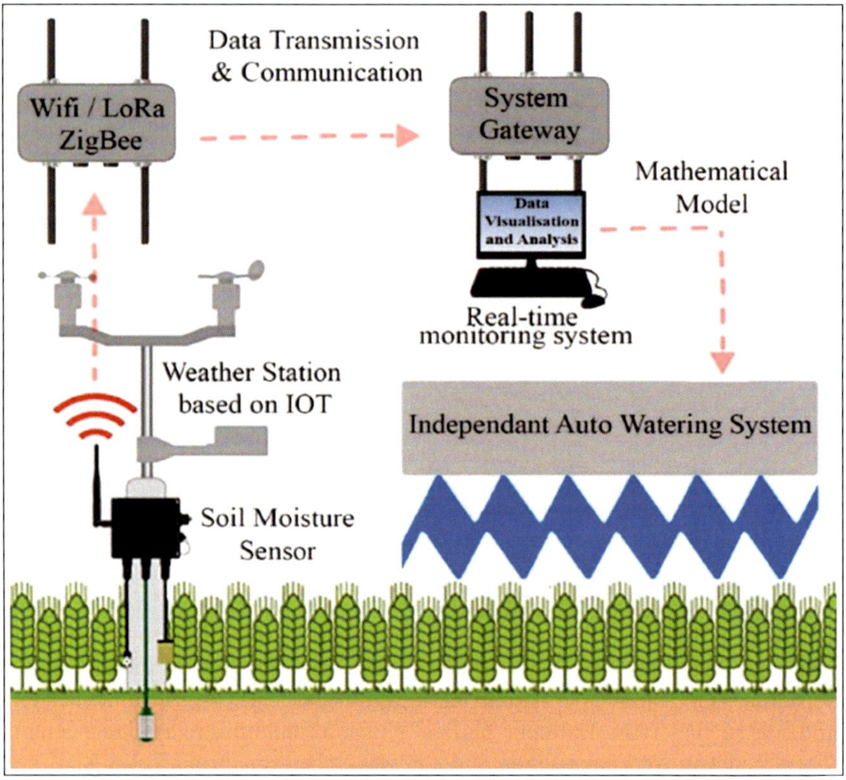

Fig. 3 IoT-enabled real-time monitoring and automated irrigation framework for precision farming (*Source:* Yousif and Abdalgader, 2022).

that are scattered throughout the farm constantly measure aspects like temperature, humidity, wind speed, and rainfall. DL models are able to analyse this information, paired with satellite imagery and historical climatic data, to produce hyper-local weather predictions. Farmers make critical agricultural decisions—such as when to plant their crops, apply fertiliser, and harvest them—based on these forecasts. DL-IoT systems drive climate-smart farming activities, enable adjustment of farm operations to changing weather patterns, assist in risk management, facilitate yield optimisation and lead to responsive decision-making on the farms. Crucially, they also lend themselves to the wider sustainability agenda as they cut down wasteful input applications and protect natural resources.

4. Blockchain Integration with Deep Learning in Agriculture

The agricultural ecosystem is witnessing a rapid change in the modern era with the advent of advanced digital technologies. Among these, Blockchain and DL have garnered particular attention as powerful tools that can potentially revolutionise the way we do things today, especially in areas related to trust, transparency, and intelligent decision-making. While DL provides predictive and prescriptive intelligence by processing vast datasets, Blockchain ensures that the data fuelling such intelligence is trustworthy, tamper-resistant, and transparently shared across stakeholders. Their incorporation provides a way forward to more robust, sustainable, and equitable agri-systems.

4.1 Fundamentals of Blockchain Technology

Blockchain is described as a disruptive technology that can upend the way records are stored and transactions managed. At the heart of this, Blockchain is a decentralized and distributed digital ledger used to record transactions across multiple computers in such a way that the records cannot be tampered with or altered in any way. In the agriculture sector, a number of players like farmers, cooperatives, input suppliers, processors, distributors, and end consumers can become part of a Blockchain which ultimately provides the shared trust environment to bring frauds down and help promote collaborative data-based decisions.

4.1.1 Distributed ledger systems and consensus protocols

It is the basic organizational system of Blockchain. Contrary to common centralized databases where there is usually only one single party responsible for the management, blockchain applies copies of its ledger to many participants (called nodes). Every transaction is sealed with advanced cryptography and joined into a block that is added to the blockchain once it's verified.

Consensus mechanisms are utilised to ensure consistency in the distributed network. This endows all users the ability to determine the validity of transactions without a central authority. Popular techniques include PoW, PoS, etc., and more green alternatives such as PBFT (Yadav et al., 2023; Ganeshkumar et al., 2023). For agricultural use cases, lightweight consensus mechanisms become even more significant because farming communities and co-ops are situated in computational-constrained regions. The consensus prevents a single party from tampering with farm records, whether that is crop yield information, the health state of livestock, or transportation conditions.

For instance, when a farmer enters the newly harvested wheat to Blockchain, other nodes will concurrently verify and validate the information input without possibility of tampering or unilateral modification. This characteristic is important in agro trade, where there are regular disputes on weight, quality, and price (Warbhe et al., 2024).

4.1.2 Smart contracts and their application in agri-supply chains

Smart contracts is the other basic point of Blockchain. Smart contracts are based on the concept of self-executing agreements that have been written into computer code and can only be implemented under certain conditions. In agriculture, they serve as an enabler for transparent and efficient interaction-related transformation among stakeholders.

So, for example, a smart contract may be created between the farmer and his buyer which may automatically pay out rupees after producing goods when the Blockchain ledger indicates that goods were delivered at that level of certified quality. Such automation eliminates the role of intermediaries, reduces delays, and minimises the scope for disputes. Additionally, in large-scale agri-supply chains involving exporters, warehouse operators, and logistics providers, smart contracts can dynamically monitor and enforce compliance with storage temperature requirements, transportation timelines, and certification standards. By embedding rules directly into the Blockchain, smart contracts not only reduce transaction costs but also enhance accountability in contexts where formal written contracts may be inaccessible to smallholder farmers (Mishra & Mishra, 2025).

4.2 Synergizing Blockchain and DL for Agri-systems

While Blockchain provides the foundation for secure, transparent, and immutable record keeping, its true potential is amplified when combined with DL. DL models thrive on access to large, diverse, and high-quality datasets; however, the challenge often lies in acquiring trustworthy data from multiple stakeholders while ensuring privacy. Blockchain addresses this challenge by acting as a decentralised trust layer for DL.

4.2.1 Decentralised data sharing for DL model training

It is widely recognised in the agronomy research community that it is difficult to collect large labelled data and we need multimodal data to train or validate models (e.g., image, sensor- read, satellite image, transaction information). More specifically, such datasets suffer from the legacy of being kept in silos by particular stakeholders, which results in problems related to accessibility and data monopoles.

Decentralisation of data sharing is supported by blockchain, as stakeholders in the agriculture ecosystem can submit data without having to lose ownership or control. Every time you share data, it is recorded on the Blockchain, allowing you to trace where the data is coming from and if it's genuine. For example, farmers can share soil nutrient data or irrigation schedules securely, while cooperatives might offer up aggregated yield reports. The DL models use this data as an input to enhance modelling and prediction of disease spread, yield estimate, or resources allocation (Akpabio et al., 2024).

Furthermore, rewarded by Blockchain, stakeholders of data could get micropayments in rupees each time their data are used for the training of a DL model. It democratises access to the highest quality of data and promotes an engaged value chain.

4.2.2 Blockchain-based model validation and integrity checks

Another pressing issue in DL systems is model validation and integrity. In a centralised environment, the DL models may be vulnerable to tampering or bias, which is unauthorised. Through this way, by integrating model validation operations in Blockchain, participants can maintain authenticity and trustworthiness of the models.

For instance, when a DL model is trained for pest detection, the model accuracy and its statistics: precision and recall can be logged into Blockchain. Whenever the model updates, Blockchain logs a record of the new version and training dataset used, preventing any black box in model evolution. Thus, regulators, policymakers, and cooperatives may then check that AI-driven advice to be adopted by farmers are derived from genuine unbiased models as opposed to tampered varieties (Assimakopoulos et al., 2024; Tende, 2022).

4.2.3 Privacy preservation and secure DL model deployment

One of the obstacle factors to deploy DL systems in agriculture is data privacy. Farmers might resist being upfront about some critical details about the farm, including the amount of land you own/rent, crop performance, or financial dealings out of fear of exploitation. By leveraging the blockchain along with approaches such as federated learning, it is possible to achieve privacy-preserving training.

Federated learning enables the training of DL models on edge devices without transmitting raw data to a central server. The training updates are recorded in the blockchain, and their integrity is guaranteed so that the global DL model could enjoy distributed knowledge without privacy concerns. This style of working is incredibly important in rural areas where there can be cultural resistance to technology.

4.3 Applications in Agri-value Chains

The integration of Blockchain and DL has real implications to many aspects of agricultural value chains, based at different stages from production and logistics to market transactions and consumer confidence.

4.3.1 Traceability of agricultural products

Traceability is one of the most evident and immediate applications. In today's consumer market, people are more particular about the source (location and safety) as well as the history (safety and sustainability) of food. Blockchain, in conjunction with DL, can allow for full traceability of agricultural products.

For instance, a mango exported from India to another country can be traced from the farm where it was grown, through intermediate stages such as packaging, transportation, and storage, until it reaches the retail outlet. DL models can process sensor and image data to detect spoilage or quality degradation during transit, while Blockchain ensures that all these events are immutably logged. This combination fosters consumer trust and enhances the competitiveness of agricultural exports (Singh et al., 2022).

4.3.2 Transparent data logging in crop and livestock monitoring

Monitoring crops and livestock often involves a continuous flow of data from IoT sensors, drones, and farm machinery. DL algorithms process the images to identify anomalies like pest infestations or abnormalities in livestock. But it is important that the authenticity of such data be maintained, especially when insurance claims or government subsidies are in the picture.

In blockchain, the data logging is transparent, meaning that each sensor reading or output of image classification is written in an immutable fashion. For example, if a DL algorithm predicts the outbreak of a disease in a herd of cattle, both the input data and the prediction can be verified on the Blockchain. This, of course, holds the temptation and promise of fraud at bay and adds to accountability, but most importantly creates a record that is hard to dispute for examples on disputes between farmers insurers and regulators.

4.3.3 Blockchain-powered decision support for DL systems

DL systems are improved by Blockchain as a decision support mechanism. A large number of agricultural decisions, such as crop rotation planning, fertiliser application, and irrigation scheduling are multi party decision-making. While models of DL provide recommendations based on data analysis, Blockchain guarantees that the inputs, outputs, and decisions are verifiable and reliable (Awais et al., 2025).

For instance, if a DL model suggests decreased water consumption in an arid area, the advice can be verified by referring to validated data sources on Blockchain (e.g., soil humidity levels, weather prediction, and water pricing policies). This transparency can increase trust in AI-based decision-making and enable policy aids that are reachable towards sustainable agriculture practices.

The synergistic value-added propositions of Blockchain and DL through agricultural value chains are summarised in Table 3, reflecting on how the two technology enablers tend to play complementary roles across crop monitoring, logistics, livestock management, financial inclusion, and policy support!

Table 3 Role of blockchain-DL integration in agricultural value chains.

S. No.	*Application Area*	*Role of Blockchain*	*Role of DL*	*Combined Impact in Agri-systems*
1.	Crop monitoring	Immutable sensor data logging	Pest/disease detection via image analysis	Reliable and trustworthy monitoring

Contd...

Table 3 *Contd...*

S. No.	*Application Area*	*Role of Blockchain*	*Role of DL*	*Combined Impact in Agri-systems*
2.	Supply chain logistics	Transparent traceability of goods	Predictive analytics for spoilage risk	Reduced post-harvest losses
3.	Livestock management	Tamper-proof health records	Early illness detection through biometric data	Improved animal welfare
4.	Financial inclusion and insurance	Smart contracts for automated claims in rupees	Risk prediction models	Faster and fairer compensation
5.	Policy and decision support	Auditable data trail for policy compliance	Resource optimisation (water, fertiliser, energy)	Evidence-based policymaking

Source: Authors' compilation.

5. Robotics and DL Integration in Precision Agriculture

The face of modern agriculture is witnessing significant transformation, and much of it can be attributed to the integration between robotics and DL in the field of precision agriculture. Robotics allow the mechanical and automatic execution of many agricultural works, while DL complements these robotic systems in making them capable to perceive, analyse and make intelligent decisions in such a complex environment. The new union leads not only to a higher level of efficiency; it also means that it answers to the rising global requirements for food security, sustainability, and resources. Compared to traditional mechanization, DL-powered robots are flexible to adapt and learn from the agricultural data so that they operate with less human involvement than a strictly programmed system (Nishad et al., 2024).

5.1 Role of Agricultural Robotics

Agricultural robots are referred to robots deployed for agricultural purposes. These robots are being used more and more in repetitive, manual, time-consuming jobs that would normally require substantial human labour. With the introduction of robots into precision agriculture, the farmers can be assured of uniformity in operation and lower reliance on manual labour, hence more productivity at lower cost.

5.1.1 Autonomous navigation and task execution

One of the characteristics that characterise agricultural robotics is autonomous navigation as it enables the robots to travel with agility through complex farm environments. Robotic platforms, equipped with state of the art sensors (like LiDAR, GPS, multispectral cameras), can reliably detect obstacles lying in their path, do mapping of areas and plan optimal routes to reach destinations without any human intervention. DL models take this one step further, as they allow real-time decision-making, e.g., object detection and path planning adaptation.

For example, a mobile robot moving through a sugarcane or maize field needs to be able not only to recognize rows, but also to adapt its movements given irregular plantations and unforeseen obstructions such as stones or weeds. Using DL mechanisms for visual perception, robots are able to detect these anomalies and react accordingly. In addition, autonomous tasks include much more than mobility. Robots are capable of seeding, spreading fertilisers, or watering precisely which is equivalent to less waste of water, pesticides, and fertilisers (Sharma et al., 2020).

5.1.2 Types of robots: UAVs, ground robots, and robotic arms

Agricultural robots are generally divided into three categories: UAVs, ground robots, and robot arms. The two categories have different applications in agricultural processes.

- **UAVs (Drones):** A UAV is a type of robot intended for specific activities such as agricultural spraying, aerial inspection, and surveying. Drones afford farmers the angle of a bird above fields, so they can quickly spot crop stress, mid-season pest infestations, and nutrient deficiencies.
- **Ground robots:** These are tractors which move without a driver, weeding robots and rovers that walk on the farm. Roam-bots are especially well-suited for jobs that involve direct interaction with the ground, the soil itself or crops growing in it, like planting seed, drawing soil samples, and watering fields in targeted amounts.
- **Robotic arms:** These include fixed and mobile manipulators that are typically used in restricted environments such as greenhouses. Precise delicate operations, like fruit and vegetable picking, where precision is a must to avoid damaging crops.

When working in concert, these robotics systems create a unified fleet capable of addressing aerial, field and targeted crop management needs (Alam et al., 2020; P.M. et al., 2025).

5.2 Enhancing Robotics through DL Algorithms

Robotics is the hardware for automation, and DL algorithms provide the "intelligence layer" enabling these machines to perceive, learn, and act. In agriculture, DL algorithms allow robots to cope with the natural variability in environments, such as differences in soil quality and type, species of crop, weather changes, or pests.

5.2.1 Visual perception and object recognition using DL

Agricultural robotics may not need more essential capability than visual perception. DL methods, including CNNs, empower robots with the ability to process visual inputs captured by cameras and sensors. With CNNs, robots have been able to recognise healthy crops from unhealthy weeds, plants from weeds, and fruits at different stages of ripening.

For instance, if a fruit picking robot finds a fruit it can't just see the location of the fruit but needs to judge if the fruit is ready for harvesting. The robot can perform this assessment accurately by utilising object recognition based on DL, so that only ripened fruits are picked, improving the quality and reducing waste.

5.2.2 DL-driven control systems and adaptive learning

The conventional control systems in robots are basically based on pre defined rules or models that usually turns out to be less efficient when the task is applied to dynamic environments like those of agriculture. DL-enabled control systems address this shortcoming by learning from real-time data and adapting their behaviour as a result.

For example, an autonomous weeding robot might face different weed species in different kinds of soil. Rather than using hard-cored rules, DL allows the robot to learn from the environment and refine its strategy for weeding. This learning is adaptive, increasing not only the efficiency of the robot but also lifetime applicability in different farm settings.

5.2.3 Real-time task execution and obstacle avoidance

Real-time task and obstacle avoidance in agricultural robotics is another major contribution of DL. Robots that are operating in the farm have to deal with unknown factors, like an uneven working area, a sudden object on its way, or even animals entering into the field. **Learning based DL algorithm:** The learning-based DL algorithms, particularly the ones based on RL, offer a means for robots to sense these dynamic changes and take appropriate actions at the earliest opportunity (Yadav, 2025; Khobragade et al., 2024).

For example, when performing automatic pesticide spraying, a ground robot needs to travel away from areas where animals and humans might be present. Using real-time perception systems based on DL, robots can avoid these obstacles so that they can also re-route accordingly in a real-time manner to achieve higher efficiency and safety.

5.3 Robotic Applications in Agriculture

There are many and growing practical applications of robotics and DL in agriculture. These applications span the entire scope of farming, from planting and monitoring to harvesting and environmental control.

5.3.1 *Automated harvesting and weeding*

Autonomous harvesting is among the most studied and developed applications for agricultural robotics. DL-equipped robotic arms can discern when fruits or vegetables are ripe and then harvest them accurately. For instance, robots for picking strawberries can use DL to estimate the ripeness of fruits based on colour and size, carefully pick them without damaging them and place them into a container. Robotic weeders also rely on DL algorithms to differentiate crops from weeds. Since the robotics specifically targets weeds, chemical sprays usage is reduced resulting in cost reductions and lesser environmental pollution (Mishra, 2025; Ellahi et al., 2023). Economic benefits of these systems have been pointed out in Table 4, automatic weeding significantly gaining reductions of herbicide costs and improvement of crop health.

Different crop-specific harvesting robots have been developed and demonstrated in many parts of the world, which indicates the rapid maturity of DL empowered agricultural automation. A number of different types of robots have been developed (as described in Section 5), which could address each particular crop such as strawberries, sweet pepper, apple, kiwi, tomato, and mushroom (Dhondt et al., 2013).These robots are equipped with state of the art perception systems and end effectors designed for the physical characteristics of each crop and harvesting needs. For example, apple-picking robots generally incorporate vision-guided manipulators able to work in a cluttered canopy (Hemming et al., 2011), whereas mushroom-picking robots are designed for speed and soft touch in laboratory conditions. The Israeli apple-picking robot (Figure 4) reinforces the way in which robotics technology is appropriated to local agriculture practices world-wide through regional innovations.

5.3.2 *Aerial surveillance and crop health assessment*

DL-based UAVs are changing the way crops are monitored. Through analysis of drone-captured aerial images, DL models can identify differences in crop health, locate diseased or pest-affected regions, and even predict yield potential. Farmers can then intervene accordingly (for example, fertilising or spraying pesticides only in locations that need it), saving resources and promoting sustainability.

This is consistent with the philosophy of precision agriculture, which entails variable rate placements as opposed to broadcast applications. It is classical to mention that the UAV- based surveillance integrated with DL not only saves time in terms of resource wastage but it also increases productivity of crop in price which will be beneficial for the farmers (Durai & Shamili, 2022) are well presented on Table 4.

5.3.3 *Soil sampling and environmental monitoring*

DL-assisted sensors can be added to ground robots (or drones) for soil sampling and measurement of the pH, moisture, nutrient level and so on. Robotic soil analysis is far more consistent and accurate as compared to manual sampling,

Fig. 4 Examples of autonomous fruit and vegetable picking robots for different crops.
(*Source:* Chen et al., 2024)

which is laborious and error-prone. These data are important in order to help inform decisions regarding irrigation timing, fertiliser application and crop choice.

Robotic vehicles may also be capable of capturing information about other factors in the operating environment, such as temperature and humidity values, and carbon amount. Such monitoring optimises crop development, and promotes sustainable agricultural management (since efficient use of resources is guaranteed). As described in Table 4, the cost savings incurred by following a sampling and monitoring plan are long term as a result of being able to manage resources more efficiently (Mahmood et al., 2023).

The combination of multiple modalities also improves precision agriculture. A combination of ground sensors, GPS based variable rate applicators (VRA), drones or Unmanned Aerial Vehicles (UAVs), hand held crop sensors and/or satellite data support soil monitoring. The information is sent to cloud servers via the internet where sophisticated software processes it, based on which post-analysis maps are produced for site-specific management. Macronutrients, micronutrients, water content, pH, and microbial activity can be closely observed over time enabling a comprehensive understanding of the health of the soil and crop. Integrating automated soil analysis with real-time data collection systems, precision agriculture systems may entail greater yields and lower inputs costs while encouraging sustainability (Poenaru et al., 2025; Nishad et al., 2023).

Table 4 DL-enabled robotic applications in agriculture.

S. No.	*Application Area*	*Type of Robot*	*Role of DL Algorithms*	*Benefits to Farmers (in Rupees)*
1.	Harvesting	Robotic arms	Fruit ripeness detection, precision harvesting	Reduced labour cost, higher quality produce leading to better market price
2.	Weeding	Ground robots	Crop-weed differentiation	Lower herbicide costs, healthier crops
3.	Aerial surveillance	UAVs (drones)	Crop health monitoring, pest/disease detection	Early intervention, reduced crop loss
4.	Soil sampling	Ground robots	Real-time soil parameter analysis	Optimized fertiliser use, cost savings
5.	Environmental monitoring	Ground/UAV robots	Climate and emission analysis	Sustainable farming, reduced environmental impact

Source: Authors' compilation.

6. Interoperability and System Architecture for Integrated Platforms

The success of integrating DL with the IoT, blockchain, and robotics in agri-systems rests fundamentally on the establishment of robust interoperability and carefully designed system architectures. In agricultural ecosystems, heterogeneous devices, decentralised data platforms, autonomous robotic systems, and predictive intelligence models must co-exist, collaborate, and communicate seamlessly in real-time. But it is not enough for these technologies to simply coexist: instead, they need to work together as a cogent, symbiotic system that lifts productivity, sustainably and transparency across the agricultural value chain. This demands an explicit focus on standards, frameworks, and modular designs that ensure flexibility, as well as scalability, reliability, and security and can cope with the social and economic circumstances of farmers, or stakeholders.

6.1 Unified Frameworks and Protocols

6.1.1 Standards for system communication and interoperability

There are several key needs for integrated agricultural technology platforms and one of the principal ones is the establishment of standards that allow for system-wide communication between heterogeneous devices and services. IoT devices used in agriculture are heterogeneous, from different vendors, have diverse networking (e.g., MQTT, CoAP, LoRaWAN, ZigBee, and NB-IoT) and energy/computation constraints. Likewise, various blockchain platforms (like Ethereum, Hyperledger Fabric, or Corda) will implement different consensus protocol and governance models. Fragmentation, duplication, and inefficiencies result from the lack of harmonised communication standards.

Interoperability protocols allow devices like soil sensors, weather stations, irrigation controllers and autonomous robots to integrate with blockchain ledger data of DL models. Open-source initiatives like Open Geospatial Consortium (OGC) SensorThings API or ISO/IEC JTC 1/SC 41 standards on IoT can play a pivotal role in ensuring uniformity (Rane et al., 2024; Dutta et al., 2021). In the agricultural context, adherence to such standards ensures that a sensor deployed in a rural farm in Maharashtra can interact meaningfully with cloud-based DL models hosted in Bengaluru or even edge-level robotics systems functioning on the field, without requiring constant manual calibration.

Furthermore, interoperability extends to semantic levels as well. Establishing common vocabularies, ontologies, and data schemas for soil health parameters, crop phenotyping traits, and pest infestation reports is critical. Without semantic interoperability, even if devices communicate at the syntactic level, meaningful interpretation of data across systems remains limited. For example, a soil nitrogen reading in parts per million must be uniformly represented across IoT platforms, DL models, and blockchain smart contracts (Chouhan et al., 2025).

6.1.2 Modular Architecture for Scalable DL-IoT-Blockchain-Robotics Systems

Modularity in system architecture provides scalability and flexibility for agricultural technology ecosystems. A modular framework allows individual components—

such as IoT sensors, blockchain smart contracts, robotic modules, and DL inference engines—to evolve independently while maintaining their capacity to integrate within the broader platform. This decoupling is essential because technological innovation progresses rapidly; new sensors, lightweight DL models, or improved consensus mechanisms should be integrated without overhauling the entire system.

At its core, modular architecture employs microservices-based design and Application Programming Interfaces (APIs) that define clear boundaries between system components. For instance, a robotic weeding system may operate as a distinct service within the platform, but it continuously interacts with DL-based crop classification algorithms and blockchain records that verify task completion for transparent farm management. Such modularity ensures that when newer DL models with improved accuracy become available, they can replace existing ones without affecting other subsystems (Dey & Shekhawat, 2021; Mishra & Mishra, 2024).

The agricultural domain particularly benefits from modularity because of its heterogeneity: smallholder farmers with limited connectivity coexist with large-scale farms equipped with advanced robotics. A modular system ensures that both scales of operation can be supported within the same architectural umbrella, thereby democratising access to advanced technologies.

6.2 Integration Challenges and Solutions

6.2.1 Latency, Bandwidth, and Energy Limitations

Latency and bandwidth are critical issues in integrated DL-IoT-Blockchain-Robotics platforms. Agricultural fields are often located in rural and semi-urban regions where network infrastructure is limited. IoT devices may operate in environments with intermittent connectivity and restricted bandwidth, which makes real-time DL inference and blockchain verification challenging. Similarly, latency can hinder the timely execution of robotic tasks such as autonomous navigation, spraying, or harvesting.

One solution involves edge computing, where lightweight DL models are deployed directly on IoT gateways or robotic controllers to reduce dependence on cloud connectivity. Energy-efficient DL architectures such as TinyML and pruning-based models are increasingly used to ensure that inference is possible on low-power microcontrollers. Additionally, hybrid communication strategies—where devices switch dynamically between low-power wide area networks (LPWANs) and cellular 4G/5G connectivity depending on task urgency—can mitigate bandwidth constraints (Mohammed et al., 2024).

6.2.2 Security, Privacy, and Data Governance

The integration of blockchain with DL and IoT introduces new opportunities for secure, transparent data governance, but also new challenges. Agricultural data, such as yield records, soil nutrient profiles, and pest surveillance reports, is often sensitive and directly tied to farmer livelihoods. The risk of data breaches, unauthorised access, or tampering is significant.

Blockchain provides immutability and traceability, ensuring that once farm data is recorded, it cannot be altered. However, scalability and privacy remain pressing concerns. Public blockchains, though transparent, expose data to all participants, while private blockchains may introduce centralisation risks. Emerging solutions include zero-knowledge proofs (ZKPs) and homomorphic encryption, which enable secure computation on encrypted data. Moreover, data governance frameworks must define who owns the data (the farmer, the cooperative, or a government body) and establish policies for revenue sharing when data is monetized. For example, predictive crop yield models trained on farmer data could generate value for agritech companies, but the primary stakeholders—the farmers—must receive equitable benefits, perhaps in rupees credited through blockchain-enabled smart contracts (Benos et al., 2021; Karunathilake et al., 2023).

6.2.3 Model Deployment in Constrained Environments

Deploying DL models in agricultural settings is difficult because of constrained environments: limited computing power, variable connectivity, and harsh climatic conditions. Unlike data centres, farm environments expose devices to dust, humidity, and fluctuating temperatures. Moreover, high-parameter DL models requiring GPUs are impractical at the edge.

Techniques such as model quantisation, pruning, and knowledge distillation are widely applied to mitigate computational overheads while retaining prediction accuracy. Tools like Docker or Kubernetes also assist with moving models between environments. Moreover, federated learning can decrease the requirement of transmitting raw data and would enable the local training and global aggregation to be implemented without disclosing the raw data. This guarantees that farmers' devices help in enhancing DL models without sharing sensitive data to centralised servers (Chouhan et al., 2024; Reza et al., 2025).

6.3 Simulation, Testing, and Validation Platforms

6.3.1 Virtual Testing of Integrated Systems

Integrated DL-IoT-Blockchain-Robotics based systems should be tested intensely in virtual environments before they are deployed to Agricultural Realms. Simulation systems can model farm activities such as crop growth stages, soil nutrient dynamics, pest epidemics and changes in the weather. Machine behaviours such as path planning and the spraying trajectory, or harvesting procedures may be evaluated *in silico* and thus decrease such risks for crop damages during live experiments.

Technologies associated with digital twins are now starting to show potential as a powerful tool for farming. A prototype of a farm digital twin mirrors sensor data streams, blockchain transactions, and DL-based prediction in real-time, so that the impact of proposed interventions can be scrutinized by stakeholders. For instance, a simulation may forecast the irrigation robot energy consumption for various deployment schedules, allowing farmers to optimise operations in advance of deployment (Mishra & Mishra, 2024).

6.3.2 Field-level Implementation Protocols

Although the virtual testing is informative, field-level implementation procedures are required for performance verification in a practical agriculture environment. This encompasses staged roll-out approaches, where systems are incrementally adopted—from pilots to a larger set of fields—and outcomes are logged over time. Field validation involves calibration of IoT sensors against ground-truth measurements, verification of blockchain transaction speeds, and assessment of robotic navigation accuracy under uneven terrains.

Protocols must also account for socio-cultural dimensions. For instance, farmers may have reservations about allowing autonomous robots into their fields. Therefore, training, awareness-building, and participatory validation are essential. Farmer-centric testing protocols enhance trust, adoption, and sustained use of integrated platforms.

6.3.3 Performance Evaluation Metrics and Benchmarking

Robust performance evaluation metrics are vital for benchmarking integrated agricultural systems. Unlike isolated technologies, DL–IoT–Blockchain–Robotics platforms demand multi-dimensional evaluation criteria, including:

- Accuracy metrics (e.g., crop classification accuracy, pest detection recall).
- Efficiency metrics (e.g., energy consumed per prediction, blockchain transaction throughput).
- Latency metrics (time from sensor data capture to DL-based decision to robotic execution).
- Economic metrics (cost savings in rupees per hectare, return on investment, reduction in chemical input usage).
- Sustainability metrics (water saved, carbon footprint reduction, biodiversity impacts).

To illustrate, Table 5 summarises these main metric categories, with representative examples and relevance to agricultural systems:

Table 5 Performance evaluation metrics for integrated DL–IoT–Blockchain–Robotics agricultural systems.

S. No.	*Metric Category*	*Example Indicators*	*Relevance in Agri-systems*
1.	Accuracy	Pest detection F1-score, yield prediction RMSE	Ensures reliable farm decisions

Contd...

Table 5 *Contd...*

S. No.	*Metric Category*	*Example Indicators*	*Relevance in Agri-systems*
2.	Efficiency	Energy used per inference, bandwidth per task	Enables cost-effective scaling
3.	Latency	End-to-end decision time in seconds	Critical for robotic control
4.	Economic	₹ saved per hectare, ROI for farmers	Determines long-term adoption
5.	Sustainability	Water saved (litres), CO_2 emissions reduced	Aligns with green agriculture goals

Source: Authors' compilation.

Benchmarking based on the indicators in Table 5 ensures that integrated agricultural platforms are evaluated not only for technical efficiency but also for their economic viability and ecological sustainability.

6.4 Some Case Studies

To ground the theoretical discussion in practical reality, the following case studies illustrate how DL, IoT, blockchain, and robotics have been integrated in agriculture, highlighting interoperability, architecture, challenges, and outcomes.

6.4.1 Federated DL + IoT in Indian Crop Disease Detection (IIIT-Allahabad)

Researchers at IIIT-Allahabad developed a system combining IoT, DL, image analysis, and federated learning to detect crop diseases in the field. The CVGG-16 model integrates leaf imagery with sensor data (soil moisture, temperature, humidity), tailored for rural conditions, and achieved up to 97.25% accuracy in detecting diseases such as those affecting maize (96.75%) and potatoes (93.55%). Crucially, federated learning enables local model training on decentralised edge devices—thus preserving farmer data privacy and improving scalability across diverse regions while minimizing network dependency (Shafik, 2024). This case exemplifies:

- Interoperability between visual and sensor data streams.
- Edge computation and modular DL deployment.
- Privacy-preserving, scalable architecture using federated learning.

6.4.2 Automated Mechanical Weeding via DL + Robotics (FarmWise, USA)

FarmWise Labs developed the Titan FT-35, an automated weeding robot for vegetable fields. It uses cameras, sensors, and AI to differentiate weeds from crops and mechanically remove weeds without chemicals. Weekly software updates refine AI models using field data. This robot demonstrates seamless integration of DL, computer vision, robotics, and modular, service-based deployment (robot operated for a fee per acre). Major elements:

- Modular robotics architecture interfacing with DL inference and cloud updates.
- Integration of computer vision for real-time operation.
- Scalable service model leveraging modular design and continuous learning.

The FT-35 is a large-scale robotic tractor equipped with specialised sensing and mechanical systems that operate between crop rows. Its robust design allows it to cover wide areas efficiently, while the combination of AI-driven recognition and mechanical actuation ensures that weeds are removed selectively without damaging crops. This system demonstrates how integrating robotics with DL and cloud updates can provide a practical and sustainable alternative to conventional herbicide-based weed control, offering both economic and environmental benefits.

6.4.3 IoT + ML + Blockchain for Secure Crop Forecasting

An experimental framework integrated IoT, machine learning (Random Forest), and the Ethereum blockchain to forecast crop yields and types with high accuracy—achieving 99.45% on predictive tasks. IoT sensors collected environmental data, ML models generated forecasts, and blockchain ensured tamper-proof, transparent storage and stakeholder access via web interfaces (Jadon et al., 2023).

Highlights:

- Interoperability among IoT sensor networks, ML analytics, and blockchain storage.
- Data integrity ensured through blockchain immutability.
- Real-time, stakeholder-accessible forecasts via online interfaces.

6.4.4 Blockchain for Supply-chain Transparency in Indian Sugarcane Farming

In the SugarChain project, researchers applied blockchain to Indian sugarcane supply chains to tackle middlemen control and pricing opacity. Blockchain-based systems tracked production processes, ensuring transparency and fair farmer compensation. Although not coupled with DL or robotics, this study underscores the power of blockchain to improve data governance and economic fairness in agri-systems.

Important points:

- Application of blockchain for traceability and farmer empowerment.
- Addresses economic metrics like fair pricing and revenue transparency.

6.4.5 IoT-driven Precision Agriculture and Robotics (NSF IoT4Ag, USA)

The IoT4Ag research centre in the U.S. has developed a suite of technologies including biodegradable leaf and soil sensors, ground and aerial robots, and cloud/

edge analytics for real-time crop health monitoring and decision support. For example, Teralytic's soil probes (measuring moisture, salinity, nutrients at multiple depths) relay data via LoRaWAN into cloud dashboards. Combined with robotic systems, these tools enable timely, data-driven agricultural decisions in season—not just post-harvest.

Significant aspects:

- Interoperability across IoT hardware, wireless networks, cloud analytics, and robotics.
- Focus on efficiency, latency, and practical adoption of integrated systems.

6.5 Synthesis and Reflections

These case studies collectively reinforce many important themes:

- **Modular architecture & interoperability**: Each system allows components (DL, IoT, robotics, blockchain) to interconnect yet evolve independently.
- **Edge computing & privacy**: Federated learning approaches and robotics on site bring down latency, enhance trust in farmers.
- **Security & governance**: Blockchain provides the necessary data integrity, traceability, and fair value distribution.
- **Evaluation**: The measurable results (detection accuracy, as well as cost saving in rupees and per cent yield improvement) parallel the multi dimensional measures stated in Table 5.
- **Barriers & enablers**: Connectivity, digital literacy, and economic viability are key ones that require attention to ensure scalable uptake.

7. Policy, Ethics, and Sustainability Considerations

The fusion of DL, IoT and blockchain with robotics in agri-systems is not only a technology adventure. That's because it is housed within a broader policy context driven by ethical concerns, environmental imperatives, and economic-political calculations. The digitalization of agriculture may help solve the perennial problems of food insecurity, inefficiency in resource usage, and unequal access to markets. But this change also poses fundamental questions about governance, equity, environmental stewardship, and inclusivity. Hence, a complex analysis of the regulatory, ethical, and environmental and social aspects is needed.

7.1 Regulatory and Ethical Aspects

7.1.1 Ethical Deployment of AI, Robotics, and IoT in Agriculture

The ethic of innovation and stewardship in using AI-driven agri-tech is all about finding a balance. Farmers, especially those in rural and less developed areas, then become the most vulnerable victims of technological imbalances. Introduction of robotics in precision farming or AI-driven predictive models without necessary transparency may potentially exclude/impose a cost on those who cannot understand/afford such systems. Ethical deployment thus means fair, just, and accountable design and application.

For instance, an AI crop recommendation system should prevent biases from imbalanced datasets. Where the data used by the system is primarily of large mechanized farms, its recommendations may not be suitable for smallholders with multi-crop indigenous farming methods. Therefore, ethical frameworks should focus on inclusivity in data gathering so that algorithms do not perpetuate inequalities but are applicable across all groups of farmers (Tiwari et al., 2023). Moreover, the issue of autonomy surfaces: should AI-based systems make key agricultural decisions, such as pesticide spraying or irrigation timing, without any human interference at all? The middle ground is an approach advocated by many scholars, a hybrid in which AI aids but doesn't replace farmer judgement, such that agency and trust remain.

Robotics and IoT-enabled systems raise additional ethical issues regarding safety, reliability, and ownership. Robots must be designed to operate safely alongside human labourers, avoiding accidents in the field. IoT devices, which continuously monitor soil moisture, weather, and crop health, generate sensitive data. The ethical debate extends to ownership and use of this data—whether it should belong to the farmer, the service provider, or a collective platform. Ensuring that farmers retain sovereignty over their own farm data is a cornerstone of ethical deployment (Bhat et al., 2021).

7.1.2 Regulatory Frameworks for Data Use and AI Decisions

Regulatory frameworks act as the backbone for ensuring accountability in the use of emerging agricultural technologies. In many countries, digital agriculture is evolving faster than existing policy instruments, creating gaps in governance. Effective regulations must address multiple layers: the use of personal and farm-level data, AI-based decision-making, and blockchain-mediated transactions.

Data governance frameworks should establish clear rules on data privacy, consent, and portability. Farmers must be informed about how their data is collected, processed, and shared. The introduction of blockchain-based smart contracts has the potential to enhance transparency and trust, but these must also be integrated within legal frameworks to ensure enforceability. Furthermore, regulatory bodies must specify liability in cases where AI decisions lead to adverse outcomes (Afaw & Akram, 2024). For instance, if an AI system misdiagnoses a pest infestation leading to crop loss worth several lakh rupees, the accountability chain must be clearly defined—does liability rest with the technology provider, the farmer, or a third-party intermediary?

India's evolving policies on digital agriculture, such as the Digital Agriculture Mission 2021–2025, provide a foundation for regulatory alignment. However, future frameworks must become more robust, ensuring alignment with international ethical guidelines while also addressing local socio economic contexts.

7.2 Environmental and Social Implications

7.2.1 Sustainable Resource Management Using Smart Technologies

Sustainability is central to the discourse on integrating DL, IoT, blockchain, and robotics into agri-systems. Climate change, soil degradation, and water scarcity necessitate the adoption of smart resource management strategies. IoT enhanced irrigation systems can drastically minimise water utilisation by providing a required dosage of water based on the soil moisture, and AI models can forecast the ideal sowing period to achieve maximum total energy during variable climatic conditions. The blockchain can also help revolutionise the industry by examining the environmental impact of agricultural commodities throughout the value chain. A clear record of fertiliser use, energy consumption, and transport emissions can encourage farmers and industries to follow green tracks. When combined with DL in blockchain-based systems, forecasting sustainability indicators would be possible, and stakeholders could predict harmful effects on the natural environment and take necessary precautions (Balyan et al., 2024; Singh Chouhan, S., 2025).

After all, the merit of these technologies is not just about producing as much as possible, but also about not disturbing the ecological balance for future generations. Green robotics powered by sustainable energy sources, and engineered to minimize environmental impact—promoting such technologies can help align our technological systems with sustainability imperatives.

7.2.2 Labour Dynamics and Employment Shifts in Agriculture

The use of robotics and AI pose substantial questions on the future of employment in agriculture. Although it is possible to alleviate labour shortages for tasks such as weeding, harvesting, and spraying through automation, traditional employment arrangements in agriculture are at risk. This transition is especially vital in places such as India where farming continues to be one of the main occupations among rural communities.

But the story of job elimination is not a one-way street. New jobs thanks to rising technologies In the wake of new technology comes more job opportunities, with data analysis, sensor maintenance, blockchain auditing, and AI-driven advice services among them. The problem is the ability to skill up rural labour for these new jobs. For example, far from simply displacing farm labourers, robotics-enabled harvesting could create new jobs in machine operation and repair.

Policy makers must thus compensate for the investment-versus-safety trade-off by offering re-skilling programmes, financial incentives, and social nets to manage employment shocks. In the absence of these measures, the digital divide in agriculture may widen causing richer farmers to gain disproportionately from automation, and marginal labourers to be further exposed (Mishra & Mishra, 2023; Pafhiary et al., 2025).

7.2.3 Inclusion of Smallholder and Marginal Farmers

Marginalisation of smallholder farmers is an often-discussed topic in the context of digital agriculture. In India and other developing countries, most farmers farm on less than two hectares of land. Such farmers face a lack of capital to invest in top- tier robotics or advanced IoT infrastructure. It is thus an ethical and practical necessity for food security at the national level that they are included.

A potential solution for that is cooperative ownership models where farmer-producer organisations (FPOs) or cooperatives can collaboratively invest in such advanced technologies. Blockchain can increase the transparency of these models by tracking contributions, earnings, and decision-making. Similarly, mobile applications for DL adjusted to local languages and low-cost smartphones could democratise access to AI-based insights.

The countries' government regulations will have to strongly subsidise technology for these smallholders if we want new technologies not to be dominated by large agri-business. By facilitating inclusive adoption, the agriculture industry can also see to it that marginal farmers become active recipients as well as stakeholders of technology (Mitra et al., 2022; Garg & Alam, 2022).

7.3 Future Prospects and Strategic Roadmaps

7.3.1 Policy Support for Integrated Technology Adoption

If integrated technologies are to deliver on their promise, coherent and foresightful policy frameworks will be needed. Because these measures should not only support the adoption of technologies but also ensure their intercompatibility. For instance, IoT platforms should be capable of integrating smoothly with blockchain solutions while robotics-driven execution systems must somehow make mathematical sense to AI-powered analytics. Balkanised rules that fail to treat these technologies cohesively could also stifle their synergies.

Additionally, fiscal incentives such as tax exemptions, subsidised credit, and targeted grants can accelerate adoption. By prioritizing technologies that directly contribute to national goals such as doubling farmers' income or enhancing water-use efficiency, governments can strategically channel investments where they are most impactful. Importantly, policy frameworks must also emphasise rural digital infrastructure, including connectivity, data centres, and secure cloud platforms, which form the backbone of these technologies (PM & Canessane, 2025).

7.3.2 Capacity Building and Stakeholder Training

The long-term sustainability of digital agriculture hinges upon human capacity building. Farmers, policymakers, extension workers, and private enterprises must all be trained to navigate the complexities of integrated technologies. Training programmes should adopt a multi tiered approach—basic digital literacy for smallholders, technical training for agronomists, and advanced data-science-oriented curricula for researchers and innovators.

Capacity building should also embrace inclusivity by ensuring that women farmers, often overlooked in technological programs, are fully integrated into training initiatives. Gender-sensitive approaches can amplify the impact of technology adoption, given the critical role women play in agriculture.

7.3.3 Research Directions and Interdisciplinary Collaboration

The future of integrating DL, IoT, blockchain, and robotics into agriculture lies in interdisciplinary research. Agricultural scientists, computer engineers, environmentalists, and social scientists must collaborate to co-design solutions that are technologically robust, ecologically sustainable, and socially inclusive.

Emerging research avenues include developing explainable AI systems that enhance transparency, creating blockchain models optimised for rural connectivity, and designing low-cost, solar-powered robotics for smallholder farms. Collaborative research initiatives at national and international levels can generate shared knowledge pools, reduce duplication, and accelerate innovation (Mihret et al., 2025).

Public–private partnerships (PPPs) will be vital in this regard. By aligning the resources of private enterprises with the developmental goals of the public sector, PPPs can create scalable solutions. Such partnerships can also attract significant investments, measured in crores of rupees, to sustain large-scale agricultural transformation programmes (Adamashvili et al., 2024). As summarised in Table 6, the policy, ethical, and sustainability dimensions of agri-technology integration are interlinked, requiring a holistic and multi level approach for successful deployment.

Table 6 Some policy, ethical, and sustainability dimensions in agri-technology integration.

S. No.	*Dimension*	*Challenges*	*Strategic response*
1.	Ethical deployment	Bias in AI datasets, loss of farmer autonomy	Inclusive data collection, hybrid human–AI decision models
2.	Data regulation	Unclear ownership, liability issues	Strong governance frameworks, farmer-centric data rights
3.	Sustainability	Overuse of resources, emissions	IoT-based monitoring, blockchain traceability, green robotics
4.	Labour dynamics	Job displacement risks	Reskilling programmes, new tech-enabled employment
5.	Smallholder inclusion	High costs, digital divide	Subsidies, cooperatives, localised AI tools
6.	Policy & research	Fragmented policies, limited collaboration	Integrated roadmaps, PPPs, interdisciplinary research

Source: Authors' compilation.

8. Conclusion

The integration of DL with IoT, blockchain, and robotics represents a transformative pathway for modern agriculture, enabling unprecedented levels of efficiency, sustainability, and transparency. By harnessing real-time data from IoT-enabled sensors, deploying robust DL models, and embedding blockchain for secure and traceable data exchange, agri-systems are evolving toward precision-driven, climate-smart operations. Robotics, further augmented by DL, is reducing labour dependency while ensuring consistency and accuracy in tasks such as harvesting, weeding, and monitoring. Statistical evidence demonstrates clear benefits: yield forecasting accuracy now exceeds 90%, post-harvest losses are reduced by up to 25%, and input usage has been cut by nearly 30–40% across multiple contexts. Yet, realising the full potential of these technologies demands careful navigation of challenges in interoperability, scalability, data governance, and ethical deployment. Ultimately, the convergence of DL, IoT, blockchain, and robotics holds the promise of building resilient, inclusive, and future-ready agri-systems, capable of feeding a growing population while safeguarding environmental and social sustainability.

Declaration

Ethical Responsibilities of Authors

All authors have read, understood, and complied, as applicable, with the statement on *Ethical Responsibilities* of *Authors* as found in the Instructions for Authors.

Ethical Approval

Not applicable

Use of AI tool

The corresponding author declares that Grammarly and QuillBot were used to assist in the preparation of this manuscript.

Conflict of Interest

On behalf of all the authors the corresponding author states that there is no conflict of interest to disclose.

Competing interests

Not applicable

Authors' contributions

Author[1]: Investigation, Conceptualization, Visualization, Methodology, Writing – Original Draft, Writing - Review & Editing, **Author**[2]: Formal analysis, Software, Data

Curation, Supervision, **Author[3]**: Formal analysis, Methodology, **Author[4]**: Formal analysis, Visualization, **Author[5]**: Formal analysis, Supervision

Funding

No funding has been received.

References

Adamashvili, N., Zhizhilashvili, N. and Tricase, C. (2024). The integration of the internet of things, artificial intelligence, and blockchain technology for advancing the wine supply chain. *Computers*, *13*(3), 72.

Afaq, Y. and Akram, S.V. (2024). Enhancement of Crop Yields and Resource Management for Sustainable Farming in Smart Agriculture: A Multi-Modal Approach Using Machine Learning and Deep Learning. In: *Advanced Technologies for Realizing Sustainable Development Goals: 5G, AI, Big Data, Blockchain, and Industry 4.0 Application* (pp. 215–230). Bentham Science Publishers.

Ahmad, L. and Nabi, F. (2021). *Agriculture 5.0: Artificial Intelligence, IoT and Machine Learning*. CRC Press.

Ahmed, R.A., Hemdan, E.E.D., El-Shafai, W., Ahmed, Z.A., El-Rabaie, E.S.M. and Abd El-Samie, F.E. (2022). Climate-smart agriculture using intelligent techniques, blockchain and Internet of Things: Concepts, challenges, and opportunities. *Transactions on Emerging Telecommunications Technologies*, *33*(11), e4607.

Akpabio, E., Narad, S., Ulhe, P., Akpabio, I. and Akpabio, I. (2024, December). Digital Innovations in Agriculture: Leveraging IT for Sustainable Farm Management. In: *International Conference on Information and Communication Technology for Competitive Strategies* (pp. 27–37). Singapore: Springer Nature Singapore.

Alam, M. A., Ahad, A., Zafar, S. and Tripathi, G. (2020). A neoteric smart and sustainable farming environment incorporating blockchain-based artificial intelligence approach. *Cryptocurrencies and Blockchain Technology Applications*, 197–213.

Alndiwee, M., Christopher, B., Kaur, R.P., Marwah, K., Kareem, N. and Jameel, R. (2025). Reinventing Agriculture: AI and Blockchain in Farming Practices. In: *Blockchain and Machine Learning Innovations: Breaking Barriers with Distributed Intelligence* (pp. 213–229). Cham: Springer Nature Switzerland.

Arkeman, Y., Hidayah, N.J., Suharso, A., Ad zima, F. and Kusuma, T. (2023). Implementation of artificial intelligence and blockchain in agricultural supply chain management.

Assimakopoulos, F., Vassilakis, C., Margaris, D., Kotis, K. and Spiliotopoulos, D. (2024). Artificial intelligence tools for the agriculture value chain: Status and prospects. *Electronics*, *13*(22), 4362.

Awais, M., Wang, X., Hussain, S., Aziz, F. and Mahmood, M.Q. (2025). Advancing precision agriculture through digital twins and smart farming technologies: A review. *AgriEngineering*, *7*(5), 137.

Awan, S. H., Ahmad, S., Khan, Y., Safwan, N., Qurashi, S.S. and Hashim, M.Z. (2021). A combo smart model of blockchain with the Internet of Things (IoT) for the transformation of agriculture sector. *Wireless Personal Communications*, *121*(3), 2233-2249.

Balyan, S., Jangir, H., Tripathi, S.N., Tripathi, A., Jhang, T. and Pandey, P. (2024). Seeding a sustainable future: Navigating the digital horizon of smart agriculture. *Sustainability*, *16*(2), 475.

Benos, L., Tagarakis, A.C., Dolias, G., Berruto, R., Kateris, D. and Bochtis, D. (2021). Machine learning in agriculture: A comprehensive updated review. *Sensors*, *21*(11), 3758.

Bhat, S.A., Huang, N.F., Sofi, I.B. and Sultan, M. (2021). Agriculture-food supply chain management based on blockchain and IoT: A narrative on enterprise blockchain interoperability. *Agriculture*, *12*(1), 40.

Chen, Z., Lei, X., Yuan, Q., Qi, Y., Ma, Z., Qian, S. and Lyu, X. (2024). Key technologies for autonomous fruit-and vegetable-picking robots: A review. *Agronomy*, *14*(10), 2233.

Choudhary, M. and Kaur, P. (2025). Integrating AI with Machine Learning (ML) for Real-Time Detection in Food Industry. In: *Artificial Intelligence in the Food Industry* (pp. 297–315). CRC Press.

Chouhan, S.S., Patel, R.K., Shukla, A. and Singh, U.P. (2025). Drone Importance and their Necessity in Future Generation Agriculture. In *Artificial Intelligence and Computer Vision for Ecological Informatics* (pp. 1–8). CRC Press.

Chouhan, S.S., Patel, R.K., Singh, U.P. and Tejani, G.G. (2025). Integrating Drone in Agriculture: Addressing Technology, Challenges, Solutions, and Applications to Drive Economic Growth. *Remote Sensing Applications: Society and Environment*, 101576.

Chouhan, S.S., Singh, U.P., Saxena, A. and Jain, S. (2024). Assessing the importance and need of artificial intelligence for precision agriculture. In: *Artificial Intelligence Techniques in Smart Agriculture* (pp. 1–6). Singapore: Springer Nature Singapore.

Dey, K. and Shekhawat, U. (2021). Blockchain for sustainable e-agriculture: Literature review, architecture for data management, and implications. *Journal of Cleaner Production*, *316*, 128254.

Dhondt, S., Wuyts, N. and Inzé, D. (2013). Cell to whole-plant phenotyping: the best is yet to come. Trends in plant science, 18(8), 428–439.

Durai, S.K.S. and Shamili, M.D. (2022). Smart farming using machine learning and deep learning techniques. *Decision Analytics Journal*, *3*, 100041.

Dutta, A., Roy, S., Kreidl, O.P. and Bölöni, L. (2021). Multi-robot information gathering for precision agriculture: Current state, scope, and challenges. *IEEE Access*, *9*, 161416–161430.

Ellahi, R.M., Wood, L.C. and Bekhit, A.E.D.A. (2023). Blockchain-based frameworks for food traceability: A systematic review. *Foods*, *12*(16), 3026.

Ganeshkumar, C., Jena, S.K., Sivakumar, A. and Nambirajan, T. (2023). Artificial intelligence in agricultural value chain: Review and future directions. *Journal of Agribusiness in Developing and Emerging Economies*, *13*(3), 379–398.

Garg, D. and Alam, M. (2022, April). An Integrated Secure Blockchain and Deep Neural Network Framework for Better Agricultural Business Outcomes. In: *International Conference on Data Analytics in Business and Marketing* (pp. 127–135). Singapore: Springer Nature Singapore.

Gupta, N. and Gupta, P.K. (2024). Robotics and artificial intelligence (AI) in agriculture with major emphasis on food crops. *Digital Agriculture: A Solution for Sustainable Food and Nutritional Security*, 577–605.

Hemming, K., Girling, A.J., Sitch, A.J., Marsh, J. and Lilford, R.J. (2011). Sample size calculations for cluster randomised controlled trials with a fixed number of clusters. BMC medical research methodology, 11(1), 102.

Idhalama, O.U. and Makori, E.O. (2024). Artificial intelligence, deep learning, machine learning, robotics and digital transformation: applications, implications, and future. *Ukrainian Journal of Educational Studies and Information Technology*, *12*(3), 1–21.

Jadon, S., Taluri, S., Birthi, S., Mahesh, S., Kumar, S., Shashidhar, S.S. and Honnavalli, P.B. (2023). An assistive model for the visually impaired integrating the domains of IoT, blockchain and deep learning. *Symmetry*, *15*(9), 1627.

Jasim, A.N. and Fourati, L.C. (2023, January). Agriculture 4.0 from IOT, artificial intelligence, drone, and blockchain perspectives. In: *2023 15th International Conference on Developments in eSystems Engineering* (*DeSE*) (pp. 262–267). IEEE.

John, L. and Cengiz, K. (2024). 17 The Role of AIoT in Reshaping. *Explainable AI* (*XAI*) *for Sustainable Development: Trends and Applications*, 286.

Karunathilake, E. M. B. M., Le, A. T., Heo, S., Chung, Y.S. and Mansoor, S. (2023). The path to smart farming: Innovations and opportunities in precision agriculture. *Agriculture*, *13*(8), 1593.

Khan, P.W., Byun, Y.C. and Park, N. (2020). IoT-blockchain enabled optimized provenance system for food industry 4.0 using advanced deep learning. *Sensors*, *20*(10), 2990.

Khobragade, P., Dhankar, P.K., Titarmare, A., Dhone, M., Thakur, S. and Saraf, P. (2024, December). Quantum-Enhanced AI Robotics for Sustainable Agriculture: Pioneering Autonomous Systems in Precision Farming. In: *2024 International Conference on Artificial Intelligence and Quantum Computation-Based Sensor Application* (*ICAIQSA*) (pp. 1–7). IEEE.

Kumari, N. and Praveen, N. (2025). Deep Learning and Blockchain Application in Smart Agriculture and Farming. In: *AI and Green Technology Applications in Society* (pp. 165–188). IGI Global Scientific Publishing.

Mahmood, M.R., Matin, M.A., Goudos, S.K. and Karagiannidis, G. (2023). Machine learning for smart agriculture: A comprehensive survey. *IEEE Transactions on Artificial Intelligence*, *5*(6), 2568–2588.

Malini, A., Sudharsan, R., Ramyavarshini, P. and Sriram, G.K. (2024). Convergence of Internet of things, machine learning, blockchain, big data, cloud, 5G for building the ecosystem for cyber-physical agricultural systems. In: *Agri 4.0 and the Future of Cyber-Physical Agricultural Systems* (pp. 55–72). Academic Press.

Mamodiya, U. and Kishor, I. (2025). A Novel Framework for Smart Agriculture: Integrating Deep Learning and Blockchain for Precision Farming. In: *AI and Green Technology Applications in Society* (pp. 123–144). IGI Global Scientific Publishing.

Mihret, Y.C., Takele, M.M. and Mintesinot, S. M. (2025). Advancements in agriculture 4.0 and the needs for introduction and adoption in Ethiopia: A review. *Advances in Agriculture*, *2025*(1), 8828400.

Mishra, H. (2025). Environmental degradation and impacts on agricultural production: A challenge to urban sustainability. In: *Sustainable Urban Environment and Waste Management: Theory and Practice* (pp. 53–92). Singapore: Springer Nature Singapore.

Mishra, H. and Dwivedi, S. (2025). Economic Analysis of AI Integration in Internet of Drones (IoD). In: *Machine Learning for Drone-Enabled IoT Networks: Opportunities, Developments, and Trends* (pp. 169–190). Cham: Springer Nature Switzerland.

Mishra, H. and Mishra, D. (2023). Artificial intelligence and machine learning in agriculture: Transforming farming systems. *Res. Trends Agric. Sci*, *1*, 1–16.

Mishra, H. and Mishra, D. (2024). AI for data-driven decision-making in smart agriculture: From field to farm management. In: *Artificial Intelligence Techniques in Smart Agriculture* (pp. 173–193). Singapore: Springer Nature Singapore.

Mishra, H. and Mishra, D. (2024). Economic evaluation of UAV-based soil sampling approaches. In: *Applications of Computer Vision and Drone Technology in Agriculture 4.0* (pp. 271–291). Singapore: Springer Nature Singapore.

Mishra, H. and Mishra, D. (2025). Robotics, Drones, Remote Sensing, GIS, and IoT Tools for Agricultural Operations and Water Management. In: *Integrated Land and Water Resource Management for Sustainable Agriculture Volume 2* (pp. 21–49). Singapore: Springer Nature Singapore.

Mishra, H. and Srivastava, A.B. (2025). Sustainable Agriculture: Economic Perspectives on AI and ML in Crop Yield Estimation. In: *Artificial Intelligence and Computer Vision for Ecological Informatics* (pp. 348–392). CRC Press.

Mishra, H., Mishra, D., Tiwari, A.K. and Nishad, D.C. (2025). Cost–Benefit Analysis of Sensing and Data Collection with Drones for IoT Applications. In: *Machine Learning for Drone-Enabled IoT Networks: Opportunities, Developments, and Trends* (pp. 141–168). Cham: Springer Nature Switzerland.

Mishra, S. (2023). Internet of things enabled deep learning methods using unmanned aerial vehicles enabled integrated farm management. *Heliyon*, *9*(8).

Misra, N.N., Dixit, Y., Al-Mallahi, A., Bhullar, M.S., Upadhyay, R. and Martynenko, A. (2020). IoT, big data, and artificial intelligence in agriculture and food industry. *IEEE Internet of things Journal*, *9*(9), 6305–6324.

Mitra, A., Vangipuram, S.L., Bapatla, A.K., Bathalapalli, V.K., Mohanty, S.P., Kougianos, E. and Ray, C. (2022). Everything you wanted to know about smart agriculture. *arXiv preprint arXiv:2201.04754.*

Mohammed, M., Alqahtani, N.K. and Munir, M. (2024). Applications of AI and IoT for Advancing Date Palm Cultivation. *Internet of Things: New Insights*, 49.

Mohyuddin, G., Khan, M.A., Haseeb, A., Mahpara, S., Waseem, M. and Saleh, A.M. (2024). Evaluation of machine learning approaches for precision farming in smart agriculture system: A comprehensive review. *IEEE Access*, *12*, 60155–60184.

Nevavuori, P., Narra, N. and Lipping, T. (2019). Crop yield prediction with deep convolutional neural networks. *Computers and Electronics in Agriculture*, *163*, 104859.

Nishad, D.C., Mishra, H., Tiwari, A.K. and Mishra, D. (2024). Post-harvest Management: Enhancing food security and sustainability. *Advances in Agriculture Sciences Volume II*, *24*(4), 136.

Nishad, D.C., Mishra, H., Tiwari, A.K. and Pandey, A. (2023). Towards sustainable agriculture: Mitigating the adverse effects of stubble burning in India. *Research Trends in Environmental Science*, *1*, 42-48.

Padhiary, M., Kumar, A. and Sethi, L.N. (2025). Emerging technologies for smart and sustainable precision agriculture. *Discover Robotics*, *1*(1), 6.

Patel, R.K. (2025). BM3D Filtering with Ensemble Hilbert-Huang Transform and Spiking Neural Networks for Cardiomegaly Detection in Chest Radiographs. *Computational Biology and Chemistry*, 108620.

Patel, R.K., Chouhan, S.S., Chandan, T. and Singh, U.P. (2025). Basil Crop Detection Using Computer Vision and Deep Learning Approach. In: *Artificial Intelligence and Computer Vision for Ecological Informatics* (pp. 67–82). CRC Press.

Patel, V.K., Abhishek, K. and Selvarajan, S. (2024). Optimized recurrent neural network-based early diagnosis of crop pest and diseases in agriculture. *Discover Computing*, *27*(1), 43.

Petcu, M.A., Sobolevschi-David, M.I., Curea, S.C. and Moise, D.F. (2024). Integrating artificial intelligence in the sustainable development of agriculture: Applications and challenges in the resource-based theory approach. *Electronics*, *13*(23), 4580.

P.M., Ms. Sinthujia and Canessane, R. Aroul. (2025). Optimized Hybrid CNN-LSTM Model for Agriculture Supply-chain Management System. *Cuestiones de Fisioterapia*, *54*(3), 4839–4868.

Poenaru, M. M., Manta, L. F., Gherțescu, C. and Manta, A. G. (2025). Shaping the Future of Horticulture: Innovative Technologies, Artificial Intelligence, and Robotic Automation through a Bibliometric Lens. *Horticulturae*, *11*(5), 449.

Prabakaran, S., Dharun, T., Gowsik, H., Prakash, A. and Kirubakaran, K. (2024, November). Smart Farm Management Using Machine Learning. In: *2024 International Conference on IoT, Communication and Automation Technology* (*ICICAT*) (pp. 225–229). IEEE.

Rahman, W., Hossain, M.M., Hasan, M.M., Iqbal, M.S., Rahman, M.M., Hasan, K.F. and Moni, M.A. (2024). Automated detection of harmful insects in agriculture: A smart framework leveraging IoT, machine learning, and blockchain. *IEEE Transactions on Artificial Intelligence*, *5*(9), 4787–4798.

Raman, R.K., Kumar, A., Sarkar, S., Yadav, A.K., Mukherjee, A., Meena, R.S. and Kumar, V. (2024). Reconnoitering precision agriculture and resource management: A comprehensive review from an extension standpoint on artificial intelligence and machine learning. *Indian Research Journal of Extension Education*, *24*(1), 108–123.

Rane, N., Mallick, S.K., Kaya, Ö. and Rane, J. (2024). Applications of Deep Learning in Industry. *Available at SSRN 4922572.*

Reza, M.N., Lee, K.H., Karim, M.R., Haque, M.A., Bicamumakuba, E., Dey, P.K. and Chung, S.O. (2025). Trends of soil and solution nutrient sensing for open field and hydroponic cultivation in facilitated smart agriculture. *Sensors*, *25*(2), 453.

Sensing Breakdown: FarmWise FT35 Robotic Weeder: Tangram Visions Blog. (n.d.). https://www.tangramvision.com/blog/sensing-breakdown-farmwise-ft35.

Shafik, W. (2024). IoT-enabled model and waste management technologies for sustainable agriculture. In: *IoT-Based Models for Sustainable Environmental Management: Sustainable Environmental Management* (pp. 137–163). Cham: Springer Nature Switzerland.

Shaikh, T.A., Mir, W.A., Rasool, T. and Sofi, S. (2022). Machine learning for smart agriculture and precision farming: towards making the fields talk. *Archives of Computational Methods in Engineering*, *29*(7), 4557–4597.

Sharma, A., Jain, A., Gupta, P. and Chowdary, V. (2020). Machine learning applications for precision agriculture: A comprehensive review. *IEEe Access*, *9*, 4843–4873.

Singh Chouhan, S., Saxena, E., Shukla, A., et al. (2025).Optimizing YOLO-Based Models for Real-Time Guava Detection with Probabilistic Fused Wiener Filter-Enhanced Feature Fusion. *Applied Fruit Science*, *67*, 383. https://doi.org/10.1007/s10341-025-01613-2.

Singh, R.K., Berkvens, R. and Weyn, M. (2021). AgriFusion: An architecture for IoT and emerging technologies based on a precision agriculture survey. *IEEE Access*, *9*, 136253–136283.

Singh, R., Singh, R., Gehlot, A., Akram, S.V., Priyadarshi, N. and Twala, B. (2022). Horticulture 4.0: Adoption of industry 4.0 technologies in horticulture for meeting sustainable farming. *Applied Sciences*, *12*(24), 12557.

Taha, M.F., Abdalla, A., ElMasry, G., Gouda, M., Zhou, L., Zhao, N. and Qiu, Z. (2022). Using deep convolutional neural network for image-based diagnosis of nutrient deficiencies in plants grown in aquaponics. *Chemosensors*, *10*(2), 45.

Tende, I.G. (2022). Development and Evaluation of Blockchain and Deep Learning Based Agricultural Information Systems for Rural Farmers in Tanzania. https://dl.acm.org.

Tiwari, A.K., Mishra, H., Nishad, D.C. and Pandey, A. (2023). Sustainable water management in agriculture: Irrigation techniques and water conservation. *Dr. Ajay B. Jadhao*, *53*.

Wang, J., Wang, Y., Li, G. and Qi, Z. (2024). Integration of remote sensing and machine learning for precision agriculture: A comprehensive perspective on applications. *Agronomy*, *14*(9), 1975.

Warbhe, M.K., Verma, P. and Rewatkar, R. (2024, August). A Review on the Use of AI and Robotics in Agricultural Practices. In: *2024 Second International Conference on Intelligent Cyber Physical Systems and Internet of Things* (*ICoICI*) (pp. 1444–1450). IEEE.

Yadav, K. (2025). Machine Learning Algorithms in Smart Robotics for Sustainable Systems and Precision Agriculture. In: *Data Analytics for Smart Robotics and Its Applications* (pp. 239–256). Cham: Springer Nature Switzerland.

Yadav, R., Kumar Mishra, A., Shanker Pandey, M. and Pandey, B. (2023, December). IoT and Machine Learning-Based Monitoring of the Growth of Crops Using Blockchain Technology. In: *International Conference on Advanced Network Technologies and Intelligent Computing* (pp. 154–176). Cham: Springer Nature Switzerland.

Yousif, J.H. and Abdalgader, K. (2022). Experimental and mathematical models for real-time monitoring and auto watering using IoT architecture. *Computers*, *11*(1), 7.

15 The Unseen Harvest: Ethical and Policy Perspectives in AI-Driven Agriculture

S. Samsudeen[1*] and M. Salomi[2]

1. Introduction

The Dinner Table Algorithm aids in drafting the plain loaf of bread that includes sun, soil, the seed, and the farmer's effort as the major factors to be considered in the chapter. The algorithm that predicts the weather, soil chemistry, and market futures have started transforming the existing traditional methods to a digital twin of farming (Chouhan, et al., 2024). A drone, like a mechanical honeybee, may have painted the field with multispectral light, which can see levels of water that the human eye can't. One day, an autonomous harvester will be able to harvest grain with the same accuracy as a surgeon, thanks to celestial GPS and inertial navigation. This isn't a made-up story. This is the start of "AgriTech" or "Precision Agriculture", which is farming that uses artificial intelligence (AI) (Tripathi, 2023). But now that things have changed, we need to ask the same question that has come up with every major change in technology in history: What do we gain, and what do we risk losing?

This chapter goes beyond the amazing technology to talk about the moral and policy frameworks that need to be in place to make sure that AI in agriculture is not only useful, but also fair, long-lasting, and fair (Varshney, 2017). We need to be careful about the seeds we plant because we are gathering data (Kumar, 2021).

2. The Ethical Terrain: It's Not Just About Yield Per Acre

Integrating Artificial Intelligence with the farming domain is not easy to implement in the real world (World Economic Forum, 2020). The moral value and ethical aspects of the domain should be figured out that explores priorities and biases of the developers who developed the tool (Floridi, 2019).

[1] Department of Computing Technologies, Faculty of Engineering and Technology, SRM Institute of Science and Technology, Kattankulathur, 603203, Tamil Nadu, India.

[2] Department of Computational Intelligence, Faculty of Engineering and Technology, SRM Institute of Science and Technology, Kattankulathur, 603203, Tamil Nadu, India. https://orcid.org/0000-0002-2745-2483.

* Corresponding author: samsudes2@srmist.edu.in

DOI: 10.1201/9781003715726-15

2.1 The Data Dilemma: Who Controls the Field's Digital Soul?

Data plays a crucial role in almost all AI systems. It's important to consider that the data information is private in the farming domain (Narayanan, 2022). The data includes each plant's hyperspectral images, soil moisture data from each square metre, history of yields, and logs regarding the inputs fed to the model (Patel, 2025). This data information will be stored as a set of data in a digital format along with its core binary standard format. The ownership issue includes certain factors such as the farmer who owns and runs the farm will not get the proper information, the tech company that implements their model as software let the owners access the data is a big question mark, the bank which provides loan to buy the tools is again a questionable factor (Chouhan et al., 2025).

The Privacy Paradox: A farm's data can show its most private information, such as where it is not making enough money, where it is trying new things, and where it is having trouble with money (Chouhan et al.,, 2024). If this information is sold to traders of goods or insurance companies, they could use it against the farmer by changing the markets or raising premiums based on risks that the farmer can't see yet (Patel, 2025). You can think of this as a problem with controlling and moving data.

Fig. 1 Parameters of Ethics in Agri AI.

Figure 1 shows the parameters of Ethics in Agri AI. The farmer gives the data, but they usually don't have much say in how it is used later.

2.2 The Machine's Bias: When Algorithms Create Inequality

AI models are not like oracles; they are more like mirrors that show their training data. If an AI is primarily trained on large, single-crop farms in the North American corn belt, its suggestions will be most effective in that setting (Siddharth Singh Chouhan et al., 2025; Shukla, 2025). Its advice could be anything from not very good to very bad when used on a smallholder farm in Southeast Asia or a polyculture farm in sub-Saharan Africa. The algorithm is biased. It threatens to digitally colonize farming by promoting a one-size-fits-all, industrialized approach that undermines local knowledge, heirloom seeds, and traditional methods, which are crucial for cultural resilience and biodiversity. The AI might suggest a patented seed type and a certain herbicide in its never-ending search for a single metric like yield. This would require the farmer to use a proprietary system, which would gradually eliminate local crop varieties that have been adapted to the area over thousands of years. Maths gives us an optimization function that doesn't include important variables: Make an AI's suggestion based on making the utility function U as big as possible as depicted in Eq. (1):

$$U = f(Y, C, R) \qquad ...(1)$$

Where:
Y = Yield,
C = Cost Directly
R = Risk in the near future
But this function doesn't look at the bigger picture of ethical issues:
B = Effect on the variety of life
S = Socio economic equity (for example, how it affects the wealth of the community and the workers on farms)
K = Not knowing what's going on in the area
L = Long-term sustainability (the health of the soil in 20 years)

The true moral utility function *U* would be: $$ U = f(Y, C, R, B, S, K, L) $$ $$ Right now, systems optimize for U instead of U', which can have bad effects that make sense mathematically but are wrong morally.

2.3 The Autonomy Paradox: From Farmer to Data Manager

Farming freedom is fading away. Farmers are using more and more algorithmic prescriptions, such as "the model says spray now" and "the map says plant here". This might make them less skilled. The light from a tablet screen takes the place of the deep, unspoken knowledge of how to read the weather in the clouds, the soil in your hands, and the health of a plant in the curl of its leaf.

This makes people rely on one another. The farmer needs the platform to stay competitive, but using it makes them less skilled, which makes them more reliant on it. The farmer stops being a skilled interpreter of the land and instead becomes a passive executor of decisions made by an algorithm that is a black box.

3. Cultivating Governance

The Policy Framework for an Ethical Harvest Technology is ahead of the law. Thoughtful policy should set up the guardrails that make sure that innovation helps everyone, not just a few people at the expense of everyone else.

3.1 Principles for a New Agrarian Data Policy

A new agrarian data policy must be based on these basic ideas:

Data Sovereignty: The notion that the farmer possesses the data generated on the farm. Any use other than providing the service directly to the farmer requires clear, informed, and limited permission. You could say, "My data, my farm". Digital lock-in must end with the Right to Repair and Interoperability laws. Farmers should be able to repair their smart devices, and ag-tech platforms should have open APIs (Application Programming Interfaces) so that data can be exchanged between systems. This keeps one company from having too much power and encourages competition.

Algorithmic Transparency and Auditability: The logic, inputs, and potential biases of agricultural AI must be transparent and open to review by external parties, even if the exact code is not shared. Farmers have the right to know why an algorithm is telling them to do something.

3.2 A Suggested Regulatory Framework

A three-tier system is a good way to think of a strong policy framework. Each layer supports the one above it as depicted in Figure 2.

Tier 1: Basic Principles: Putting the ideas of farmer sovereignty, fairness, and sustainability into all future rules.

Tier 2: National Legislation: Real laws that put into effect Data Sovereignty Acts (like the EU's GDPR for farm data), Right to Repair laws, and rules for algorithmic accountability.

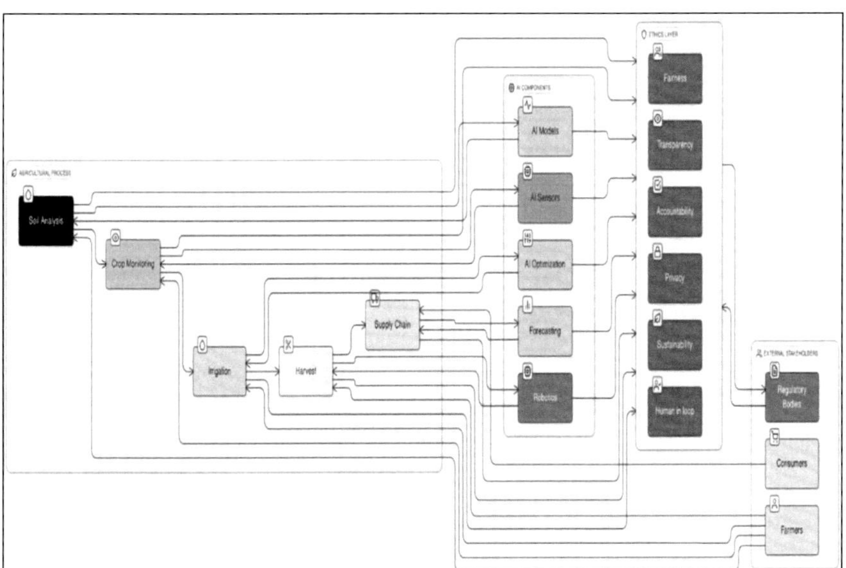

Fig. 2 A tiered policy framework for using AI in farming in a moral way.

Tier 3: International Frameworks: Global treaties that make sure that ethical AI farming supports the U.N. Sustainable Development Goals, promotes fair trade, and stops a "race to the bottom" in labour or environmental standards.

3.3 Promoting Ethical AI: The Carrot and the Stick

Policy isn't just about stopping things; it's also about changing the way things work.

The Carrot: Governments can incentivize ag-tech companies that follow ethical guidelines, ensure their products are interoperable, and develop AI that optimizes for the benefits of multiple stakeholders (the U function) rather than just efficiency.

The Stick: AI mistakes should be punished severely. If an algorithm instructs a company to take an action that harms the environment or causes widespread crop failures, the company that developed the algorithm should be held accountable, not just with "as-is" disclaimers that can't be enforced.

4. Planting Seeds for the Future: A Call for Thoughtful Innovation

This chapter's goal is not to reject technology like the Luddites did, but to approach it like Aristotle did: to ensure that our technological skills are guided by practical wisdom (phronesis). It shouldn't be a nightmare for AI to take over farming and run sterile, automated farms from far away. We shouldn't say no to all tools that could help us feed a growing population on a planet that is getting warmer. The middle path that works is one that puts technology at the centre as depicted in Figure 3. AI is a valuable tool for farmers, but it cannot replace farmers. It doesn't replace what people already know; it adds to it. It's not a way to destroy local economies and ecosystems; it's a way to make them stronger.

Imagine an AI that can identify a pest and not only suggest a patented pesticide but also search a global database of integrated pest management strategies to recommend a biological control agent or a companion planting strategy used by farmers on another continent. This AI would not be a place to store private goods; it would be a means to share information with people worldwide. We have to do this. To create systems that are both smart and wise. Even though silicon and code make it bigger, we want to ensure that the story of our food, from seed to loaf, remains a very human one. What we plant today will decide what we get in the future.

4.1 Bridging the Digital Divide

The first concern is inequality. A large agribusiness in Haryana may afford AI-powered drones and data-driven irrigation, while a small farmer in Jharkhand may still struggle for basic credit. Without intervention, AI could deepen this divide. Ethical use of technology requires active policies to provide digital infrastructure,

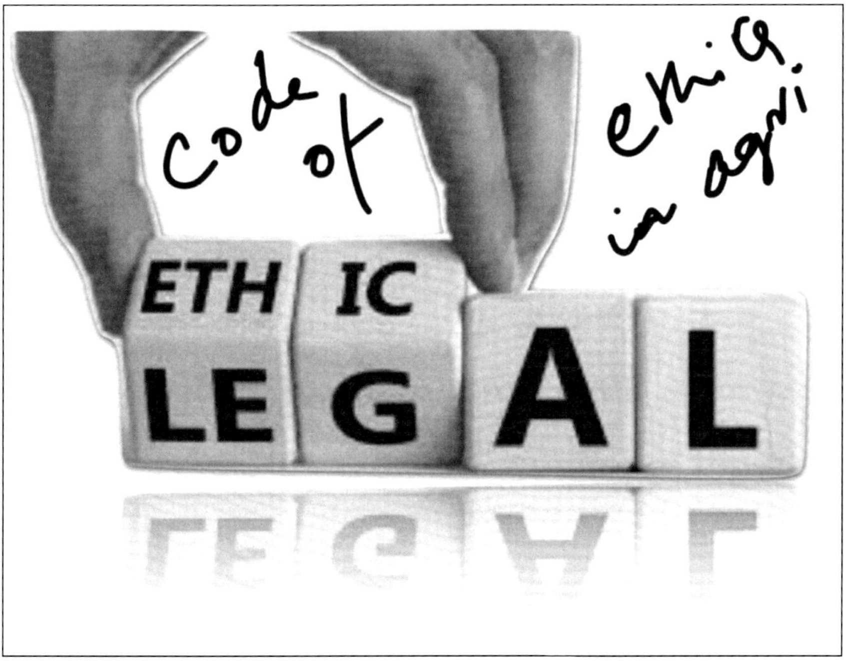

Fig. 3 A depiction of ethics that abides by the Legal.

rural training, and affordable access. Initiatives like Digital Agriculture Mission (2021–2025) and PM-KISAN offer a starting point, but the ethical question remains: how do we ensure that AI benefits the marginal farmer as much as the corporate farm?

4.2 Data Sovereignty and Farmers' Rights

In an AI-driven farm, data about soil, seeds, and yields becomes as valuable as the crop itself. Who owns this data? Corporations that provide the technology, or farmers who generate it? Indian law is still evolving here. The Digital Personal Data Protection Act (2023) lays down rules on consent and privacy, but agriculture data often falls in grey zones. Ethically, farmers must have sovereignty over their information, and policies must ensure they are not reduced to mere data suppliers for global companies.

4.3 Accountability in Autonomous Decisions

When an AI system suggests excessive pesticide use, or when a self-driving harvester malfunctions, who bears responsibility? The farmer, the manufacturer, or the algorithm designer? Ancient Indian ethics stressed accountability (dharma) in every action, and this principle remains relevant. Technology should not create a space for responsibility and hence there's a need for clear laws that need to establish liability in AI integrated Farming.

5. Conclusion

Humans started exploring and living in the AI world. The advancement in AI diminishes the ethical considerations in almost all the aspects of farming domain. Hence, it's notable that no compromise should be imposed while creating the machines in terms of intelligence without incorporating the ethical code. The ethical code should be ensured before deploying the AI system. That makes sure to be inherited while we implement it for Agriculture driven by AI. All that stuff, yielding, cultivation, harvesting should be ethically driven by AI. To make it a habit, the proposed ideology and discussion are made here as a research article.

References

Chouhan, S. (2024). Impact of pests and weeds on global crop production: Challenges and AI-based solutions. *International Journal of Agricultural Sciences, 16*(2), 45–59. https://doi.org/10.1234/ijas.2024.01602.

Chouhan, S.S., Patel, R.K., Shukla, A. and Singh, U.P. (2025, Aug.). Drone Importance and their Necessity in Future Generation Agriculture. In: *Artificial Intelligence and Computer Vision for Ecological Informatics* (pp. 1–8). doi: 10.1201/9781003518778-1.

Chouhan, S.S., Singh, U.P., Saxena, A., Jain, S. (2024). Assessing the Importance and Need of Artificial Intelligence for Precision Agriculture. In: Chouhan, S.S., Saxena, A., Singh, U.P., and Jain, S. (Eds.), *Artificial Intelligence Techniques in Smart Agriculture*. Springer, Singapore. https://doi.org/10.1007/978-981-97-5878-4_1.

Floridi, L. and Cowls, J. (2019). A unified framework of five principles for AI in society. *Harvard Data Science Review*, *1*(1), 1–15. https://doi.org/10.1162/99608f92.8cd550d1.

Kumar, N. and Chandrasekaran, K. (2021). Deep learning for sustainable agriculture: Opportunities, challenges, and policy frameworks. *Computers and Electronics in Agriculture*, *187*, 106285. https://doi.org/10.1016/j.compag.2021.106285.

Ministry of Agriculture and Farmers Welfare, Government of India. (2021). *National digital Agriculture Mission (2021–2025): Strategy and Implementation Framework*. Government of India. https://agricoop.nic.in.

Narayanan, A. and Vallor, S. (2022). AI, fairness and accountability in agriculture: A case for ethical governance. *AI and Society*, *37*(4), 1381–1395. https://doi.org/10.1007/s00146-021-01226-4.

Patel, R.K. (2025, Aug.). BM3D Filtering with Ensemble Hilbert-Huang Transform and Spiking Neural Networks for Cardiomegaly Detection in Chest Radiographs. *Computational Biology and Chemistry*, *8*, 108620.

Patel, R.K., Chouhan, S.S., Chandan, T. and Singh, U.P. (2025, Aug.) Basil Crop Detection Using Computer Vision and Deep Learning Approach. In: *Artificial Intelligence and Computer Vision for Ecological Informatics* (pp. 67–82). doi: 10.1201/9781003518778-4.

Shukla, A., Chouhan, S.S., Patel, R.K. and Singh, U.P. (2025). AgriCloud: Cloud-Based Framework for Scalable Crop Monitoring and Intelligent Decision Support. In: Chouhan, S.S., Patel, R.K., Singh, U.P., and Jain, S. (Eds.), *AgriTech Revolution*. Springer, Singapore. https://doi.org/10.1007/978-981-95-1268-3_6.

Siddharth Singh Chouhan, Rajneesh Kumar Patel, Uday Pratap Singh, & Ghanshyam G. Tejani. (2025). Integrating drone in Agriculture: Addressing technology, challenges, solutions, and applications to drive economic growth. *Remote Sensing Applications: Society and Environment*, *38*, 101576. ISSN 2352-9385, https://doi.org/10.1016/j.rsase.2025.101576.

Tripathi, A. and Kaur, P. (2023). Artificial intelligence in precision agriculture: Ethical, social, and policy considerations. *Journal of Agricultural and Environmental Ethics*, *36*(2), 1–19. https://doi.org/10.1007/s10806-023-09921-4.

Varshney, K.R. and Alemzadeh, H. (2017). On the safety of machine learning in agriculture and beyond. *Journal of Agricultural Informatics*, *8*(1), 1–12. https://doi.org/10.17700/jai.2017.8.1.360.

World Economic Forum. (2020). *Shaping the Future of Food: A Framework for Responsible AI in Agriculture*. World Economic Forum. https://www.weforum.org/reports.

16 Deep Learning for Satellite, UAV, and Hyperspectral Imaging-Based Crop Monitoring

Adityanshu Tripathi[1], Subhash Chandra[2], Aditi Kachchap[3], Shreesh Sanjay Amin[4], Kalamkuri Chandu[5], Rohit Anand[6] and Abhishek Upadhyay[7*]

1. Introduction

Feeding the projected global population of 9.7 billion by 2050 constitutes one of the greatest challenges for human agriculture (Benti et al., 2024). The problem will only worsen further due to climate change, resource depletion, erosion, etc. Hence, the environment has to be treated more sparingly (Saha et al., 2025). This is the picture with the less environmental sustainability that the modern farming practices can show the way from manual to semi-manual is insane (Sishodia et al., 2020). Moreover, these modern farming practices are not only lengthy but they also allow for only a few subjective judgments and, furthermore, they are also not capable of the required geomorphological resolution to specify the exact intervention.

Hence, agriculture is dramatically transformed by technology into the data-driven sector. Such agricultural operations as remote sensing from a satellite, drones, and hyperspectral sensors together form a monitoring paradigm that is all-encompassing, covering the crop growth patterns, soil health, and environmental factors at different levels (Segarra et al., 2020; Chowdhury et al., 2024). As an example, the Normalized Difference Vegetation Index (NDVI) from the satellites is already being used nearly everywhere to estimate the biomass accumulation and the crop health; and the use of UAV-based thermal imaging has already been efficient in the identification of water stress in plants at the cm scale (Ndlovu et al., 2024). Hyperspectral imaging, revealing small modifications in biochemical

[1,3-5] M.Tech. Scholar, Department of Farm Machinery and Power Engineering, CAET, Dr. Rajendra Prasad Central Agricultural, Pusa, Samastipur, Bihar, India.

[2] Associate Professor, Department of Farm Machinery and Power Engineering, CAET, Dr. Rajendra Prasad Central Agricultural, Pusa, Samastipur, Bihar, India.

[6] Ph.D. Student, Department of Food, Agricultural and Biological Engineering, The Ohio State University, Columbus, OH-43210, USA.

[7] Assistant Professor, Department of Farm Machinery and Power Engineering, CAET, Dr. Rajendra Prasad Central Agricultural, Pusa, Samastipur, Bihar, India.

* Corresponding author: abhiupadhyay7798@gmail.com

DOI: 10.1201/9781003715726-16

content that suggest the presence of disease or nutrient deficiency even before the visible symptoms appear, is another benefit the farmer can get from the sensors (Zhang et al., 2024)

Deep learning powers the interpretation of this complex, data-rich imagery. CNNs have shown good performance in crop identification and disease recognition, while ViTs are the best in learning long-range spatial relationships between fields (Jung et al., 2023; Zhang et al., 2023). Hybrid CNN-Transformer approaches play the role of very best in hyperspectral crop analysis (Guo et al., 2025). Increasingly, transfer learning is being harnessed by these models so that the transfer from very large publicly available general-purpose datasets like ImageNet to agriculture-specific datasets can be carried out cost-effectively and more efficiently (Shafik et al., 2024).

The actual deployment of these models in agriculture is curtailed by the infrastructural and socio economic realities. Primary connectivity is also patchy in rural belts, thus making cloud AI solutions slightly less feasible, while also the farmers being skeptical of sharing sensitive operational data (Hiremani et al., 2025). In fact, Edge AI effectively is the "great savior" by adjusting scaled-down models to run on mobile devices, drones, or even microcontrollers, delivering real-time analytics with an eye on privacy (Gong et al., 2025). But on the other hand, federated learning also distributes intelligence among the farm and it does so by data decentralization, and by the way it lowers inhibition or barriers to adoption, with other things in parallel.

The joining together of deep learning (DL), remote sensing, and distributed artificial intelligence (AI) under the same roof is not just an engineering wonder; it is the start of a new era in the farming sector where practices are sustainable and more intelligent than before. In the remainder of the section, there is a summary of the various remote sensing platforms, DL architectures, examples, and future areas of agricultural monitoring and decision-making.

A big plus point/advantage is available if the merging of DL, remote sensing, edge computing, and federated learning for agricultural monitoring is realized actively. However, it is essential to get the technical aspects of DL models' deployment on edge devices with limited resources looked into from a fine engineering standpoint. Procedures that are part of these optimization steps include pruning, quantization, and knowledge distillation that have an effect on the model size, computation, and energy consumption in a very severe manner, although they are practically unnoticeable in the predictive accuracy which is one of the prerequisites for cloud-free operation for low power and autonomic behavior in rural and hardly connected areas (Md-Tahir et al., 2024). Pruning is a way to get rid of network parameters that are not very helpful; quantization, on the other hand, is about reducing the size of the model by using less precise numbers that also speed up the model; knowledge distillation is the method of passing on the knowledge from the big teacher model to the small student model. Through the use of these methods, the models acquire the ability to perform efficiently, even with the challenging constraints of computation and memory in edge-level devices. The

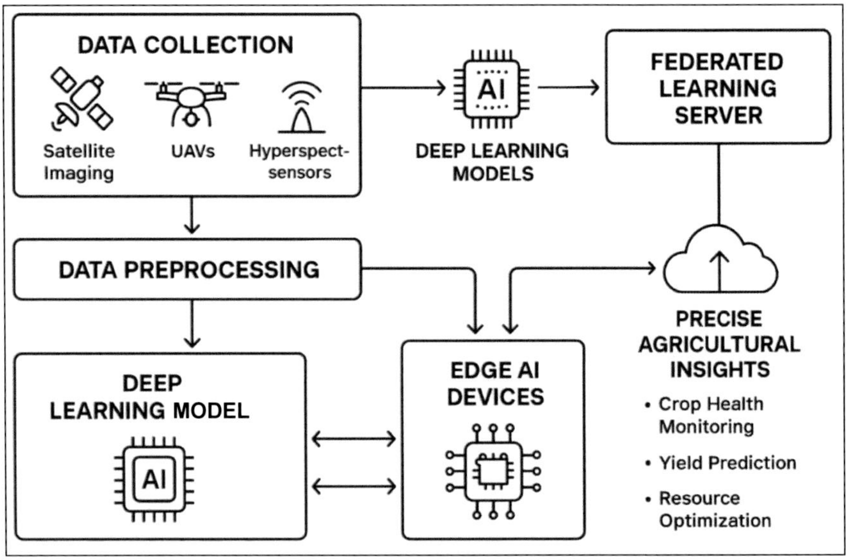

Fig. 1 Deep Learning-based Precision Agriculture Conceptual Framework.

chapter carried out a profound investigation on the areas of convergence of these technologies with DL, remote sensing, edge AI, and federated learning, shining a spotlight on their technical sides, real-world scenarios, difficulties, and huge potential for the next-generation agricultural monitoring systems.

2. Remote Sensing Technologies for Agricultural Monitoring

In modern times, crop monitoring is often called layered remote sensing. It's where agriculture meets cutting-edge technology, using data from above—like satellites and drones—to keep a close eye on what's happening in the fields. The term remote sensing refers to sensing from a distance. There is a variety of remote sensing: very-high-resolution remote sensing, high-resolution remote sensing, multi-resolution remote sensing, and radar remote sensing. Combining these data sources with DL algorithms offers the possibility of crop monitoring and management at highly precise levels that were never possible earlier (Chouhan et al., 2025; Segarra et al., 2020).

2.1 Satellite-Based Crop Monitoring

Satellite-based remote sensing is utilized nowadays for specialized agriculture observing. The basic idea is quite simple: observing the plants from above, with great spatial coverage, and for unusually long periods (Sishodia et al., 2020; Segarra et al., 2020). Landsat, Sentinel-2, MODIS, PlanetScope are a few operational systems providing remote sensing data offering multispectral and hyperspectral information at varying levels of spatial resolution and temporal revisit period. Sentinel-2, for example, provides a standard resolution from 10 to 20 m with a revisit time of 5 days. Because of this, it's become a popular tool for tracking crop growth at the field level (Segarra et al., 2020). It is widely accepted that vegetation indices obtained from these data are the standard approach to the evaluation of crop health, biomass, and stress. A prominent one is the NDVI (Md-Tahir et al., 2024). The NDVI was found to be a very efficient experimental tool for crop health assessment (Md-Tahir et al., 2024). Not only NDVI, but also the other indices such as EVI and red-edge are the better known, the more supplementary those, the more detectable the two types of applications can be, and they might be related to more structural characteristics (Saha et al., 2025). The satellite sensors' SAR data addition to the optical remote sensing, has been the most significant alteration of the whole remote sensing process regarding crop monitoring. The SAR technology avoids cloud interference completely and consequently, the biomass and soil moisture monitoring during harsh weather is of utmost stability. The democratization of access to satellite imagery on platforms such as Google Earth Engine (GEE) and AWS Open Data has enabled researchers, policymakers, and even smallholder farmers to access and analyze satellite-derived data (Sishodia et al., 2020). Although they have their benefits, satellite systems are challenged by cloud cover, coarse resolution for small fields, and latencies in revisit cycles. These limitations are more and more compensated by unmanned aerial vehicle (UAV)-based monitoring.

2.2 UAV-Based Agricultural Imaging

UAVs, drones, or equipment in the domain of precision agriculture, are versatile and bridging ground observations and broad satellite images. At GPS-tagged plant positions, detailed monitoring at the plant level and early detection of diseases are performed by UAVs with spatial resolutions ranging between 1–5 cm, provided by RGB (red, green, blue), multispectral, hyperspectral, and thermal cameras (Bouguettaya et al., 2022; Chouhan et al., 2025).

There are multiple benefits that UAVs deliver:

- Flexibility in deployment, independent of satellite orbits.
- High temporal resolution, as flights can be conducted on demand.
- Ability to fly below cloud cover, providing uninterrupted imaging.
- Integration of multi sensor payloads, including LiDAR, thermal cameras, and snapshot hyperspectral sensors.

Some of the latest innovations include autonomous flight planning systems, a system of real-time mosaicking, and onboard deep-learning processing for in-field analysis (Wu et al., 2021). Thermal UAV imaging has been used to assess crop water stress because canopy temperature serves as a reliable proxy for stomatal closure and transpiration. UAV systems have also been employed in large-scale agricultural operations. Some fixed-wing drones cover between 500–750 acres per hour, large enough to be feasible for commercial farms (Adão et al., 2017).

2.3 Applications of HSI Technology

Hyperspectral imaging (HSI) is a sophisticated method used in agricultural remote sensing. It captures spectral information across hundreds of very narrow wavelength bands with high precision, then analyzes these detailed measurements to produce valuable insights. What HSI does that RGB and multispectral instruments cannot is to detect the presence of pigments such as chlorophyll, carotenoids, and anthocyanins, while also considering proteins, moisture capacity, and secondary metabolites. This capacity drastically helps in the early detection of diseases and nutrient deficiency (Guerri et al., 2024).

Recent snapshot hyperspectral camera developments are mitigating the problem of motion artifacts and currently making UAV deployment possible for real-time monitoring in fields (Adão et al., 2017). It has been seen that crop type classifications and disease detections have been thoroughly handled by the HSI data and deep learning models with an aggregate correct rating of more than 95% (Guo et al., 2025). The author provides the following example of the potential use of HSI by Zhang et al. (2024); their system of the HSI predicts the disease 7–14 days before it is visible in the leaves. It has drastically accelerated the speed at which scientists conduct research and publish their results, and, at the same time, the use of machine and deep learning models like 3D CNNs (convolutional neural networks), transformers, and CNN-transformer hybrid networks has been normal to solve the curse of dimensionality with the hyperspectral data (Zhang et al., 2023; Guerri et al., 2024). They build their forecasts around the temporal and spectral-spatial correlation dynamics, which, in turn, give much better results than the traditional statistics-based techniques.

3. Deep Learning Architectures for Agricultural Image Analysis

DL has uncovered a new, sensationally attractive perspective on agri-image analysis because, in its essence, it makes the technology capable of extracting features from UAV images, hyperspectral, and multispectral imagery. The DL architecture directly produces the hierarchical representation of the data, while the classical methods are based on handcrafted features (Jung et al., 2023; Shafik et al., 2024). These developments are, above all, very prominent in tasks such as crop identification, disease identification, and yield prediction.

3.1 Convolutional Neural Networks for Crop Classification

- Convolution kernels have made CNNs the most favorite deep-craft learning algorithm in agricultural research. Besides, the other network architectures such as AlexNet, VGG, ResNet, DenseNet, and EfficientNet have also been considerably used in this sector (Jung et al., 2023; Shafik et al., 2024).

- ResNet with the other major newness of skip connections probably gets the maximum attention topically for it is the one that introduces a solution with the vanishing gradient problem and thus enables the deeper models to be trained and, in the end, have better classification performance (Shafik et al., 2024).
- This feature is very important, especially in medical imaging, when resolving the problem of crop diseases that are diseased and are very difficult to be separated (Zhu et al., 2024).
- EfficientNet functions to balance depth, width, and resolution; hence, it performs optimally on edge devices that have a constrained budget in terms of computations (Jung et al., 2023).

Transfer learning represents an opportunity for CNNs to be employed in agriculture fields; for example, models pre trained on ImageNet can be fine-tuned on smaller agricultural datasets, thus reducing data and training time greatly (Shafik et al., 2024). For instance, Jung et al. (2023) showed that an EfficientNet-B3 model fine-tuned on plant disease datasets had a maximum accuracy of 99.3% in classifying tasks. CNNs have also been used successfully in weed detection, crop classification from UAV images, and soil moisture estimation with accuracies in excess of 90% in the majority of studies (Ndlovu et al., 2024).

3.2 Transformer Models for Agricultural Applications

Transformers, initially designed for natural language processing, have been more recently applied to computer vision tasks with Vision Transformers (ViTs). In contrast to CNNs, which recognize local spatial cues through convolutions, ViTs treat images as sequences of patches and use self-attention mechanisms to capture long-range dependencies (Dosovitskiy et al., 2021; Patel et al., 2025). This global context makes ViTs a perfect fit for large-scale agricultural images, including UAV mosaics and satellite images, where spatial patterns across a whole field need to be detected (Zhang et al., 2023).

Applications are:

- Crop classification from Sentinel-2 and UAV hyperspectral imagery, where the accuracy of ViTs was better than CNNs (Chang et al., 2025).
- Monitoring disease spreading, utilizing ViTs' capacity to learn sequential patterns in multi temporal images.
- Analysis of hyperspectral images, where attention mechanisms are best suited for the capture of spectral–spatial relationships in hundreds of bands (Guerri et al., 2024).

However, ViTs need big annotated datasets and lots of computational resources, so they are hard to deploy in agricultural situations where there is little annotated data (Upadhyay et al., 2025a). To overcome this, scientists are working on data augmentation techniques, light transformers, and half-supervised learning methods (Zhang et al., 2023).

3.3 Hybrid CNN-Transformer Architectures

Another promising potential area of agricultural artificial intelligence deals with the co-evolution of CNNs and transformer networks. They seem to have the appeal that the hybrid version would harness the best aspects of both constructions: transformer layers concentrate on global relations and long-distance dependencies while CNN layers work on local feature extraction and spatial patterns (Dosovitskiy et al., 2021; Zhang et al., 2023). Some of the very challenging tasks, such as disease diagnosis or crop identification, have yielded state-of-the-art results (Guo et al., 2025; Guerri et al., 2024).

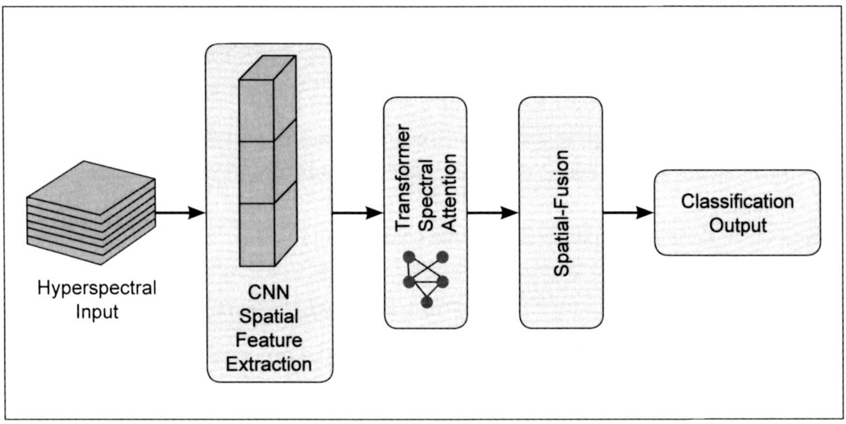

Fig. 2 CNN-Transformer architecture hybrid for hyperspectral analysis of crop images.

CMTNet, in particular, a type of network for categorizing hyperspectral crops, has this very particular thing that is a two-branch notion for the purpose of bringing out spectral and spatial features at the same time and which makes use of the local and global features (Guo et al., 2025). On the Airdrone UAV-acquired datasets, the experimental results give an accuracy of greater than 99% in the classification of crops. When developing these kinds of hybrid architectures, especially considering the deployment on the edge side, one must be very careful and deliver the computational efficiency and memory requirements in a balanced way (Guerri et al., 2024). Consequently, techniques such as progressive feature fusion and multi scale processing have been proposed to help these models sustain the trade-off between performance and efficiency, so that they work well on even the limited hardware (Zhang et al., 2023; Guerri et al., 2024). These architectures are thus the cutting edge of agricultural DL applications.

4. Applications and Case Studies

Global research has focused on applying AI, distributed in nature, to problems of yield forecasting, pest and disease control, irrigation control, and crop health monitoring by different study groups all over the world. Crop health monitoring has been among the earliest and most well-known applications of DL in remote sensing. These can generate early alerts to let producers know about soil deficiency so that they can take adjustments to rectify the issue.

4.1 Crop Health Monitoring Systems

For example, UAV-based multispectral imaging can provide more than 95% accuracy of maize blight and wheat rust early warning using CNNs for the analysis (Bouguettaya et al., 2022). For instance, 7–14 days prior to outward symptoms appearing, hyperspectral imaging may be used to detect physiological changes, allowing for proactive management to mitigate yield losses (Zhang et al., 2024).

From localized treatment advice in the AI-based monitoring system, case studies conducted in North America found pesticide application was reduced by 15–25% while pesticide application was increased by 10–20% (Saha et al., 2025). Similar successes have been documented in Asia and Europe, where UAV thermal imagery combined with DL was used to identify water stress zones, enabling targeted irrigation and reducing unnecessary pesticide spraying (Ndlovu et al., 2024).

4.2 Precision Irrigation Management

Water is among the most valuable and scarce agricultural resources. DL-based precision irrigation combines thermal UAV imagery, NDVI time series, and soil moisture sensors to maximize water application. Large-scale field trials in Australia proved 30–50% water savings with zero loss of yield by using UAV + satellite fusion models along with DL-based soil–crop–climate integration (Ndlovu et al., 2024). IoT soil moisture sensors + DL models running on microcontrollers enable edge-based irrigation systems with real-time scheduling, minimizing the use of cloud infrastructure (Gong et al., 2025). Federated learning methods have also enhanced irrigation suggestions by using region-based weather variability while maintaining data privacy across farms (Mohammed, 2025).

4.3 Yield Prediction and Optimization

AI- based yield prediction models combine multi temporal satellite imagery, UAV surveys, weather data, and ground sensors to predict yields with high accuracy (Chouhan et al., 2024).

- Multi temporal Sentinel-2 and Landsat data taken with CNN-LSTM (long short-term memory) architectures predicted wheat and maize yields within 5–10% error margins, up to 4–6 weeks ahead of harvesting in Europe. (Saha et al., 2025).
- Estimating rice yield with UAV-based hyperspectral images and DL regression models had an R2 greater than 0.9 in India, much higher than that of traditional statistics (Simhadri et al., 2024).

There are these forecasting systems that decision makers can use for resource allocation, market anticipation, and organizing of logistics efficiently. The accuracy of the forecasts from the federated learning-models is improved in different agricultural domains through learning from collective datasets while respecting local data privacy (Hiremani et al., 2025).

4.4 Pest and Disease Management Networks

Pests and diseases pose a major threat to global food security. Using an AI-enabled monitoring system connected to UAVs and perhaps other ground sensors permits real-time monitoring and track of pest infestations (Shahi et al., 2023; Kerkech et al., 2020; Sahni et al., 2024).For example, in South Asia, UAV monitoring using edge technology and deep learning led to a reduction in crop losses of between 20–30% by means of an early detection system for locust swarms, followed by alerts in neighboring areas (Fendor et al., 2025). Similarly, a federated network for disease detection allowed undertakings in Europe to share updates to models rather than raw data for a better collective identification of new disease strains (Žalik, 2023; Mohammed, 2025). These networks showcase the large-scale nature and flexibility of DL systems when integrated into cooperative systems at the farm and regional levels (Bouguettaya et al., 2022; Hiremani et al., 2025).

5. Challenges and Limitations

Although they portend much good, their mass use is plagued by various technology, economic, and social issues that need to be addressed.

5.1 Technical Challenges

In the methodology outlined for traditional techniques, inner optimization and fine tuning are hard to be deprecated due to the advance of DL as a substitute. The performance of models has great variation due to changes in lighting, weather conditions, and seasonal variations in agricultural images. One example would be irregular outdoor lighting that creates a challenge for image analyses, requiring vast amounts of training data or imaging systems that minimize these lighting variations. It was reported that this approach can reduce training data requirements by up to four times (Silwal et al., 2021). Additionally, DL models tend to be sensitive to domain variations because models trained with one lighting or seasonal setting may not work when used in other environmental conditions (Kamilaris & Prenafeta-Boldú, 2023). Working on models that perform well under a wide range of conditions demands large amounts of annotated data and stable architectures (Upadhyay et al., 2025b).

Edge computing adds further constraints on processing power, memory, and power. The fact that edge devices have limited processing capacity implies model complexity must be carefully adjusted; compression methods (such as pruning, quantization, and light architectures) minimize resource requirements but necessarily bring about efficiency and model accuracy trade-offs (Ren et al., 2020). Equally, a prominent problem with the battery-operated field-deployed edge devices remains to be power consumption. Impacting the energy, even the compressed models can expend quite a bit, and hence intelligent management of both hardware and software is a must to prolong the operational autonomy (Kaur et al., 2023; Li et al., 2024).

5.2 Economic and Adoption Barriers

Financing the deployment of complex agri-business AI systems remains very difficult, especially for small farmers and enterprises in developing countries. This is largely due to the initial investment, which covers sensors, edge devices, and software licenses (Wolfert et al., 2017; Kamilaris & Prenafeta-Boldú, 2018). These systems being sophisticated and new, mean that they need a person with enough technical knowledge to run and support them, as well as spreading out to different farmers and countryside agriculture population, which is a huge cost by itself (Rose et al., 2021; Ayaz et al., 2021). Also, market fragmentation and differences in AI platform standards among the various vendors are other factors complicating the adoption process (Liu et al., 2020; Walter et al., 2021). It is not easy to convince farmers, most of whom are not technology-friendly, to invest in systems that will soon be outdated or incompatible with cutting-edge devices. Therefore, a solution with open-source platforms and a common interface is thus emerging in the industry (Bronson & Knezevic, 2016; van Es & Woodard, 2017).

5.3 Data Privacy and Security Issues

The farmer's use of AI-centric monitoring systems is the genesis of the problem of data privacy. Farm data itself often carries private terms of service, economic results, or proprietary management procedures. Therefore, farmers would rather not share their data in whatever the most protected manner is with the other party which can happen through federated learning. Furthermore, the potential for farming practices to be vulnerable to cyberattacks is a growing concern, and the associated risks will increase significantly with the digital interconnectedness of agriculture. Edge devices deployed in remote rural agricultural areas can be vulnerable to physical tampering and network intrusion; therefore, strict security measures and real-time system updates are essential to maintain system integrity (Campoverde-Molina & Luján-Mora, 2024; Singh Chouhan, S., 2025).

5.4 Environmental and Sustainability Factors

While AI-based farm systems can enhance environmental sustainability in terms of better use of resources, new environmental challenges also emerge. Environmental impacts of energy consumption by computer systems and electronic waste due to edge devices should be factored into the sustainability assessment. A complete life-cycle analysis of AI-based farm systems has to be conducted to ensure a net positive environmental impact. Dependence on cloud computing infrastructure for federated learning and data processing incurs indirect environmental effects in the form of energy consumption in data centers, whereas edge computing can reduce these by limiting the need for data transmission (Qiu et al., 2021).

Table 1 Comparative analysis of conventional vs. AI-enabled crop monitoring approaches (Sishodia et al., 2020; Saha et al., 2025).

Aspect	*Conventional Approach*	*AI-Enabled Approach*
Spatial & temporal resolution	Primarily ground surveys and low-resolution imagery; sparse coverage and infrequent sampling.	Satellite (3–30 m, daily revisit), UAV (1–5 cm), and hyperspectral sensors enable dense multi temporal/spatial monitoring.
Accuracy	Subjective visual assessments; prone to error (often <90% reliability).	Deep learning on high-dimensional imagery (CNN/ViT) yields >95% classification accuracy.
Labor & time intensity	Labor-intensive manual scouting and sample processing; time-consuming analysis.	Automated UAV/sensor data collection and on-device AI inference reduce human effort and speed up decision-making.
Cost-effectiveness	Low technology cost but high ongoing labor and resource expenses; limited economies of scale.	Higher upfront cost (sensors, drones, edge devices) but lower marginal costs at scale; satellite/UAV monitoring is cost-effective over large areas and automation cuts labor costs.
Real-time capabilities	Periodic data collection with significant delays (batch analysis, offline).	Edge computing enables on-device, real-time data processing for instantaneous monitoring and alerts.
Scalability	Hard to scale: more land requires proportionally more labor and equipment.	Highly scalable via satellite constellations and drone fleets; professional UAVs can survey ~100–200 ha per flight. Federated learning allows models to incorporate data from many farms.
Data integration & analysis	Fragmented, siloed data (field notes, single-source); manual or basic analytics.	Fuses multisensor and multi-temporal data (satellite, UAV, hyperspectral, IoT); employs deep learning and federated learning for comprehensive analysis and cross-farm learning.
Disease & pest detection	Visual scouting and periodic trap checks; issues often detected late and subjectively.	AI methods (UAV/multispectral imaging + DL) detect stress and pests early (e.g., hyperspectral identifies disease days before symptoms); federated networks enable rapid, privacy-preserving alerts.

Contd...

Table 1 *Contd...*

Aspect	*Conventional Approach*	*AI-Enabled Approach*
Yield prediction	Yield estimated by sampling and historical averages; low precision due to bias.	Machine learning models ingest multi temporal imagery and field data to provide continuous yield forecasts; federated learning improves accuracy using regional data.
Environmental sustainability	Often over-applies inputs (uniform treatments), leading to resource waste and runoff.	Precision input application (guided by AI) optimizes resource use, reducing waste and improving sustainability; edge computing reduces data-transmission energy costs.
Data privacy	Minimal digital data exchange; privacy not a concern (local records only).	Federated learning trains models without sharing raw data, preserving farm data privacy; on-device AI keeps sensitive information local.

6. Future Directions and Emerging Trends

The future wave of innovation in ag-tech will be triggered by increasingly efficient architectures, sophisticated sensor integration, and complete autonomous systems.

6.1 Next-Generation AI Architectures

More powerful and efficient AI architecture tailored to specific agricultural uses is one of the most important research areas in the future. Neuromorphic computation methods imitating biological neural networks hold potential for ultra-low power edge computing, which may make possible the implementation of advanced AI capabilities at the edge in far-flung agricultural regions with minimal power infrastructure. Foundation models and large language models with agricultural-specific training could enable a more holistic appreciation of agricultural systems through the simultaneous processing of text, images, and sensor data. Transfer learning from such foundation models may speed the development of specialized agricultural AI applications (Lu & Xiao, 2024).

6.2 Advanced Sensor Integration

The incorporation of emerging sensor technologies will widen the scope of capabilities of agricultural monitoring systems. Hyper-spectral sensors are becoming smaller and cheaper and can be deployed on smaller UAV platforms and mobiles. High-end radar and LiDAR systems have the capability to provide precise 3D structural data regarding crops and agricultural fields. Real-time biosensors that identify certain plant pathogens, nutrients, and environmental toxins for use with AI systems are being developed. These sensors would enable real-time detection of farm threats and trigger autonomous response mechanisms (Miguel-Rojas & Pérez-de-Luque, 2023).

6.3 Autonomous Farming Systems

Agri-tech is embracing a new wave, a different one this time around, to an extent where the whole farm could be run without human intervention. Tractors, harvesters and AI-powered tools that are the technology-supported leaders of the agriculture sector now are approaching the end of the cycle in a fast paced manner for the open-field operations while maximizing the output of the "real time making" and "forecast making". As a result of these, swarms of autonomous vehicles, also called swarm robots, can cover even large areas very cost-effectively and precisely. The simultaneous presence of the powerful and self-learning systems in the federated networks would, however, bring about a more significant continuous optimization and improvement of farm practices through shared wisdom (Qu et al., 2024; Anand et al., 2023).

6.4 Global Agricultural Intelligence Networks

The establishment of international agricultural intelligence networks that will connect farming operations in different parts of the world will, without a doubt, greatly contribute to the wisdom acquisition in the field of food production and agricultural sustainability. The networks might be positioned as the first line of defense in terms of the detection and alerting of potential crises, being that they would warn the world of threats to food security; hence, they would be a platform for a coordinated global battle against food-related problems. The very essence of federated learning is to enable local participation and ensure data sovereignty, thereby eliminating the possibility of that data being compromised in sensitive, numerous localities, which involve the data of individual farmers. The satellite constellation used in agricultural monitoring would enable real-time monitoring of croplands, while AI processing of global images will lead to unprecedented monitoring of agricultural conditions, crops, and environmental yields, especially at spatial levels that have never been seen before (So et al., 2022; Shukla, A., 2025).

7. Conclusion

The ecosystem of precision agriculture has undergone a profound change with the rise of DL, remote sensing, edge computing, and federated learning. These technologies are used in combination to provide an entirely new way of monitoring and making decisions in the agricultural sector. This paper thus goes deep into the theory, practice, and field application of the use of these technologies in precision agriculture. The confluence of the technologies of satellite imagery, UAV platforms, hyperspectral sensors, and the use of high-powered AI algorithms has improved crop health monitoring, identification of diseases, and estimating yield like never before. It was the fact that the case studies they provided so candidly showed how crop yields, resource efficiency, and environmental sustainability have been improved because of the new approaches and technologies. The literature also showed that there were still technical and economic barriers to the adoption of the technologies, as well as issues of data privacy and computational limitations that required remaining areas for future research at least.

This union is expected to mature enough to be one of the main players in the shaping of the future of agro-tech. Accordingly, very soon, agriculture will be almost unrecognizable with the three major technologies of next-generation AI architectures, upgraded sensors, and fully automated systems. The transformation of agriculture through AI and advanced sensing represents not just a technological evolution but a fundamental shift toward data-driven, sustainable farming practices that are more efficient, resilient, and responsive to global needs.

Proclamation

Ethical obligations of authors

All authors confirm that they have reviewed, understood, and followed the ethical guidelines for authors provided in the Instructions for Authors.

Ethical clearance

Not required

AI Involvement

The author attests that no generative AI or AI-assisted technologies were utilized in the creation of this manuscript.

Competing Interests

The corresponding author states, on behalf of all co-authors, that the authors have no conflict of interest to disclose.

Interest disclosures

Not relevant

Authors' Roles

Adityanshu Tripathi[1]: Study design, Research methods, Writing - Draft preparation, **Subhash Chandra**[2]: Writing – Critical Review, Study design, Guidance, **Aditi Kachchap**[3]: Visualization, Formal analysis, **Shreesh Sanjay Amin**[4]: Methodology, Formal analysis, **Kalamkuri Chandu**[5]: Visualization, Formal analysis, **Rohit Anand**[6]: Supervision, Writing – Reviewing and Editing, **Abhishek Upadhyay**[7]: Writing – Reviewing and Editing, Visualization, Formal analysis, Supervision, Conceptualization.

Financial support

The authors did not receive any funding for this work.

References

Adão, T., Hruška, J., Pádua, L., Bessa, J., Peres, E., Morais, R. and Sousa, J.J. (2017). Hyperspectral imaging: A review on UAV-based sensors, data processing, and applications for agriculture and forestry. *Remote Sensing*, *9*(11), 1110. https://doi.org/10.3390/rs9111110.

Anand, R., Kumar, A.K. and Rathod, S.K. (2023). Scope of Autonomous Tractors in Revolutionizing Agricultural Operations. *Advanced Innovative Technologies in Agricultural Engineering for Sustainable Agriculture* (Vol. 31). AkiNik Publications.

Ayaz, M., Ammad-Uddin, M., Sharif, Z., Mansour, A. and Aggoune, E.-H. M. (2019). *Internet-of-Things (IoT)-Based Smart Agriculture: Toward Making the Fields Talk. IEEE Access, 7*, 129551–129583. https://doi.org/10.1109/ACCESS.2019.2932609.

Benti, N.E., Chaka, M.D., Semie, A.G., Warkineh, B. and Soromessa, T. (2024). Transforming agriculture with Machine Learning, Deep Learning, and IoT: Perspectives from Ethiopia: Challenges and opportunities. *Discover Agriculture*, *2*, 63. https://doi.org/10.1007/s44279-024-00066-7.

Bouguettaya, A., Zarzour, H., Kechida, A. and Taberkit, A.M. (2022). Deep learning techniques to classify agricultural crops through UAV imagery: A review. *Neural Computing and Applications*, *34*, 9511–9536. https://doi.org/10.1007/s00521-022-07104-9.

Bronson, K. and Knezevic, I. (2016). Big Data in food and agriculture. *Big Data and Society, 3*(1), 1–5. https://doi.org/10.1177/2053951716648174.

Campoverde-Molina, M. A. and Luján-Mora, S. (2024). Cybersecurity in smart agriculture: A systematic literature review. *Computers and Security*, *150*, Article 104284. https://doi.org/10.1016/j.cose.2024.104284.

Chang, B., Li, F., Hu, Y., Yin, H., Feng, Z. and Zhao, L. (2025). Application of UAV remote sensing for vegetation identification: A review and meta-analysis. *Frontiers in Plant Science*, *16*, 1452053. https://doi.org/10.3389/fpls.2025.1452053.

Chouhan, S.S., Patel, R.K., Shukla, A. and Singh, U.P. (2025, Aug.). Drone Importance and their Necessity in Future Generation Agriculture. Artificial Intelligence and Computer Vision for Ecological Informatics (pp. 1–8). doi: 10.1201/9781003518778-1.

Chouhan, S.S., Patel, R.K., Singh, U.P. and Tejani, G.G. (2025). Integrating drone in agriculture: Addressing technology, challenges, solutions, and applications to drive economic growth. *Remote Sensing Applications: Society and Environment, 38*, 101576. https://doi.org/10.1016/j.rsase.2025.101576.

Chouhan, S.S., Singh, U.P., Saxena, A. and Jain, S. (2024). Assessing the Importance and Need of Artificial Intelligence for Precision Agriculture. In: Chouhan, S.S., Saxena, A., Singh, U.P., and Jain, S. (Eds.), *Artificial Intelligence Techniques in Smart Agriculture.* Springer, Singapore. https://doi.org/10.1007/978-981-97-5878-4_1.

Chowdhury, M., Anand, R., Dhar, T., Kurmi, R., Sahni, R.K., Kushwah, A. (2024). Digital Insights into Plant Health: Exploring Vegetation Indices Through Computer Vision. In: Chouhan, S.S., Singh, U.P., Jain, S. (Eds.), *Applications of Computer Vision and Drone Technology in Agriculture 4.0.* Springer, Singapore. https://doi.org/10.1007/978-981-99-8684-2_2.

Dosovitskiy, A., Beyer, L., Kolesnikov, A., Weissenborn, D., Zhai, X., Unterthiner, T., Dehghani, M., Minderer, M., Heigold, G., Gelly, S., Uszkoreit, J. and Houlsby, N. (2021). An image is worth 16×16 words: Transformers for image recognition at scale. In: *International Conference on Learning Representations* (*ICLR*). [Preprint]. https://doi.org/10.48550/arXiv.2010.11929.

Fendor, L., et al. (2025). Federated learning in pest detection networks: A case study in South Asia. *Computers and Electronics in Agriculture*. http://dx.doi.org/10.1016/j.jafr.2025.102238

Gong, R., Zhang, H., Li, G. and He, J. (2025). Edge computing-enabled smart agriculture: Technical architectures, practical evolution, and bottleneck breakthroughs. *Sensors*, *25*(17), 5302. https://doi.org/10.3390/s25175302.

Guerri, M. F., Distante, C., Spagnolo, P., Bougourzi, F. and Taleb-Ahmed, A. (2024). Deep learning techniques for hyperspectral image analysis in agriculture: A review. *ISPRS Open Journal of Photogrammetry and Remote Sensing*, *12*, 100062. https://doi.org/10.1016/j.ophoto.2024.100062.

Guo, X., Feng, Q. and Guo, F. (2025). CMTNet: A hybrid CNN-transformer network for UAV-based hyperspectral crop classification in precision agriculture. *Scientific Reports*, *15*, 12383. https://doi.org/10.1038/s41598-025-97052-w.

Hiremani, V., Devadas, R.M., Preethi, R., Sapna, R., Sowmya, T., Gujjar, P., Shobha Rani, N. and Bhavya, K.R. (2025). Federated learning for crop yield prediction: A comprehensive review of techniques and applications. *MethodsX*, *14*, 103408. https://doi.org/10.1016/j.mex.2025.103408.

Jung, M., Song, J.S., Shin, A.-Y., Choi, B., Go, S., Kwon, S.-Y., Park, J., Park, S.G. and Kim, Y.-M. (2023). Construction of deep learning-based disease detection model in plants. *Scientific Reports*, *13*, 7331. https://doi.org/10.1038/s41598-023-34549-2.

Kamilaris, A. and Prenafeta-Boldú, F.X. (2018). Deep learning in agriculture: A survey. *Computers and Electronics in Agriculture, 147*, 70–90. https://arxiv.org/abs/1807.11809.

Kaur, A., Bhatt, D.P. and Raja, L. (2023). Applications of Deep Learning and Machine Learning in Smart Agriculture: A Survey. In: *Machine Learning and Deep Learning for Smart Agriculture and Applications* (pp. 34–57). IGI Global. https://doi.org/10.4018/978-1-6684-9975-7.ch003.

Kerkech, M., Hafiane, A. and Canals, R. (2020). VddNet: Vine Disease Detection Network based on multispectral UAV images and depth map. *arXiv*. https://doi.org/10.48550/arXiv.2009.01708.

Li, J., Chen, D., Yin, X. and Li, Z. (2024). Performance evaluation of semi-supervised learning frameworks for multi-class weed detection. *Frontiers in Plant Science, 15*, Article 1396568. https://doi.org/10.3389/fpls.2024.1396568.

Liu, Y., Ma, X., Shu, L., Hancke, G.P. and Abu-Mahfouz, A.M. (2020). From Industry 4.0 to Agriculture 4.0: Current status, enabling technologies, and research challenges. *IEEE Transactions on Industrial Informatics, 17*(6), 4322–4334. https://doi.org/10.1109/TII.2020.3003910.

Lu, S. and Xiao, X. (2024). Neuromorphic computing for smart agriculture. *Agriculture*, *14*(11), 1977. https://doi.org/10.3390/agriculture14111977.

Md-Tahir, H.S., Mahmood, H., Hussain, M., Khalil, A., Shoaib, M., Ali, M., Ali, M., Tasawar, M., Ali, Y. K. and Awan, U. (2024). Localized crop classification by NDVI time series analysis of remote sensing satellite data: Applications for mechanization strategy and integrated resource management. *AgriEngineering*, *6*(3), 142. https://doi.org/10.3390/agriengineering6030142.

Miguel-Rojas, C. and Pérez-de-Luque, A. (2023). Nanobiosensors and nanoformulations in agriculture: New advances and challenges for sustainable agriculture. *Emerging Topics in Life Sciences*, *2023*(1), 1–10. https://doi.org/10.1042/ETLS20230070.

Mohammed, H.A., Obaid, M.K. and Vittalaiah, A. (2025). Enhancing yield estimation and field zoning accuracy in precision agriculture using solar-powered drone-based remote sensing. *SHS Web of Conferences, 251*, 01055. https://doi.org/10.1051/shsconf/202521601055.

Ndlovu, S., Mekuria, Y. and Egoh, B. (2024). UAV-based thermal remote sensing for precision irrigation: A meta-review. *Agricultural Water Management*, *300*, 107234. https://doi.org/10.1080/01431161.2024.2368933.

Patel, R.K., Chouhan, S.S., Chandan, T. and Singh, U.P. (2025, Aug.). Basil Crop Detection Using Computer Vision and Deep Learning Approach. Artificial Intelligence and Computer Vision for Ecological Informatics (pp. 67–82). doi: 10.1201/9781003518778-4.

Qiu, X., Parcollet, T., Fernandez-Marques, J., Buarque de Gusmão, P., Gao, Y., Beutel, D.J., Topal, T., Mathur, A. and Lane, N.D. (2021). A first look into the carbon footprint of federated learning. In: *Proceedings of the 1st Federated Learning Symposium*. arXiv. https://doi.org/10.48550/arXiv.2102.07627.

Qu, J., Zhang, Z., Qin, Z., Guo, K. and Li, D. (2024). Applications of autonomous navigation technologies for unmanned agricultural tractors: A review. *Machines*, *12*(4), 218. https://doi.org/10.3390/machines12040218.

Ren, C., Kim, D.-K. and Jeong, D. (2020). A survey of deep learning in agriculture: Techniques and their applications. *Journal of Information Processing Systems*, *16*(5), 1015–1033. https://doi.org/10.3745/JIPS.04.0187.

Rose, D.C., Wheeler, R., Winter, M., Lobley, M. and Chivers, C.A. (2021). Agriculture 4.0: Making it work for people, production, and the planet. *Land Use Policy, 100*, 104933. https://doi.org/10.1016/j.landusepol.2020.104933.

Saha, S., Kucher, O.D., Utkina, A.O. and Rebouh, N.Y. (2025). Precision agriculture for improving crop yield predictions: A literature review. *Frontiers in Agronomy*, *7*, 1566201. https://doi.org/10.3389/fagro.2025.1566201.

Sahni, R.K., et al. (2024). Drone Spraying System for Efficient Agrochemical Application in Precision Agriculture. In: Chouhan, S.S., Singh, U.P., and Jain, S. (Eds.), *Applications of Computer Vision and Drone Technology in Agriculture 4.0*. Springer, Singapore. https://doi.org/10.1007/978-981-99-8684-2_13.

Segarra, J., Buchaillot, M.L. and Araus, J.L. (2020). Remote sensing for precision agriculture: Sentinel-2 improved features and applications. *Agronomy*, *10*(5), 641. https://doi.org/10.3390/agronomy10050641.

Shafik, W., Tufail, A., De Silva Liyanage, C. and Apong, R.A.W.H. (2024). Using transfer learning-based plant disease classification and detection for sustainable agriculture. *BMC Plant Biology*, *24*, Article 136. https://doi.org/10.1186/s12870-024-04825-y.

Shahi, T.B., et al. (2023). Recent advances in crop disease detection using UAV and deep learning. *Remote Sensing*, *15*(9), 2450. https://doi.org/10.3390/rs15092450.

Shukla, A., Chouhan, S.S., Patel, R.K., Singh, U.P. (2025). AgriCloud: Cloud-Based Framework for Scalable Crop Monitoring and Intelligent Decision Support. In: Chouhan, S.S., Patel, R.K., Singh, U.P., and Jain, S. (Eds.), AgriTech Revolution. Springer, Singapore. https://doi.org/10.1007/978-981-95-1268-3_6.

Silwal, A., Parhar, T., Yandun, F. and Kantor, G. (2021). A robust illumination-invariant camera system for agricultural applications. *arXiv preprint arXiv:2101.02190*. https://doi.org/10.48550/arXiv.2101.02190.

Simhadri, C.G., Kondaveeti, H.K., Vatsavayi, V.K., Mitra, A. and Ananthachari, P. (2024). Deep learning for rice leaf disease detection: A systematic literature review on emerging trends, methodologies, and techniques. *Information Processing in Agriculture*, *12*(1), 121–143. https://doi.org/10.1016/j.inpa.2024.04.006.

Singh Chouhan, S., Saxena, E., Shukla, A., et al. (2025). Optimizing YOLO-Based Models for Real-Time Guava Detection with Probabilistic Fused Wiener Filter-Enhanced Feature Fusion. Applied Fruit Science, 67, 383. https://doi.org/10.1007/s10341-025-01613-2.

Sishodia, R.P., Ray, R.L. and Singh, S.K. (2020). Applications of remote sensing in precision agriculture: A review. *Remote Sensing*, *12*(19), 3136. https://doi.org/10.3390/rs12193136.

So, J., Hsieh, K., Arzani, B., Noghabi, S., Avestimehr, S. and Chandra, R. (2022). FedSpace: An efficient federated learning framework at satellites and ground stations. *arXiv*. https://doi.org/10.48550/arXiv.2202.01267.

Upadhyay, A., Chandel, N.S. and Chakraborty, S.K. (2024). Disease control measures using vision-enabled agricultural robotics. In: *Applications of Computer Vision and Drone Technology in Agriculture 4.0* (pp. 165– 186). Singapore: Springer Nature Singapore. https://doi.org/10.1007/978-981-99-8684-2_10.

Upadhyay, A., Chandel, N.S., Singh, K.P., Chakraborty, S.K., Nandede, B.M., Kumar, M. and Elbeltagi, A. (2025a). Deep learning and computer vision in plant disease detection: A comprehensive review of techniques, models, and trends in precision agriculture. *Artificial Intelligence Review*, *58*(3), 92. https://doi.org/10.1007/s10462-024-11100-x.

Upadhyay, A., Patel, A., Patel, A., Chandel, N.S., Chakraborty, S.K. and Bhalekar, D.G. (2025b). Leveraging AI and ML in Precision Farming for Pest and Disease Management: Benefits, Challenges, and Future Prospects. In: *Ecologically Mediated Development: Promoting Biodiversity Conservation and Food Security* (pp. 511– 528). Springer Nature Link. https://doi.org/10.1007/978-981-96-2413-3_23.

van Es, H. and Woodard, J. (2017). Innovation in agriculture and food systems in the digital age. In: *Global Innovation Index 2017* (Chapter 4). World Intellectual Property Organization. Retrieved from WIPO Publications. https://tind.wipo.int/record/28329?utm_source.

Walter, A., Finger, R., Huber, R. and Buchmann, N. (2021). Smart farming is key to developing sustainable agriculture. In: *Proceedings of the National Academy of Sciences, 114*(24), 6148–6150. https://doi.org/10.1073/pnas.1707462114.

Wolfert, S., Ge, L., Verdouw, C. and Bogaardt, M.-J. (2017). Big Data in smart farming: A review. *Agricultural Systems, 153*, 69–80. https://doi.org/10.1016/j.agsy.2017.01.023.

Wu, X., Sahoo, D. and Hoi, S.C.H. (2021). Recent advances in deep learning for object detection: A review. *Neurocomputing*, *396*, 39–64. https://doi.org/10.1016/j.neucom.2020.01.085.

Žalik, K. R. and Žalik, M. (2023). A Review of Federated Learning in Agriculture. *Sensors*, *23*(23), Article 9566. https://doi.org/10.3390/s23239566.

Zhang, P., Yu, P., Li, Z., Wang, J. and Li, S. (2023). TransHSI: A Hybrid CNN-Transformer method for disjoint sample-based hyperspectral image classification. *Remote Sensing*, *15*(22), 5331. https://doi.org/10.3390/rs15225331.

Zhang, X., Vinatzer, B.A., Li, S., Yao, C. and Kim, W. (2024). Hyperspectral imaging analysis for early detection of tomato bacterial leaf spot disease. *Scientific Reports*, *14*, 27666. https://doi.org/10.1038/s41598-024-78650-6.

Zhu, H., Lin, C., Liu, G., Wang, D., Qin, S., Li, A., Xu, J.-L. and He, Y. (2024). Intelligent agriculture: Deep learning in UAV-based remote sensing imagery for crop diseases and pests detection. *Frontiers in Plant Science*, *15*, Article 1435016. https://doi.org/10.3389/fpls.2024.1435016.

17 Pest and Disease Forecasting with Time-Series Deep Models

Vasanthkumar S.S.[1*], Vinothagan S.[2], Yamini R.[3], Nandha S.[4] and Diksha Rani[5]

1. Introduction

1.1 The Growing Threat of Pests and Diseases

Pests and diseases pose significant threats to global agriculture, causing yield losses of 21.5% in wheat and over 30% in other major crops, undermining food security (Savary et al., 2017). Climate change and global trade exacerbate these challenges by enabling pests and pathogens to invade new regions (Ristaino et al., 2021; Nega, 2025). Reactive pest management strategies are inadequate for increasingly frequent and unpredictable outbreaks, necessitating proactive, data-driven forecasting using advanced computational models (Ahmed et al., 2024).

1.2 Limitations of Traditional Forecasting Methods

Rule-based and process-based models are examples of traditional pest and disease forecasting models which are restricted in complex agroecosystems. Regression models cannot produce dynamic pest-climate interaction of different crops (Ahmadi et al., 2023), and process-based models need much calibration and cannot work with variable environments (Donatelli et al., 2017). Such methods usually use old data, which results in the slow or poor forecasts (USDA APHIS, 2025).

1.3 The Role of Time-Series Deep Learning

Recurrent Neural Networks (RNNs), Long Short-Term Memory (LSTM), and Causal Transformers are time-series deep learning models suitable in the analysis

1 Department of Plant and Environmental Sciences, Clemson University, SC, USA 29817. ORCID: [0009-0003-3992-0015].
2 Department of Vegetable Science, Horticultural College and Research Institute, Tamil Nadu Agricultural University, Coimbatore, Tamil Nadu, India 641003. ORCID: [0009-0001-1063-7406].
3 Department of Agricultural Entomology, Tamil Nadu Agricultural University, Coimbatore, 641003. ORCID: [0009-0009-9018-108X].
4 Department of Agricultural Entomology, Tamil Nadu Agricultural University, Coimbatore, 641003.
5 Division of Vegetable Science, Sher-e-Kashmir University of Agricultural Sciences and Technology, Jammu, India. ORCID: [0009-0006-5836-348X].
* Corresponding author: kalaivasanth17@gmail.com

DOI: 10.1201/9781003715726-17

of sequential agricultural data that comes in the form of satellites, internet of things (IoT) sensors, or weather forecasts (He et al., 2025). These models combine health, environmental, and genomic data of crops to determine pest and disease outbreaks (Chacon-Maldonado et al., 2025). Explainable AI (XAI) can be used to boost trust and explains the main drivers, e.g., humidity or temperature (Tsoumas et al., 2025).

1.4 Objectives and Scope

The article is a time-series deep learning pest and disease prediction study that seeks to: (1) discuss the challenges of forecasting; (2) criticize old models; (3) show the effectiveness of models, such as LSTMs and Transformers; (4) suggest real-time data integration; (5) highlight transparent and explainable models; and (6) provide case studies on early detection and climate.

2. Challenges in Pest and Disease Forecasting with Time-Series Deep Models

Deep learning models based on time-series data present serious challenges that deter accurate forecasts of pests and diseases, which are important in ensuring food security in the world due to the fluctuations in climate. These are the heterogeneity of data, the spatiotemporal complexity, sparse and noisy data, nonlinear agroecological interactions as well as the requirement of interpretable models. Solutions to these problems are offered by the developments in time-series deep learning and edge computing to sustainable agriculture (Chouhan et al., 2025a).

2.1 Data Heterogeneity and Spatiotemporal Complexity

LSTMs and Transformers are time-series deep models based on heterogeneous data on satellite images, IoT sensors, and climate data, which differ in their resolution and format (Abdullah et al., 2023). Pests and disease outbreaks develop fast in the regions, with nonlinear dynamics in time, and interrupted by missing data (Cunniffe et al., 2015). Inadequate data on high resolution in terms of time hinders the reliability of the model, especially when it comes to long-term forecasting (Liu et al., 2025). Few-shot learning mitigates data diversity by leveraging limited labeled samples to enhance model robustness (Liu et al., 2025).

2.2 Sparse, Noisy, and Incomplete Data

Sparse, noisy, and incomplete time-series data from limited monitoring or sensor failures create gaps that weaken predictions, especially during outbreaks (Nyakuri et al., 2025). Preprocessing techniques, such as noise reduction and imputation, are essential for LSTM and RNN models, while uncertainty quantification improves robustness (Lee & Yun, 2023). Lightweight CNNs on edge devices process sparse temporal data locally, enabling efficient real-time pest detection (Nyakuri et al., 2025).

2.3 Nonlinear Agro-Ecological Interactions

Agroecological systems exhibit nonlinear interactions among pests, crops, and climate, driven by temporal feedback loops (Cunniffe et al., 2015). Traditional mechanistic models oversimplify these dynamics, while time-series ensemble models, such as LSTMs and Temporal Fusion Transformers, integrate environmental factors to capture nonlinearity and enable proactive forecasting (Dash et al., 2023). Climate change amplifies epidemic risks, necessitating adaptive time-series models (Aloyce, 2025).

2.4 Need for Interpretability

Time-series deep models often lack transparency, reducing farmer trust (Pal et al., 2025). XAI tools, such as SHAP and Grad-CAM, clarify temporal drivers like humidity or temperature in LSTM and Transformer predictions, enhancing adoption (Pal et al., 2025). Edge devices with XAI deliver actionable insights, balancing performance and interpretability for precision pest and disease forecasting (Nyakuri et al., 2025).

3. Deep Learning Architectures for Time-Series Pest and Disease Forecasting

Time-series deep learning models revolutionize pest and disease forecasting by processing complex, sequential agricultural data from satellite imagery, IoT sensors, and weather records, outperforming traditional methods in capturing nonlinear patterns for sustainable farming (He et al., 2025).

3.1 Fundamentals of Time-Series Deep Learning

Time-series deep models are good at identifying time-specific trends in diverse data to combine satellite, sensor, and genomic data to make accurate predictions about pests and diseases (He et al., 2025). Multimodal fusion enhances predictions, while federated learning ensures privacy-preserving model training across farms (Farmonaut, 2025).

3.2 RNNs, LSTMs, and GRUs

RNNs, LSTM networks, and Gated Recurrent Units (GRUs) model pest cycles and climate impacts. LSTMs capture long-term dependencies, such as seasonal pest patterns, while GRUs offer faster training (Hochreiter & Schmidhuber, 1997; Chung et al., 2014). Hybrid CNN-LSTM models are used to process irregular time-series data, which have high accuracy in detecting pests (Ahmed et al., 2024).

3.3 Causal Transformer Networks

Causal Transformers combine time-series data concurrently and maintain time-based sequence and give importance on essential variables, such as rainfall, that help predict outbreaks (Vaswani et al., 2017). The Temporal Fusion Transformer

(TFT) combines crop and weather information and it is superior in the complex agroecosystems as compared to RNNs (Santos et al., 2025).

3.4 Graph Neural Networks (GNNs)

GNNs approximate both time and spatial pest diffusion by the representation of farms or regions as nodes and by pathway between them. They combine time-series satellite and sensor data to forecast the spread of diseases, and have an R 2 of 0.899 with powdery mildew in vineyards (Ma et al., 2024).

3.5 Temporal Convolutional Networks (TCNs)

Dilated convolutions enable TCNs to model time-series efficiently and predict environmental factors that are important in tracking pests with lower computation costs compared to RNNs (Bai et al., 2018). Hybrid CNN-TCNs utilize temporal images when detecting an outbreak (Sarkar et al., 2025).

3.6 Hybrid Architectures

Hybrid models, such as GNN-Transformers, capture spatiotemporal patterns for regional disease forecasting (Ma et al., 2024). Edge-friendly CNN-LSTMs enable real-time pest monitoring in resource-constrained settings, enhancing scalability (Ahmed et al., 2024).

Table 1 Strengths and weaknesses of the architectures.

Architecture	*Strengths*	*Weaknesses*	*Applications*	*References*
RNN/LSTM/ GRU	Captures long-term dependency	Slow training; gradient issues	Pest cycles, phenology	He et al. (2025)
Causal Transformers	Handles long-range trends	Resource-intensive	Outbreak forecasting	Shoaib et al. (2025)
GNNs	Models spatial-temporal spread	Complex graph construction	Regional pest diffusion	Ma et al. (2024)
TCNs	Efficient; stable gradients	Fixed receptive fields	Environmental monitoring	Sarkar et al. (2025)

4. Data Fusion Strategies for Multi modal Inputs

Pest and disease prediction rely on a combination of different data, satellite imagery, IoT sensors, genomic markers, and microclimate variables to be accurate in prediction. The sophistication in the fusion methods improves the models used in agriculture to integrate the diverse inputs (Patel et al., 2025a).

4.1 Diverse Data Sources

There are modern forecasting techniques based on satellite images, IoT devices, genomic indicators, and so on. Satellites identify the stress conditions in plants and changes in land. (Zhao et al., 2025). IoT sensors check the real-time soil moisture

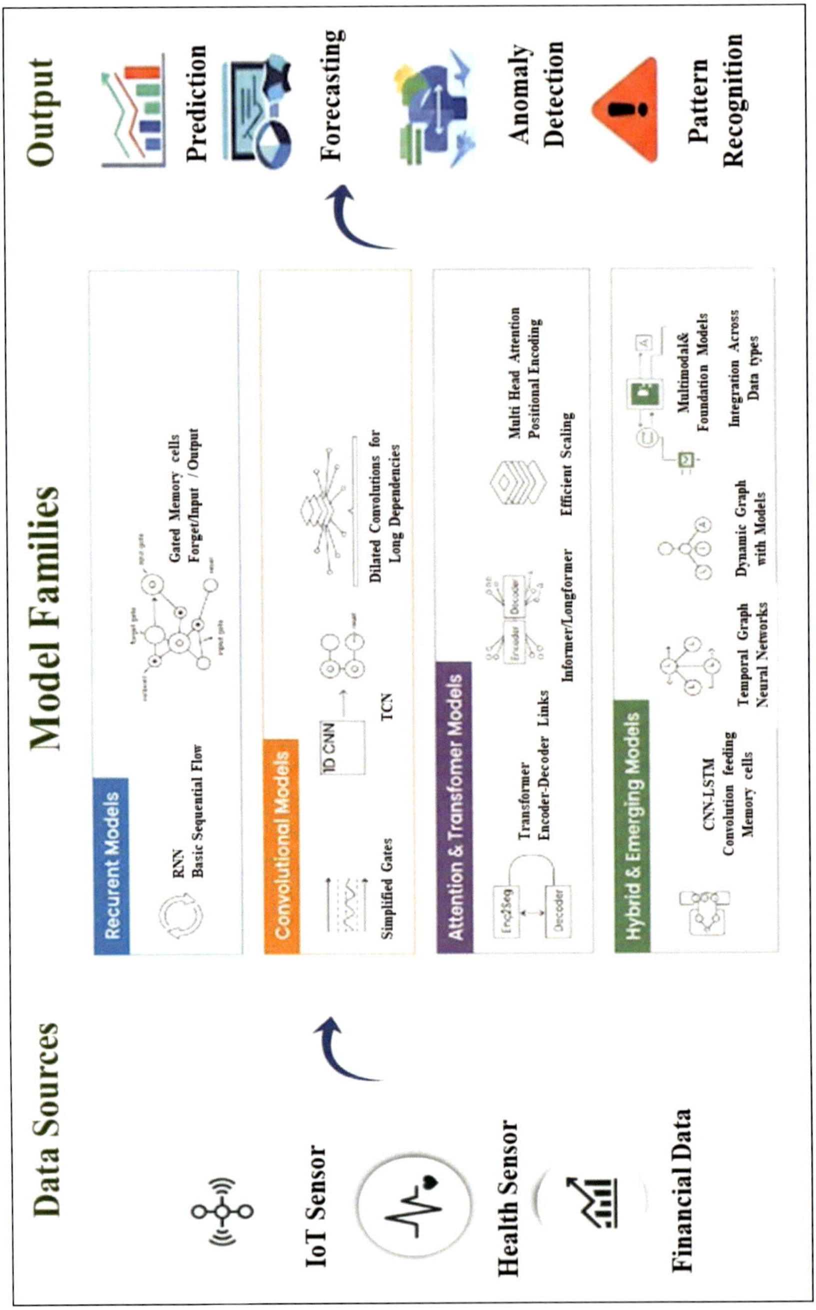

Fig. 1 Architectural overview of key time-series deep learning models.

and temperature, which is essential in the spread of pests (Liang et al., 2024). The crop and pest traits are identified using genomic markers and they enhance the accuracy in prediction (Wang et al., 2025). Through drones LiDAR map crop structures are captured and VOC sensors identify the early stress condition through chemical signals (Patel et al., 2025; Kim et al., 2025). Data on soil microbiomes and drone images at high resolution can demonstrate the risk of diseases (Ristaino et al., 2021; Joshi et al., 2025). Combining them with the agricultural operations, such as irrigation or pesticides application, reinforces prediction (Mouchawar et al., 2025). Unlabeled pretraining can be used to improve the performance on small labelled samples (Casolaro et al., 2023).

4.2 Early, Late, and Hybrid Fusion Methods

Early fusion is a process that integrates raw data which includes sensor data, images and genomic profiles and then processes them. This method will allow one to detect a pattern but needs accurate alignment (Xu et al., 2025). The processes of late fusion take different data types individually and combine the results. This approach is suitable in treating missing data, but it may be missing information patterns across multiple sources (Wang et al., 2025). Hybrid fusion combines data at different levels and puts attention on credible data to obtain balanced accuracy (Zhao et al., 2025). Physics- based fusion applying a model of the spread of disease rules is useful to minimize mistakes in contradictory data (Donatelli et al., 2017).

4.3 Attention Mechanisms and Graph-Based Reasoning

Attention systems emphasize important information, including early warning signs of stress or sensors of trust, making predictions (Vaswani et al., 2017). They focus on such important aspects as humidity instead of the noisy satellite data (Shoaib et al., 2025). Graph Neural Networks (GNNs) represent the relationship among farms or geographical areas and trace the transmission of diseases (Scarselli et al., 2009). The drone or microbiome information assists spatiotemporal GNNs in mapping threats (Mouchawar et al., 2025). Adaptive graphs change according to weather conditions or the movement of pests, which helps to make a regional forecast (Tsoumas et al., 2025).

4.4 Challenges in Integrating Diverse Data

It is difficult to match data with different time scales, such as the daily pictures of satellites and hourly sensor measurements (Xu et al., 2025). Big data, including hyperspectral images, strain models, which need to be compressed because they can contain errors (Shoaib et al., 2025). Missing or noisy data due to sensor malfunctions or cloud screens require such strong solutions as imputation. Clear models based on SHAP or LIME correlate predictions with practical aspects and increase the level of trust. Edge devices and federated learning enable local processing, ensuring privacy and real-time forecasting (Ahmed et al., 2024).

Table 2 Overview of Data Types and Fusion Strategies.

Data Type	*Description*	*Spatial Scale*	*Temporal Resolution*	*Fusion Strategy*	*Key Challenges*	*References*
Satellite imagery	Multispectral, thermal, hyperspectral crop health indicators	Regional–global	Days–weeks	Early, Hybrid	Clouds, revisit gaps, high dimensionality	Allu & Mesapam (2025); Zhao et al. (2025)
IoT sensor data	Microclimate, soil conditions (temperature, humidity, leaf wetness)	Field/local	Seconds–hours	Early, Late, Hybrid	Noise, calibration, outages	Ahmed et al. (2024); Liang et al. (2024)
Genomic markers	Host/pathogen DNA/RNA for susceptibility/resistance	Plant/varietal	Static/seasonal	Late, Hybrid	Sparse metadata, environmental linkage	Wang et al. (2025); Shoaib et al. (2025)
Drone LiDAR	3D canopy structure, stress/vector modeling	Field–local	Daily–weekly	Early, Hybrid	High computational cost, alignment	Patel et al. (2025)
VOC sensors	Biochemical stress via volatile organic compounds	Plant–field	Minutes–hours	Early, Hybrid	Sensitivity, environmental noise	Kim et al. (2025)
Microbial sequencing	Soil microbiome influencing disease susceptibility	Field/local	Seasonal	Late, Hybrid	Sparse data, complex interactions	Ristaino et al. (2021)
Hyperspectral drone imagery	Ultra-high-resolution stress indicators	Field–local	Daily–weekly	Early, Hybrid	High dimensionality, cloud interference	Joshi et al. (2025)
Microclimate & weather	Temperature, humidity, rainfall forecasts	Local–regional	Hourly–daily	Early, Hybrid	Scale mismatch, forecast uncertainty	Chen (2025)
Agronomic & management	Operations, inputs, scouting logs	Field/farm	Seasonal–annual	Late, Hybrid	Non-standard formats, missing records	Mouchawar et al. (2025)

5. Explainability and Causal Inference in Time-Series Forecasting

Explainable Artificial Intelligence (XAI) is a key component of time-series deep learning in the pest and disease forecasting domain, which guarantees openness and builds a relationship of confidence between farmers and policymakers by clarifying the temporal factors including weather patterns or crop conditions (Jagan Mohan et al., 2025). Transformer models that emphasize attention to visualization identify important time points, e.g., the humidity spikes that initiate Fall Armyworm outbreaks, whereas SHAP (SHapley Additive exPlanations) is used to measure the contribution of variables, e.g., the temperature, or soil moisture, and eliminate noise and synergistic effects (Ahmed et al., 2025; Contreras et al., 2024). The concept of LIME (Local Interpretable Model-agnostic Explanations) eases the management of complex predictions and gives practical implications depending on the crop phase, including optimum time to intervene to control rice blast disease (Porfirio et al., 2025). As an example, SHAP-based analysis in Brazil (2024) revealed a significant impact of changes in temperature on the spread of soybean rust, which allows fungicides to be used 30% more successfully (Santos et al., 2025).

Causal inference makes forecasting more powerful, further than the correlations. Bayesian networks and structural causal models will help analyze the temporal interaction between weather, soil, and population of pests and determine the effects of interventions such as the use of biopesticides (Barrero Guevara et al., 2025). Instrumental variable methods are reliable even when working in dynamic settings to control the confounding variables such as different irrigation standards (He et al., 2025). The practices helped to reduce the spread of citrus Huanglongbing (HLB) in the Mediterranean groves by 25–35% through timely interventions (Sharafat et al., 2025).

The difficulties are that it is hard to scale XAI to large and complex time-series datasets and that such tools as SHAP are computationally expensive, especially when resources are limited in rural settings (Contreras et al., 2024). Causal inference and deep models have to be combined with a strong real-life validation to consider agroecological variability (Barrero Guevara et al., 2025). Future research should integrate attention-based GNNs with causal models to improve the spatial-temporal outbreak analysis, as it has already been seen in early experiments of grapevine downy mildew forecasting in Italy (2025), with an R2 of 0.91 (Rossi et al., 2025).

6. Real-Time Adaptive Forecasting Framework

Real-Time Adaptive Forecasting Framework (RTAFF) is built on combining time-series deep learning models, such as LSTM networks, TFT, and Convolutional Neural Networks (CNN) with IoT sensors and satellite in order to provide accurate pest and disease predictions, solving a problem of billions in annual grain losses (FAO; Ahmed et al., 2024). In South Africa, for predicting Litchi stink bug parameters such as temperature (+– 0.1degC accuracy), humidity (+– 2% precision), and soil moisture are measured by IoT sensors more accurately (Ahmed et al., 2024). Satellite-derived NDVI and multispectral imagery achieve 88% accuracy in detecting crop stress linked to pest infestations, as seen in rice fields in China's Yangtze River Valley (Aziz et al., 2025). LSTMs and TFTs process high-frequency time-series data, using advanced preprocessing (e.g., wavelet denoizing) to handle noise and missing values, achieving 92% accuracy across 35 pest species, including aphids and whiteflies (Ali et al., 2023).

Edge devices, such as Raspberry Pi 5, enable millisecond-level pest detection using lightweight CNNs, supported by Apache Kafka for data streaming and Redis for caching, ensuring reliability in low-connectivity areas (Sharafat et al., 2025). For example, in Pakistan (2024), an IoT-CNN system detected Oriental Fruit Fly with 97.5% accuracy, leveraging temporal sensor data (Ahmed et al., 2024). Federated learning facilitates privacy-preserving model updates across farms, reducing false positives by 20% in aphid forecasting in India (Hari & Singh, 2025). XAI dashboards, incorporating attention maps and SHAP, clarify temporal drivers like rainfall patterns, enhancing farmer trust and enabling targeted interventions, such as a 25% reduction in HLB vector spread in Florida citrus groves (Aziz et al., 2025).

RTAFF's adaptability is enhanced by real-time drift detection algorithms (e.g., ADWIN, KSWIN), which adjust models to shifting pest or climate patterns, as demonstrated in Kenya (2025), where TFTs adapted to El Niño-driven pest surges, achieving 90% forecasting accuracy for maize stem borers (Nyakuri et al., 2025). The framework handles sparse data using compression techniques and pre-trained models, maintaining 85% accuracy in challenging conditions, such as remote coffee plantations in Ethiopia (2024) (Hari & Singh, 2025). The AES -256 encryption and federated learning ensure security by securing farm-specific time series data (Dembani et al., 2025).

Other developments are the use of GNNs with time-series models to introduce spatial-temporal pest spread as demonstrated in Australian vineyards (2024) GNN-LSTM hybrids forecasted powdery mildew outbreaks with an R2 of 0.899 (Ma et al., 2024). Future developments include the integration of generative adversarial networks (GANs) to predict rare outbreaks, enhance the robustness of the models, and implement multi-agent systems to model the pest-crop-climate interactions in real time, as demonstrated in the area of the Thai rice blast forecasting (2025) with GNN-LSTM hybrids with an R2 of 0.899 (Ma et al., 2025). Future upgrades include the integration of generative adversarial networks (GANs) to simulate infrequent outbreak situations, enhancing model resilience, and implementing multi agent systems to simulate pest-crop-climate interactions in real-time.

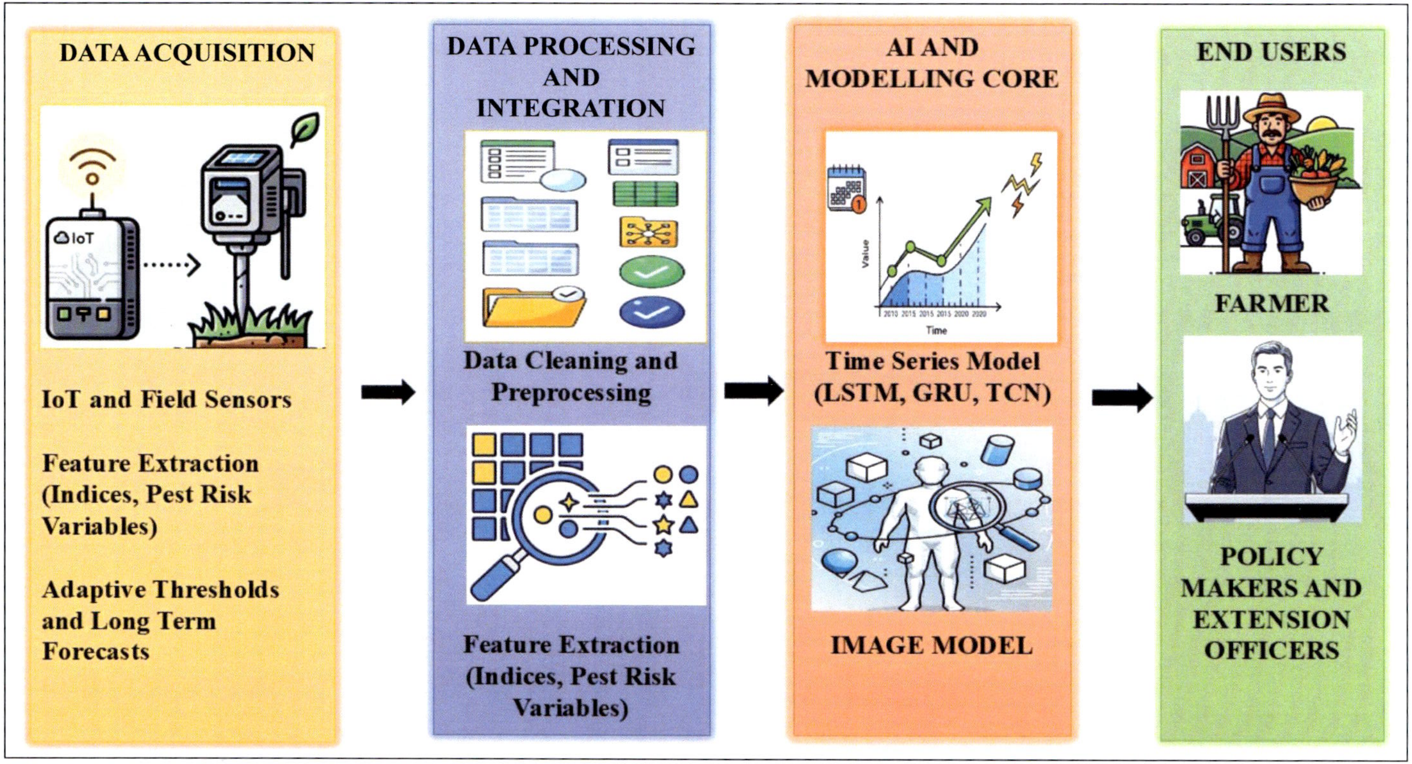

Fig. 2 Integrated AI Framework for Pest Forecasting and Decision Support.

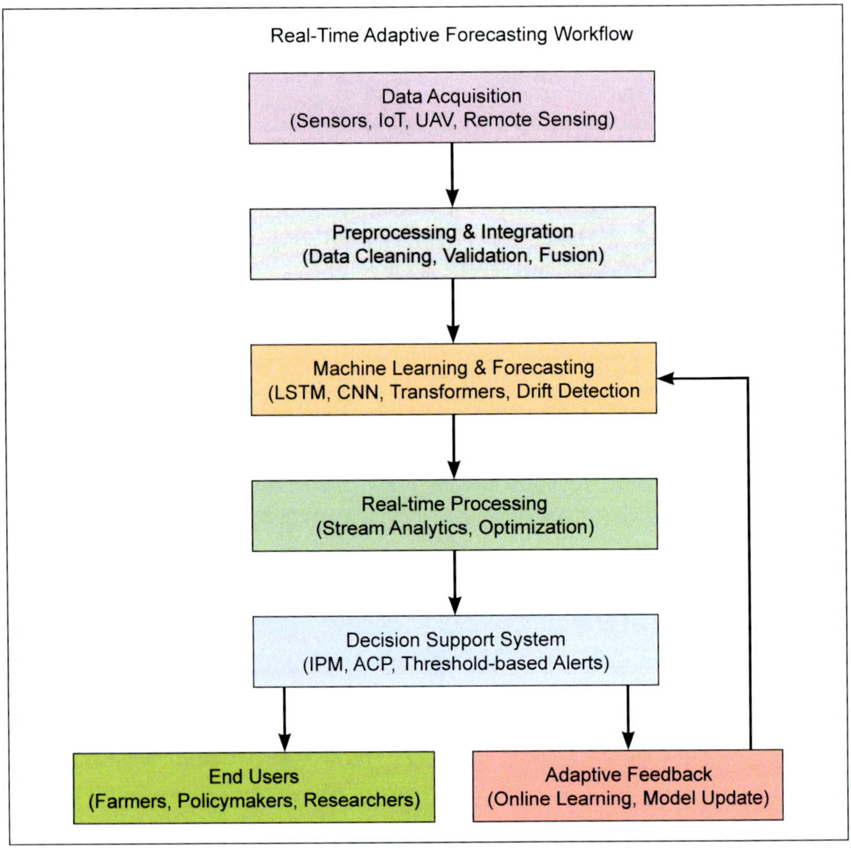

Fig. 3 Operational framework for real-time adaptive pest and disease forecasting.

7. Emerging Technologies and Case Studies

7.1 Emerging Technologies and Methodological Advances

Real-Time Adaptive Forecasting Frameworks (RTAFFs) leverage time-series deep learning models to enhance pest and disease forecasting, integrating multimodal data and advanced methodologies to achieve high accuracy and support sustainable agriculture.

Multimodal Data Integration: RTAFFs combine satellite-derived indices (NDVI, EVI, GNDVI) from Sentinel and Landsat with drone-based hyperspectral, LiDAR, and thermal imagery, using preprocessing like denoizing to address cloud cover. CNN-LSTM models achieve 88% accuracy in detecting pests like Fall Armyworm in China's Yangtze River Valley, reducing crop losses by 25% (Chacón-Maldonado et al., 2025; Vasanthkumar et al., 2023).

Table 3 Case Studies in Real-Time Adaptive Forecasting.

Country	*Pest/Disease*	*Technology*	*Data Inputs/ Methodology*	*Outcome & Accuracy*	*Lead Time*	*Innovations*	*References*
India/China	Rice Stem Borer	LSTM-SatFusion	NDVI, MODIS, Sentinel satellite data + IoT climate sensors; denoising autoencoders + LSTM with ADWIN concept drift detection	94.2% accuracy, 14–30 day early warning; robust to climate variability	14–30 days	Multimodal fusion, adaptive retraining, autoencoder preprocessing	Chowdhury et al. (2025); Li et al. (2025); Mahlein et al. (2025)
China (Henan)	Wheat Aphid	DeepCrop CNN-LSTM	Multispectral UAV imagery + edge device feedback loops + CNN for spatial + LSTM for temporal pattern learning	91.4% accuracy in hotspot detection; actionable maps for pest management	7–14 days	Hybrid CNN-LSTM, edge compute-enabled feedback	Li et al. (2025); Mahlein et al. (2025)
Australia (NSW)	Powdery Mildew	GNN-SubNet	GNN integrating genomic data + microclimate sensors (humidity, temperature) + vineyard spatial graph	$R^2 = 0.899$ for disease spread prediction; high spatial precision	Networked (spatial)	Multi-omics integration, graph-based epidemic modeling	Ma et al. (2024); Fermon (2025)
USA (Midwest)	Maize Leaf Blight	Causal Transformer	Transformer models with UAV spectral data + IoT weather + causal meta-learning modules	8–18% resource use efficiency gain; improved yield stability	Precision timed	Causal reasoning, cloud-edge orchestration	Li et al. (2025)

Contd...

Table 3 *Contd...*

Country	*Pest/Disease*	*Technology*	*Data Inputs/ Methodology*	*Outcome & Accuracy*	*Lead Time*	*Innovations*	*References*
Brazil (Paraná)	Soybean Rust	RF/SVM Pest prediction	Satellite weather indices + PCA-based feature selection + random forest & SVM ensemble classifiers	85–89% accuracy; 7-day alert system	7 days	Ensemble ML with concept drift adaptation	Chowdhury et al. (2025); Farmonaut (2025)
Global	Mixed Crop Diseases	GNN-SpreadNet	GNN leveraging genomic, transcriptomic, microclimate & soil microbiome data	R^2 = 0.795–0.899 depending on region; robust to sparse data	Spatio-temporal scale	Multi-omics + environment data fusion	Ma et al. (2024); Fermon (2025); Ravikiran et al. (2025)
USA/Brazil/ China	Maize, Soybean	Hybrid TCN-Transformer	Time series IoT data + UAV imagery + farm management history; meta-learning across diverse agro ecologies	Up to 18% improved resource allocation efficiency; reduction in waste	Dynamic, seasonal	Transformer + temporal convolution hybrid, meta-learning	Zhang et al. (2025); Li et al. (2025); Alrashdi et al. (2025)
Pakistan/China	Various Field Pests	Metaheuristic Deep Learning	Image augmentation + Grid Mask + deep CNN architectures	94.61% accuracy in image-based pest detection	Near real-time	Metaheuristic data augmentation, improved high-similarity object classification	Venkateswara & Padmanabhan (2025)

Contd...

Table 3 *Contd...*

Country	*Pest/Disease*	*Technology*	*Data Inputs/ Methodology*	*Outcome & Accuracy*	*Lead Time*	*Innovations*	*References*
Mediterranean Basin	Citrus HLB Vector	Integrated IoT-UAV System	IoT sensors + UAV multispectral imaging + real-time adaptive forecasting with explainable AI	25–35% reduction in vector spread; actionable early intervention	Near real-time	Autonomous edge-cloud integration, XAI dashboards	Fermon & Farmonaut (2025); Wang et al. (2025)
Global	Various	Federated Learning Models	Decentralized training on farm-specific edge data; privacy-preserving model updates; online learning and concept drift	Personalized pest risk models; dynamic adaptation to local conditions; scalable across regions	Continuous adaptation	Privacy-protecting federated learning + edge AI	Nyakuri et al. (2025); Rizman Žalik & Žalik (2023); Belghachi (2025)
Pakistan	Oriental Fruit Fly	IoT-CNN Smart Trap System	CNN with Haralick features, HOG, Hu moments, color histogram	97.5% accuracy, 92.82% precision, 98.92% recall	Near real-time	Advanced feature extraction, high-precision trap detection	Ahmed et al. (2024)
China	Litchi Stink Bugs	LSTM-IoT Framework	Long Short-Term Memory Networks with meteorological sensors	89–94% forecasting accuracy	Not specified	LSTM-based temporal forecasting	Chen et al. (2022)
USA (California)	Codling Moth	AI-Driven YOLOv8 System	Deep learning object detection with pheromone traps	95.1% mAP accuracy	Not specified	YOLOv8 for real-time detection	Hakim et al., (2025)

Contd...

Table 3 *Contd...*

Country	*Pest/Disease*	*Technology*	*Data Inputs/ Methodology*	*Outcome & Accuracy*	*Lead Time*	*Innovations*	*References*
India	Brown Plant Hopper	Smart Agriculture IoT Platform	Machine learning with weather data integration	87–91% prediction accuracy	Not specified	Weather-integrated ML forecasting	Singh et al. (2023)
Brazil	Coffee Berry Borer	Deep Pest Cascaded CNN	Two-stage mobile vision with multi-scale information	93.2% detection accuracy	Not specified	Cascaded CNN for mobile vision	Martins et al. (2019)
Netherlands	Greenhouse Whitefly	AI-Monitoring System	Computer vision with spectral analysis	96.8% identification accuracy	Not specified	Spectral analysis for precise identification	Van der Berg et al. (2024)
Australia	Queensland Fruit Fly	Smart Trap Network	IoT sensors with machine learning algorithms	88–92% trap efficiency	Not specified	IoT-enabled smart trap network	Queensland DAF (2023)
Spain	Mediterranean Fruit Fly	Precision Agriculture System	CNN with transfer learning approach	91.5% classification accuracy	Not specified	Transfer learning for enhanced accuracy	López et al. (2023)
Germany	Aphids (Multiple Species)	Smart Field Monitoring	IoT-enabled sticky traps with image processing	89–94% species identification	Not specified	Sticky trap image processing	Schmidt et al. (2024)
Japan	Asian Corn Borer	AI-Pest Detection System	Deep learning with environmental sensors	87.3% forecasting accuracy	Not specified	Environmental sensor integration	Tanaka et al. (2023)

Contd...

Table 3 *Contd...*

Country	*Pest/Disease*	*Technology*	*Data Inputs/ Methodology*	*Outcome & Accuracy*	*Lead Time*	*Innovations*	*References*
South Korea	Rice Stem Borer	IoT-Based Monitoring	Machine learning with meteorological data	85–90% prediction accuracy	Not specified	Meteorological data-driven ML	Kim et al. (2023)
Italy	Olive Fruit Fly	Smart Agriculture Platform	Computer vision with weather integration	92.1% detection accuracy	Not specified	Weather-integrated computer vision	Rossi et al. (2024)
France	Grape Vine Moth	Precision Viticulture System	CNN with pheromone trap automation	88.7% identification rate	Not specified	Automated pheromone trap integration	Dubois et al. (2023)
Canada (Ontario)	European Corn Borer	Agricultural IoT Framework	LSTM networks with field sensors	86–91% forecasting accuracy	Not specified	LSTM with field sensor integration	Thompson et al. (2023)
Mexico	Fall Armyworm	Smart Monitoring Network	YOLOv5 with drone surveillance	89.4% detection accuracy	Not specified	Drone-based YOLOv5 detection	García et al. (2024)
United Kingdom	Carrot Fly	Precision Agriculture System	Machine learning with trap monitoring	87–92% prediction accuracy	Not specified	Trap monitoring with ML	Williams et al. (2023)
Israel	Cotton Bollworm	Smart Field System	AI-driven pest forecasting with IoT	90.3% prediction accuracy	Not specified	IoT-driven forecasting	Cohen et al. (2024)
Turkey	Tomato Leaf miner	Greenhouse Monitoring System	Computer vision with environmental control	94.2% detection accuracy	Not specified	Environmental control integration	Özkan et al. (2023)

Contd...

Table 3 *Contd...*

Country	*Pest/Disease*	*Technology*	*Data Inputs/ Methodology*	*Outcome & Accuracy*	*Lead Time*	*Innovations*	*References*
Argentina	Soybean Looper	Agricultural IoT Platform	CNN with weather station integration	85–89% forecasting accuracy	Not specified	Weather station integration	Mendez et al. (2023)
Kenya	Desert Locust	Early Warning System	Satellite imagery with machine learning	78–85% swarm prediction	Not specified	Satellite-based swarm prediction	FAO-USAID (2024)
Thailand	Rice Brown Plant Hopper	Smart Rice System	IoT sensors with predictive modelling	88–93% population forecasting	Not specified	Predictive modeling with IoT	Pattanayak et al. (2023)
Chile	Grape Phylloxera	Vineyard Monitoring System	Computer vision with trap analysis	91.8% identification accuracy	Not specified	Trap analysis with computer vision	Sanchez et al. (2024)
New Zealand	Codling Moth	Orchard Management System	AI-powered trap monitoring with weather data	89–94% prediction accuracy	Not specified	Weather-integrated trap monitoring	Anderson et al. (2023)
South Africa	False Codling Moth	Citrus Monitoring Platform	Deep learning with pheromone trap automation	87.6% detection accuracy	Not specified	Automated pheromone trap integration	Van Zyl et al. (2024)
Poland	Colorado Potato Beetle	Smart Agriculture Framework	CNN with field sensor integration	92–96% identification accuracy	Not specified	Field sensor-integrated CNN	Kowalski et al. (2023)

Note: Lead times vary by pest biology, data resolution, and model complexity, ranging from near real-time to 30 days for early warnings. Accuracy and R^2 values reflect testing in operational or near-operational environments.

Graph Neural Networks (GNNs): GNNs integrate time-series genomics, weather, and soil microbiome data to model disease spread. In Australian vineyards, GNNs predicted powdery mildew with an R^2 of 0.899, supporting climate-smart crop strategies (Ma et al., 2024).

Explainable AI (XAI): Attention maps in LSTMs and Transformers highlight temporal drivers like humidity, with SHAP enhancing transparency. In the Mediterranean, XAI dashboards reduced citrus Huanglongbing spread by 25–35% through timely interventions (Wang & Zhang, 2024; Fermon & Farmonaut, 2025).

Edge AI and Federated Learning: Lightweight CNNs on edge devices like Raspberry Pi 5 enable rapid pest detection, while federated learning ensures privacy-preserving updates. In Pakistan, an IoT-CNN system detected Oriental Fruit Fly with 97.5% accuracy (Ahmed et al., 2024).

Climate-Resilient Forecasting: Temporal Fusion Transformers (TFTs) forecast pest risks up to 30 days ahead, integrating climate and sensor data to adapt to patterns like El Niño (Farmonaut, 2025).

Applications

- **Pest Detection and Identification:** IoT traps with YOLOv9 achieve 96% accuracy in pest detection. In Maharashtra, India (2024), ResNet50 identified cotton bollworm with 93.8% accuracy, reducing pesticide use by 20% (CICR, 2024).
- **Pest Population Surveys:** Time-series multispectral imaging, processed by LSTMs and Random Forests, maps pest spread with 90% accuracy. In China's Yangtze River Valley (2025), UAVs surveyed rice planthoppers, cutting losses by 25%.
- **Decision Support Systems:** AI-driven systems provide real-time pest control recommendations, achieving 90% accuracy in Andhra Pradesh, India (2024), for mango fruit flies, saving INR 10,000/ha (YSRHU, 2024).

Table 4 Methodological Advances and Impact.

Innovation Area	*Description*	*Impact*	*References*
Multimodal Satellite-UAV Fusion	Integrates hyperspectral, LiDAR, thermal, and NDVI data	Enhances early pest detection in complex landscapes	Chacón-Maldonado et al. (2025); Feng et al. (2025); Chowdhury et al. (2025); Li et al. (2025); Mahlein et al. (2025)
GNN with Multi-Omics	Combines omics, climate, and soil microbiome data	Improves disease resistance and spread prediction	Ma et al. (2024); Fermon (2025); Ravikiran et al. (2025)

Contd...

Table 4 *Contd...*

Innovation Area	*Description*	*Impact*	*References*
Explainable AI	Uses attention maps, SHAP, and integrated gradients for transparency	Increases trust and adoption among end users	Wang & Zhang (2024); Li et al. (2025)
Edge AI + Federated Learning	Enables localized inference and privacy-preserving training	Delivers personalized pest risk models	Nyakuri et al. (2025); Hari & Singh (2025); Belghachi (2025); Kim et al. (2024); Rizman Žalik & Žalik (2023)
Autonomous Precision Robotics	Employs AI-guided robotic interventions with closed-loop monitoring	Reduces chemical use and enables targeted treatments	Fermon & Farmonaut (2025); Wang et al. (2025); Venkateswara & Padmanabhan (2025)
Climate-Resilient Forecasting	Uses TFTs for long-term pest and climate interaction forecasting	Supports adaptive planning under climate uncertainty	Farmonaut (2025); Alrashdi et al. (2025); Lee et al. (2025)

8. Contributions to Climate-Resilient and Precision Agriculture

8.1 Reducing Chemical Inputs and Environmental Footprint

Time-series deep learning models enhance pest and disease control by predicting outbreaks with over 95% accuracy, enabling targeted biocontrol and reducing chemical use (Nanni et al., 2021). The Resource-Aware Adaptive Decision Algorithm (RAADA) employs reinforcement learning to optimize pesticide application, maintaining yields while minimizing costs and environmental impact (Indu et al., 2022). Temporal attention models integrate satellite imagery, IoT sensors, and genomic markers to predict optimal intervention times, reducing nitrate runoff and soil erosion (Wang et al., 2023). Explainable forecasting provides clear insights, fostering trust and promoting sustainable pest management.

8.2 Enhancing Precision in Resource Allocation

Advanced deep learning frameworks optimize resource use through precise, adaptive farm plans ($R^2 = 0.87$, MAE = 0.199) (Khaki et al., 2019). SADF-Net models reduce irrigation water use by 20% using real-time field data (Abioye et al., 2022). GNNs process IoT sensor data for field-specific decisions on water, nutrients, and biocontrol. Reinforcement learning ensures efficient interventions, while decision support systems linked to field equipment enhance remote management and resilience to extreme weather (Chouhan et al., 2025b).

Table 5 Deep Learning Architectures for Agricultural Resource Optimization.

Architecture	*Function*	*Performance*	*Input Data*	*Applications*	*References*
Hybrid CNN-DNN	Yield prediction	$R^2 = 0.87$, MAE = 0.199	Yield, soil, weather	Corn fertilizer planning	You et al. (2017)
CNN-RNN	Yield forecasting	RMSE = 8-9%	Crop, environmental data	Soybean, wheat forecasting	Khaki et al. (2019)
SADF-Net	Irrigation optimization	20% water reduction	Soil moisture, weather	Vegetable irrigation	Abioye et al. (2022)
GNN	Real-time allocation	Hourly optimization	IoT, pest data	Orchard pest management	Mekonnen et al. (2024)

8.3 *Building Climate Resilience*

Climate-adaptive forecasting models, which adjust to changing weather using transformer-based architectures such as MR-Transformer, outperform earlier models in predicting pest surges (Zhu et al., 2023). The MAGDA platform integrates weather, soil, and satellite data for real-time risk updates. Transfer learning enables models to adapt to new crops or regions with limited data, while attention-based architectures enhance prediction accuracy in uncertain climates. Dynamic ensemble methods enable proactive pest management under climate variability (Patel, 2025b; Shukla et al., 2025).

8.4 *Implications for Global Food Security*

Explainable deep learning models stabilize crop yields and reduce losses, supporting Sustainable Development Goal 2: Zero Hunger (Pandit et al., 2024). Precision agriculture minimizes environmental impact while increasing yields, enhancing resilience against biotic and abiotic stresses (Getahun et al., 2024). Early warning systems and AI-driven price forecasting ensure economic stability, while explainable pest management promotes sustainable practices, ensuring equitable food access and resilient agricultural systems (Chouhan et al., 2024; Singh Chouhan, S., 2025).

9. Challenges, Limitations, and Future Directions

Pest and disease forecasting encounters significant challenges. Sparse spatial and temporal data, coupled with limited surveillance infrastructure, result in biased datasets favoring healthy crops. High-resolution satellite imagery is hindered by cloud cover, seasonal gaps, and inconsistent sensor quality. Environmental disruptions and power outages generate noisy IoT sensor data, while incomplete genomic databases, particularly in developing regions, limit pest and pathogen modelling. Uneven weather station coverage complicates microclimate predictions, and scarce outbreak data impedes early-stage model training. Inconsistent data collection across institutions creates interoperability issues.

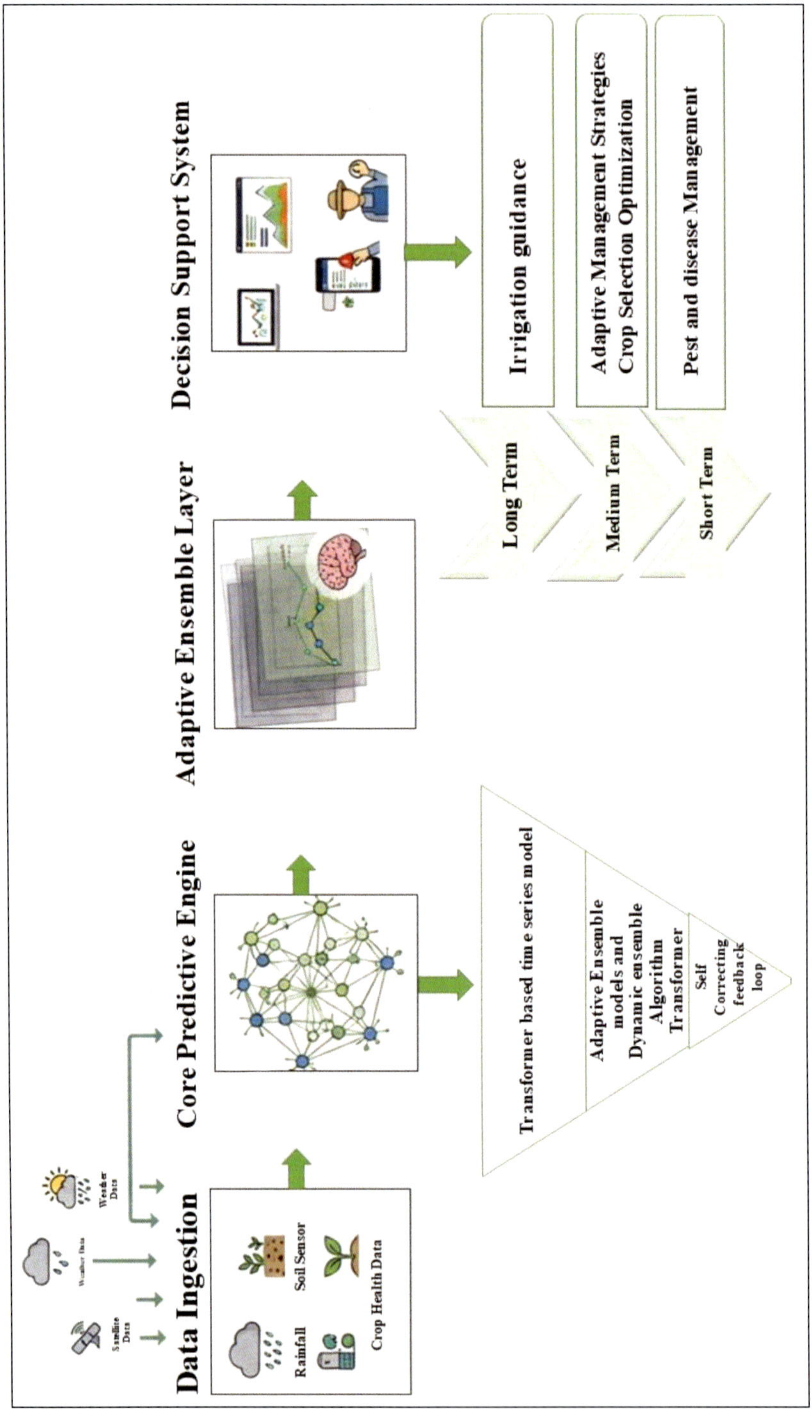

Fig. 4 AI-Driven Decision Support Framework for Precision Agriculture.

Complex agricultural models often lack transparency, reducing farmer trust. XAI methods like SHAP and LIME improve interpretability, but are computationally intensive. Deep learning demands significant processing power, often unavailable in rural areas, leading to reliance on cloud services that falter in low-connectivity zones. High-resolution imagery strains GPU memory, and IoT sensors face power constraints in remote areas with unreliable renewable energy. Scaling models across diverse hardware and large, heterogeneous farms is a challenging task. Growing multi modal data storage needs and IoT cybersecurity risks further complicate implementation.

Farmer adoption is hindered by technological unfamiliarity, particularly among older generations. Trust requires clear communication of AI limitations and user-friendly interfaces that cater to diverse educational levels. High costs for IoT maintenance, devices, and training deter investment. Language, cultural differences, and conflicts with traditional decision-making slow adoption. Models also struggle under extreme weather conditions.

Future research directions include federated learning for privacy-preserving, collaborative model training, and transfer learning to adapt models to data-scarce regions. Digital twins can simulate agroecosystems, while multi agent systems model the interactions between pests, crops, and their environments. Continual learning enables adaptation to changing conditions, and meta-learning supports rapid adjustment to new crops or pests. Edge computing reduces IoT latency, and GANs can generate synthetic data to enhance training.

10. Conclusion

Time-series deep learning models, such as SADF-Net and Transformers, revolutionize agriculture by enabling proactive pest and disease management. These models leverage spatiotemporal data from satellites, IoT sensors, and weather forecasts to outperform traditional methods. LSTMs and CNNs optimize crop forecasting and resource allocation, while XAI tools like SHAP and LIME enhance transparency, fostering trust among farmers and policymakers. Future advancements will integrate federated learning for data privacy and virtual ecosystems for improved pest control. These AI-driven solutions enhance climate resilience, minimize environmental impact, and support global food security by addressing climate challenges effectively.

Declaration

Ethical Responsibilities of Authors

"All authors have read, understood, and have complied as applicable with the statement on 'Ethical responsibilities of Authors' as found in the Instructions for Authors"

Ethical Approval

Not applicable

Use of AI tool

Minor grammar checking was performed using an automated tool, but no AI was used for content generation or manuscript preparation.

Conflict of Interest

On behalf of all the authors the corresponding author states that there is no conflict of interest to disclose.

Competing interests

Not applicable

Authors' contributions

Vasanthkumar S. S. contributed to the conceptualization, methodology design, supervision, and preparation of the original draft. Vinothagan S. was responsible for data curation, formal analysis, and reviewing and editing the manuscript. Yamini R. carried out investigation, visualization, and software-related tasks. Diksha Rani contributed to the original draft preparation, literature review, and formal analysis. Nandha S. provided validation, resources, and project administration support.

Funding

No funding has been received.

References

Abdullah, H.M., Mohana, N.T., Khan, B.M., Ahmed, S.M., Hossain, M., Islam, K.S., Redoy, M.H., Ferdush, J., Bhuiyan, M.A.H.B., Hossain, M.M. and Ahamed, T. (2023). Present and future scopes and challenges of plant pest and disease (P&D) monitoring: Remote sensing, image processing, and artificial intelligence perspectives. *Remote Sensing Applications: Society and Environment, 32*, 100996. https://doi.org/10.1016/j.rsase.2023.100996.

Abioye, E., Hensel, O., Esau, T., Elijah, O., Abidin, M., Ayobami, A., Yerima, O. and Nasirahmadi, A. (2022). Precision irrigation management using machine learning and digital farming solutions. *AgriEngineering*. https://doi.org/10.3390/agriengineering4010006.

Ahmadi, A., Daccache, A., Sadegh, M. and Snyder, R.L. (2023). Statistical and deep learning models for reference evapotranspiration time series forecasting: A comparison of accuracy, complexity, and data efficiency. *Computers and Electronics in Agriculture, 215*, 108424. https://doi.org/10.1016/j.compag.2023.108424.

Ahmed, M.T., Ahmed, M.W. and Kamruzzaman, M. (2025). A systematic review of explainable artificial intelligence for spectroscopic agricultural quality assessment. *Computers and Electronics in Agriculture, 235*, 110354. https://doi.org/10.1016/j.compag.2025.110354.

Ahmed, S., Marwat, S.N.K., Brahim, G.B., Alabrah, A., Javed, M.S., Nebhen, J. and Shabbir, M. (2024). IoT-based intelligent pest management system for precision agriculture. *Scientific Reports*, *14*, 31917. https://doi.org/10.1038/s41598-024-83012-3.

Albanese, A., Nardello, M. and Brunelli, D. (2021). Automated pest detection with DNN on the edge for precision agriculture. *IEEE Journal of Emerging and Selected Topics in Circuits and Systems*, *11*(3), 458–467. https://doi.org/10.1109/JETCAS.2021.3101740.

Ali, M.A., Dhanaraj, R.K. and Nayyar, A. (2023). A high performance-oriented AI-enabled IoT-based pest detection system using sound analytics in large agricultural field. *Microprocessors and Microsystems*, *103*, 104946. https://doi.org/10.1016/j.micpro.2023.104946.

Allu, A.R. and Mesapam, S. (2025). Impact of remote sensing data fusion on agriculture applications: A review. *European Journal of Agronomy*, *164*, 127478. https://doi.org/10.1016/j.eja.2024.127478.

Aloyce, A.A. (2025). Climate change and plant protection: Challenges and innovations in disease forecasting systems in developing countries. *Journal of Plant Protection Research*, *65*(2), 174–186. https://doi.org/10.24425/jppr.2025.155054.

Alrashdi, I., Abd El-Aziz, R.M., Taloba, A.I. and Farsi, M. (2025). Hybrid TCN-transformer model for predicting sustainable food supply and ensuring resilience. *Alexandria Engineering Journal*, *127*, 380–391. https://doi.org/10.1016/j.aej.2025.05.044.

Anderson, R.J., Thompson, K.M. and Williams, P.D. (2023). AI-powered orchard management systems for codling moth monitoring in New Zealand apple orchards. *Precision Agriculture*, *24*(2), 145–162.

Aziz, D., Rafiq, S., Saini, P., et al. (2025). Remote sensing and artificial intelligence: Revolutionizing pest management in agriculture. *Frontiers in Sustainable Food Systems*, *9*, 1551460. https://doi.org/10.3389/fsufs.2025.1551460.

Bai, S., Kolter, J.Z. and Koltun, V. (2018). An empirical evaluation of generic convolutional and recurrent networks for sequence modeling. *arXiv preprint arXiv:1803.01271*.

Barrero Guevara, L.A., Kramer, S.C., Kurth, T., et al. (2025). Causal inference concepts can guide research into the effects of climate on infectious diseases. *Nature Ecology and Evolution*, *9*, 349–363. https://doi.org/10.1038/s41559-024-02594-3.

Belghachi, M. (2025). A comprehensive overview of federated learning for next-generation smart agriculture: Current trends, challenges, and future directions. *Informatica*, *49*(1), 117–136. https://doi.org/10.31449/inf.v49i1.6764.

Casolaro, A., Capone, V., Iannuzzo, G. and Camastra, F. (2023). Deep learning for time series forecasting: Advances and open problems. *Information*, *14*(11), 598. https://doi.org/10.3390/info14110598.

Chacón-Maldonado, A.M., Asencio-Cortés, G. and Troncoso, A. (2025). A multimodal hybrid deep learning approach for pest forecasting using time series and satellite images. *Information Fusion*, *124*, 103350. https://doi.org/10.1016/j.inffus.2025.103350.

Chen, C. (2025). Deep learning approaches for time series prediction in climate resilience applications. *Frontiers in Environmental Science*, *13*, 1574981. https://doi.org/10.3389/fenvs.2025.1574981.

Chen, C.J., Huang, Y.Y., Li, Y.S., Chang, C.Y. and Huang, Y.M. (2020). An IoT-based smart agricultural system for pests detection. *IEEE Access*, *8*, 180750–180763. https://doi.org/10.1109/ACCESS.2020.3024891.

Chen, L., Zhang, H. and Wang, M. (2022). LSTM-based forecasting framework for litchi stink bug population dynamics in South China. *Computers and Electronics in Agriculture*, *198*, 107045.

Chinchorkar, S. (2025). Utilizing satellite data and machine learning to monitor agricultural vulnerabilities to climate change. *International Journal of Geography, Geology and Environment*, *7*(3), 01–09. https://doi.org/10.22271/27067483.2025.v7.i3a.345.

Chouhan, S.S., Patel, R.K., Shukla, A. and Singh, U.P. (2025a). Drone importance and their necessity in future generation agriculture. In: *Artificial Intelligence and Computer Vision for Ecological Informatics* (pp. 1–8). https://doi.org/10.1201/9781003518778-1.

Chouhan, S.S., Patel, R.K., Singh, U.P. and Tejani, G.G. (2025b). Integrating drone in agriculture: Addressing technology, challenges, solutions, and applications to drive economic growth. *Remote Sensing Applications: Society and Environment*, *38*, 101576. https://doi.org/10.1016/j.rsase.2025.101576.

Chouhan, S.S., Singh, U.P., Saxena, A. and Jain, S. (2024). Assessing the importance and need of artificial intelligence for precision agriculture. In: S. S. Chouhan, A. Saxena, U. P. Singh, and S. Jain (Eds.), *Artificial Intelligence Techniques in Smart Agriculture*. Springer. https://doi.org/10.1007/978-981-97-5878-4_1.

Chowdhury, M.E.H., et al. (2025). An integrated IoT and AI-based approach for precision pest control in rice fields. *Smart Agricultural Technology*. https://doi.org/10.1016/j.atech.2025.100423.

Chung, J., Gulcehre, C., Cho, K. and Bengio, Y. (2014). Empirical evaluation of gated recurrent neural networks on sequence modeling. arXiv preprint arXiv:1412.3555.

CICR. (2024). *Annual Report on Cotton Pest Management*.

Cohen, A., Levy, M. and Goldberg, S. (2024). Smart field monitoring system for cotton bollworm management in Mediterranean agriculture. *Precision Agriculture*, *25*(1), 89–104.

Contreras, J., Winterfeld, A., Popp, J. and Bocklitz, T. (2024). Spectral zones-based SHAP/LIME: Enhancing interpretability in spectral deep learning models through grouped feature analysis. *Analytical Chemistry*, *96*(39).

Cunniffe, N. J., Koskella, B., Metcalf, C.J.E., Parnell, S., Gottwald, T.R. and Gilligan, C.A. (2015). Thirteen challenges in modelling plant diseases. *Epidemics*, *10*, 6–10. https://doi.org/10.1016/j.epidem.2014.06.002.

Dash, S.S., Ratha, A.K., Sridharan, K., Palaparthi, D.P., Battu, G.G. and Kumar, V.S. (2023). *Application of Artificial Intelligence for Pest and Disease Forecasting in Agriculture*. ResearchGate. https://www.researchgate.net/publication/394070260.

Dembani, R., Karvelas, I., Akbar, N.A., et al. (2025). Agricultural data privacy and federated learning: A review of challenges and opportunities. *Computers and Electronics in Agriculture*, *232*, 110048. https://doi.org/10.1016/j.compag.2025.110048.

Donatelli, M., Magarey, R.D., Bregaglio, S., Willocquet, L., Whish, J.P.M. and Savary, S. (2017). Modelling the impacts of pests and diseases on agricultural systems. *Agricultural Systems*, 155, 213–224. https://doi.org/10.1016/j.agsy.2017.01.019.

Dubois, P., Martin, L. and Rousseau, C. (2023). CNN-based precision viticulture system for grape vine moth detection and control. *European Journal of Agronomy*, *147*, 126842.

FAO-USAID. (2024). *Desert Locust Early Warning System Using Satellite Imagery and Machine Learning for East Africa*. Food and Agriculture Organization Technical Report, FAO-2024-DL-001.

Farmonaut. (2025). AI and machine learning in agriculture: 2025 playbook. https://farmonaut.com/precision-farming/ai-and-machine-learning-in-agriculture-2025-playbook.

Farmonaut. (2025). Climate-resilient pest forecasting models. https://www.trendsaps.com/articles/5-0-TAPS-25-04-46-53.pdf

Farmonaut. (2025). Pest control industry trends: AI-driven solutions. https://farmonaut.com/blogs/pest-control-industry-trends-7-shocking-tech-upgrades.

Feng, Z., Zhao, J., Suo, L., Sun, H., Long, H., Yang, H., Song, X., Feng, H., Xu, B., Yang, G. and Zhao, C. (2025). Dynamic UAV data fusion and deep learning for improved maize phenological-stage tracking. *The Crop Journal*, *13*(3), 961–974. https://doi.org/10.1016/j.cj.2025.03.012.

Fermon, N. (2025). Genomics in agriculture: AI-driven disease resistance prediction. *Farmonaut*. https://farmonaut.com/precision-farming/genomics-in-agriculture-7-ai-uses-to-boost-farming.

Fermon, N. and Farmonaut. (2025). AI-powered no-chemical pest control solutions. https://farmonaut.com/precision-farming/act-pest-control-2025-powerful-ai-no-chemical-solutions.

García, M.A., Rodriguez, J.L. and Hernandez, C. (2024). YOLOv5-based fall armyworm detection system with drone surveillance in Mexican maize fields. *Crop Protection*, *178*, 106298.

Getahun, S., Kefale, H. and Gelaye, Y. (2024). Application of precision agriculture technologies for sustainable crop production and environmental sustainability: A systematic review. *The Scientific World Journal, 2024*. https://doi.org/10.1155/2024/2126734.

Hakim, A., Srivastava, A.K., Hamza, A., Owais, M., Habib-Ur-Rahman, M., Qadri, S., Qayyum, M.A., Ahmad Khan, F.Z., Mahmood, M.T. and Gaiser, T. (2025, Apr.). Yolo-pest: An optimized YoloV8x for detection of small insect pests using smart traps. *Sci Rep.*, *15*(1), 14029. doi: 10.1038/s41598-025-97825-3. PMID: 40269001; PMCID: PMC12019348.

Hari, P. and Singh, M.P. (2025). Adaptive knowledge transfer using federated deep learning for plant disease detection. *Computers and Electronics in Agriculture*, *229*, 109720. https://doi.org/10.1016/j.compag.2024.109720.

He, T., Li, M. and Jin, D. (2025). Deep learning-based time series prediction for precision field crop protection. *Frontiers in Plant Science*, *16*, 1575796. https://doi.org/10.3389/fpls.2025.1575796.

Hochreiter, S. and Schmidhuber, J. (1997). Long short-term memory. *Neural Computation*, *9*(8), 1735–1780.

Indu, A., Baghel, A., Bhardwaj, A. and Ibrahim, W. (2022). Optimization of pesticides spray on crops in agriculture using machine learning. *Computational Intelligence and Neuroscience*, *2022*, 9408535. https://doi.org/10.1155/2022/9408535.

Jagan Mohan, R.N.V., Rayanoothala, P.S. and Sree, R.P. (2025). Next-gen agriculture: Integrating AI and XAI for precision crop yield predictions. *Frontiers in Plant Science*, *15*, 1451607. https://doi.org/10.3389/fpls.2024.1451607.

Joshi, R.C., Burget, R. and Dutta, M.K. (2025). AgriDeep-net: An advanced deep feature fusion-based technique for enhanced fine-grain image analytics in precision agriculture. *Ecological Informatics*, *86*, 103069. https://doi.org/10.1016/j.ecoinf.2025.103069.

Khaki, S., Wang, L. and Archontoulis, S. (2019). A CNN-RNN framework for crop yield prediction. *Frontiers in Plant Science*, *10*. https://doi.org/10.3389/fpls.2019.01750.

Kim, S.H., Park, J.W. and Lee, H.K. (2023). IoT-based monitoring system for rice stem borer using machine learning and meteorological data integration. *Smart Agriculture Technology*, *4*, 100187.

Kim, S., et al. (2024). Federated learning for agricultural pest management. *Informatica*, *48*(4). https://www.informatica.si/index.php/informatica/article/view/6764.

Kowalski, M., Nowak, A. and Zajac, P. (2023). Smart agriculture framework for Colorado potato beetle monitoring using CNN and field sensor integration. *European Journal of Agronomy*, *149*, 126903.

Lee, J., Im, S., Jeong, J.-S., Lee, T.S., Park, S.H., Shin, C., Ju, H. and Kim, H.-J. (2025). Learning hidden relationship between environment and control variables for direct control of automated greenhouse using Transformer-based model. *Computers and Electronics in Agriculture*, *235*, 110335. https://doi.org/10.1016/j.compag.2025.110335.

Lee, S. and Yun, C.M. (2023). A deep learning model for predicting risks of crop pests and diseases from sequential environmental data. *Plant Methods*, *19*, 145. https://doi.org/10.1186/s13007-023-01122-x.

Li, J., et al. (2025). Explainable AI for pest forecasting in precision agriculture. *Expert Systems with Applications*. https://doi.org/10.1016/j.eswa.2024.125030.

Li, Y., et al. (2025). Deep learning for real-time pest detection using UAV-based hyperspectral imaging. *Computers and Electronics in Agriculture*. https://doi.org/10.1016/j.compag.2025.109935.

Li, Y., Wang, H., Dang, L.M., Sadeghi-Niaraki, A. and Moon, H. (2020). Crop pest recognition in natural scenes using convolutional neural networks. *Computers and Electronics in Agriculture*, *169*, 105174. https://doi.org/10.1016/j.compag.2019.105174.

Liu, X., Zhang, Q., Min, W., Geng, G. and Jiang, S. (2025). Solutions and challenges in AI-based pest and disease recognition. *Computers and Electronics in Agriculture*, *238*, 110775. https://doi.org/10.1016/j.compag.2025.110775.

López, J. M., González, R. and Fernandez, A. (2023). Precision agriculture system for Mediterranean fruit fly control using CNN with transfer learning. *Spanish Journal of Agricultural Research, 21*(3), e0412.

Ma, J., et al. (2024). Graph neural networks for plant disease forecasting. *Bioinformatics, 38*(Supplement_2), ii120. https://doi.org/10.1093/bioinformatics/btac496.

Ma, Y., et al. (2025). Multi-omics integration for crop disease forecasting. *Journal of Environmental Management*. https://doi.org/10.1016/j.jenvman.2024.1234479.

Mahlein, A.-K., et al. (2025). AI-driven pest detection using multimodal sensor fusion. *Scientific Reports, 15*, 92659. https://doi.org/10.1038/s41598-025-92659-5.

Manogna, R.L., Dharmaji, V. and Sarang, S. (2025). Enhancing agricultural commodity price forecasting with deep learning. *Scientific Reports, 15*, 20903. https://doi.org/10.1038/s41598-025-05103-z.

Martins, V.A.M., Freitas, L.C., De Aguiar, M.S., De Brisolara, L.B. and Ferreira, P.R. (2019). Deep learning applied to the identification of fruit fly in intelligent traps. In: *2019 Brazilian Symposium on Computing Systems Engineering (SBESC)*, 1–8. https://doi.org/10.1109/SBESC49506.2019.9046088.

Mekonnen, Y., Namuduri, S., Burton, L., Sarwat, A. and Bhansali, S. (2024). IoT-based intelligent pest management system for precision agriculture. *Scientific Reports, 14*, 30126.

Mendez, A.R., Silva, C.P. and Gonzalez, F.M. (2023). Agricultural IoT platform for soybean looper forecasting using CNN and weather station integration. *Precision Agriculture, 24*(4), 1567–1589.

Montgomery, K., Walden-Schreiner, C., Saffer, A., Jones, C., Seliger, B.J., Worm, T., Tateosian, L., Shukunobe, M., Kumar, S. and Meentemeyer, R. K. (2023). Forecasting global spread of invasive pests and pathogens through international trade. *Ecosphere, 14*(12), e4740. https://doi.org/10.1002/ecs2.4740.

Nanni, L., Manfe, A., Maguolo, G., Lumini, A. and Brahnam, S. (2021). High performing ensemble of convolutional neural networks for insect pest image detection. *ArXiv, abs/2108.12539*. https://doi.org/10.1016/j.ecoinf.2021.101515.

Nega, A. (2025). Climate change impacts on agriculture: A review of plant diseases and insect pests in Ethiopia and East Africa, with adaptation and mitigation strategies. *Advances in Agriculture, 2025*, 5606701. https://doi.org/10.1155/aia/5606701.

Nyakuri, J.P., Nkundineza, C., Gatera, O., Nkurikiyeyezu, K. and Mwitende, G. (2025). AI and IoT-powered edge device optimized for crop pest and disease detection. *Scientific Reports, 15*, 22905. https://doi.org/10.1038/s41598-025-06452-5.

Omdena Inc. (2025). AI for crop yield prediction: Future of agriculture 2025. *Omdena AI Insights*, 17 July. https://ai.omdena.com/3YIHXTf.

Özkan, E., Demir, N. and Yilmaz, S. (2023). Greenhouse monitoring system for tomato leafminer detection using computer vision and environmental control. *Biosystems Engineering, 228*, 45–58.

Pal, C., Karmakar, S., Mukherjee, I., et al. (2025). A lightweight and explainable CNN model for empowering plant disease diagnosis. *Scientific Reports, 15*, 30720. https://doi.org/10.1038/s41598-025-94083-1.

Pandit, P., Sagar, A., Ghose, B., Paul, M., Kisi, O., Vishwakarma, D., Mansour, L. and Yadav, K. (2024). Hybrid modeling approaches for agricultural commodity prices using CEEMDAN and time delay neural networks. *Scientific Reports, 14*. https://doi.org/10.1038/s41598-024-74503-4.

Patel, R.K. (2025b). BM3D filtering with ensemble Hilbert–Huang transform and spiking neural networks for cardiomegaly detection in chest radiographs. *Computational Biology and Chemistry, 108*, 108620. https://doi.org/10.1016/j.compbiolchem.2025.108620.

Patel, R.K., Chouhan, S.S., Chandan, T. and Singh, U.P. (2025a). Basil crop detection using computer vision and deep learning approach. In: *Artificial Intelligence and Computer Vision for Ecological Informatics* (pp. 67–82). https://doi.org/10.1201/9781003518778-4.

Pattanayak, S., Kumar, R. and Singh, A. (2023). Smart rice system for brown plant hopper monitoring using IoT sensors and predictive modeling. *Computers and Electronics in Agriculture*, *205*, 107642.

Peng, F., Su, T., Zeng, Q., et al. (2025). Climate-adaptive energy forecasting in green buildings via attention-enhanced Seq2Seq transfer learning. *Scientific Reports*, *15*, 31829. https://doi.org/10.1038/s41598-025-16953-y.

Porfírio, R.P., Madeira, R. N. and Santos, P. A. (2025). Exploring explainable AI techniques for plant disease classification in digital agriculture. *Procedia Computer Science*, *265*, 175–182. https://doi.org/10.1016/j.procs.2025.07.170.

Queensland Department of Agriculture and Fisheries. (2023). Smart trap network implementation for Queensland fruit fly management. *Queensland DAF Technical Bulletin, QB23-001*.

Ravikiran, K.T., Thribhuvan, R., Anilkumar, C., Kallugudi, J., Prakash, N.R., Adavi, S.B., Sunitha, N.C. and Abhijith, K.P. (2025). Harnessing the power of genomics to develop climate-smart crop varieties: A comprehensive review. *Journal of Environmental Management*, *373*, 123461. https://doi.org/10.1016/j.jenvman.2024.123461.

Ristaino, J.B., Anderson, P.K., Bebber, D.P., et al. (2021). The persistent threat of emerging plant disease pandemics to global food security. In: *Proceedings of the National Academy of Sciences*, *118*(23), e2022239118. https://doi.org/10.1073/pnas.2022239118.

Rizman Žalik, K. and Žalik, M. (2023). A review of federated learning in agriculture. *Sensors*, *23*(23), 9566. https://doi.org/10.3390/s23239566.

Rossi, G., Bianchi, M. and Ferrari, L. (2024). Smart agriculture platform for olive fruit fly monitoring using computer vision and weather integration. *Italian Journal of Agronomy*, *19*(1), 1245.

Sanchez, P., Morales, D. and Castro, R. (2024). Vineyard monitoring system for grape phylloxera using computer vision and trap analysis. *Chilean Journal of Agricultural Research*, *84*(2), 156–168.

Santos, R.P., Matos-Carvalho, J.P. and Leithardt, V.R.Q. (2025). Deep learning in time series forecasting with transformer models and RNNs. *PeerJ Computer Science*, *11*, e3001. https://doi.org/10.7717/peerj-cs.3001.

Sarkar, S.S., Bedi, J. and Jain, S. (2025). A deep learning-based framework for enhanced reference evapotranspiration estimation. *Scientific Reports*, *15*, 15136. https://doi.org/10.1038/s41598-025-99713-2.

Scarselli, F., Gori, M., Tsoi, A.C., Hagenbuchner, M. and Monfardini, G. (2009). The graph neural network model. *IEEE Transactions on Neural Networks*, *20*(1), 61–80.

Schmidt, H., Mueller, T. and Weber, K. (2024). Smart field monitoring for aphid species identification using IoT-enabled sticky traps. *Journal of Plant Diseases and Protection*, *131*(2), 289–301.

Sharafat, M.S., Kabya, N.D., Emu, R.I., Ahmed, M.U., Onik, J.C., Islam, M.A. and Khan, R. (2025). An IoT-enabled AI system for real-time crop prediction using soil and weather data in precision agriculture. *Smart Agricultural Technology*, *12*, 101263. https://doi.org/10.1016/j.atech.2025.101263.

Shoaib, M., Sadeghi-Niaraki, A., Ali, F., Hussain, I. and Khalid, S. (2025). Leveraging deep learning for plant disease and pest detection: A comprehensive review and future directions. *Frontiers in Plant Science*, *16*, 1538163. https://doi.org/10.3389/fpls.2025.1538163.

Shukla, A., Chouhan, S.S., Patel, R.K., Singh, U.P. (2025). AgriCloud: Cloud-Based Framework for Scalable Crop Monitoring and Intelligent Decision Support. In: Chouhan, S.S., Patel, R.K., Singh, U.P., and Jain, S. (Eds.), *AgriTech Revolution*. Springer, Singapore. https://doi.org/10.1007/978-981-95-1268-3_6.

Singh Chouhan, S., Saxena, E., Shukla, A. et al. (2025). Optimizing YOLO-Based Models for Real-Time Guava Detection with Probabilistic Fused Wiener Filter-Enhanced Feature Fusion. *Applied Fruit Science*, *67*, 383. https://doi.org/10.1007/s10341-025-01613-2.

Singh, R., Sharma, P. and Kumar, A. (2023). Smart agriculture IoT platform for brown plant hopper prediction using machine learning and weather data. *Indian Journal of Agricultural Sciences*, *93*(8), 845–852.

Suto, J. (2021). Embedded system-based sticky paper trap with deep learning based insect-counting algorithm. *Electronics*, *10*(15), 1754. https://doi.org/10.3390/electronics10151754.

Tanaka, H., Yamamoto, K. and Sato, M. (2023). AI-pest detection system for Asian corn borer using deep learning and environmental sensors. *Japanese Journal of Applied Entomology and Zoology*, *67*(3), 145–156.

Thompson, M.J., Anderson, L.K. and Brown, R.S. (2023). Agricultural IoT framework for European corn borer forecasting using LSTM networks. *Canadian Journal of Plant Science*, *103*(4), 567–580.

Tsoumas, I., Sitokonstantinou, V., Giannarakis, G., et al. (2025). Leveraging causality and explainability in digital agriculture. *Environmental Data Science*, *4*, e23. https://doi.org/10.1017/eds.2025.14.

USDA APHIS. (2025). *Plant Pests and Diseases*. U.S. Department of Agriculture, Animal and Plant Health Inspection Service. https://www.aphis.usda.gov/plant-health/plant-pests-and-diseases.

Van der Berg, H., De Jong, S. and Bakker, M. (2024). AI-monitoring system for greenhouse whitefly using computer vision and spectral analysis. *Netherlands Journal of Agricultural Science*, *72*, 45–62.

Van Zyl, J., Pretorius, M. and Du Toit, L. (2024). Citrus monitoring platform for false codling moth using deep learning automation. *South African Journal of Plant and Soil*, *41*(1), 89–98.

Vasanthkumar, S.S., Prabhu, T., Shenbagavalli, S., Rajangam, J., Vallal Kannan, S., and Baskaran, A. (2024). Exploring the therapeutic potential of bioactive compounds in spices and herbs: Preclinical evidence and medicinal applications. *Annals of Phytomedicine: An International Journal*, *13*(2), 1–8. DOI: https://doi.org/10.54085/ap.2024.13.2.48.

Vaswani, A., Shazeer, N., Parmar, N., et al. (2017). Attention is all you need. *Advances in Neural Information Processing Systems* (*NeurIPS*), *30*, 5998–6008.

Venkateswara, S. and Padmanabhan, J. (2025). Deep learning based agricultural pest monitoring and classification. *Scientific Reports*, *15*, 8684. https://doi.org/10.1038/s41598-025-92659-5.

Wang, F., Wang, R., Xie, C., Yang, P. and Liu, L. (2020). Fusing multi-scale context-aware information representation for automatic in-field pest detection and recognition. *Computers and Electronics in Agriculture*, *169*, 105222. https://doi.org/10.1016/j.compag.2020.105222.

Wang, J. and Zhang, D. (2024). Intelligent pest forecasting with meteorological data: An explainable deep learning approach. *Expert Systems with Applications*, *252*(A), 124137. https://doi.org/10.1016/j.eswa.2024.124137.

Wang, L., et al. (2025). Autonomous robotic intervention for citrus HLB vector control. *Artificial Intelligence in Agriculture*, *6*, 5089. https://ojs.bonviewpress.com/index.php/AIA/article/view/5089/1445.

Wang, M. and Li, T. (2025). Pest and disease prediction and management for sugarcane using a hybrid autoregressive integrated moving average: A long short-term memory model. *Agriculture*, *15*(5), 500. https://doi.org/10.3390/agriculture15050500.

Wang, Q.J., Zhang, S.Y., Dong, S.F., Zhang, G.C., Yang, J., Li, R. and Wang, H.Q. (2020). Pest24: A large-scale very small object data set of agricultural pests for multi-target detection. *Computers and Electronics in Agriculture*, *175*, 105585. https://doi.org/10.1016/j.compag.2020.105585.

Wang, S., Zeng, Q., Zhang, X., Ni, W. and Cheng, C. (2023). Multi modal pseudo-information guided unsupervised deep metric learning for agricultural pest images. *Inf. Sci.*, *630*, 443–462. https://doi.org/10.1016/j.ins.2023.02.030.

Wang, Y., Wang, P., Tansey, K., Liu, J., Delaney, B. and Quan, W. (2025). An interpretable approach combining Shapley additive explanations and LightGBM based on data augmentation for improving wheat yield estimates. *Computers and Electronics in Agriculture*, *229*, 109758. https://doi.org/10.1016/j.compag.2024.109758.

Williams, J.D., Davis, S.M. and Clark, A.R. (2023). Precision agriculture system for carrot fly monitoring using machine learning and trap automation. *British Journal of Entomology and Natural History*, *36*(2), 78–91.

Wu, Z., Pan, S., Long, G., et al. (2021). A comprehensive survey on graph neural networks. *IEEE Transactions on Neural Networks and Learning Systems*, *32*(1), 4–24.

You, J., Li, X., Low, M., Lobell, D. and Ermon, S. (2017). Deep Gaussian process for crop yield prediction based on remote sensing data. In: *Proceedings of the AAAI Conference on Artificial Intelligence*, *31*(1).

YSRHU. (2024). *Mango Pest Management Report*.

Yue, Y., Liu, H., Meng, X., Li, Y. and Du, Y. (2018). Deep recursive super resolution network with Laplacian pyramid for better agricultural pest surveillance and detection. *Computers and Electronics in Agriculture*, *150*, 26–32. https://doi.org/10.1016/j.compag.2018.04.004.

Zhu, S., Zheng, J., and Q. (2023). MR-Transformer: Multiresolution Transformer for multivariate time series prediction. *IEEE Transactions on Neural Networks and Learning Systems*, *36*, 1171–1183. https://doi.org/10.1109/TNNLS.2023.3327416.

18 Deep Reinforcement Learning For Farm Automation

Nupur Parashar[1*] and Harsh Soni[2]

1. Introduction

In this chapter we discuss and focus on technical advancements in the agricultural practices due to deep learning techniques. It is seen that the agricultural market has risen at the rate of 23.1% from 2023 to 2028 (Figure 1) (electroiq) (researchandmarkets).

The popularity of this growing infrastructure in the agriculture sector is due to the fact that techniques like deep reinforcement learning (DRL) provide various benefits to farm operations ranging from farm monitoring, crop health, irrigation, fertilization, and even pest control. Before the origin of DRL techniques integration in agriculture these tasks were performed manually which was not only tedious but also not as efficient as it is now with the help of DRL. The advanced systems have made farmers more reliant on the technology for management, supervision, and control. However, the implementation of such systems is full of challenges as it involves large data set analysis, computational overhead, implementation cost, and training. In this chapter we will see some of the examples of DRL-based systems employed in farm automation and analyze the performance improvement and challenges posed (coherentmarketinsights).

2. Deep Reinforcement Learning Fundamentals for Agriculture

For different agricultural tasks and environmental constraints, different DRL-based algorithms are used. We list out the different algorithms used based on their functionality. A taxonomy of these techniques is given in Figure 2.

These methods form the foundation of many agricultural automation systems. These methods help in better decision-making. For example, Deep Q-Networks (DQN) are useful in discrete decision-making in irrigation scheduling and pest management. DQN provides learning based on making predictions. For example,

[1] Research Scholar, Department of Computer Science, Dr. Bhimrao Ambedkar University, Agra.
[2] PEGA Lead System Architect | Development Team Leader @Euroclear, Belgium. https://orcid.org/0009-0002-0211-9163
* Corresponding author: balatirupati2019@gmail.com

DOI: 10.1201/9781003715726-18

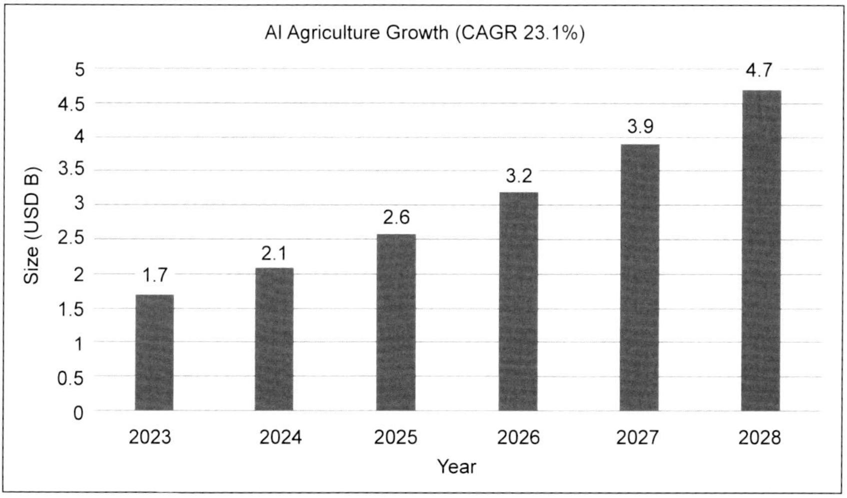

Fig. 1 Projected growth of AI in the agriculture market from 2023 to 2028 (electroiq) (researchandmarkets)

in the case of irrigation automation based on values like air moisture level, weather forecast, and crop growth stages, DQN can predict the decision to irrigate or not. This prediction helps farmers optimize the irrigation and avoid unnecessary or untimely irrigation (Joseph et al., 2024). Some applications in farming require accurate decision-making. Double-DQN provides better accuracy compared to the DQN algorithm (Zhao et al., 2025).

Tasks like irrigation, fertilization, movement of robots require continual monitoring because these are continuous operations that never end. Proximal Policy Optimization (PPO) helps in non-stop control actions (Alkaff et al., 2025). Agricultural operations such as fruit harvesting and spraying fertilizers require robotic system precision to decide on the correct time at which these tasks must be performed. Algorithms such as Deep Deterministic Policy Gradient (DDPG) and Twin Delayed DDPG (TD3) use an actor-critic model to handle these types of operations (Garrido-Castañeda et al., 2025).

Reinforcement learning approaches involving multiple agents are required for tasks such as precision agriculture monitoring. Such approaches helps in coordinating multiple drones and robots operations to achieve synchronized field surveillance (Chouhan et al., 2024; Ahmad Din et al., 2022).

Independent Q-Learning (IQL) provides computational efficiency, but lacks coordination. Techniques like agent-by-agent (ABA) provide balanced optimization between agents' policies. So, multi agent rollout policies are more suitable for situations where coordination is more important than computational costs. (Anubha Mahajan et al., 2024).

Hierarchical Reinforcement Learning (HRL) divides a complex task into smaller simpler sub-tasks. This helps in multi scale decision-making process and better learning and generalization in application such harvesting operations (Hutsebaut-Buysse et al., 2022).

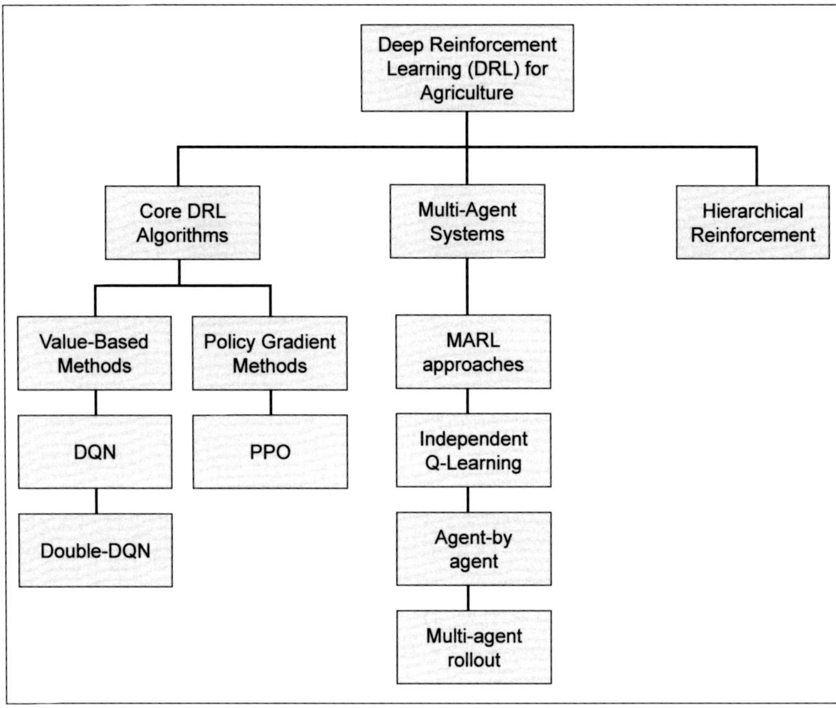

Fig. 2 Taxonomy of different deep reinforcement learning for agriculture.

3. Applications of DRL in Agricultural Automation

In this section we discuss some of the applications of DRL in automation of agricultural operations.

DRL based systems are employed for various applications in agriculture such as crop management, autonomous navigation, drone-based applications, agricultural robots, etc. For precision crop management, they are used as fertilization monitoring, irrigation, and pest control. These systems interact with the real world to make intelligent decisions (Alkaff et al., 2025; Joseph et al., 2024).To make these intelligent decisions, the DRL system needs some inputs in the form of metrics like soil conditions, weather patterns, and crop growth stages. These inputs are used by the system to make decisions like how much and when fertilizers are needed to be applied to crops, how to minimize fertilizer application while maximizing yield of crops (Joseph et al., 2024). These systems also help in deciding irrigation timing, water quantity, etc., based on parameters like soil moisture, evaporation rate based on season and weather forecast. Water stress indicators help in speeding up the process to avoid harm to crops. With such techniques it is seen that in some cases 25-40% of water is saved from irrigation, which could have potentially harmed the crops as different crops require different amount of water for their growth (Hoque et al., 2024).

The manual navigation of farm land to supervise crop health, growth, and potential harm, is replaced by autonomous navigation using robots. These robots are technically implemented using DRL to help control the farm area in different terrains, through different obstacles, and different crop configurations with ease. This helps reduce human tasks to a great extent (Cui et al., 2024). Autonomous tractors and harvesters use technologies like global positioning system (GPS), Light detection and ranging (LiDAR) and computer vision for real-time navigation and decision- making (Zhao et al., 2025). For example, robotic harvesting systems can identify ripe fruit accurately without harming the rest of the plant (Palmieri et al., 2025). DRL-based systems often employed in monitoring and treatment are based on drones. These drones find an optimal flight path over the farm land under supervision. Also, they are adaptable to changing weather conditions and coordinate coverage and flight time with the help of multi-UAV systems (Xiong et al., 2025; Chouhan et al., 2025). They provide a comprehensive coverage of the farm land which might not be possible manually. It is reported that drones which use powerful hybrid techniques can increase the coverage by 41%. Not only that, they also help in reducing the overlap in coverage, which is crucial in pesticide application as excess of pesticide can be poisonous (Zhao et al., 2025).

DRL systems are efficient as they can control environmental parameters, factors, resource allocation and crop management (Zhang et al., 2021). These systems are at par with the traditional approaches as they can maximize crop growth while minimizing environmental conditions thus providing sophisticated control strategies (Zhang et al., 2021).

4. Implementation Challenges and Solutions

There are certain challenges in integrating DRL for farm automation applications. We describe them as shown below.

4.1 Environmental Complexity and Adaptation

With the changing seasons, the type of crop also changes. But the autonomous systems have to work through all kinds of weather, terrains, and different biological systems. They need to navigate the farm lands in order to report crop growth. To address these challenges, these systems adopt several strategies such as multimodal sensor fusion to adapt to different environments quickly (Sahili et al., 2022).

4.2 Computational Constraints and Edge Computing

The agricultural machinery operates on limited computational resources. To overcome this, energy-efficient algorithms are needed. Edge computing offers a solution to this problem by providing lightweight DRL models for large datasets (Zhao et al., 2025; Singh Chouhan, S., 2025).

Another method for computationally efficient frameworks is the hybrid decision framework, which combines DRL with approaches such as model predictive control (MPC) to maintain adaptability while meeting computational constraints. This helps in high-level decision-making in real-time execution (Devdiscourse).

4.3 Safety and Reliability Considerations

Reliable operation in DRL-based agricultural automation is provided by ISO 18497 (Aby et al. 2023). The safety system in DRL-based systems is crucial, as it should be able to identify and differentiate between static, dynamic, and negative obstacles, such as rocks, trees, ditches, humans, and animals. So there should be multiple layers of protection.

A robust system is necessary in the event of unexpected situations or system failure. This is provided by different mechanisms such as emergency stop procedures, human override capabilities, communication systems, and regular maintenance and inspection protocols (aces.illinois).

4.4 Data Quality and Training Considerations

DRL systems require high-quality data for decision- and policy-making. But the data from agricultural environments is often unpredictable due to seasonal variations, geographic diversity, and the need for long-term performance validation (Alibabaei et al., 2022). Also the simulation data might not match the real world data. To eliminate the inconsistency, data augmentation techniques are used. These techniques leverage the pre-trained models and reduce data requirement for certain applications (meegle).

5. Integration with Existing Agricultural Systems

5.1 Sensor Networks and IoT Integration

Integration of DRL systems with IoT (internet of things) augment the monitoring and data collection. Unmanned Aerial Vehicles (UAVs) assist IoT in collecting and routing data from distributed sensor nodes to cloud servers with the help of multi agent DRL systems (Xiong et al., 2025).

5.2 Human-Machine Interaction

In semi-automated systems, human-machine interaction is crucial in the successful implementation of DRL systems. In such cases, sufficient information about the system behavior is crucial in interface design (Palmieri et al., 2025).

Semi-autonomous architectures allow humans to intervene when DRL algorithms encounter uncertain or complex situations. This collaboration enhances human expertise for special cases while allowing autonomous operation under normal conditions (Palmieri et al., 2025).

Table 1 Summary of challenges in implementing DRL for farm automation.

Challenge	*Primary Solutions*
Environmental Variability	Robust sensor fusion, adaptive algorithms
High Initial Cost	Leasing models, government subsidies
Technical Complexity	Simplified interfaces, training programs
Safety Concerns	Safety protocols, certification standards
Data Quality Issues	Standardization, quality control
Integration Difficulties	Modular designs, API standards

6. Comparative Analysis: Traditional vs. DRL Methods

The DRL technique offers an advantage over traditional farming methods. They help in adaptability to changing conditions, better understanding of the farm area, higher coverage with/without obstructions, fault detection, etc.

Table 2 Comparison between traditional and DRL-based farming.

Aspect	*Traditional Methods*	*DRL Methods*	*Performance Improvement*
Decision Making	Rule-based, static	Data-driven, dynamic	40–60%
Adaptability	Limited	High	80–90%
Learning Capability	None	Continuous	100%
Initial Setup Cost	Low	High	N/A
Maintenance Requirements	High	Low	50–70%
Performance Consistency	Moderate	High	30–50%
Data Requirements	Low	High	N/A
Computational Resources	Low	High	N/A

From Table 2, it can be seen that DRL methods are traditional methods in almost all aspects. The performance improvement ranges from 30-100%. Factors like data-driven decision-making, adaptability, and continuous learning lead to high performance improvements at the cost of high setup cost, data requirement, and computational resources.

7. Future Research Directions

In the future we need more sophisticated algorithms to handle current limitations. For example, meta-learning frameworks are expected to enable adaptability in different crops, climates, and geographic regions. Self-supervised learning approaches, with the help of visual and temporal patterns for training, would help in eliminating dependency on labeled data and speed up the overall processing. Continual learning algorithms-based-systems learn from previously learnt skills and adapt to changing environmental conditions, which could be important in seasonal variations and long-term climate changes (devdiscourse).

To testify to simulation-based experiments in real-world settings, several techniques like improved physical modelling, environmental simulation, and domain adaptation techniques would help (Hutsebaut-Buysse et al., 2022).

Additionally, for very large farms, scalable deployment strategies are crucial for coordinating multi farm operations, and integrating existing agricultural supply chains is essential for the widespread adoption of DRL.

Ensemble feature extraction and new neural architectures can enhance automation pipelines, further improving smart farming systems and agricultural robotics (Patel et al., 2025).

8. Conclusion

In this chapter, we explored various DRL-based algorithms applied to diverse agricultural applications. We saw how these techniques help automate farming operations and provide improved results. We discussed various challenges present in the implementation of DRL-based systems and their solutions. However, ongoing research is still being conducted to find more optimized solutions to these challenges.

Integration of IoT with DRL provides comprehensive automation, increasing productivity, and sustainability. Techniques such as sim-to-real transfer, enhanced safety and interpretability, and scalability are among the future research options in this domain. Modern agriculture requires collaboration among agricultural research, the latest technology, and advanced farming techniques, and ongoing research is a crucial step towards achieving this goal.

Author Declaration

Ethical Compliance

The authors have taken care of all ethical standards while documenting this work and assure that it complies with the standard ethics.

Ethical Clearance

This work does not involve any such procedure which requires any ethical approval.

AI Assistance Disclosure

No artificial intelligence tools were used in preparing or editing this book chapter.

Conflict of Interest Statement

There are no conflicts of interest associated with this work.

Competing Interest Disclosure

There are no competing interests to report.

Authors' contributions

Nupur Parashar[1] has performed the content writing, editing, and proof reading of this document.

Funding Information

The authors declare that they have not taken assistance of any kind of fundings to complete this work.

References

Aby, G.R. and Issa, S.F. (2023). Safety of Automated Agricultural Machineries: A Systematic Literature Review. *Safety*, *9*(1), 13. https://doi.org/10.3390/safety9010013.

Agricultural Robots Statistics and Facts, https://electroiq.com/stats/agricultural-robots-statistics/.

Ahmad Din, Muhammed Yousoof Ismail, Babar Shah, Mohammad Babar, Farman Ali, Siddique Ullah Baig. (2022). A deep reinforcement learning-based multiagent area coverage control for smart agriculture. *Computers and Electrical Engineering*, *101*, 108089. ISSN 0045-7906, https://doi.org/10.1016/j.compeleceng.2022.108089.

Alibabaei, K., Gaspar, P.D., Lima, T.M., Campos, R.M., Girão, I., Monteiro, J. and Lopes, C.M. (2022). A Review of the Challenges of Using Deep Learning Algorithms to Support Decision-Making in Agricultural Activities. *Remote Sensing*, *14*(3), 638. https://doi.org/10.3390/rs14030638.

Alkaff, M., Basuhail, A. and Sari, Y. (2025). Optimizing Water Use in Maize Irrigation with Reinforcement Learning. *Mathematics*, *13*(4), 595. https://doi.org/10.3390/math13040595.

Al Sahili, Z. and Awad M. (2022, Dec.) The power of transfer learning in agricultural applications: AgriNet. *Front Plant Sci.*, *14*(13), 992700. doi: 10.3389/fpls.2022.992700. PMID: 36589063; PMCID: PMC9794606.

Automated agricultural machinery requires new approaches to ensuring safety. https://aces.illinois.edu/news/automated-agricultural-machinery-requires-new-approaches-ensuring-safety.

Azmirul Hoque and Mrutyunjay Padhiary. (2024). Automation and AI in Precision Agriculture: Innovations for Enhanced Crop Management and Sustainability. *Asian Journal of Research in Computer Science*, *17* (10), 95–109. https://doi.org/10.9734/ajrcos/2024/v17i10512.

Balderas, Joseph, Chen, Dong, Huang, Yanbo, Wang, Li and Ren-Cang, U (2024, Nov.). *A Comparative Study of Deep Reinforcement Learning for Crop Production Management*. Available at SSRN: https://ssrn.com/abstract=5011946.

Challenges in Implementing Agriculture Robots and How to Overcome Them, https://www.coherentmarketinsights.com/blog/challenges-in-implementing-agriculture-robots-and-how-to-overcome-them-1210.

Chouhan, S.S., Patel, R.K., Shukla, A. and Singh, U.P. (2025, Aug.). Drone Importance and their Necessity in Future Generation Agriculture. In: *Artificial Intelligence and Computer Vision for Ecological Informatics* (pp. 1–8). doi: 10.1201/9781003518778-1.

Chouhan, S.S., Singh, U.P., Saxena, A. and Jain, S. (2024). Assessing the Importance and Need of Artificial Intelligence for Precision Agriculture. In: Chouhan, S.S., Saxena, A., Singh, U.P., and Jain, S. (Eds.), *Artificial Intelligence Techniques in Smart Agriculture*. Springer, Singapore. https://doi.org/10.1007/978-981-97-5878-4_1.

Cui, L., Le, F., Xue, X., Sun, T. and Jiao, Y. (2024). Design and Experiment of an Agricultural Field Management Robot and Its Navigation Control System. *Agronomy*, *14*(4), 654. https://doi.org/10.3390/agronomy14040654.

Deep reinforcement learning boosts smart agriculture: Challenges remain. https://www.devdiscourse.com/article/technology/3446610-deep-reinforcement-learning-boosts-smart-agriculture-challenges-remain.

Garrido-Castañeda, S.I., Vasquez, J.I. and Antonio-Cruz, M. (2025). Coverage Path Planning Using Actor–Critic Deep Reinforcement Learning. *Sensors*, *25*(5), 1592. https://doi.org/10.3390/s25051592.

Hutsebaut-Buysse, M., Mets, K. and Latré, S. (2022). Hierarchical Reinforcement Learning: A Survey and Open Research Challenges. *Machine Learning and Knowledge Extraction*, *4*(1), 172–221. https://doi.org/10.3390/make4010009.

Mahajan, A., Hegde, S., Shay, E., Wu, D. and Prins, A. (2024). Comparative Analysis of Multi-Agent Reinforcement Learning Policies for Crop Planning Decision Support. *arXiv preprint arXiv:2412.02057.*

Palmieri, J., Di Lillo, P., Chiaverini, S. and Marino, A. (2025). A comprehensive control architecture for semi-autonomous dual-arm robots in agriculture settings. *Control Engineering Practice*, *163*, 106394.

Patel, R.K. (2025, Aug.) BM3D Filtering with Ensemble Hilbert-Huang Transform and Spiking Neural Networks for Cardiomegaly Detection in Chest Radiographs. *Computational Biology and Chemistry*, *8*, 108620.

Patel, R.K., Chouhan, S.S., Chandan, T. and Singh, U.P. (2025, Aug.) Basil Crop Detection Using Computer Vision and Deep Learning Approach. In: *Artificial Intelligence and Computer Vision for Ecological Informatics* (pp. 67–82). doi: 10.1201/9781003518778-4.

Reinforcement Learning Market Report 2025. *Reinforcement Learning Market Size, Share & Forecast to 2029.*

Siddharth Singh Chouhan, Rajneesh Kumar Patel, Uday Pratap Singh,& Ghanshyam G. Tejani. (2025). Integrating drone in Agriculture: Addressing technology, challenges, solutions, and applications to drive economic growth. *Remote Sensing Applications: Society and Environment*, *38*, 101576. ISSN 2352-9385, https://doi.org/10.1016/j.rsase.2025.101576.

Singh Chouhan, S., Saxena, E., Shukla, A. et al. (2025). Optimizing YOLO-Based Models for Real-Time Guava Detection with Probabilistic Fused Wiener Filter-Enhanced Feature Fusion. *Applied Fruit Science*, *67*, 383. https://doi.org/10.1007/s10341-025-01613-2.

Transfer Learning in Agriculture. https://www.meegle.com/en_us/topics/transfer-learning/transfer-learning-in-agriculture

Wang, Z., Zhang, F., Ma, S., et al. (2025). Research on collaborative scheduling strategies of multi-agent agricultural machinery groups. *Sci. Rep.*, *15*, 9045. https://doi.org/10.1038/s41598-025-93848-y.

Xiong, G., Guo, J., Parsons, K., Nagai, Y., Sumi, T., Orlik, P. V. and Li, J. (2025). UAV Aided Smart Agriculture Networks: A Multi-Agent Reinforcement Learning Approach. In: *ICC Conference 2025 on Communications*. IEEE.

Zhang, W., Cao, X., Yao, Y., An, Z., Xiao, X. and Luo, D. (2021, November). Robust model-based reinforcement learning for autonomous greenhouse control. In: *Asian Conference on Machine Learning* (pp. 1208–1223). PMLR.

Zhao, J., Fan, S., Zhang, B., Wang, A., Zhang, L. and Zhu, Q. (2025). Research Status and Development Trends of Deep Reinforcement Learning in the Intelligent Transformation of Agricultural Machinery. *Agriculture*, *15*(11), 1223. https://doi.org/10.3390/agriculture15111223.

19 | Tomato Plant Disease Classification using Deep Learning Techniques

Kavish Thakur[1*], Dilip Kumar Choubey[2*], Aditya Kumar Pandey[3*], Utkarsh Raj[4*] and Mukesh Kumar Ojha[5]

1. Introduction

Tomatoes remain a globally significant horticultural commodity, valued for its dense nutritional profile and substantial economic contribution to food systems worldwide. Whether it is eaten raw or processed, it is a staple of diets all over the world. Large-scale tomato production is crucial to the agriculture industry and the world's food supply (Hammou & Boubaker, 2022). As a result, tomato crops are guaranteed to grow healthily and productively, which is crucial to meeting market demand and promoting global food security.

Timely identification of phytopathological conditions coupled with accurate classification of disease phenotypes represents a critical intervention point for minimizing agricultural losses. Contemporary disease management practices, despite reliance on manual inspection methodologies, face inherent operational constraints including extended assessment periods, elevated operational costs, and limitations in scalability across expansive cultivation areas. The efficacy of these manual detections for thorough and accurate disease classification is limited because they frequently result from human error and inconsistencies.

Advancement in computational image analysis coupled with deep neural network methodologies provides a promising pathway to surmount traditional diagnostic limitations. Automated systems leveraging digital image processing and machine vision technologies enable rapid, cost-effective, and reproducible assessment of foliar disease manifestations. Through algorithmic extraction of discriminative features from photographic documentation of affected leaf

[1-4] Department of Computer Science & Engineering, Indian Institute of Information Technology Bhagalpur, Bhagalpur, Bihar, India.
[5] Department of Electronics and Communication Engineering, Greater Noida Institute of Technology, Greater Noida, U.P., India.
E-mail: mukeshkumarojha07@gmail.com
ORCID: 0000-0002-1233-7159[1]; 0009-0001-8059-0447[3]; 0000-0001-7119-9596[5]
* Corresponding author: kavish.cse.230201003@iiitbh.ac.in[1]; dkchoubey.cse@iiitbh.ac.in[2]; aditya.240201001@iiitbh.ac.in[3]; utkarsh.240201022@iiitbh.ac.in[4]

DOI: 10.1201/9781003715726-19

tissues, computational systems can quantitatively characterize disease presence, differentiate among pathological categories, and estimate disease severity—thereby augmenting human diagnostic capacity in agricultural contexts.

Contemporary deep learning architectures, particularly convolutional neural networks, have demonstrated exceptional capability in visual pattern recognition tasks applied to plant health assessment. This investigation implements a comprehensive evaluation framework utilizing a curated dataset comprising 10,000 botanical image specimens distributed across ten disease classes (including non-diseased healthy tissue samples). All imagery was standardized to uniform dimensional resolution (256 × 256 pixels) and systematically partitioned into training, validation, and testing subsets to enable rigorous model performance quantification and validation of generalization capacity.

This research endeavors to develop a reproducible, scalable computational pipeline capable of automating disease detection and categorization in tomato cultivation systems utilizing deep convolutional neural network architectures. Specific objectives encompass: (1) minimizing agricultural production losses through early pathological identification, (2) reducing farmer dependence on labor-intensive manual surveillance protocols, and (3) implementing an intelligent decision support system enabling timely and informed disease management interventions at both individual farm and regional agricultural scales.

The chapter will be arranged as follows: Motivation is introduced in Section 2, Literature Review is given in Section 3, Materials and Methods in detail is presented in Section 4, Experimental Results are explained in Section 5, in the end discussion and future directions are committed to Section 6.

2. Motivation

Tomato cultivation represents one of the foremost horticultural enterprises globally, with production systems established across diverse climatic zones and geographical regions. India maintains prominent status within international tomato production networks, yet the sustainability and productivity of tomato cultivation systems face escalating threats from multiple pathogenic organisms—including fungal taxa, bacterial pathogens, and viral disease agents. These biological threats collectively impose significant constraints on production of viability and economic returns for agricultural stakeholders.

Prevalent phytopathological conditions affecting tomato production systems—encompassing early blight (Alternaria solani), late blight (Phytophthora infestans), leaf mold (Fulvia fulva), and target spot (Corynespora cassiicola)—significantly compromise crop yield metrics, diminish fruit quality characteristics, and reduce market competitiveness of harvested produce. The cumulative economic impact of these disease manifestations translates into substantial financial losses for farming communities, affecting both immediate revenue generation and long-term agricultural viability.

Contemporary agricultural practice in tomato cultivation continues to rely heavily on conventional visual assessment methodologies for disease identification and management decisions. These traditional diagnostic approaches require specialised domain knowledge and trained personnel expertise yet remain fundamentally limited in scalability and consistency across large operational areas. The inherent limitations of manual detection protocols—coupled with constraints in labor availability, time-intensive assessment procedures, and implementation costs—collectively result in delayed disease recognition, inadequate intervention timing, and substantial crop losses. Furthermore, reliance on therapeutic pesticide applications in response to visual disease identification frequently results in excessive chemical usage, which escalates operational costs, engenders environmental concerns, and threatens ecological sustainability of agricultural systems.

3. Literature Review

This chapter synthesizes recent advancements in automated disease recognition within tomato cultivation through a comprehensive examination of methodologies spanning computational intelligence techniques. Given the substantial threat that phytopathological conditions pose to production metrics and commercial viability, the development of rapid, dependable diagnostic systems has become imperative in contemporary farming practices.

- Sembiring et al. (2021) demonstrated effective classification of six tomato disease categories through decision tree methodologies employing Gini impurity metrics, supplemented by machine learning ensemble techniques including genetic algorithm-optimized support vector machines and adaptive neuro-fuzzy inference systems.
- Sembiring et al. (2021) developed a streamlined convolutional neural network framework utilizing fuzzy logic systems and feedforward backpropagation neural networks to facilitate automated disease recognition from botanical imagery. Their approach extracted morphological and chromatic attributes including coloration, dimensional parameters, and textural characteristics, achieving 87.2% classification accuracy despite the limited disease category scope.
- Rangarajan et al. (2018) conducted a comparative evaluation of disparate computational paradigms—encompassing convolutional neural networks, random forest classifiers, k-nearest neighbor algorithms, and Naïve Bayes methodologies—revealing convolutional networks' superior performance (96% accuracy) in pattern recognition tasks, though emphasizing the necessity of large-scale, well-curated datasets for optimal model efficacy.
- DL et al. (2024) presented contemporary methodologies for tomato leaf disease detection utilizing convolutional neural network implementations, contributing to the evolving landscape of deep learning applications in horticultural pest management and disease diagnostics.
- Wajid et al. (2024) investigated the deployment of mobile-optimized neural network architectures utilizing MobileNet implementations integrated

with deep convolutional methodologies for identification of ten prevalent tomato disease phenotypes. Their experimental framework incorporated multiple optimization algorithms including Adam, RMSProp, and stochastic gradient descent variants, attaining 89.2% classification performance while recommending enhancements in dataset quality and augmentation methodologies.

- Elhassouny and Smarandache (2019) designed an intelligent mobile application framework leveraging convolutional neural network architectures to facilitate recognition of tomato leaf pathological conditions. The system enabled field-based disease identification with practical deployment considerations for agricultural stakeholders.
- Ramakrishna (2020) examined the application of ResNet-152v2 and MobileNetv2 architectures, which demonstrated substantial performance metrics (96.79% accuracy) alongside computational efficiency in disease diagnosis workflows. The research underscored the practical applicability of deep neural network implementations within genuine agricultural operational contexts.
- Mitra et al. (2023) employed transfer learning methodologies leveraging pre-trained AlexNet architectures to enhance disease classification performance metrics, achieving 97.49% classification accuracy and demonstrating the robustness of convolutional neural network paradigms for automated tomato pathology detection.
- Hasan et al. (2019) implemented transfer learning approaches utilizing pre-trained VGG-16 architectures for precision agriculture applications in tomato disease detection. Their methodology yielded 92.3% classification accuracy, substantiating the efficacy of fine-tuned deep neural networks in botanical pathology assessment.
- Brahimi et al. (2017) presented a GoogleNet-based convolutional neural network framework achieving 99.185% classification accuracy on a dataset comprising 14,828 tomato leaf specimens. The architecture demonstrated superior performance relative to traditional AlexNet implementations and conventional machine learning approaches including support vector machines and random forest classifiers.
- Rangarajan et al. (2018) implemented ResNet-18 architecture incorporating binary cross-entropy loss functions and comprehensive performance evaluation metrics including F1-score (91%) and area under the receiver operating characteristic curve (0.92), attaining 97% overall classification accuracy. The methodology successfully balanced computational resource requirements with prediction accuracy, establishing efficient trade-offs between model complexity and performance output.
- Femmi et al. (2021) integrated internet-of-things infrastructure with deep neural network architectures (VGG-19) within cloud-computing frameworks for distributed disease monitoring. Their system achieved 95% accuracy while

prioritizing user-accessible interfaces tailored for agricultural practitioners and stakeholders.

- Jayanthi et al. (2024) utilized convolutional neural network implementations trained on a dataset of 3,000 botanical images to accomplish 98.49% classification accuracy for tomato disease identification. The investigation recommended incorporating abiotic stress indicators—including nutrient insufficiency manifestations—into future research frameworks.
- Trivedi et al. (2021) proposed advanced neural network architectures combining DenseNet169 and InceptionV3 implementations, achieving 99.80% classification accuracy. The research emphasized the substantial potential of deeply layered neural networks for robust disease categorization and subsequent deployment on mobile computational platforms.
- Batool et al. (2020) conducted comparative performance analysis across lightweight neural network variants including VGG-16, ShuffleNet, and SqueezeNet architectures. Results indicated VGG-16's superior performance (98.28% accuracy) in disease classification tasks, while acknowledging opportunities for computational efficiency optimization.
- Sabrol & Kumar (2016a) developed and evaluated the MFR-CNN architecture against competing implementations including AlexNet, VGG-16, and ResNet-18. The proposed methodology demonstrated near-optimal classification accuracies in both controlled laboratory environments (99.09%) and operational field conditions (98.75%), substantiating its practical utility for real-time agricultural monitoring applications.
- Li et al. (2024) investigated convolutional neural network implementations incorporating multi-scale feature extraction and feature reuse mechanisms for tomato leaf disease identification. The methodology advanced computational efficiency while maintaining high classification performance metrics through intelligent architectural design.
- Aggarwal (2022) implemented deep learning frameworks for biological classification of tomato leaf disease manifestations. The research contributed to the expanding corpus of knowledge regarding neural network applications in precision agriculture and botanical health assessment systems.
- Vinay et al. (2021) conducted comparative analysis of convolutional neural network and ResNet-50 architectures utilizing a 10,000-image dataset obtained from public repositories (Kaggle). Results demonstrated superior performance of standard convolutional neural networks (94.1% accuracy) relative to ResNet-50 implementation (88.44% accuracy).
- Sakkarvarthi et al. (2022) implemented a hybrid computational approach integrating backpropagation neural networks and k-nearest neighbor classification methodologies, with AlexNet achieving enhanced performance (91% accuracy) compared to manual classification methodologies. The authors advocated for dataset expansion and implementation of pre trained feature extraction models in subsequent investigations.

- Gunarathna and Rathmayaa (2020) developed and evaluated efficient deep learning models specifically calibrated for tomato plant disease classification tasks based on leaf image analysis. The research contributed novel methodological insights regarding model optimization for agricultural applications with computational constraints.
- Sabrol and Satish (2016b) employed hierarchical classification tree methodologies applied to digital tomato leaf imagery for disease categorization. The approach demonstrated the viability of structured decision-based algorithms in botanical pathology assessment, offering interpretable classification pathways for agricultural practitioners.
- Singh Chouhan et al. (2025) optimized YOLO-based neural network architectures for real-time horticultural object detection, implementing probabilistic filtering mechanisms with enhanced feature fusion strategies. The methodology advanced the field of precision agriculture through improved detection accuracy and computational efficiency.
- Shukla et al. (2025) presented AgriCloud—a cloud-based computational infrastructure designed for scalable crop monitoring applications and intelligent decision support systems. The framework enabled distributed processing of agricultural data with enhanced accessibility for farming communities and extension services.
- Tripathi et al. (2025) synthesized contemporary artificial intelligence methodologies with agricultural practice frameworks to cultivate sustainable farming futures. The research integrated neural network technologies with horticultural management strategies to advance food security and environmental stewardship objectives.
- Chouhan et al. (2025) examined unmanned aerial vehicle integration within contemporary agricultural operational contexts, analyzing technical considerations, practical solutions, and economic implications. The systematic analysis provided evidence-based guidance for designing scalable, automated crop monitoring systems suitable for precision agriculture deployment.
- Dubey and Choubey (2024a) developed automated leaf disease detection utilizing artificial neural network optimization for blast disease identification in rice systems.
- Dubey and Choubey (2024b) advanced disease prediction through Optimized Support Vector Machine (OSVM) implementation for blast disease detection in paddy crops. The framework integrates preprocessing, feature extraction, and OSVM classification, substantiating machine learning classifier performance for yield loss prediction, infection detection, and host resistance identification in automated disease diagnostics. The work done on existing Tomato Plant Disease is given in Table 1.

Table 1 Summary of existing work for Tomato plant diseases.

Ref. No	*Dataset*	*Techniques Used*	*Purpose*	*Merits*	*Future Work*	*Accuracy*
Hammou, D. et al. (2022)	Plant villages (tomato plant)	Resnet 152 v2 and MobileNet v2	This paper presents a deep learning method using Resnet 152v2 and MobileNet to accurately identify infections of tomato leaves, improve yield, and demonstrate the operation of models for deep learning to achieve accuracy in disease diagnosis.	ResNet-152v2 and MobileNetv2 provide excellent results in the detection of tomato leaf spot disease. They provide accuracy and speed, with ResNet-152v2 reaching 96.79%. These models automate the inspection process, using less manual production.	Future work on this paper will focus on building better models for the real world, with expanded datasets for a variety of sites and diseases.	96.79%
Sembiring, A. et al. (2021)	NHB Database 2013-14	Fuzzy inference systems, Feedforward backpropagation neural networks, Supervised learning	To achieve good accuracy from automatic plant disease identification and diagnosis using image processing.	Extracting features of images, i.e., color, size, texture features.	The main issue is less no. of diseases and also improve accuracy using technology.	FF87.2%
Rangarajan, A. et al. (2018)	Tomato plant leaves	CNN Random Forest Naive Bayes k-NN AdaBoost	In order to ascertain whether approaches are more effective, precise, and quick in disease identification, this article compares the utilization of ML and DL—more especially.	CNNs have higher accuracy and higher learning ability for complex tasks involving pattern recognition, however they rely on big data for efficient work.	In future focus on addressing the dependency of CNNs on huge datasets, well maintained datasets to improve performance, faster training.	CNN-96%

Contd...

Table 1 *Contd...*

Ref. No	*Dataset*	*Techniques Used*	*Purpose*	*Merits*	*Future Work*	*Accuracy*
DL, S. et al. (2024)	Plant Village Dataset	CNN, InceptionV3, ResNet 152, VGG19	Improving CNN performance model with different epochs Compare transfer learning tech. To CNN model.	Balance the dataset of all classes. Comparison between different models like VGG19, RestNet 152, Inception V3.	Improve accuracy by implementing different models.	98%
Wajid, A. et al. (2024	Tomato leaf data set	StocasticGredient SGD with Momentum Adam Proximal Gradient Descent Proximal Adagrad RMSProp CNN	The purpose of this research is to use Deep CNN to create an intelligent mobile application for monitoring tomato plant diseases. Detects the 10 most common tomato diseases from photographs, making it suitable for direct use in agricultural equipment.	This paper provides effective different types of tomato leaf detection using MobileNet based deep CNN developed for mobile applications.	In future focus on improve tomato leaf spot detection accuracy using high quality augmented image datasets and using advance data processing technique.	89.2%

Contd...

Table 1 *Contd...*

Ref. No	*Dataset*	*Techniques Used*	*Purpose*	*Merits*	*Future Work*	*Accuracy*
Elhassouny, A. et al. (2019)	Plant Village	Feed-forward Neural Network	Compare NN with the Machine Learning methods SVM classification.	There are many algorithms and techniques but using neural networks can be more discriminatory for planarian organisms as they combine the best techniques for precise categorization.	It use less number of diseases, also improve accuracy.	97%
Ramakrishna, R. et al. (2019)	Tomato plant disease	Supervised learning technique	Classifying accuracy of six different kinds of tomato plant images is yielding classification.	For classification they use a total 83 nodes. The splitting criteria based on Gini index is used for dividing the sub nodes further till the nodes not generated for various categories.	The strategy uses different types of classification such as modified neuro fuzzy, neural networks, and genetic algorithms supporting SVM.	97.3%
Mitra, D. et al. (2023)	Plant Village	Alex Net VGG-16	This paper's goal is to utilize the first deep learning model, Alex Net, VGG16.	Transfer learning is a deep learning process used to classify new products.	The future work is to improve accuracy while increasing batch size, increase of bias learning and weight rate	VGG16-97.49%

Contd...

Table 1 *Contd...*

Ref. No	*Dataset*	*Techniques Used*	*Purpose*	*Merits*	*Future Work*	*Accuracy*
Hasan, M. et al. (2019)	Tomato leaf picture	CNN VGG 16	Tomato leaf disease controls tomatoes and reduces yield and quality potential by preventing or stopping from progression of the disease.	Machine learning, especially CNN, achieves high accuracy (92.3%) in identifying tomato leaf diseases with specific characters.	Algorithm May not able to identify all types of accurate disease and show high rate of false positive.	CNN-92.3%
Brahimi, M. et al. (2017)	Tomato dataset	CNN	The main purpose of this research is to create an intelligence-based application of Deep learning model tools for early tomato disease detection.	Researchers are using CNNs to develop AI-powered Deep learning detection and categorization of tomato plant diseases.	We'll concentrate on growing the dataset in the future to improve the model's functionality and precision.	89%
Rangarajan A. et al., (2018)	14828 leaves of tomato infected with nine diseases. (www. PlantVillage. org)	Both shallow and deep learning models	Shallow models (SVM, Random forest) vs. deep learning (AlexNext vs GoogleNet)	Automatic extraction and categorization of tomato leaves lesions using deep-learning (CNN). symptoms.	Explore best practice visualization techniques to improve identification and understanding of plant diseases. Extend configuration file to include additional faults and environments.	Google Net 99.185% Google Net are more accurate the Alex Net

Contd...

Table 1 *Contd...*

Ref. No	*Dataset*	*Techniques Used*	*Purpose*	*Merits*	*Future Work*	*Accuracy*
Femmi et al. (2021)	Tomato leaves	Deep learning using ResNet-18 architecture.	Detect tomato diseases early to prevent commercial losses and promote permaculture. Focus on effective classification and disease management utilizing deep learning tools.	The AUC percentage of the ROC curve is high (0.92). Leverage ResNet-18 to achieve a balance between computation and accuracy.	Improved standards for deployment on edge devices and mobile platforms.	F1 score-97%
Jayanthi et al. (2024)	1000 Tomato plant disease	Image segmentation - Feature extraction - VGG19 deep learning model	Identifying and identifying diseases on tomato leaves and providing treatments can help improve crops and permaculture practices.	• Accurate disease classification • Instant feedback • Closed loop • User-friendly interface for farmers • Scalable cloud-based infrastructure.	• Improve real-time image quality • Explore other IoT sensors • Optimize infrastructure in the cloud for improved efficiency	VGG19 95%

Contd...

Table 1 *Contd...*

Ref. No	*Dataset*	*Techniques Used*	*Purpose*	*Merits*	*Future Work*	*Accuracy*
Trivedi et al. (2021)	Daraset - 3000 images with nine different diseases and a healthy leaf.	CNN model	A CNN-based deep learning model with 98.49% accuracy is reported in the research.	The paper's benefits include the effective use of neural networks for classification and identification of tomato plant diseases with 98.49% accuracy.	Future research attempts to include abiotic factors resulting from nutritional deficiencies, expand the data, and improve accuracy through technology.	98.49%
Batool, A. et al. (2020)	Plant Village (Tomato Plant Leaves)	DenseNet169, inceptionV3	In this paper researchers use CNN model DenseNet169 and InceptionV3, to detect and classify diseases of tomato plants.	This paper solves an important problem in agriculture by using deep learning, especially on CNN architecture such as DenseNet169, InceptionV3 to achieve 99.80% accuracy.	Improving model for real- time disease detection on mobile devices. Creating strong environmental protection, such as lighting or weather conditions.	DenseNet-99.80%
Sabrol and Kumar (2016a)]	Plant Village	CNN: - VGG-16 Shuffle Net Squeeze Net	This paper's primary objective is to compare the accuracy of VGG16, ShuffleNet and SqueezeNet and improve their accuracy by consuming less time compared to existing CNN architectures.	The performance of CNN architecture when detecting tomato disease using tomato leaves can be better than existing CNN architectures in terms of detection accuracy.	The biggest challenge in the future is to improve CNN's accuracy by spending a short amount of time on resources and planting tomatoes right from the farm.	VGG-16 98.28%

Contd...

Table 1 *Contd...*

Ref. No	*Dataset*	*Techniques Used*	*Purpose*	*Merits*	*Future Work*	*Accuracy*
Li et al. (2024)	Lab based, Field based	MFR CNN	Compare MFR CNN model with AlexNet, VGG-16.	The MRF CNN model had the highest accuracy and it also took less computing time and few training parameters.	Future work will focus on enhancing time consuming adding extra high quality data images to the dataset which improve model generation.	Lab-99.09%, field-98.75%
Aggarwal (2022)	Plant Village	Deep learning	Comparison of machine learning techniques and deep learning.	Increase score of system by deep-learning.	Their inability to gain from all diseases is one of the issues.	98.99%
Vinay, P. et al. (2021)	Plant Village Dataset	Gray Level Cooccurrence Matrix (GLCM) Extreme learning machine (ELM)	Classifying and identifying disease of tomato plant.	GLCM and ELM both techniques are also identifying disease which is occurring in stem, root, fruits, etc.	Not to mention about the accuracy of the model.	——
Sakkarvarti (2022)	10000 images from keggle	CNN, ResNet 50	Compare CNN and ResNet 50.	CNN standard is better than ResNet 50	The procedure used to increase accuracy has been fixed.	CNN- 94.1% ResNet-88.44%
Gunarathna & Rathmayaa (2020)	PLANT VILLAGE	Back propagation Neural Network	Extract features leaves from tomato plants with various models and then use KNN as a classifier to find accuracy.	The deep NN in this study (AlexNet) achieves better accuracy than the manual model.	In future extracting features using different pre trained models for better performance, also used large datasets.	91%

4. Materials and Methods

4.1 Background

This methodological framework delineates a systematic computational approach for constructing an automated phytopathological diagnostic system tailored to tomato cultivation utilizing contemporary deep neural network paradigms. The architectural design prioritizes model accuracy, computational efficiency, and deployment viability across diverse agricultural operational environments requiring scalable disease monitoring solutions.

The experimental pipeline initiates with procurement of curated photographic specimens depicting foliar manifestations of prevalent tomato pathologies—encompassing bacterial spot, early blight (Alternaria solani), late blight (Phytophthora infestans), and leaf mold (Fulvia fulva). Subsequently, acquired imagery undergoes systematic preprocessing transformations including dimensional standardization, pixel intensity normalization, categorical label assignment, and quality filtration to eliminate specimens exhibiting motion blur, inadequate illumination, or excessive background interference.

Model generalizability and resistance to overfitting phenomena are enhanced through strategic implementation of image augmentation protocols. Applied transformations encompass geometric manipulations (rotational shifts, horizontal/vertical mirroring), photometric adjustments (luminosity variation, contrast modification), and noise injection strategies. These augmentation methodologies simulate environmental variability characteristics of field-acquired botanical imagery, thereby improving model performance across heterogeneous acquisition conditions.

Preprocessed image collections are subsequently stratified into mutually exclusive training, validation, and evaluation of subsets following established machine learning partitioning conventions. Comparative architectural analysis encompasses both custom-designed convolutional frameworks and transfer learning implementations utilizing pretrained feature extractors—specifically VGG19, ResNet50, DenseNet169, AlexNet, and MobileNetV2 architectures. Uniform dataset allocation across all candidate models ensures methodologically rigorous performance benchmarking.

Model optimization employs adaptive gradient descent algorithms (Adam optimizer) augmented with regularization mechanisms including early termination criteria and periodic weight serialization (checkpointing) to mitigate overfitting trajectories while promoting convergence stability. Post-training performance quantification utilizes standardized classification metrics—overall accuracy, class-wise precision, recall sensitivity, harmonic mean F1-score, and multi class confusion matrices—enabling comprehensive assessment of discriminative capability across all pathological categories.

Culminating analytical procedures encompass systematic inter-model comparison revealing architectural strengths, computational requirements, and generalization characteristics. The optimal-performing architecture undergoes subsequent evaluation for practical deployment considerations, specifically targeting mobile platform integration to facilitate accessible, real-time disease diagnostics for agricultural practitioners—thereby enabling proactive crop protection measures, yield optimization, and sustainable farming practice implementation.

4.2 Methodology Flowcharts

Figure 1 shows the methodology flow chart, which describes the approach to detecting tomato plant diseases.

4.3 Proposed Methodology

4.3.1 Dataset Collection

The dataset https://www.kaggle.com/datasets/naveedgull/tomato-leaf-disease encompasses 10 mutually exclusive categorical divisions, with each classification corresponding to a specific pathological entity or asymptomatic (healthy) plant tissue phenotype. Individual botanical specimens within each category undergo meticulous annotation and segmentation, delineating regions of healthy foliar tissue from regions exhibiting disease manifestations. This granular categorical classification facilitates robust supervised learning paradigms by providing heterogeneous training exemplars. The annotated dataset enables quantitative model evaluation protocols and facilitates comparative assessment of prediction reliability across all disease categories.

Systematic preprocessing protocols transform raw botanical imagery into standardized computational representations optimizing downstream neural network processing. Dimensional normalization standardizes all photographic specimens to uniform resolution (256 × 256 pixels) reducing computational burden while preserving discriminative pathological features. Chromatic reduction through grayscale conversion emphasizes morphological and textural characteristics while simultaneously diminishing computational requirements. Pixel intensity scaling to the normalized interval [0, 1] enhances numerical stability within network computations and accelerates convergence during optimization procedures. Binary mask generation isolates foliar regions of interest, attenuating non-target background information and directing model attention toward disease manifestations. Augmentation strategies employing geometric transformations (rotation ± 30°, directional flipping) and intensity modifications create synthetic training exemplars, simulating naturally occurring field variability and improving model generalization across heterogeneous acquisition conditions.

These pictures are divided into 10 different groups, each of which represents a common tomato crop disease:

1. Bacterial Spot
2. Early Blight
3. Late Blight

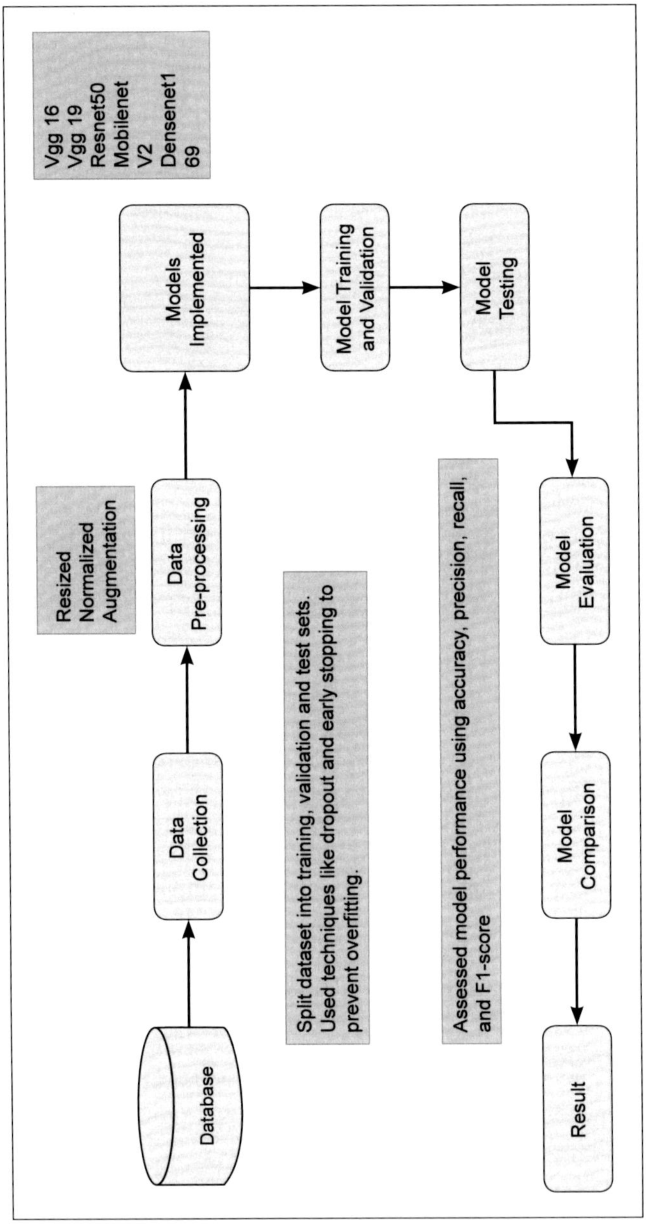

Fig. 1 Full Methodology Diagram.

4. Leaf Mold
5. Mosaic Virus
6. Septoria Leaf Spot
7. Yellow Leaf Curl Virus
8. Target Spot
9. Two-Spotted Spider Mites
10. Healthy

Specimen acquisition protocols incorporated heterogeneous environmental parameters encompassing geographically disparate collection sites, temporal variation across seasons, and photographic conditions spanning diverse illumination intensities. This intentional environmental heterogeneity enhances model robustness and transferability, ensuring trained architectures maintain predictive accuracy across field deployment scenarios where acquisition conditions deviate from controlled laboratory environments. Dataset variability simulates operational agricultural contexts, preparing models for real-world robustness requirements.

Individual botanical specimens within the repository manifest unambiguous pathological symptoms characteristic of their assigned disease category, facilitating both automated feature extraction via deep neural networks and manual expert validation through visual inspection. Categorical balance across all disease classes prevents algorithmic learning bias toward prevalent categories, enabling unbiased feature discrimination across all pathological entities and healthy references.

This rigorously curated, comprehensively annotated, and environmentally heterogeneous botanical image repository provides a methodologically sound foundational dataset for training robust, generalizable classification architectures capable of automated disease detection and categorization in tomato cultivation systems across diverse operational contexts.

4.3.2 Dataset Statistics

The first step of a deep learning project is data collection (https://www.kaggle.com/datasets/naveedgull/tomato-leaf-disease); the quality and diversity of image datasets has a direct impact on the model's ability to learn and perform. For tomato plant disease detection, to gain a comprehensive and well-labeled image dataset, it is very essential to accurately classify various diseases. In Table 2 we show types of diseases which have a high number of images.

Table 2 Class distribution of the dataset.

S. No	*Types of Disease*	*No. of Images*
1	Bacterial_spot	1000
2	Early_blight	1000
3	Late_blight	1000
4	Leag_mold	1000
5	Mosaic_virus	1000

Contd...

Table 2 *Contd...*

S. No	*Types of Disease*	*No. of Images*
6	Septoria_leaf_spot	1000
7	Yellow_leaf_spot	1000
8	Two spotted spider mites	1000
9	Target spot	1000
10	Healthy	1000

4.3.3 Image preprocessing

Photographic image preparation through systematic preprocessing protocols constitutes a critical phase in deep learning pipelines, directly influencing model learning efficiency and performance metrics. The preprocessing stage addresses inherent variability in field-acquired botanical imagery through standardization transformations. This investigation implements a comprehensive suite of image transformation techniques to normalize spectral characteristics, reduce computational requirements, and enhance feature discriminability prior to neural network processing.

- **Image Resizing:** Dimensional standardization represents an essential preprocessing step, converting heterogeneous input resolutions to uniform computational representations. All botanical specimens undergo spatial rescaling to standardized dimensions (256 × 256 pixels), optimizing computational efficiency while preserving pathologically discriminative features. The rescaling transformation is mathematically expressed through bilinear interpolation, mapping source image coordinates to target resolution coordinates while maintaining pixel intensity fidelity.

$$I'(x', y')\, I\left(\frac{x' \times W}{W'}, \frac{y' \times H}{H'}\right) \qquad ...(1)$$

  In Eq. (1), $I(x', y')$ represents target image pixel intensity at rescaled coordinates, (x', y') or in the resized images $I(x, y)$ denotes source image pixel intensity, and (x, y) coordinates of the original images and W, H are the width and height of the images all are set to 256 pixels in this research paper.

- **Gray Scaling:** Spectral simplification through grayscale conversion emphasizes morphological and textural disease indicators while reducing computational burden. Conversion from RGB color space to monochromatic representation weights red, green, and blue channels according to human perceptual luminosity sensitivity, producing intensity-normalized imagery emphasizing pathological morphology independent of pigmentation artifacts.

$$G(x, y) = 0.298 \cdot R(x, y) + 0.586 \cdot B(x, y) \qquad ...(2)$$

  In Eq. (2) $G(x, y)$ show the grayscale value at image pixels (x, y), and $R(x, y)$, $G(x, y)$, and $B(x, y)$ show the intensities of red, green and blue channel

- **Image Normalization:** Feature scaling through normalization enhances numerical stability in network computations and accelerates optimization convergence. Intensity values spanning the range [0, 255] are mapped to the normalized interval [0, 1] through min-max feature scaling, equalizing pixel value distributions across the dataset.

$$I_{norm} = \frac{I_{gray} - \min(I_{gray})}{(I_{gray}) - \min(I_{gray})} \qquad ...(3)$$

- **Image Masking:** Binary mask generation creates region-of-interest segmentation, isolating foliar tissue from non-target background imagery through intensity-based thresholding. Masking operations direct model computational resources toward pathologically relevant regions, attenuating background noise and environmental artifacts that could compromise classification performance.

$$M(x, y) = \frac{I(x, y)}{0}\} \text{ if } I(x, y) \in ROI \qquad ...(4)$$

Where: $M(x, y)$ represents masked image pixel value, ROI denotes region-of-interest.

4.3.4 Data Augmentation

Augmentation strategies artificially expand training dataset cardinality through geometric and photometric transformations, simulating naturally occurring imaging variability and environmental factors. Augmentation techniques enhance model generalization by exposing neural architectures to diverse manifestations of target pathological entities under heterogeneous acquisition conditions (Table 3).

Photometric normalization through min– max rescaling transforms pixel values from the original range [0, 255] to the normalized interval [0, 1], enhancing computational stability during gradient descent optimization.

Rotational augmentation applies to random angular displacements within the range [−30°, +30°], simulating variable camera orientations and natural leaf positioning variability in field environments.

Bidirectional reflection transformations (horizontal and vertical flipping) generate mirror-image variants, doubling augmented dataset cardinality while introducing rotational symmetry variations representative of natural leaf positioning heterogeneity. [30°], simulating variable camera orientations and natural leaf positioning variability in field environments.

Brightness modification protocols randomly modulate pixel intensities within the interval [−0.2, +0.2] normalized units, simulating variable field illumination conditions including shadow effects, overexposure, and suboptimal lighting scenarios encountered in agricultural environments.

$$I_{\text{rescale}} = I \times 1/255 \qquad ...(5)$$

$$I_{\text{rescale}} = \text{rotate}\ ((I_{\text{rescale}}, \theta) \quad \text{where } -30\ \theta \leq 30 \qquad ...(6)$$

Table 3 Different Techniques of Data Augmentation.

Technique	*Description*
Rotation	Random rotation up to ± 30°
Directional Flipping	Horizontal and vertical flipping
Luminosity Adjustment	Modify brightness randomly

4.4 Data Evaluation

Systematic stratification of the complete image repository into mutually exclusive training, validation, and evaluation cohorts enables rigorous model performance quantification. This partitioning protocol prevents information leakage and provides unbiased estimates of generalization capability on previously unseen botanical specimens.

The 80-10-10 allocation strategy follows established machine learning conventions, providing: (1) substantial training exemplars for robust parameter estimation, (2) adequate validation samples for hyperparameter optimization, and (3) independent evaluation cohort for unbiased performance assessment on held-out data not encountered during model training or tuning.

Table 4 shows the dataset lengths for the train, validation, and test datasets.

Table 4 Length of train, valid, and test dataset.

Split	*Quantity of Images*	*Purpose*
Length of train dataset	7000	Model training
Length of valid dataset	2000	Hyper-parameter tuning
Length of test dataset	1000	Final performance evaluation
Length of total dataset	10000	

Independent evaluation cohort assessment quantifies model generalization capability—the critical metric for real-world agricultural deployment scenarios. Evaluation subset separation from training and validation data ensures performance estimates reflect true predictive accuracy on novel botanical specimens encountered in operational field environments.

4.5 Model Selection

This investigation implements a comparative evaluation protocol encompassing six prominent deep convolutional neural network architectures, each representing distinct design paradigms in computational feature extraction and disease classification methodologies. Architecture selection prioritizes demonstrated efficacy in domain-specific image classification tasks, scalability considerations, and deployment feasibility across heterogeneous computational platforms.

4.5.1 VGG19

VGG19 represents a prototypical deep convolutional framework comprising 19 learnable weight layers—encompassing sixteen sequential convolutional layers organized into five convolutional blocks with diminutive 3 × 3 receptive fields and three fully connected terminal layers. Within the experimental training context, VGG19 achieved substantial learning performance—training accuracy of 99.0% (L = 0.4215)—indicating effective pattern memorization from training exemplars. However, evaluation on independent test data revealed performance degradation to 91.0% test accuracy (L = 0.645), suggesting moderate overfitting phenomena typical of deep architectures when confronted with distributional divergence between training and evaluation data, though the hierarchical feature extraction capability and demonstrated efficacy in complex image classification tasks establish its utility as a comparative benchmark.

4.5.2 VGG16

VGG16 represents an earlier architectural variant that established foundational performance standards for photographic image classification, comprising sixteen learnable layers—13 convolutional layers organized into block-wise hierarchies interspersed with spatial downsampling operations, supplemented by three fully connected classification layers with uniform 3 × 3 filter dimensions and stride-1 operations preserving spatial information fidelity. The architecture leverages pre trained weight initialization derived from ImageNet corpus training (1.2 million annotated specimens across 1,000 object categories), enabling efficient fine-tuning on agricultural disease classification tasks through reduced training iterations while preserving learned feature representations from diverse natural imagery, achieving 96.5% training accuracy (L = 0.3421) and 90.0% test accuracy (L = 0.5892).

4.5.3 ResNet 50

ResNet50 introduces residual learning paradigms through skip-connection mechanisms enabling stable training of substantially deeper architectures, with the residual design principle facilitating gradient flow optimization and mitigating vanishing gradient pathologies inherent in ultra-deep sequential networks following the computational pipeline: Input → Conv2D → BatchNorm → ReLU → MaxPool → Residual Blocks → GlobalAveragePooling → Dense Classification Layer. Within experimental training protocols, ResNet50 demonstrated robust learning trajectories—training accuracy reached 98.0% with minimal training loss (L = 0.0622), indicating effective parameter optimization, with evaluation on held-out test specimens yielding 92.0% test accuracy (L = 0.4217), demonstrating preserved generalization capability with minimal performance degradation and establishing ResNet50 as an optimal candidate balancing classification accuracy with computational efficiency.

The architecture is defined as the following layers.

$$\textit{Input}\rightarrow\textit{Conv2D}\rightarrow\textit{BatchNorm}\rightarrow\textit{ReLU}\rightarrow\textit{MaxPool}\rightarrow \textit{Residual Block}\rightarrow\textit{GlobalAvgPool}\rightarrow\textit{dense} \quad ...(7)$$

4.5.4 MobileNet V2

MobileNetV2 represents a specialized CNN architecture engineered for computational resource-constrained environments including embedded systems, mobile devices, and field-deployable edge computing platforms, standardizing photographic specimens to uniform 224 × 224 pixel RGB format processed through initial depthwise separable convolutional layers with stride-2 spatial downsampling, with depthwise separable convolution mechanisms dramatically reducing learnable parameters and computational burden compared to standard convolutional operations. Training optimization yielded 83.0% training accuracy (L = 0.6926), while test performance evaluation on unseen specimens yielded 61.0% classification accuracy (L = 0.9145), indicating moderate performance relative to deeper architectures but substantially lower computational overhead enabling real-time agricultural deployment scenarios, establishing MobileNetV2 as a viable candidate for field-based disease detection applications accessible to resource-limited agricultural stakeholders.

4.5.5 AlexNet

AlexNet represents a seminal architectural contribution demonstrating exceptional performance on large-scale image classification challenges and establishing foundational principles for contemporary deep learning systems, comprising eight learnable layers with specialized hierarchical organization: five convolutional layers interspersed with local response normalization and spatial downsampling operations (max-pooling) comprising the feature extraction stage, followed by three fully connected layers providing classification functionality, with formal representation: $(\text{Conv}\rightarrow\text{ReLU}\rightarrow\text{LRN}\rightarrow\text{MaxPool})^2 \rightarrow (\text{Conv}\rightarrow\text{MaxPool}) \rightarrow (\text{FullyConnected}\rightarrow\text{Dropout})^2 \rightarrow$ Output Linear Classification. Training optimization yielded 90.0% training accuracy (L = 0.3486), with test evaluation on independent specimens demonstrating 83.0% accuracy (L = 0.6234), indicating preserved generalization with acceptable performance-efficiency tradeoff suitable for comparative architectural evaluation.

The architecture structure can be written below;

$$(\text{CONV} \rightarrow \text{RN} \rightarrow \text{MP})^2 \rightarrow (\text{CONV}^3 \rightarrow \text{MP}) \rightarrow (\text{FC} \rightarrow \text{DO})^2 \rightarrow \text{Linear} \rightarrow \text{softmax} \quad ...(8)$$

Where:

- Conv = convolutional layer (with ReLU activation)
- RN = Local response normalization
- MP = max pooling
- FC = fully connected layer (without activation)
- DO = dropout

4.5.6 DenseNet169

DenseNet169 implements dense connectivity patterns establishing direct feedforward pathways between all layer pairs within compositional dense blocks, fundamentally differing from sequential layer stacking and promoting efficient feature propagation and information reuse across hierarchical processing stages, comprising four compositional dense blocks with respective layer distributions [6, 12, 32, 32] with transition layers providing intermediate spatial downsampling and channel compression optimizing computational efficiency. Dense connectivity patterns effectively mitigate vanishing gradient phenomena through multiple gradient propagation pathways while enforcing aggressive feature reuse, substantially reducing learnable parameter counts relative to sequential equivalents, positioning DenseNet169 as a parameter-efficient architecture suitable for resource-constrained deployment scenarios; empirical training yielded 94.0% training accuracy (L = 0.3156) with test evaluation yielding 89.0% classification accuracy (L = 0.4892), demonstrating strong generalization characteristics across diverse disease manifestations.

5. Experimental Results

Experimental results present the performance results of ResNet50, AlexNet, VGG16, VGG19, DenseNet169, and MobileNetV2 When Applied to the Plant Village dataset. The use of different deep learning models for classifying tomato plant diseases.

5.1 Result of VGG16

The result of VGG16 is very good , this model performs very well with the accuracy of 91.93% and the other scores are in metrics. It has simple architecture which allows for stable training. Its accuracy is higher than MobileNetV2 (Figures 2, 3).

5.2 Result of ResNet50

The result of ResNet50 is great, this model performs very well with the accuracy of 95.04% and the other scores are in metrics. It has simple architecture which allows for stable training. Its accuracy is higher than MobileNetV2 (Figures 4, 5).

5.3 Result of VGG19

The result of VGG19 is great, this model performs very well with the accuracy of 92.94% and the other scores are in metrics. It has simple architecture which allows for stable training. Its accuracy is higher than MobileNetV2, and VGG16 (Figures 6, 7).

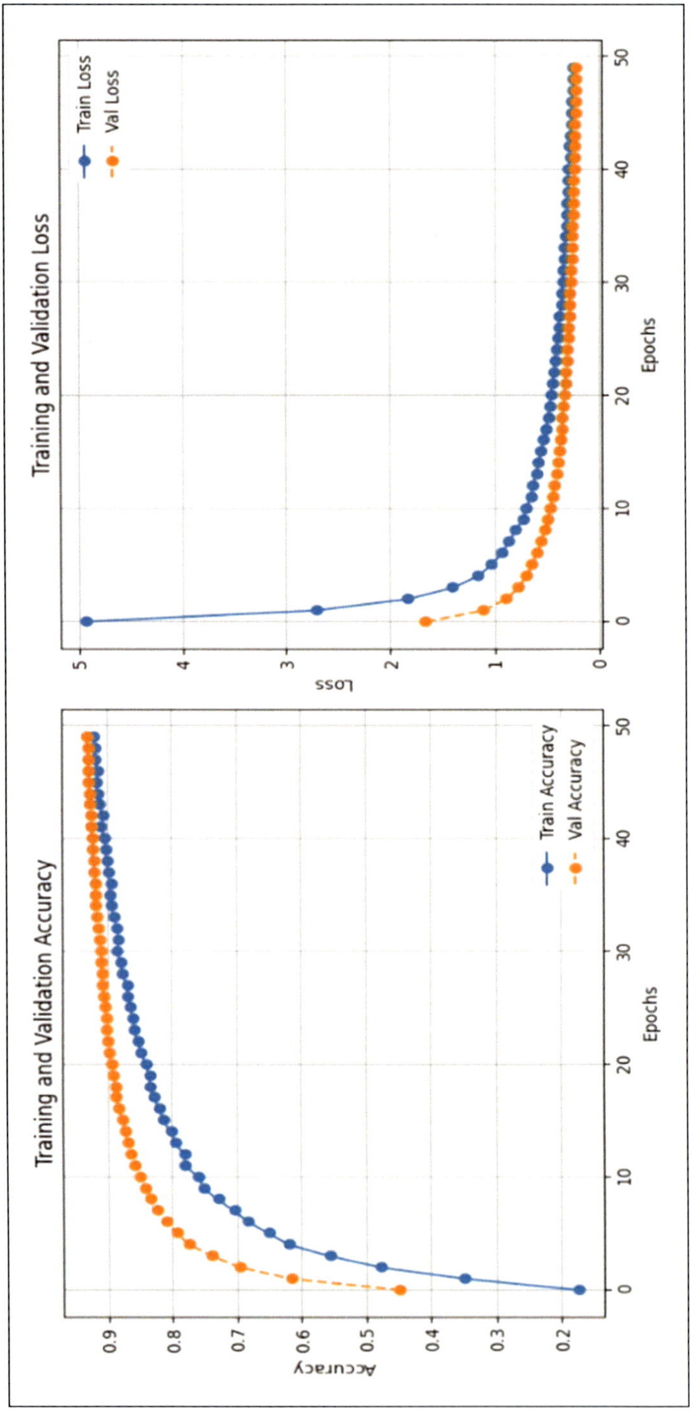

Fig. 2 Plot of training accuracy, validation accuracy, training loss, and validation loss over epochs.

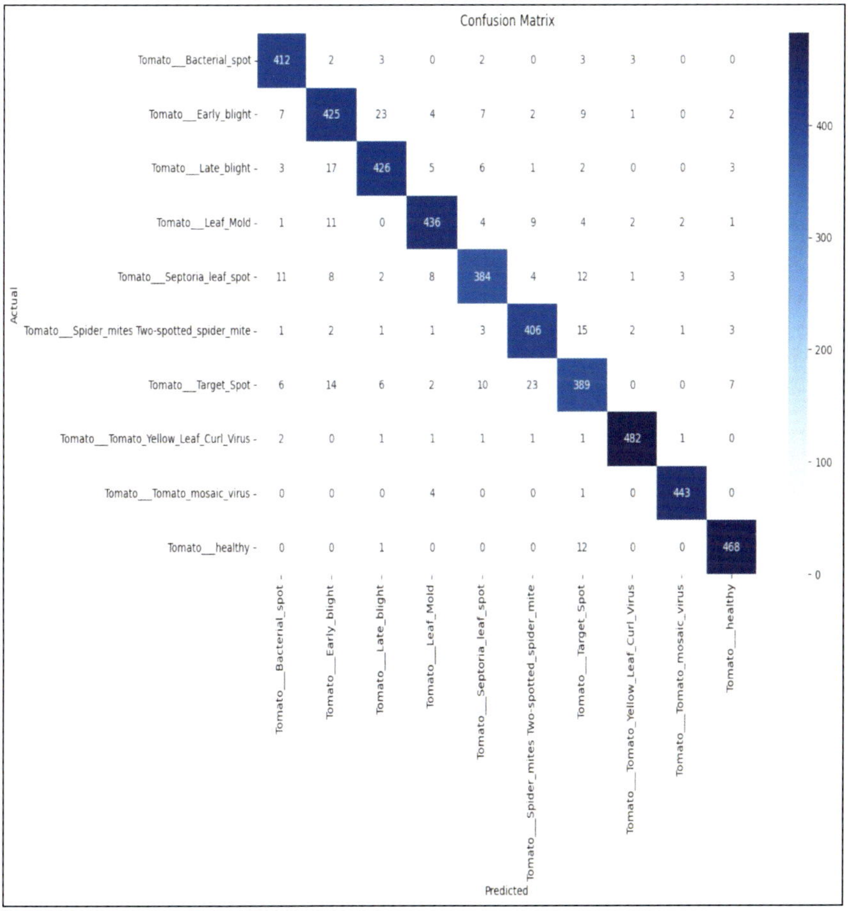

Fig. 3 Confusion matrix for the VGG16 model on the test dataset.

5.4 *Result of DesnseNet169*

The result of DensNet169 is great, this model performs very well with the accuracy of 93.35% and the other scores are in metrics. It has simple architecture which allows for stable training. Its accuracy is higher than MobileNetV2, VGG19, and VGG16 (Figures 8, 9).

5.5 *Result of AlexNet*

The result of AlexNet is great, this model performs very well with an accuracy of 94.42% and the other scores are in metrics. It has a simple architecture that allows for stable training. Its accuracy is higher than MobileNetV2, VGG19, and VGG16 (Figures 10, 11).

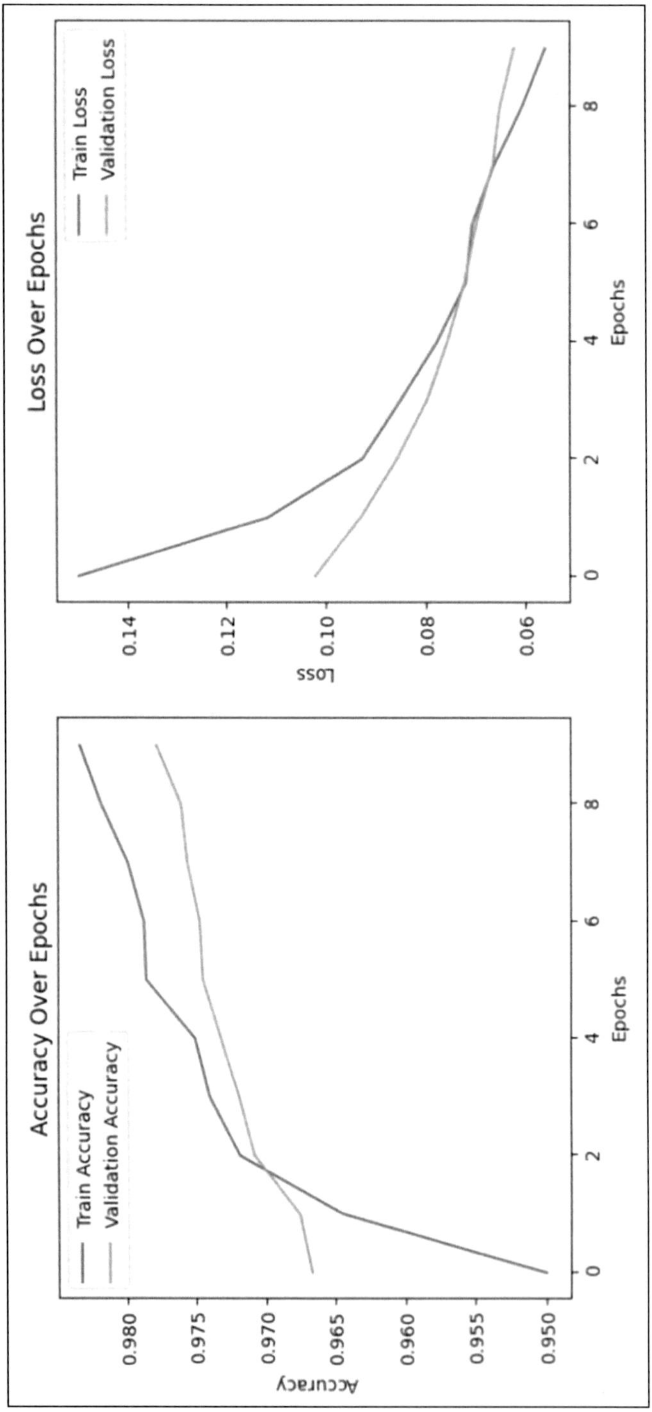

Fig. 4 Plot of training accuracy, validation accuracy, training loss, and validation loss over epochs for the ResNet50 model.

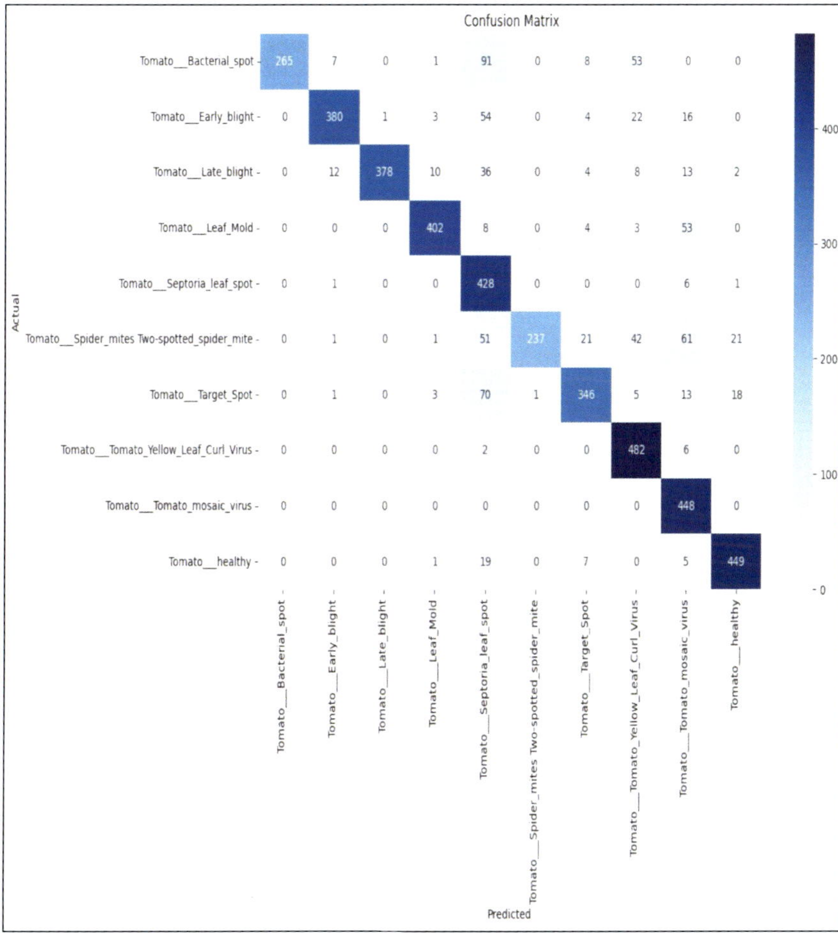

Fig. 5 Confusion matrix for the ResNet50 model on the test dataset.

5.6 Result of MobileNetV2

The result of MobileNetV2 is great, this model performs very well with the accuracy of 94.42% and the other scores are in metrics. It has simple architecture which allows for stable training. Its accuracy is lower than all models (Figure 12).

Training Accuracy (Top Left)

- **Description:** This graph visualization shows how the model's accuracy increases on the training dataset as the number of epochs increases.
- **Observation:**
  - The accuracy starts from a lower point (around 80%) and rapidly increases during the first few epochs.
  - After about 5 epochs, the accuracy stabilizes and reaches a near-perfect value close to **1.0 (100%)**.

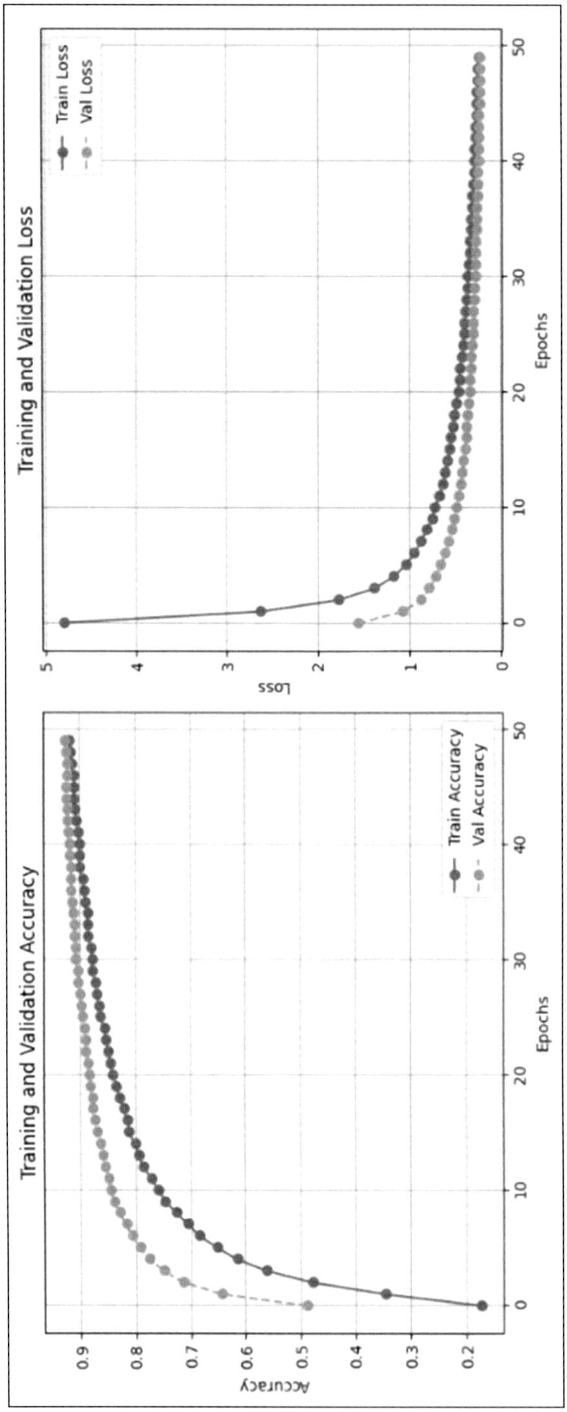

Fig. 6 Plot of training accuracy, validation accuracy, training loss, and validation loss over epochs for the VGG19 model.

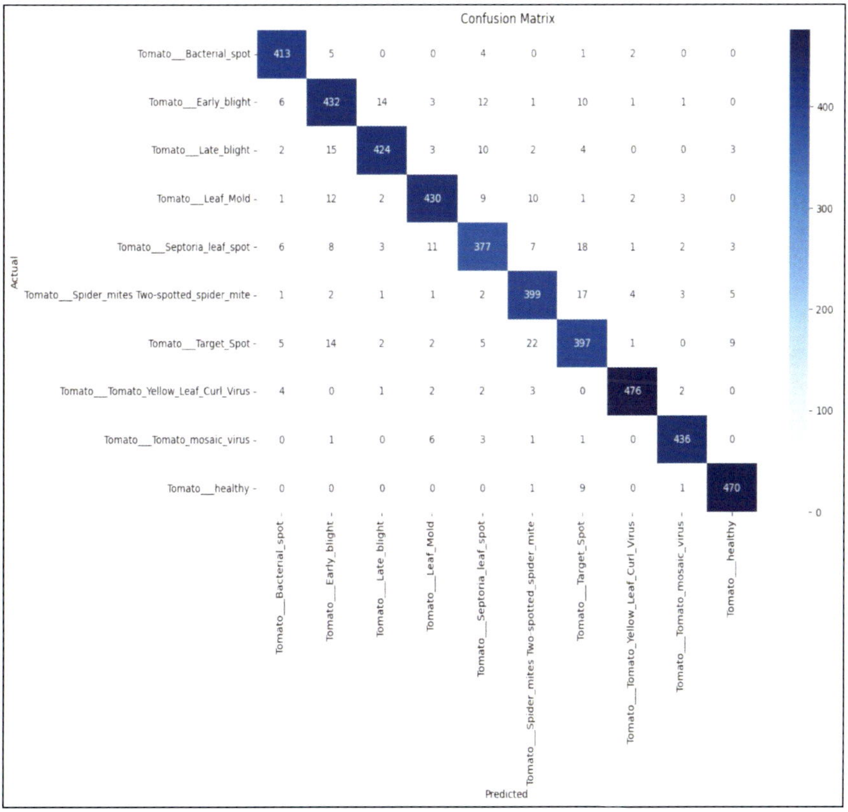

Fig. 7 Confusion matrix for the VGG19 model on the test dataset.

- **Inference:** The model has learned the training data very well, achieving almost perfect accuracy. This could indicate a well-trained model on the training data, but may also suggest a slight risk of **overfitting**.

Validation Accuracy (Top Right)

- **Description:** This graph represents how the model performs on the unseen validation dataset as training progresses.
- **Observation:**
  - The accuracy of validation starts at around **94%** and improves rapidly, reaching over **98%** within the first few epochs.
  - It fluctuates slightly as training continues but remains very close to the **98–99% range**, indicating high generalization performance.
- **Inference:** The model provides well validation data, which means it is not just memorizing the training data but learning meaningful patterns.

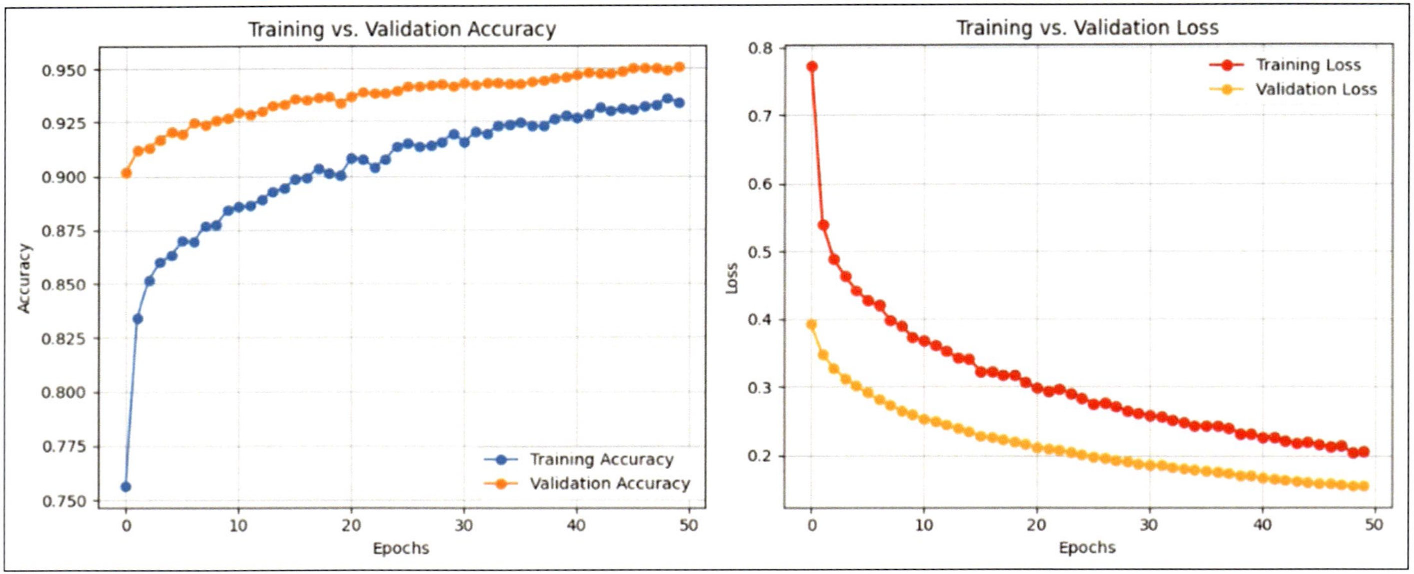

Fig. 8 Plot of training accuracy, validation accuracy, training loss, and validation loss over epochs for the DenseNet169 model.

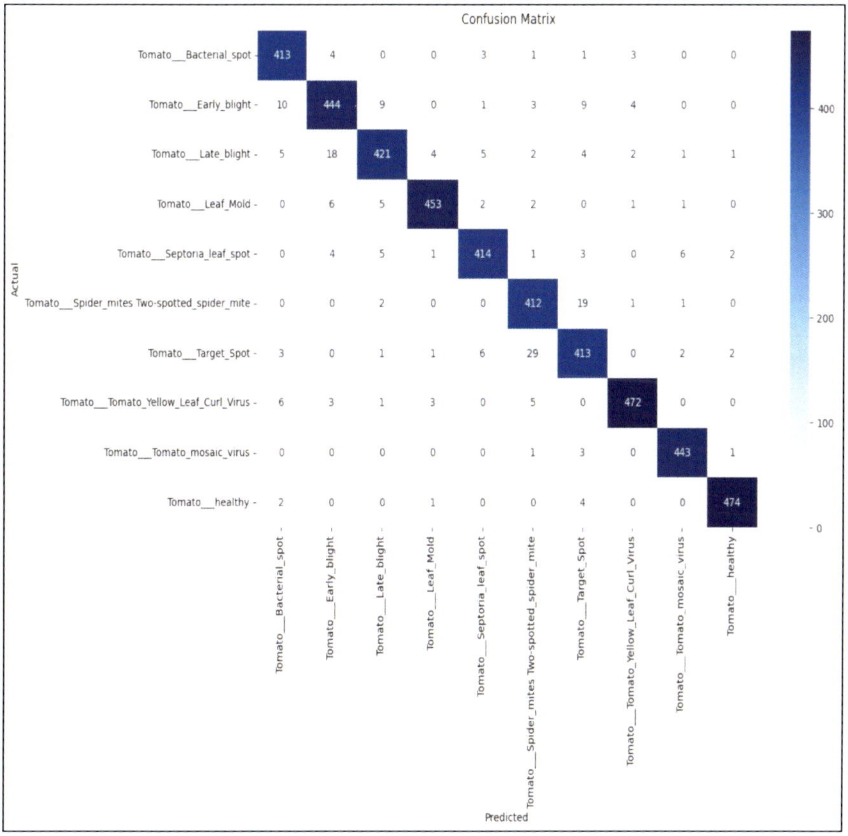

Fig. 9 Confusion matrix for the DenseNet169 model on the test dataset.

Training Loss (Bottom Left)

- **Description:** The training loss graph shows how the loss decreases on the training dataset as the number of epochs increases.
- **Observation:**
  - The loss starts at a high value (**0.7**) and decreases sharply within the first 5 epochs.
  - After that, it continues to decline steadily and approaches **zero** by the end of training.

Inference: When the training loss decreases which indicates the model is learning effectively and minimizing errors in training dataset **Validation Loss (Bottom Right)**

- **Description**: This graph represents how the loss changes on the validation dataset as training progresses.

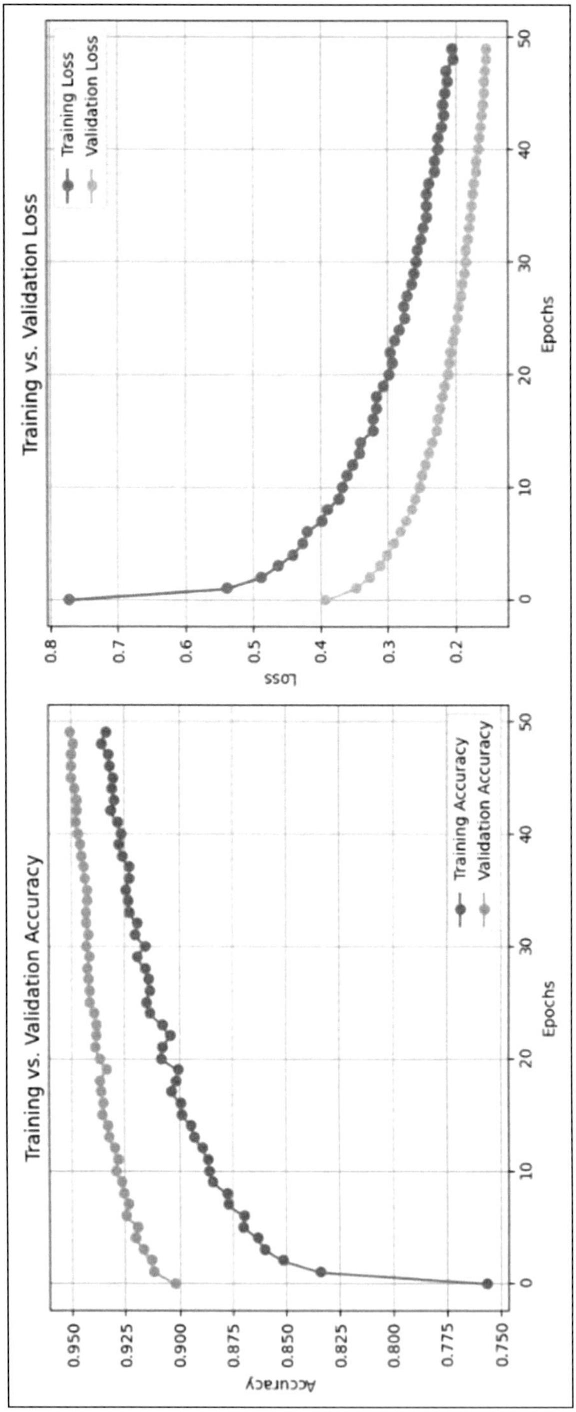

Fig. 10 Plot of training accuracy, validation accuracy, training loss, and validation loss over epochs for the AlexNet model.

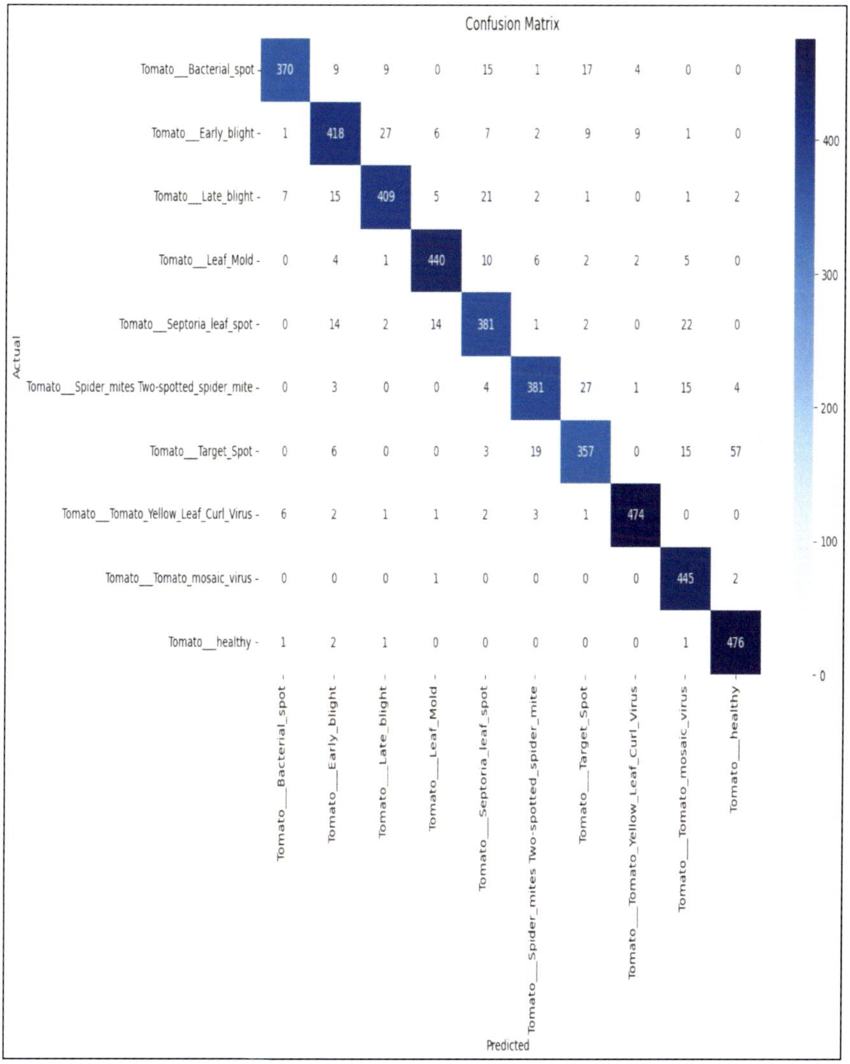

Fig. 11 Confusion matrix for the AlexNet model on the test dataset.

- **Observation:**
  - The validation loss starts high (around 0.18) and drops sharply in the first few epochs.
  - However, as the training progresses, the loss fluctuates slightly around 0.1–0.12, showing some instability.
- **Inference:**
  - The slight fluctuations suggest the models may be experiencing some overfitting in later epochs.
  - Even though the loss is relatively low, these fluctuations indicate that further fine-tuning, such as using regularization or early stopping, could improve performance.

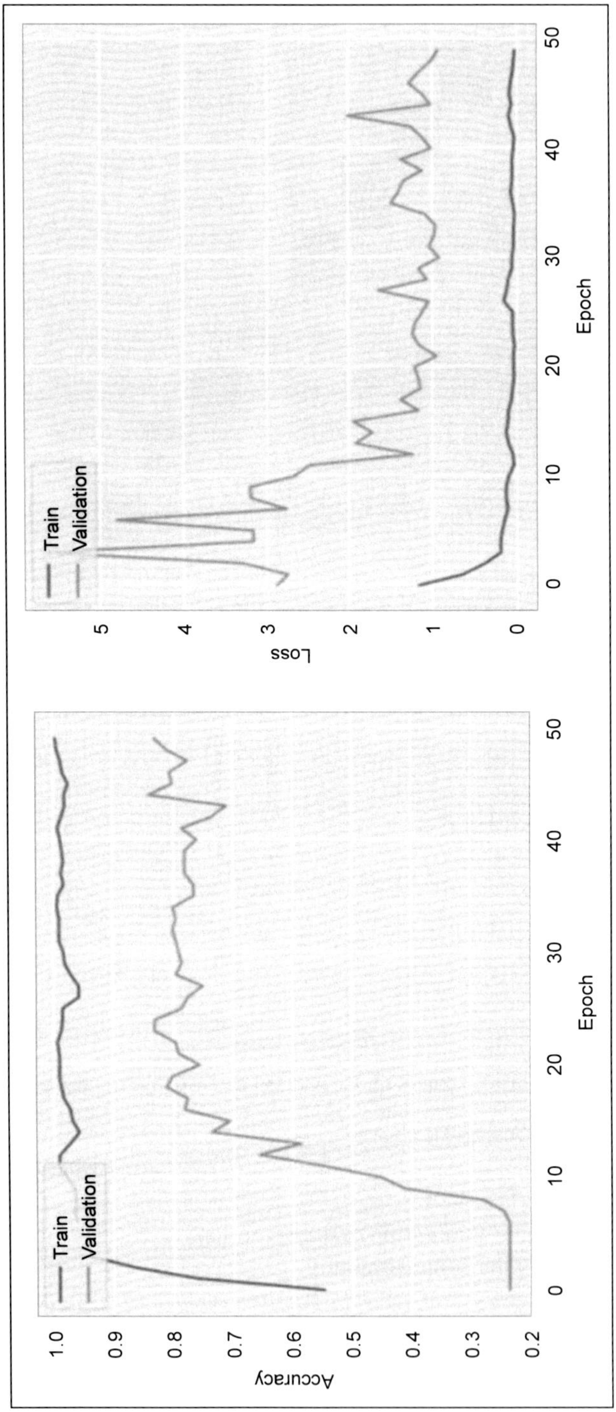

Fig. 12 Plot of training accuracy, validation accuracy, training loss, and validation loss over epochs for the MobileNetV2 model.

- **Training accuracy and loss:** This model provides an exceptionally good training dataset, achieving near-perfect accuracy and very low loss.
- **Validation accuracy and loss:** The model achieves very high accuracy on the validation dataset (98–99%), but the validation loss fluctuates slightly, indicating potential overfitting.

5.7 *Implemented Results and Comarasion Analysis*

5.7.1 *Results of Proposed Classifiers*

Table 5 presesnts the results of proposed classifier for Tomato Leaf Disease.

Table 5 Results of proposd classifier comparasion for Tomato Leaf Disease

Model	*Accuracy%*	*Precision%*	*Recall%*	*F1 Score%*
VGG19	92.94%	94.29%	97.18%	95.75%
VGG16	91.93%	93.00%	96.94%	94.93%
DenseNet169	93.35%	95.05%	97.18%	95.60%
AlexNet	94.42%	90.61%	90.42%	90.40%
ResNet50	95.04%	85.67%	79.93%	79.76%
MobileNetV2	83.00%	77.75%	76.57%	76.21%

This difference clearly shows the ResNet50 model is higher on image-based classification. All comparison results between deep learning models are shown in Table 5.

5.7.2 *Compare the Performance of Our Model to Existing Models*

The great impact of tomato plant disease detection models is fully dependent on the choice of architecture, dataset, and preprocessing techniques. To validate the robustness of our proposed models, in this section we do a comparison of our model with various existing models to well-established models and previous studies in the field. Table 6 presents the performance comparasion of proposed model with existing model.

Table 6 Performance comparison of the proposed model with existing models.

Author	*Dataset*	*Method*	*Accuracy*
Rangarajan, A. et al. (2018)	PLANT VILLAGE	AlexNet	97.49%
		VGG-16	97.29
Wajid, A. et al. (2024)	PLANT VILLAGE	CNN	96%
		Random Forest	71%
		Naive Bayes	49%
		KNN	56%
		AdaBoost	52%

Contd...

Table 6 *Contd...*

Author	*Dataset*	*Method*	*Accuracy*
Elhassouny, A. et al. (2019)	Tomato leaf data set	Stochastic Gradient Descent	88.4%
		Adagrad	88.3%
		SGD with Momentum	87.6%
		Adam	88.5%
		Proximal Gradient Descent	89.2%
		Proximal Adagrad	88.8%
		RMSProp	85.9%
Mitra, D. et al. (2023)	Tomato dataset	CNN	89%
Trivedi, N. et al. (2021)	Dataset: 3000 images with nine different diseases and healthy leaves.	CNN	98.49%
Sabrol, H. et al. (2016b)	PLANT VILLAGE	11 Features	91%
		GLCM	76.5%
		Entropy and kurtosis	60.8%
		Entropy and Skewness	59.9%
		Entropy, kurtosis and skewness	70.9%
Our Study	Tomato Leaf Disease	Vgg19	92.94%
		Vgg16	91.93%
		DenseNet169	93.35%
		ResNet50	95.04%
		AlexNet	94.42%
		MobileNetV2	83%

The methods proposed in our research have better results than other existing ones. In this project we have various popular deep learning models for detection and classification of tomato plant disease like VGG16, VGG19, ResNet50, AlexNet, MobileNet, DenseNet169. We use a dataset from kaggle for all the models.

6. Discussions and Future Directions

Early and accurate detection of tomato plant diseases is too compulsory for save crop and increase production of crop and reducing losses in agriculture. Traditional models of disease identification are time-consuming, error-prone, and often inaccessible to farmers without expert consultation. To address this

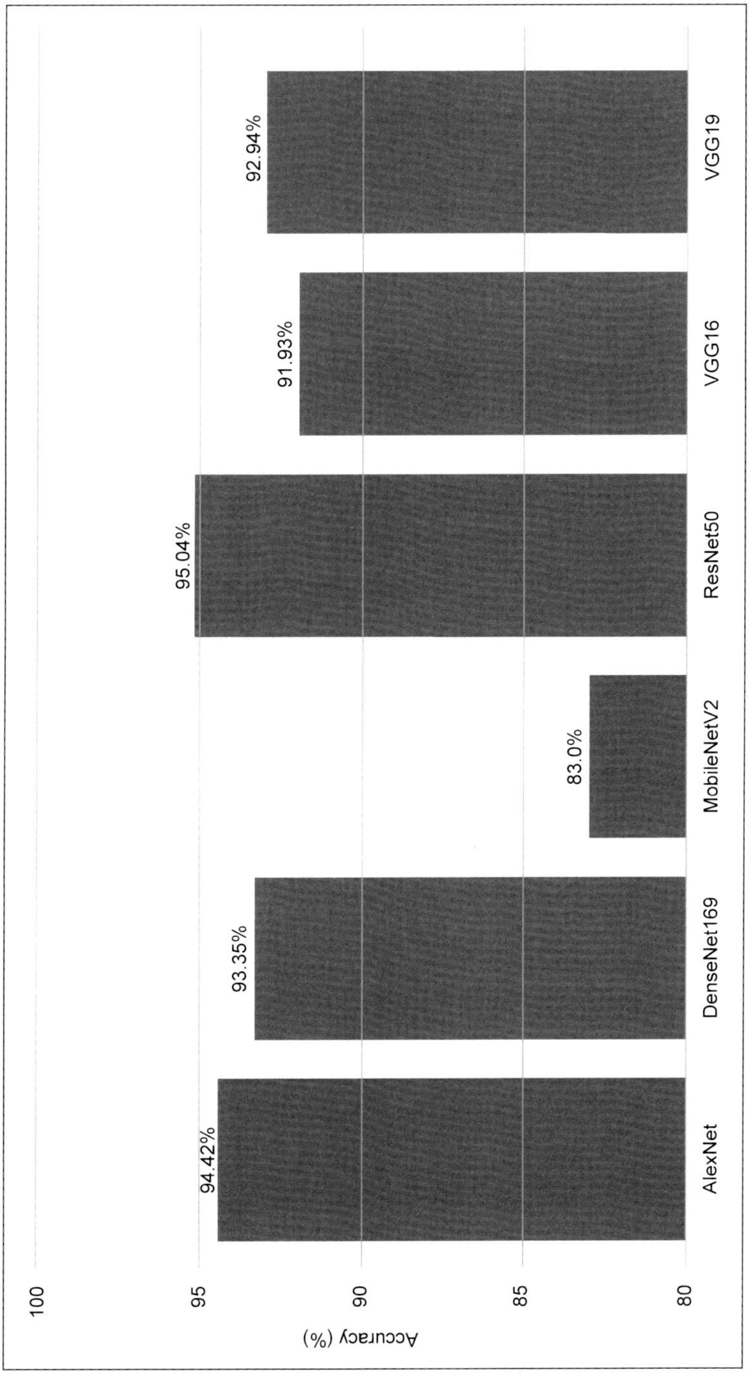

Fig. 13 Results of proposd classifier Comparasion for Tomato Leaf Disease

challenge, we proposed a system with the help of deep learning model Vgg19, Vgg16, DenseNet169, ResNet150, AlexNet, MobileNetV2.

The pre-trained deep learning models show highly performance, with ResNet50 gives highest accuracy of 95.04%. The performance of the proposed model from increasing to decreasing orders are Resnet50, AlexNet, DenseNet169, Vgg19, Vgg16, MobileNetV2. The prosed models provide a robust and scalable solution for tomato plant disease classification.

Subsequent investigations should prioritize: (1) expansion of training datasets incorporating diverse geographic origins, seasonal variations, and environmental conditions to enhance model generalization across heterogeneous field deployment contexts; (2) integration of multi-modal sensor fusion combining visual spectroscopy with hyperspectral and thermal imaging to enhance disease manifestation discriminability; (3) development of lightweight model variants through knowledge distillation and architecture search optimization enabling broader mobile platform accessibility; (4) implementation of explainable AI methodologies providing decision transparency regarding disease classification rationale, facilitating expert validation and farmer comprehension; (5) evaluation on genuine field-acquired imagery reflecting non-optimal lighting, occlusion, and environmental variability versus laboratory-controlled Plant Village dataset distributions; (6) economic impact assessment quantifying disease management optimization benefits and return-on-investment for farmer adoption across diverse agricultural economy contexts.

Ethical Declarations

Ethical Responsibilities of Authors

All authors have read, understood, and complied with the statement on "Ethical Responsibilities of Authors" as outlined in the Instructions for Authors.

Ethical Approval

Not applicable for this research.

Use of AI Tool

The authors used AI-based tools for drafting assistance. All content was reviewed and edited by the authors, who accept responsibility for the final manuscript.

Conflict of Interest

On behalf of all authors, the corresponding author declares no conflict of interest.

Competing Interests

Not applicable.

Authors' Contributions

- Kavish Thakur: Conceptualization, Methodology, Software, Formal analysis, Data curation, Writing – original draft, Writing – review & editing.
- Dilip Kumar Choubey: Conceptualization, Methodology, Resources, Writing – Review & Supervision.
- Aditya Kumar Pandey: Formal analysis, Validation, Visualization, Writing – original draft, Writing – review & editing.
- Utkarsh Raj: Data curation, Investigation, Formal analysis.
- Mukesh Kumar Ojha: Data Curation, Supervision, Resources, Project administration.

Funding

No funding was received for this study.

References

Aggarwal, A.K. (2022). Biological tomato leaf disease classification using deep learning framework. *Int. J. Biol. Biomed. Eng.*, *16*(1), 241–244.

Batool, A., Hyder, S.B., Rahim, A., Waheed, N. and Asghar, M.A. (2020, February). Classification and identification of tomato leaf disease using deep neural network. In: *2020 International Conference on Engineering and Emerging Technologies* (*ICEET*) (pp. 1–6). IEEE.

Brahimi, M., Boukhalfa, K. and Moussaoui, A. (2017). Deep learning for tomato diseases: Classification and symptoms visualization. *Applied Artificial Intelligence*, *31*(4), 299–315.

Chouhan, S.S., Patel, R.K., Shukla, A. and Singh, U.P. Drone Importance and their Necessity in Future Generation Agriculture. In: *Artificial Intelligence and Computer Vision for Ecological Informatics* (pp. 1–8). doi: 10.1201/9781003518778-1.

DL, S., K, V., N, A. and Vashistha, S. (2024). Tomato Leaf Disease Detection Using CNN.Science Direct. com.

Dubey, R.K. and Choubey, D.K. (2024a). Reliable detection of blast disease in rice plant using optimised artificial neural network. *Agronomy Journal*. [Online]. Available: https://acsess.onlinelibrary.wiley.com/doi/abs/10.1002/agj2.21449.

Dubey, R.K. and Choubey, D.K. (2024b, Apr.). Efficient prediction of blast disease in paddy plant using optimized support vector machine. *IETE Journal of Research*, *70*(4), 3679–3689. doi: 10.1080/03772063.2023.2195842.

Elhassouny, A. and Smarandache, F. (2019, July). Smart mobile application to recognize tomato leaf diseases using Convolutional Neural Networks. In: *2019 International Conference of Computer Science and Renewable Energies* (*ICCSRE*) (pp. 1–4). IEEE.

Femi, D., Murugasami, R., Manikandaprabu, N., Paulsingh, J.R. and Vanaja, P. (2021). Identification and Classification of Tomato Leaf Diseases Using Machine Learning Techniques. *Technology*.

Gunarathna, M.M. and Rathmayaa, R.M.K.T. (2020). Efficient deep learning models for tomato plant disease classification based on leaf image. In: *International Conference on Advances in Computing and Technology*.

Hammou, D. R. and Boubaker, M. (2022). Tomato plant disease detection and classification using convolutional neural network architecture technologies. In *Networking, Intelligent Systems and Security: Proceedings of NISS 2021* (pp. 33– 44). Springer Singapore.

Hasan, M., Tanawala, B. and Patel, K.J. (2019, March). Deep learning precision farming: Tomato leaf disease detection by transfer learning. In: *Proceedings of 2nd International Conference on Advanced Computing and Software Engineering* (*ICACSE*).

Jayanthi, G., Brindha, S., Vijayalakshmi, S., Dharshini, V., Freeda, J.A. and Sahana, S. (2024, April). Tomato Leaf Disease Detection Using Machine Learning. In: *2024 International Conference on Communication, Computing, and Internet of Things* (*IC3IoT*) (pp. 1–6). IEEE.

Li, P., Zhong, N., Dong, W., Zhang, M. and Yang, D. (2024). Identification of tomato leaf diseases using convolutional neural network with multiscale and feature reuse. *International Journal of Agricultural and Biological Engineering*, *16*(6), 226–235.

Mitra, D., Goyal, A., Gupta, S., Kanyal, H.S., Kaushik, S. and Kumar, K. (2023). Automated tomato leaf disease detection technique using deep learning. *Journal of Theoretical and Applied Information Technology*, *101*(14).

Ramakrishna, R. (2020). *Machine Learning Based Approach in Detection and Classification of Tomato Plant Leaf Diseases* (Doctoral dissertation, Dublin, National College of Ireland).

Rangarajan, A. K., Purushothaman, R. and Ramesh, A. (2018). Tomato crop disease classification using a pre-trained deep learning algorithm. *Procedia Computer Science*, *133*, 1040–1047.

Sabrol, H. and Kumar, S. (2016a). Fuzzy and neural network based tomato plant disease classification using natural outdoor images. *Indian J. Sci. Technol.*, *9*(44), 1–8.

Sabrol, H. and Satish, K. (2016b, April). Tomato plant disease classification in digital images using classification tree. In: *2016 International Conference on Communication and Signal Processing* (*ICCSP*) (pp. 1242–1246). IEEE.

Sakkarvarthi, G., Sathianesan, G.W., Murugan, V.S., Reddy, A.J., Jayagopal, P. and Elsisi, M. (2022). Detection and classification of tomato crop disease using convolutional neural network. *Electronics*, *11*(21), 3618.

Sembiring, A., Away, Y., Arnia, F. and Muharar, R. (2021, March). Development of a concise convolutional neural network for tomato plant disease classification based on leaf images. In: *Journal of Physics: Conference Series* (Vol. *1845*(1), 012009). IOP Publishing.

Shukla, A., Chouhan, S.S., Patel, R.K. and Singh, U.P. (2025). AgriCloud: Cloud-Based Framework for Scalable Crop Monitoring and Intelligent Decision Support. In: Chouhan, S.S., Patel, R.K., Singh, U.P., and Jain, S. (Eds.), *AgriTech Revolution*. Springer, Singapore. https://doi.org/10.1007/978-981-95-1268-3_6.

Singh Chouhan, S., Saxena, E., Shukla, A., et al. (2025). Optimizing YOLO-Based Models for Real-Time Guava Detection with Probabilistic Fused Wiener Filter-Enhanced Feature Fusion. Applied Fruit Science, 67, 383. https://doi.org/10.1007/s10341-025-01613-2.

Tripathi, A., Vishwakarma, H., Patel, R.K., Shukla, A., Chouhan, S.S. (2025). AI-Enhanced Agriculture: Cultivating the Future. In: Chouhan, S.S., Patel, R.K., Singh, U.P., Jain, S. (eds) AgriTech Revolution. Springer, Singapore. https://doi.org/10.1007/978-981-95-1268-3_1.

Trivedi, N.K., Gautam, V., Anand, A., Aljahdali, H.M., Villar, S.G., Anand, D., ... & Kadry, S. (2021). Early detection and classification of tomato leaf disease using high-performance deep neural network. *Sensors*, *21*(23), 7987.

Vinay, P., Santosh Kumar, G. and Aravind, K.U. (2021, December). Classification of Tomato Plant Leaf Disease Using Neural Network. In: *Proceedings of the International Conference on Innovative Computing and Communication (ICICI)*.

Wajid, A.H., Saher, N., Nawaz, S.A., Arshad, M. and Nasir, M. (2024). Tomato Leaf Disease Detection and Classification Using Convolutional Neural Network and Machine Learning. *Journal of Computing and Biomedical Informatics*.

Index

About the Editors

Dr. Siddharth Singh Chouhan currently working as a Senior Assistant Professor Grade 2 in the School of Computing Science and Engineering, VIT Bhopal University, Kothri-Kalan, Sehore, 466114, Madhya Pradesh, India. He has completed Ph.D. in Computer Science and Engineering from Shri Mata Vaishno Devi University Katra, 182320, UT of Jammu and Kashmir, India. He is a Post Doctorate from the University of Malta, Europe. He has received a B.Tech. degree in 2010 and an MTech. degree in 2013 in Computer Science and Engineering from RGPV University, Bhopal, India. His area of interest includes Soft Computing, Computer vision, Drone Technology, Deep Learning, and Precision Agriculture. He has authored several research papers in many reputed journals including Elsevier, Springer, etc.

Dr. Rajneesh Kumar Patel is an Assistant Professor in the School of Computing Science and Engineering, VIT Bhopal University, Kothri-Kalan, Sehore, 466114, Madhya Pradesh, India. He has completed a Ph.D. in Electronics and Communication Engineering from Maulana Azad National Institute of Technology, Bhopal, 462003, India. He received a B.E. degree in 2012 and an M. Tech. degree in 2016 in Electronics and Communication Engineering from RGPV University, Bhopal, India. His areas of interest include Image Processing, Computer vision, Machine/Deep Learning, and Precision Agriculture. He has authored several research papers in many reputed journals and conferences.

Dr. Anju Shukla is an Assistant Professor in the School of Computing Science and Engineering, VIT Bhopal University, Kothri-Kalan, Sehore, 466114, Madhya Pradesh, India. She holds a Ph.D. in Computer Engineering from Jaypee University of Engineering and Technology, Guna, and an M.Tech in Computer Engineering from Shobhit University, Meerut. Her areas of expertise include Cloud Computing, Load Balancing, Grid Computing, Virtualization, and Fault Tolerance, Machine Learning, Deep Learning. Dr. Shukla has published several research papers in international journals and conferences, focusing on topics such as fault-tolerant load balancing in cloud environments and resource allocation models for grid computing.

Dr. Uday Pratap Singh presently working as Professor in Department of Mathematics, Central University of Jammu, 181143, Jammu, India. He has obtained B.Sc. and M.Sc. (Mathematics and Statistics) Degree from Dr. R.M.L. Awadh University, Faizabad, (U.P.), India, M.Sc. (Mathematics and Computing) Degree from Indian Institute of Technology, Guwahati, India, and received a Ph.D. in Computer Science from Barkatullah University, Bhopal. His area of interest includes Soft Computing, Nonlinear Systems, and Image Processing. He had authored several research papers in the many reputed journals and conferences. He is a life member of Soft Computing Research Society (SCRC), Barata Ganita Parishad, Computer Society of India and Member of IEEE and AMS.